Introduction to Nano Solar Cells

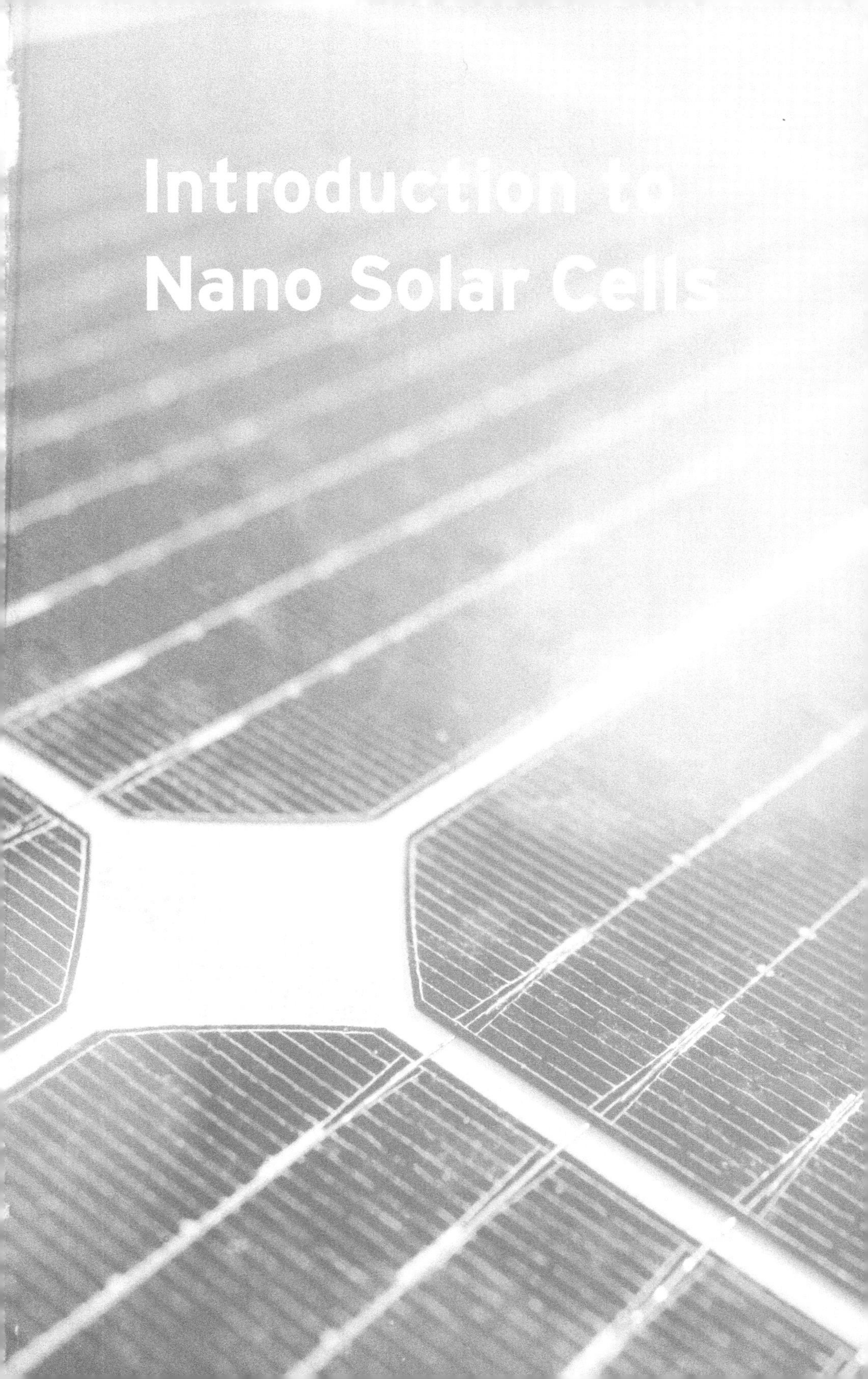

Introduction to Nano Solar Cells

Ning Dai

Jenny Stanford Publishing

Published by

Jenny Stanford Publishing Pte. Ltd.
101 Thomson Road
#06-01, United Square
Singapore 307591

Email: editorial@jennystanford.com
Web: www.jennystanford.com

British Library Cataloguing-in-Publication Data
A catalogue record for this book is available from the British Library.

Introduction to Nano Solar Cells

ISBN 978-981-4877-49-7 (Hardcover)
ISBN 978-1-003-13198-4 (eBook)

Contents

Preface xv

1. Electrons in Semiconductors **1**

1.1 Introduction 1
1.2 Fundamental Properties of Semiconductors 3
1.2.1 Crystal Structures of Semiconductors 3
1.2.2 Electronic Resistivity and Optical Absorption Edge 5
1.3 Electrons and Holes in Semiconductors 8
1.3.1 Atomic Orbits and Energy Bands 9
1.3.2 Fermi–Dirac Statistics 13
1.3.3 *P*-Type and *n*-Type Doping 15
1.3.4 Transport Process of Electrons and Holes 19
1.3.5 Defects in Semiconductors 22
1.4 Built-in Electric Field 23
1.4.1 *p*–*n* Junction 23
1.4.2 Heterojunction 27
1.4.3 Semiconductor–Metal Contact 30
1.4.4 Semiconductor–Insulator Interface 32
1.5 Nanomaterials 34
1.5.1 Quantum Confinement and Quantum Confinement States 36
1.5.2 Moving Electrons in Quantum Confined System 36
1.5.3 Density of States of Quantum Confined System 47
1.5.4 Surface Modification of Nanomaterials 53
1.5.5 Preparation of Nanomaterials 56

1.6 Interaction between Light and Matter 62
1.6.1 Some Facts about Solar Radiation 62
1.6.2 Optical Transition Probability 64
1.6.3 Optical Transition in Low-Dimensional Semiconductors 67

2. How Solar Cells Work 75

2.1 Photovoltaic Effect 75
2.1.1 The Physical Process of Photovoltaic Effect 76
2.1.2 The Photoelectric Conversion Efficiency 77
2.1.3 Modeling a Conventional Semiconductor Solar Cell 79
2.1.4 Charge Separation in Solar Cells 91
2.2 The Major Losses in Solar Cells 93
2.2.1 The Loss Mechanisms 94
2.2.2 Optical Losses 94
2.2.3 Electric Losses 96
2.2.4 Thermal Losses 98
2.3 Photovoltaic Devices: Solar Cells 99
2.3.1 Wafer-Based Si Solar Cells 100
2.3.2 Important Parameters of Solar Cells 101
2.3.3 Thin-Film Solar Cells 104
2.3.4 Nanomaterial-Based Solar Cells 114
2.4 Factors That Affect the Efficiency of Nano Solar Cell 119
2.4.1 Special Characteristics of Nano Photovoltaic Materials and Devices 119
2.4.2 Existing Problems That Affect Solar Cell Efficiency 121
2.4.3 Some Fundamental Rules in Designing a Solar Cell 125
2.5 Existing Problems in Solar Cells Used for Power Generation 127

3. Nanomaterials and Structures for Photon Trapping **131**

3.1 Wave-Particle Duality of Light and Light Reflection, Diffraction, and Refraction 131

3.1.1 Light Reflection and Diffraction at Interface between Two Dielectrics 134

3.1.2 Rayleigh Scattering and Mie Scattering 135

3.1.3 Lambertian Scattering 137

3.2 Light Trapping Methods 138

3.2.1 Light Trapping in Conventional Crystalline Si Solar Cells 139

3.2.2 Light Trapping in Nano Solar Cells 142

3.2.3 Light Trapping Using NW Arrays 143

3.2.4 Suppression of Light Reflection Using Tapered Two-Dimensional Gratings 145

3.2.5 Enhancement of Light Absorption Reflection by Periodic Arrays of Nanowires 147

3.2.6 Enhancement of Light Absorption by Periodic Arrays of Nanoholes 152

3.2.7 Light Trapping Using Step-Like Nanocone Array 153

3.3 Light Trapping Using Plasmonic Technique 157

3.3.1 Localized Surface Plasmons 157

3.3.2 Light Trapping Effect Using Plasmons 160

3.3.3 Nanoparticle/Dielectric/Metal Structures for Light Trapping 163

3.3.4 Plasmonic Light Trapping Used for Solar Cells 165

3.3.5 Light Trapping Using High-Index Nanostructures 170

3.4 Photon Trapping Using Photonic Crystals 173

3.4.1 Photonic Crystals for the Manipulation of Light Propagation 174

3.4.2 One-Dimensional Photonic Crystals 179

3.4.3 Light Trapping Using 2D Photonic Crystals 185
3.4.4 Light Trapping Using 3D Photonic Crystals 186
3.4.5 Preparation of Photonic Crystal Used for Light Trapping 188
3.5 Some Applications 192
3.5.1 Plasmonic Light Trapping for Thin-Film a-Si:H Solar Cells 192
3.5.2 Light Trapping Layers Fabricated by Nano-Imprint Lithography 195

4. Transparent Conducting Electrodes and Dye-Sensitized Solar Cells 205

4.1 Dye-Sensitized Solar Cells 206
4.1.1 Photoanodes of Dye-Sensitized Solar Cells 207
4.1.2 Transparent Conducting Electrodes 209
4.1.3 Nanomaterial Scaffold 213
4.1.4 Examples of Photoanodes 214
4.2 Carbon-Based Transparent Conducting Electrodes 222
4.2.1 Carbon Nanotubes 222
4.2.2 Graphene 224
4.2.3 Polymer Transparent Conducting Electrodes 227
4.3 Synthesis of TCF 227
4.3.1 Wet-Chemical Preparation 227
4.3.2 Anodic Oxidation 230
4.3.3 Vapor Transport Synthesis 232
4.3.4 Hydrothermal Process 237
4.4 Sensitizers 239
4.4.1 Ruthenium(II) Sensitizers 239
4.4.2 Metal-Free Tetrathienoacene Sensitizers 240

4.4.3 Porphyrin-Based D–π–A Sensitizers 242
4.4.4 Natural Dyes 245
4.5 Counter Electrodes 246
4.5.1 Metal Nanoparticles 246
4.5.2 Carbon Nanomaterials 247
4.5.3 TiN Nanotube Arrays 249
4.5.4 Porous Silicon 250
4.5.5 $Cu_2ZnSn(S_{1-x}Se_x)_4$ 253

5. Quantum Dot Solar Cells 265

5.1 General Properties of Semiconductor Quantum Dots 267
5.1.1 Quantum Dot Composite Thin Films 267
5.1.2 Electronic Properties of Quantum Dots 269
5.1.3 Optical Properties of Quantum Dots 272
5.2 Growth of Quantum Dot 274
5.2.1 Hot-Injection Synthesis 275
5.2.2 Non-Injection Heat-Up Synthesis 276
5.2.3 Flow Reactor Method 280
5.3 Thin-Film Preparation 283
5.3.1 Spin Coating 283
5.3.2 Dip Coating 285
5.3.3 Drop Casting 286
5.3.4 Spray Coating and Inkjet Printing 286
5.4 Surface Ligand Exchanges and Shell Layer Growth 287
5.4.1 CdSe QDs Capped with Fullerene Derivatives 289
5.4.2 Air Stabilization of Colloidal Quantum Dots in Solid Matrixes 290
5.4.3 CdS Quantum dots with Tunable Surface Composition 291
5.5 Device Architectures of QD Solar Cells 293
5.5.1 Schottky Junction 293
5.5.2 Depleted Heterojunction 297

5.5.3 Nanoheterojunction Colloidal QD Solar Cells 301
5.5.4 Quantum Dot–Sensitized Cells 304
5.5.5 Other Quantum Dot Solar Cell Configurations 306
5.5.6 Extremely Thin Absorber Cells 308
5.5.7 Tandem Nano Solar Cells 309
5.6 Recent Work in Quantum Dot Solar Cells 310
5.6.1 QD Solar Cells with Conversion Efficiency Over 12% Due to Improved Photoanodes and Counter Electrode 311
5.6.2 High Conversion Power Achieved Using Perovskite QDs 312
5.6.3 QD Solar Cells with Improved Quality of QD Films 315
5.7 Existing Problems for Quantum Dot Solar Cells 317

6. Solar Cells Based on One-Dimensional Nanomaterials 325

6.1 Fundamental Material Properties of Nanowires 326
6.1.1 Basic Theory 326
6.1.2 Electrical Properties of Semiconductor Nanowires 329
6.1.3 Optical Properties of Semiconductor Nanowires 331
6.2 Growth of 1D Nanomaterials 333
6.2.1 Vapor–Liquid–Solid Synthesis of III–V Nanowires 334
6.2.2 Vapor–Liquid–Solid Synthesis of Si Nanowires 337
6.2.3 Growth of Core/Shell NWs 339
6.2.4 Growth of NW with Axial Junction 340
6.2.5 Catalyst-Free Growth 341
6.2.6 Top-Down Etching 342
6.3 Device Architectures of Nanowire Solar Cells 345
6.3.1 Charge Generation and Separation in NWs 345
6.3.2 Charge Collection 351

6.3.3 Device Architecture of Nanowire Solar Cells 353

6.4 Some Nanowire Solar Cells 360

6.4.1 Group IV Nanowire Solar Cells 360

6.4.2 III–V Nanowire Solar Cells 365

6.4.3 II–VI Nanowire Solar Cells 370

6.4.4 Single Nanowire Solar Cells 374

6.5 Some Important Work Review 381

6.5.1 Solar Cells Based on Solution-Processed Core–Shell Nanowires 381

6.5.2 Si Nanowire Solar Cells with Axial and Radial *p*–*n* Junctions 382

6.5.3 GaAs Nanowire Array Solar Cells with Axial *p*–*i*–*n* Junctions 385

6.5.4 InP Nanowire Array Solar Cells with 13.8% Efficiency 388

6.5.5 Three-Dimensional Nanopillar-Array Solar Cells Based on *n*-Type CdS Nanopillars Embedded in *p*-Type CdTe 390

6.5.6 ZnO Core/Shell Nanowire Solar Cells 393

6.6 Existing Problems 394

7. Hybrid Nano Solar Cells 405

7.1 About Hybrid Solar Cells 405

7.2 Fundamental Material Properties 407

7.2.1 Optical and Electrical Properties of Polymers 408

7.2.2 Some Electron and Hole Transport Polymers 409

7.3 Device Architectures and Working Principles 415

7.3.1 Device Work Principle 415

7.3.2 Junction Potentials in Devices 419

7.4 Device Structure of Hybrid Solar Cells 423

7.4.1 3D Bulk Heterojunctions 423

7.4.2 Double Heterojunctions 424

7.4.3 Ordered Lamellar Architecture 425
7.4.4 Ordered Nanowire Structures 426
7.5 Some Hybrid Solar Cells 427
7.5.1 A Hybrid Cell with Si Nanowires on Pyramid-Textured Surface 427
7.5.2 Nano-CdSe/Polymer Hybrid Cells 429
7.5.3 Hybrid Solar Cells Using GaAs Nanopillars 434
7.5.4 A Hybrid Tandem Solar Cell Consisting of a-Si:H Top Cell and a Dye-Sensitized Bottom Cell 437
7.5.5 A Hybrid Tandem Cell Consisting of an a-Si:H Front Cell and a Polymer-Based Organic Back Cell 440
7.6 Several Hybrid Nano Solar Cells with Efficiency ~10% 445
7.6.1 Polymer/Nano-Si Hybrid Cells 445
7.6.2 Efficient Hybrid Heterojunction Solar Cells Containing Perovskite Compound and Polymeric Hole Conductors 453
7.6.3 Highly Efficient Si-Nanorods/Organic Hybrid Core–Sheath Heterojunction Solar Cells 457
7.7 The Existing Problems 460

8. Some Advanced Ideas for Enhancing the Conversion Efficiency 467

8.1 Overview of Strategies for Improving Photovoltaic Cell Efficiency 467
8.2 Multi-Junction Tandem SCs 469
8.2.1 Principle of Multi-Junction Solar Cells 470
8.2.2 Thin-Film Multi-Junction Tandem Solar Cells 471
8.2.3 Nano Multi-Junction Tandem Solar Cells 477
8.2.4 Spectrum Splitting 484
8.3 Hot-Carrier Capture 490
8.3.1 Hot-Carrier Generation in Semiconductors 491

8.3.2 Device Structures of Hot Carrier Solar Cells 492
8.3.3 Hot Carrier Solar Cells Based on Metal–Insulator–Metal and Metal–Semiconductor Structures 495
8.3.4 Hot Carrier Solar Cells Based on QDs 496
8.4 Multiple Exciton Generation or Downconversion 498
8.4.1 Physical Background of Multiple Exciton Generation 498
8.4.2 Multiple Exciton Generation in QDs 500
8.4.3 Solar Cell Architectures with Multiple Exciton Process 501
8.5 Upconversion 503
8.5.1 Principle of Spectral Upconversion 503
8.5.2 Solar Cells with a Spectral Upconverter 506
8.5.3 An a-Si:H Thin-Film Solar Cell Attached with an Upconverter 510
8.6 Plasmonic Solar Cells 511
8.6.1 The Plasmon Effect 511
8.6.2 Working Principle of Plasmonic Solar Cells 513
8.6.3 A Plasmonic Solar Cell with the $NaYF_4$:(Er^{3+}, Yb^{3+}) Upconverter 514
8.7 Intermediate Band Solar Cells 515
8.7.1 Working Principle of Intermediate Band Solar Cells 515
8.7.2 Methods of Introducing Intermediate Bands 517
8.7.3 Some Intermediate Band Solar Cells 521

Index 529

Preface

The energy radiated by the Sun to the Earth every year is enormous, more than the total power generated by one million Three Gorges hydropower stations. Photovoltaic devices are the keys to converting solar radiation into electrical energy. To achieve large-scale use, photovoltaic devices (solar cells) should have high photoelectric conversion efficiency and low price. Being high-efficiency (with a theoretical maximum efficiency of 33%) and high-cost, the photovoltaic devices of the first generation use pure silicon wafers and are based on a *p–n* junction for energy extraction from photons. The second-generation photovoltaic devices refer to thin-film solar cells that are low-efficiency and low-cost; these devices use minimal materials (thin-film amorphous silicon, CdTe, etc.) and inexpensive manufacturing processes such as fast printing. The third-generation photovoltaic devices use nanomaterials and nanostructures, such as nano quantum dots and nanowires, aimed at achieving high efficiency and low cost.

The purpose of the book is to describe how photovoltaic devices composed of nano forms of materials and nanostructure work to convert solar radiation into electrical power, focusing on the typical device structures and the advantages and limitations for achieving high conversion efficiencies. The book also introduces typical works in constructing nano photovoltaic devices. Chapter 1 presents the electronic properties of semiconductors and nanomaterials and quantum mechanical description of the photon–electron interaction, fundamental for the understanding of the photovoltaic process. Chapter 2 elaborates working principles and device models (equivalent circuits) of solar cells and important device parameters for the characterization of solar cells. Various loss factors that limit the performance of solar cells and the currently existing problems affecting the efficiency of nano solar cells are discussed.

At the first step in the photovoltaic process, it is necessary to allow as much incident light as possible to enter the photovoltaic cell. Light trapping principle and nanomaterials and nanostructures for light trapping are elaborated in Chapter 3, which includes nanowire patterns, two-dimensional gratings, nanocone array, and photonic crystals and surface plasmonic resonance. Chapter 4 is devoted to transparent contacting electrodes, which are important components in various nano solar cells and have a critical impact on device performance. The function of transparent conductive electrodes in photovoltaic cells was explained in detail by means of dye-sensitized solar cells. Chapters 5, 6, and 7 deal with nano solar cells based on quantum dots, nanowires, and organic-inorganic hybrid structures, respectively. The photoelectric characteristics and typical growth methods of the nanomaterials are presented, and the corresponding device architectures and photovoltaic processes are discussed. These chapters also introduce some examples of breakthrough works. Finally, in Chapter 8, some important ideas for achieving high power conversion efficiency are simply reviewed.

Chapter 1

Electrons in Semiconductors

1.1 Introduction

Human beings are facing the challenge of increasing use of energy. The traditional fossil energy resource, the major fraction of our total energy usage, is drying up, while the environmental problem is becoming increasingly prominent. The sun, an ordinary star in the universe, releases a huge amount of energy to the earth through solar radiation. The astonishing amount of radiation power to the earth, total $\sim 1.7 \times 10^{17}$ W or $\sim 1.4 \times 10^{3}$ W/m^2, is more than the total power generated by 1 million Three Gorges hydropower stations. The earth receives a huge amount of solar radiation energy per year!

The use of solar energy largely relies on the photovoltaic (PV) effect, a process in which sunlight is converted directly into electrical energy. The process is completed by the photoelectronic device, termed as solar cells that collect solar radiation and output electricity based on the photovoltaic effect. In principle, a semiconductor with a given bandgap absorbs all photons with energies equal to and large than its bandgap. The performance of a photovoltaic device (a solar cell) is characterized by the solar power conversion efficiency. Thus, an efficient solar cell should be well-designed to convert solar radiation energy into electric power as much as possible. Unfortunately the power conversion efficiency is below 45% for precisely designed multiple junction solar cells whose operation requires an additional light

Introduction to Nano Solar Cells
Ning Dai

ISBN 978-981-4877-49-7 (Hardcover), 978-1-003-13198-4 (eBook)
www.jennystanford.com

condensing and intricate solar radiation tracking system. The most widely used wafer-based Si solar cells can only convert ~20% of the collected solar radiation energy. In other words, 80% of the solar radiation energy is currently unconvertible due both to fundamental limits and to imperfection in device structure and materials. Low conversion efficiency of the solar cells partially leads to expensive cost of photovoltaic power generation. Nanomaterials and nanostructures, powered by quick development of nanotechnology, are highly expected to raise the photoelectric power conversion efficiency and, at the same time, to reduce cost of photovoltaic devices and the cost of solar power generation. Conversion efficiency and cost are currently the two major concerns for solar cells used for power generation. It is expected that, in the future, direct solar power will become the major source of energy that our humankind will rely on.

The construction of efficient nano solar cells relies on the material, structure, and device architecture. The main advantage of nanomaterials and nanostructures lies in their tunability with respect to physical properties and diversity in materials and structures, making them widely suitable for the use in solar cell fabrication. Practical photovoltaic devices use semiconductors, insulators and metals but semiconductors (including organic semiconductors) play a major role in the device for photon absorption and photocurrent generation that are the major steps in a photovoltaic process. The properties of semiconductors, together with device architectures, are thus critical to device performance.

The unique properties of nano semiconductor materials and structures are the key for the fabrication of next-generation solar cells and some of them still need to be explored. The optical and electrical properties of nano semiconductors are quite different from those of their bulk. The distinct difference stems from the boundary and the size of nano-sized semiconductors. Boundaries and sizes are ignorable for bulk materials but extremely important for nano semiconductors in determining material properties. In other words, the properties of nano semiconductors can be "tuned" from bulk by introducing boundary and size effects. It is thus necessary to review some fundamental properties of bulk semiconductors to better understand those of nanomaterials.

1.2 Fundamental Properties of Semiconductors

Semiconductor materials include elementals like Si and Ge and compounds like binary GaAs, InP, InSb, CdTe, and ZnTe, as well as many ternaries such as $Al_{1-x}Ga_xAs$, $InAs_{1-x}Sb_x$, $Zn_{1-x}Cd_xTe$. The compound materials can alloy among themselves or with other elements to form quaternaries that display properties of semiconductors. Semiconductors come with a large family and, among them, element Si and Ge, as well as compound GaAs are the most popular ones used for optoelectronic devices.

1.2.1 Crystal Structures of Semiconductors

The atoms in a semiconductor crystal are held together by bonding of the valence electrons [1–2]. The bonding and crystalline structure are, in general sense, determined in such a way that the internal free energy of the crystal is minimized. This condition is generally satisfied when each atom has eight valence electrons to complete its full orbit. However, an independent atom does not have eight electrons so that the out-shell orbit is incomplete. When the atoms are brought together to form a crystal, an atom becomes strongly interacted with its neighboring atoms, bonds forming to minimize the crystalline energy. However, its neighboring atoms are all in short of electrons to complete a full valence shell. A semiconductor crystal is structured, in which valence electrons are shared among atoms through bonding, and the bonds are named as covalent bonds. In elemental semiconductors like Si and Ge, each atom contributes four valence electrons to share with its four nearest neighbor atoms and gains four electrons from its four nearest neighbor atoms to complete a close valence shell, although the four contributed electrons and the four gained ones are all shared. For compound semiconductors like GaAs and CdTe, for instance, the bonds are not completely covalent, the bonds being partially ionic due to different Coulomb interactions of electron–cation and electron–anion. The crystalline structure of a semiconductor relies largely on the bonding energy and bonding geometry.

The unit cell of Si has the diamond structure and those of GaAs, InAs, and CdTe have the zinc blende one. CdS and some others crystallize to form the close-packed hexagonal structure. Figure 1.1b depicts the unit cells of diamond, zinc blende, and close-packed hexagonal structures.

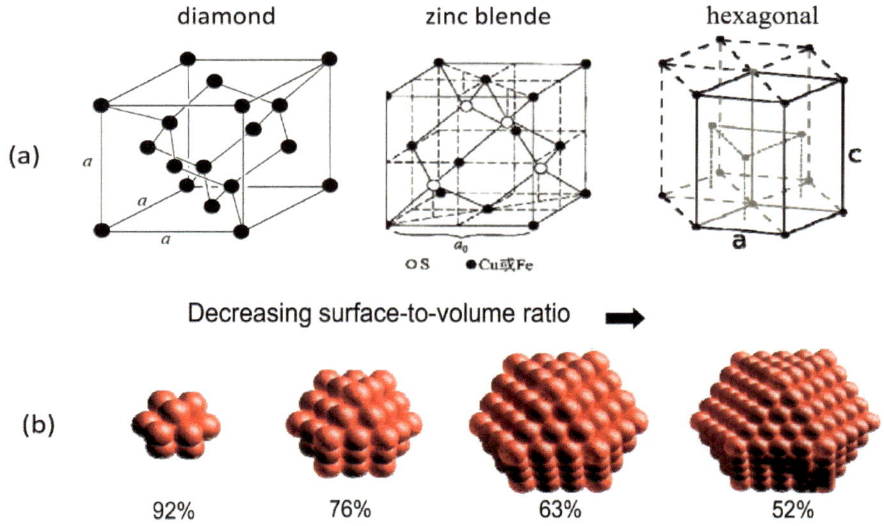

Figure 1.1 (a) Unit cells of diamond, zinc blende, and close-packed hexagonal structures. (b) Schematic illustration: decreasing the size of the nanocrystal gives rise to increasing fraction of surface atoms.

Nanomaterials are substances with their size less than ~100 nm at least in one dimension. Nanomaterials are classified as zero-dimensional (0D), one-dimensional (1D), two-dimensional (2D) nanostructures, corresponding to quantum confinement in all three directions, two directions, and one direction, respectively. Thus, nanomaterials could be 2D quantum wells or nano films, 1D nanowires, or 0D nanoparticles (often termed as quantum dots for semiconductors). In addition, nanomaterials may appear in single, fused, or aggregated forms. They can be further classified according to their shapes. 0D nanomaterials, for instance, could have the spherical, triangle, or cubic shapes. Nano-sized crystals have far larger surface areas than the same masses of bulk crystals, leading to a large fraction of atoms to be the surface atoms. Figure 1.1b shows that the fraction of surface atoms becomes larger and larger with decrease of the nanoparticle (0D) size.

On surface, the unit cells are not perfect and the symmetry of the crystal is broken. The imperfection of the surface atom arrangement introduces high density of the surface states acting as carrier trapping centers. They are troublesome for device operation.

1.2.2 Electronic Resistivity and Optical Absorption Edge

The resistivity of semiconductors lies in between conductors such as silver and insulators such as glass quartz, so that they are called semiconductors. Semiconductors play the key role in microelectronics for wide range of applications stemming from their unique properties, characterized by their very strong optical and electrical non-linearities. As typical examples, the active layers in many optoelectronics devices are made of semiconductors. In a low-temperature range (below 100 K), the conductivity of semiconductors increases with temperature so that semiconductors have a negative temperature coefficient of resistance in contrast to that of a metal, as shown in Fig. 1.2 [3, 4]. The increase of conductivity is due to thermal excitation of carriers, which becomes increasingly remarkable as the rise of temperature. The very low conductivity of an insulator is essentially temperature independent. The conductivity of semiconductors can be varied easily by controlled impurity element doping into the materials. The conductivity of silicon, for instance, could increase a few thousand even million times just by doping a small amount of trivalent elements like indium and gallium, or pentavalent elements like antimony and phosphorus. The doped semiconductors are called extrinsic ones.

Electronic states in semiconductors are characterized by energy bands which can be filled with electrons. At zero temperature, the lowest empty band is termed as the conduction band and the highest filled band is the valence band. The conduction band and the valence band are separated by a forbidden band with bandgap energy of certain energy for a given semiconductor. Ideal semiconductors are not conductive at zero temperature, since there are no electrons in the conduction band. At finite temperature, some electrons are excited from the valence band into the conduction band where electrons are conductive.

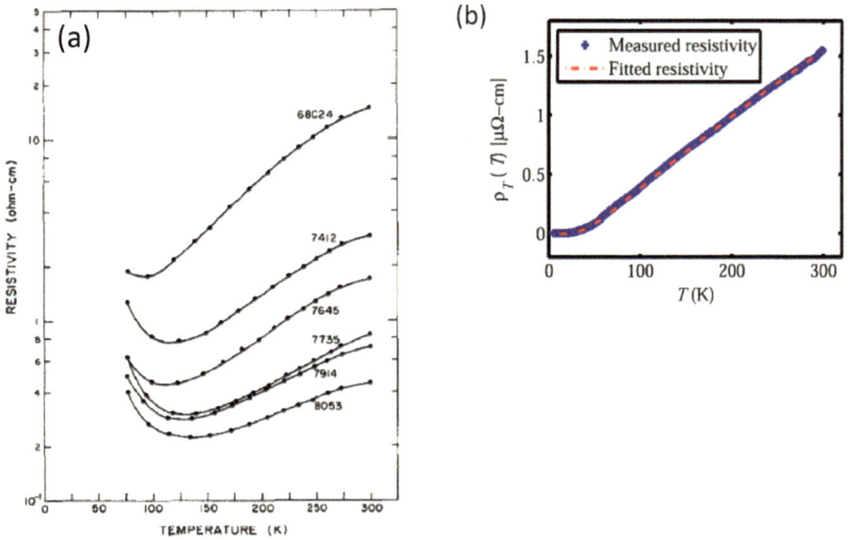

Figure 1.2 (a) Temperature dependence of resistivity of Si and (b) the measured and fitted resistivity (using the Bloch–Grüneisen formula) of a 100 nm Cu film [3, 4]. In a low-temperature range (below 100 K), the resistivity of Si decreases with temperature, a typical characteristic of a semiconductor. Reproduced with permission from [3, 4].

Optically, semiconductors present a turn-on characteristic of optical absorption at the photon energy equal to the bandgap, as shown in Fig. 1.3a. A semiconductor absorbs photons with energies equal to or larger than its bandgap, while the material is transparent to photons with lower energies [5]. On the other hand, a semiconductor might emit photons with energies equal to the bandgap roughly, when electrons in the conduction band combine with holes in the valence band as shown by the main peak on the photoluminescence (PL) spectrum in Fig. 1.3b. Note that the PL spectrum is also characterized by satellite peaks due to defects and phonons (see the peaks highlighted by dashed rectangles in Fig. 1.3b). The electrons can be excited into the conduction band by photons (photon excitation) and phonons (thermal excitation), or through electron injection via externally applied electric field [6]. The rich photoelectric properties make semiconductors excellent candidates for their use as photodetectors and light-emitting diodes.

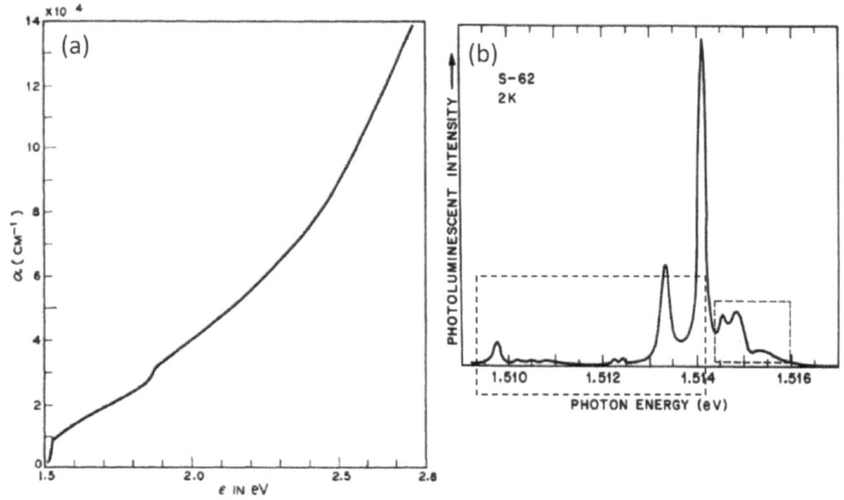

Figure 1.3 Optical absorption (a) and photoluminescence (b) curves of GaAs. The PL curve is characterized by the main peak (band-to-band emission) and the satellite peaks due to the defects and phonons as highlighted in dashed rectangles. Reproduced with permission from [6].

When the size of semiconductors reaches the nanometer level, these excellent properties of the semiconductor material become tunable within a certain range. Nanomaterials are useful and have great potential in many wide-range applications, including electronics and biology [7, 8]. In nanometer-sized semiconductors, many important material parameters, such as conductivity, electron mobility, electron Bohr radius, etc., are modified due to the boundary and the size of nanomaterials, since electron waves are reflected by the potential discontinuity on the boundary. The values of those parameters depend how and/or how much the boundary and the size of nanomaterials are varied. Due to quantum confinement, the electronic eigen states in nanomaterials are quantized, i.e., the density of states (DOS) is modified. The eigen states are also affected by the environment (absorbed foreign atoms, for instance), since the electron wavefunction is likely to extend into the environment outside nanomaterials. Thus, optical absorption and luminescence spectra of nano-materials are tunable through varying the size. As depicted in Fig. 1.4, variation of the size of the CdSe quantum dots leads to shift of the absorption edge and the luminescence peak (the longer the growth time, the larger the size) [9]. The unique property makes it possible to fabricate photoelectronic devices like solar cells to realize power conversion for the full solar spectrum, using just a single semiconductor material with different particle sizes.

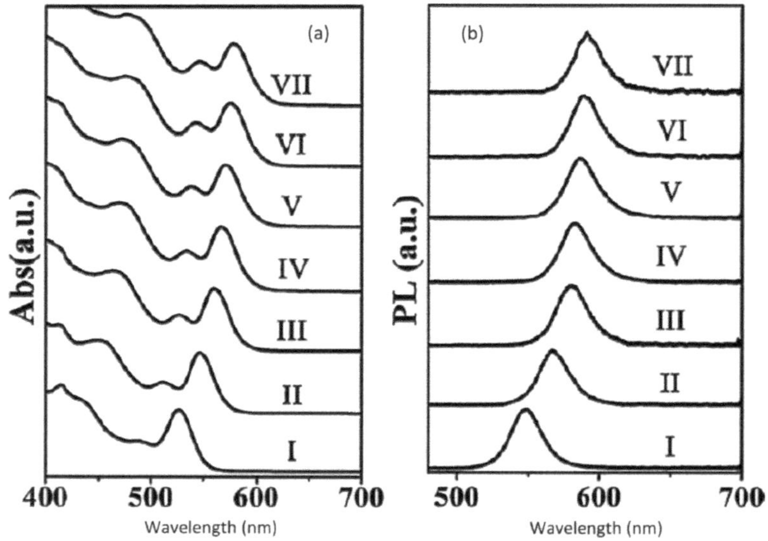

Figure 1.4 Optical absorption (a) and photoluminescence (b) spectra of CdSe nanocrystals that are grown in 10 s, 20 s, 30 s, 40 s, 50 s, 60 s and 70 s at the reaction temperature of 260 °C, corresponding to curves I–VII, respectively. Reproduced with permission from [9].

Some semiconductors also show excellent thermoelectric properties and they are useful in fabricating thermoelectric devices. Semiconductors with high thermoelectric figures of merit have been demonstrated for making thermoelectric coolers.

1.3 Electrons and Holes in Semiconductors

An electron is a charged particle that moves in the lattice background and receives scattering from the periodic lattice potential and other charged particles. In semiconductors, an electron behaves like a Bloch particle that moves on certain energy levels.

1.3.1 Atomic Orbits and Energy Bands

In crystalline semiconductors, insulators, and metals, atoms are arranged in a periodic order with certain crystalline symmetry, while in vapor and liquid phase materials atoms or molecules are randomly distributed. Like electrons in inner shell orbits, the energy levels of valence electrons in an isolated atom are discrete. When the atoms are brought closer and closer to each other, the interaction between valence electrons of the neighboring atoms is enhanced continuously. Note that the interaction between valence electrons of neighboring atoms is much stronger than that between inner shell electrons, characterized by strong wavefunction overlapping of the valence electrons. The originally degenerate energy levels of valence electrons in isolated atoms then split into separated levels due to the atomic interaction. Since the energy spacing of split levels is much smaller than k_BT, the thermal energy at conventional temperature, in the case of a large number of atoms, the collection of the closely spaced energy levels may be viewed as a continuous band of eigen states, as shown schematically in Fig. 1.5 [1, 10]. The energy bands dominantly formed by the ground and the excited states of the valence electrons are called the valence and the conduction bands, respectively. Similar to the valence shell electrons in isolated atoms, electrons in the valence band are the most active ones. The interaction of valence electrons between neighboring atoms dominantly determines the physical and chemical properties of the formed crystal.

There is no allowed electronic state in a forbidden gap (energy gap, bandgap) that separates the valence and the conduction bands. The forbidden gap forms as a result of strong interference between incident and reflective electron waves on the Brillouin boundaries. The interference leads to the formation of standing waves at the boundaries of Brillouin zone where the group velocity of electrons is zero. This is clearly demonstrated based on a nearly free electron model described in many semiconductor textbooks. The forbidden gap, characterized by the energy difference between the valence band maximum and the conduction band minimum in energy, is in the range of 0–3 eV for semiconductors. The bandgap energy E_g is a very important parameter for a semiconductor.

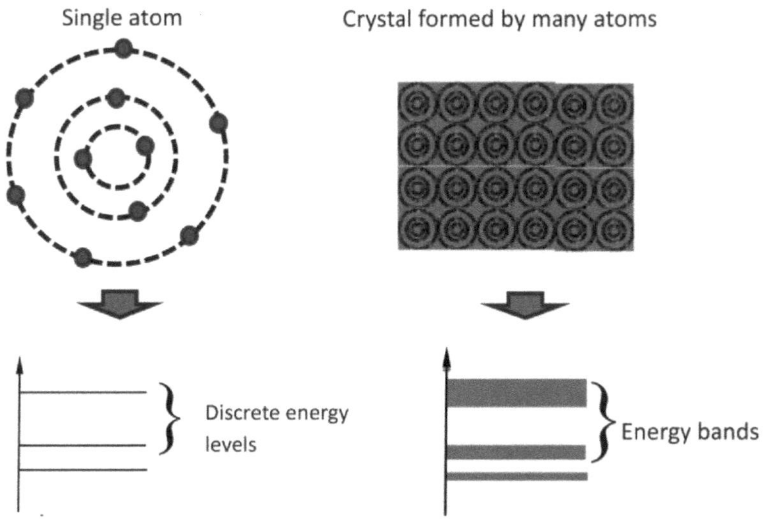

Figure 1.5 Schematic diagram of energy band formation when atoms are brought together to form a crystal.

For semiconductors, many important electrical and optical properties are determined by their bandgaps. The semiconductor bandgaps are typically below 3 eV, while the bandgaps of insulator are above 3 eV, although there is not a cutting line between a semiconductor and an insulator. In metals, however, the highest occupied band is half-filled, rendering them very different properties from semiconductors and insulators. As a result, the properties of very narrow-bandgap semiconductors are close to those of metals and the wide-bandgap semiconductors are similar to insulators. When we say semiconductors, insulators, and metals, we refer to those typical ones. Figure 1.6 presents the schematics of energy band and electron filling for a metal, a semiconductor, and an insulator.

Energy bands in solids are formed when individual atoms are brought together to form crystals. Therefore, the energy band formation is due to many body interactions in a semiconductor—a multiple particle system. For an accurate description of electronic states in such a system, one has to solve for the eigen value problem of a Schrödinger equation with the form of a Hamiltonian describing interactions among all particles:

$$\mathrm{H} = \sum_{I} \frac{\hbar^2 \nabla_I^2}{2M_I} + \sum_{i} \frac{\hbar^2 \nabla_i^2}{2m_i} + \frac{1}{2} \sum_{I' \neq I} \frac{Z_I Z_{I'} e^2}{4\pi\varepsilon_0 |R_I - R_{I'}|}$$

$$+ \frac{1}{2} \sum_{i' \neq i} \frac{e^2}{4\pi\varepsilon_0 |R_i - R_{i'}|} - \sum_{I,i} \frac{Z_I e^2}{4\pi\varepsilon_0 |r_i - R_I|}, \qquad (1.1)$$

where, on the right side of (1.1), the first (second) term is kinetic energy operators for ions (electrons), the third (forth) term the potential energy operators for ion–ion (electron–electron) interaction, and the fifth term the potential energy operators for ion-electron interaction. M_I (m_i) represents the mass of the I-th ion (i-th electron), R_I (r_i) is the position coordinate of the I-th ion (i-th electron), Z_I the atomic number of I-th nucleus. Diagonalization of the Hamiltonian in (1.1) for a system with 10^{22} electrons/cm^3 and 10^{22} ions/cm^3 is definitely a formidable work, which is an impossible task even with the use of a fastest modern computer.

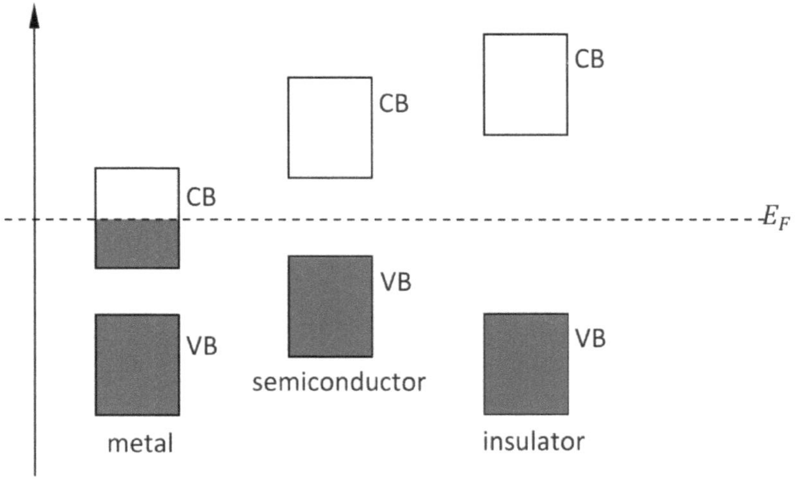

Figure 1.6 Energy band and electron filling for a metal, a semiconductor, and an insulator. The conduction band of a metal is partly filled. For semiconductors and insulators, their valence bands are filled and the conduction bands are empty at zero temperature. The shaded areas represent filled states and the dashed line is the Fermi level that lies roughly in the middle of the bandgap for a semiconductor and an insulator.

Very drastic approximations have been adopted, in which the many-particle problems are simplified to the one-electron framework [1, 11]. The single-electron framework assumes that each electron sees only an average potential distribution due to the fixed charges (positive ions) and other electrons in the crystal. In other words, the movement of each electron is only dependent on the average potential distribution of ions and other electrons throughout the crystal. Therefore, the coordinates of ions and other electrons are absent in Schrödinger equation in the single-electron framework. Under the single-electron approximation that is also termed as mean-field approximation, the solution for the eigen value problem for a many body system is reduced to find adequate single-electron eigen functions and eigen values for a crystal considered. In single-electron approximation, the Schrödinger equation takes the form of

$$H\psi(\boldsymbol{r}) = \left(\frac{\hbar^2\nabla^2}{2m} + V(\boldsymbol{r})\right)\psi(\boldsymbol{r}) = E\psi(\boldsymbol{r}), \tag{1.2}$$

where $H = \left(\frac{\hbar^2\nabla^2}{2m} + V(\boldsymbol{r})\right)$ is the single-electron Hamiltonian in which $V(\boldsymbol{r})$ is the average potential, $\psi(\boldsymbol{r})$ and E are eigen function and eigen value of the electron considered, respectively.

A number of methods have been developed to calculate the energy band structures for metals, semiconductors, and insulators. Those methods for energy band calculation differ in selection of the forms of periodic potentials and the set of base functions. The Bloch wavefunction of an electron is linearly expanded in terms of the set of base functions. In the near free electron framework, the total wavefunctions of electrons are simulated by a linear combination of a set of plane waves and the weak crystal potential is treated as perturbation [12]. The near free electron approximation is adequate choice for metals and narrow-bandgap semiconductors in which electrons are nearly free. In contrast, the tight-binding approximation is useful for crystals in which the valence electrons are tightly bounded [13, 14], so that the method is suitable for insulators and wide-bandgap semiconductors. Other frameworks often used for energy band calculation include psudo-potential and $\boldsymbol{k}\cdot\boldsymbol{p}$ models [14–16]. Although it is not a global description of electron energy bands for the whole Brillouin zone, the $\boldsymbol{k} \cdot \boldsymbol{p}$ method is particularly powerful in revealing details of high-symmetry points of band dispersion by offering analytical expressions of band dispersion and important parameters like effective masses. Therefore, the $\boldsymbol{k} \cdot \boldsymbol{p}$ model is conveniently used to explain optical transitions.

Energy band theory is one of the most successful one that has greatly promoted the advances of microelectronic technology. The general properties and fundamental nature of metals, semiconductors, and insulators can be functionally described using energy band theory. Figure 1.7 presents calculated energy bands of Ge, Si, and GaAs, as representatives of semiconductors [16].

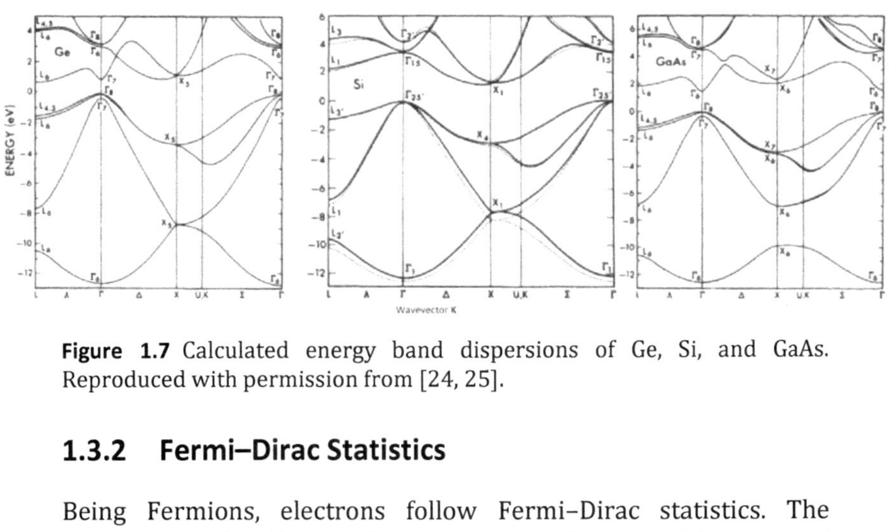

Figure 1.7 Calculated energy band dispersions of Ge, Si, and GaAs. Reproduced with permission from [24, 25].

1.3.2 Fermi–Dirac Statistics

Being Fermions, electrons follow Fermi–Dirac statistics. The Fermi–Dirac distribution function for an electron with energy E takes the form of [1]

$$f_0(E) = \frac{1}{e^{\frac{E-\mu}{k_B T}} + 1}, \tag{1.3}$$

where μ is the chemical potential, the Gibbs free energy for a single electron and the subscript "0" represents the system under zero external fields. $f_0(E)$ is the probability of an eigen state at the eigen energy E being occupied at thermal equilibrium. At not very high temperature, chemical potential μ is approximately equal to Fermi energy E_f as a good approximation (chemical potential = Fermi energy at 0K). Thus, we have

$$f_0(E) = \frac{1}{e^{\frac{E-E_f}{k_BT}} + 1}. \quad (1.4)$$

Figure 1.8 presents calculated $f_0(E)$ at several temperatures, showing that, at low temperature, $f_0(E)$ drops to 1 from 0 quickly with the increase of E going over E_f. The energy dependence of $f_0(E)$ becomes more gradual at higher temperature. The electron concentration in semiconductors is calculated by integrating over the density of states (DOS) multiplied by the distribution function, i.e., $D(E)f(E)$ with respect to energy:

$$n = \int_{-\infty}^{+\infty} D(E)f(E)dE \quad (1.5)$$

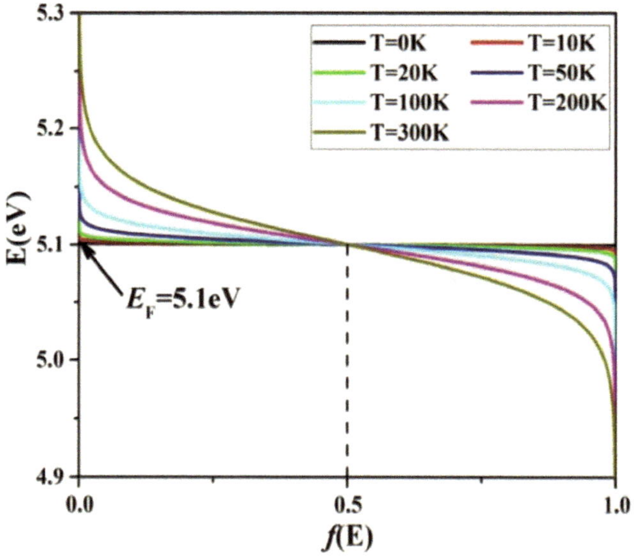

Figure 1.8 Fermi distribution function at 0, 10, 20, 50, 100, 200, and 300 K.

The total number of electrons in the conduction band is

$$n = \int_{0}^{+\infty} D(E)f(E)dE \quad (1.6a)$$

where zero energy is set at the Fermi level. For the same reason, the total number of unoccupied states in the valence band is then

$$p = \int_{-\infty}^{0} D(E)[1 - f(E)]dE \tag{1.6b}$$

Theoretically, a lot of important semiconductor parameters can be calculated, if the density of states $D(E)$ is known, by integrating a function containing the integrand $D(E)f(E)$ over energy E.

Fermi level is a description of electron occupancy on energy levels. Fermi levels of intrinsic semiconductors and insulators are located nearly in the middle of the bandgap. At zero temperature, the valence band is fully occupied by electrons and the conduction band is empty. For pure semiconductors, exciting an electron from the valence band into the conduction band has to overcome the bandgap energy. For semiconductors like Si with a bandgap of 1.12 eV, thermally exciting an electron into the conduction band is almost impossible, on considering the characteristic energies of acoustic phonons ($\sim k_D T$, typically a few meV) and optical phonons ($\sim$30 meV). However, optical excitation of electrons is possible for photon energy larger than 1.12 eV.

In narrow-bandgap semiconductors, however, thermal excitation generates a number of electrons, since phonons could offer enough energy for electrons to overcome the thermal barrier—bandgap. As a result, the intrinsic carrier concentration in narrow gap semiconductors could be quite high, in contrast to that in wide-bandgap semiconductors.

1.3.3 *P*-Type and *n*-Type Doping

There are two conduction mechanisms in semiconductors: electrons and holes. A hole is a missing electron (an unoccupied state) in the valence band. In a volume of 1 cm^3, a solid state material has $\sim 10^{22}$ atoms. In a metal, each atom contributes one or more free electrons. Therefore, the free electron concentration in a metal is at least in the order of 10^{22} cm^{-3}. In a "pure" semiconductor, the electron concentration is around 10^{10} to 10^{15} cm^{-3}, caused by unintentional doping, defects, and thermal excitation. Commercially available very pure Si has an electron concentration of only 10^{11} cm^{-3}. By doping the pure Si with

0.001% of arsenic, the impurities contribute 10^{17} free electrons per cm^3, leading to an increase in conductivity by a factor of 10^6!

The most fascinating property that a semiconductor differs from a metal and an insulator is their doping characteristics —the conductivities might increase millions times when the semiconductor is doped with 1/millions concentration of dopants [1, 17]. Doping results in increase of the number of charged carriers via donating electrons or holes by impurity atoms. For moderate and heavy doping, the carrier concentration in a semiconductor is dominantly determined by the amount of dopants. With a precise control over the amount of dopants, the electrical properties of a semiconductor can be modulated in a wide range. A semiconductor is called "p-type" if its dominant carriers are positively charged holes, or it is called "n-type" if carriers are mainly consisted of negatively charged electrons. A crystal is considered a good semiconductor if it can be easily doped with both p-type and n-type dopants of controllable carrier concentrations. Precise control over the carrier concentration is a prerequisite condition for the materials used for electronic devices. The kinds of carriers rely on doping elements. In the case of Si, for instance, the good p-type dopants are the group-III atoms like B and Ga, and the n-type dopants are some group-V elements like P and As.

Si is a group-IV element and each Si atom has four valence electrons. In Si single crystal, a silicon atom is covalently bonded to its four nearest neighbors. Each covalent bond is composed of two valence electrons and each valence electron is shared by two nearest Si atoms. Those tetrahedral covalent bonds hold the Si atoms together, forming a *diamond cubic* crystal structure (see Fig. 1.1a). The single crystal Si is very stable since, as shown in Fig. 1.9, each Si atom has an eight-electron close outer shell, although each electron is shared by two neighboring Si atoms. In a perfect Si crystal at zero temperature, all the valence electrons occupy the valence band states so that there is no conductive electron. At non-zero temperature, some valence electrons could be thermally excited into the conduction band where they become nearly free electrons. The thermal excitation level is, however, very low for pure Si at conventional temperature.

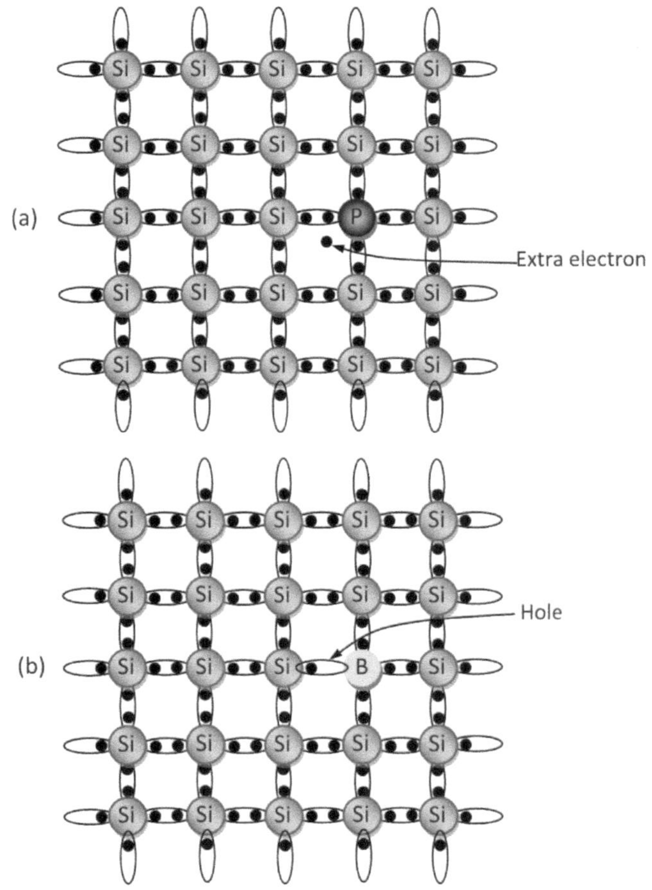

Figure 1.9 Crystalline Si when doped with phosphorous (a) and boron (b).

Elements can be doped into Si crystals as substitutional dopants that replace Si atoms at lattice points, or interstitial dopants that take the interstitial sites. A group-V phosphorus (P) atom has five valence electrons and, if a Si atom in Si crystal is replaced by a P atom as shown in Fig. 1.9a, four of the five valence electrons become covalently bonded to four nearest neighbor Si atoms. The fifth valence electron is extra with respect to the covalent bonding and free to move in the semiconductor upon P atom ionization. The ionization energy of the extra electron is very low so that the ionization is likely to occur at room temperature. Indeed, in P-doped Si crystals, the extra electrons donated by P atoms are located at shallow donor impurity levels and they are thermally excited into the conduction band. A P atom in Si crystal is therefore called a donor impurity, since it donates a free electron for the material conduction. If a Si atom is replaced by a group-III atom like boron (B) (see Fig. 1.9b), each boron atom, having three valence electrons, is in short of one electron for the formation of four covalent bonds with its four nearest Si neighbors. A hole is thus bound in the vicinity of the boron atom. The boron atom can capture an electron from a neighboring Si atom, generating a hole in the valence band so that a boron atom (or other group III atoms) acts as an acceptor impurity. Holes are positively charged and can move freely in the crystal.

The *n*-type doping and *p*-type doping enhance conductivity significantly, since the *n*-type dopants contribute a large number of free electrons nearly equal to the number of *n*-type impurities and the *p*-type dopants lead to a large number of free holes nearly equal to the number of *p*-type impurities, as shown in Fig. 1.10a,b, respectively. However, doped Si remains charge neutral, though there are a large number of free electrons in *n*-type Si and free holes in *p*-type Si. Strictly speaking, there always exist a small number of free holes (electrons) in an *n*-type (a *p*-type) semiconductor. They are called the minority carriers, in contrast to majority ones.

Electrons in the conduction band and hole in the valence behave quite similar to free electrons and free holes—free to move, since the carriers in the bands receive only very weak scattering due mainly to the translational invariant of the periodic crystal potential, as well as screening effect of core electrons. Doping in nano semiconductors, however, is rather difficult [18, 19]. First, controlling the doping concentration is a big challenge since in quantum dots, for instance, a doping concentration of 10^{17} cm^{-3} corresponds to one or two impurities in each quantum dot. Secondly, the large surface to bulk ratio of atom numbers, together with high density of defects and dislocation, make it extremely difficult for an impurity atom to sit at a right site. In addition, electrons and holes contributed by impurities are likely to be captured by defects and surface states of nanomaterials. Although there exist many technical difficulties, some progress has been made in doping nanomaterials [20, 21].

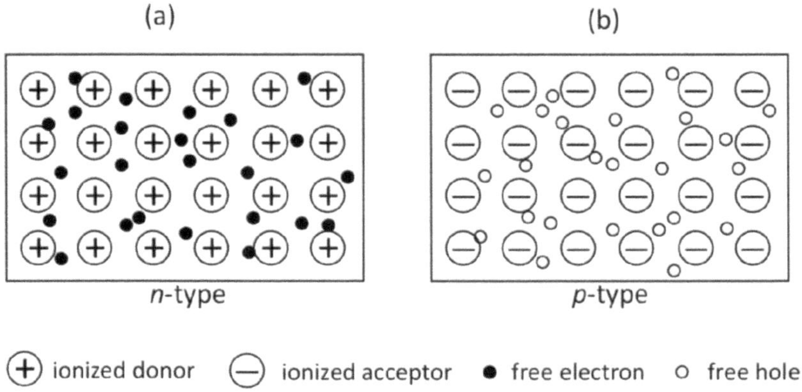

Figure 1.10 Free carriers in *n*-type (a) and *p*-type (b) doped Si, where ionized donors and acceptors are fixed charges.

1.3.4 Transport Process of Electrons and Holes

One has to rely on quantum mechanics to understand the transport properties of carriers in a semiconductor. Intuitively, an electron (a hole) traveling in a crystal experiences a great deal of scattering from ions and other electrons, which should significantly slow down the movement of electrons. However, electrons move in a crystal nearly like Bloch-like free electrons with long mean free paths. In fact, electrons in crystals are well-described by the semiclassical model in which an electron interacts with a periodic potential of fixed ions under the screening of other electrons. According to the model, the mean velocity of a Bloch electron can be written as

$$v(\boldsymbol{k}) = \frac{1}{\hbar}\frac{\partial \mathcal{E}(\boldsymbol{k})}{\partial \boldsymbol{k}}, \tag{1.7}$$

where $\varepsilon(\boldsymbol{k})$ is the eigen value of electrons obtained by solving the Schrödinger equation in terms of a Hamiltonian with periodic potential. Thus, the non-vanishing $v(\boldsymbol{k})$ contains the contribution of the scattering due to periodic potential. In an ideal crystal, an externally applied electric field accelerates electrons to an infinitive speed. However, defects, impurities, and other imperfections act as scattering centers to slow down electrons, leading to limited carrier mobility and conductivity. For high-purity and perfect materials, the boundary scattering due to surface and interface could dominate the scattering to electrons. Under zero electric field, the Fermi ball in 3D or Fermi circle in the 2D case is symmetric, so that macroscopically average electric current is zero. The average zero-current is well understood by the fact that for any given electron with a momentum of $\boldsymbol{k}$, one could always find another electron with a momentum of $-\boldsymbol{k}$, as shown by the dashed circle in Fig. 1.10a. This is made clear by the Fermi distribution function under zero-field, f_0, depicted by the dashed line shown in Fig. 1.10b.

The zero-current situation is changed by applying an electromagnetic field to the system. In an electromagnetic field, electrons are driven by the Lorenz force and the equation of motion for electrons is described by

$$\frac{1}{m^*}\frac{d\boldsymbol{p}}{dt} = \frac{1}{\hbar}\frac{d\boldsymbol{k}}{dt} = -e\left[\boldsymbol{E}(\boldsymbol{r},t) + \boldsymbol{v}(\boldsymbol{k}) \times \boldsymbol{B}(\boldsymbol{r},t)\right], \tag{1.8}$$

where $\boldsymbol{p}$ is the electron momentum, m^* the mass of the electron, $\boldsymbol{k}$ the wavevector, $\boldsymbol{E}$ the electric field and $\boldsymbol{B}$ the magnetic field. Equation (1.8) indicates that electrons are accelerated by $\boldsymbol{E}$ and $\boldsymbol{B}$ toward a certain direction determined by the orientations and magnitudes of $\boldsymbol{E}$ and $\boldsymbol{B}$. According to (1.8), electron momentum could become huge with prolonged exertion of an electric field, since no damping has been considered. Equation (1.8) could be modified using relaxation time approximation, so that

$$\frac{1}{\hbar}\frac{d\boldsymbol{k}}{dt} - \frac{m^* v}{\tau} = \frac{dv}{dt} - \frac{m^* v}{\tau} = -e\left[\boldsymbol{E}(\boldsymbol{r},t) + \boldsymbol{v}(\boldsymbol{k}) \times \boldsymbol{B}(\boldsymbol{r},t)\right] \tag{1.9}$$

where τ is relaxation time. The applied electro-magnetic field causes the shift of the Fermi circle and change of Fermi–Dirac distribution function, as shown by the solid lines in Figs. 1.11a and 1.11b, respectively. As a result, an electron with a momentum $\boldsymbol{k}_1$, for instance, cannot find its counterpart electron with a momentum equal to $-\boldsymbol{k}_1$ in the solid Fermi circle of Fig. 1.11a. This leads to non-zero average electron momentum. Correspondingly, there is a macroscopic electric current passing through the crystal. Assuming that an external electric field is applied in the **x** direction, we could rewrite (1.8) to obtain

$$\frac{dv_x}{dt} - \frac{m^* v_x}{\tau} = -eE_x \tag{1.10}$$

We can then get $v_x = -\frac{eE_x\tau}{m^*}$ or $\Delta k_x = -\frac{eE_x\tau}{\hbar}$, using $v_x = \frac{\hbar}{m^*}\frac{dk_x}{dt}$. Generally in an externally applied electric field $\boldsymbol{E}$, the Fermi distribution function is

$$f_E(\boldsymbol{k}, t) = f_0(\boldsymbol{k}) + \frac{e\tau(\boldsymbol{k})}{\hbar}\boldsymbol{E}\cdot\nabla_k f(\boldsymbol{k},t) \cong f_0\left(\boldsymbol{k} + \frac{e\tau(\boldsymbol{k})}{\hbar}\boldsymbol{E}\right) \tag{1.11a}$$

$$f_E(\boldsymbol{k}, t) - f_0(\boldsymbol{k}) = \frac{e\tau(\boldsymbol{k})}{\hbar}\boldsymbol{E}\cdot\nabla_{\boldsymbol{k}} f(\boldsymbol{k}, t). \tag{1.11b}$$

$f_E(\boldsymbol{k}, t) - f_0(\boldsymbol{k})$ is plotted in Fig. 1.11b, indicating the effect of applied field on Fermi distribution function. For detailed analysis on transport properties of materials, one has to use the Boltzmann equation.

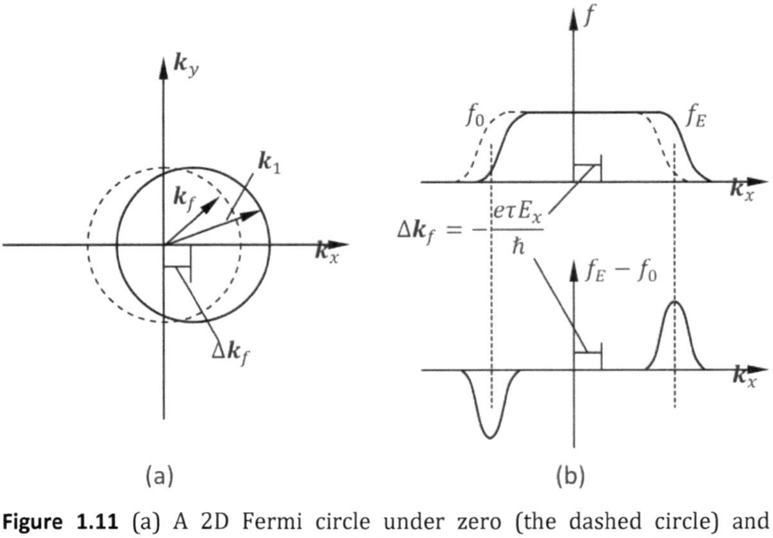

Figure 1.11 (a) A 2D Fermi circle under zero (the dashed circle) and non-zero (the solid circle) external electric field in *x* direction; (b) electron distribution function under zero field (f_0) and non-zero electric field in *x* direction.

1.3.5 Defects in Semiconductors

There are always various defects in crystals [22, 23]. Defects in crystals are categorized into two types: intrinsic and extrinsic ones. Intrinsic defects are missing atoms, or atoms in semiconductors, in geometrical sense, sit at wrong sites with respect to those of atoms in a perfect crystal. Intrinsic defects include point defects, dislocations, micro-twins, etc, falling into the categories of point defects, line defects, face defects, respectively. Extrinsic defects are caused by intentional and unintentional doping with foreign atoms in semiconductors taking either the interstitial or the substitutional positions. Some defects are even mobile in crystals. Electrically, the defects could correspond to either deficit or extra electrons, so that defects could behave as donors or acceptors. If defect density is not very high, the electronic states of the defects are localized. For those defects with their electronic states in the bandgap but very close to the bottom of the conduction band or to the top of the valence band, they are the shallow donor states or shallow acceptor states, respectively.

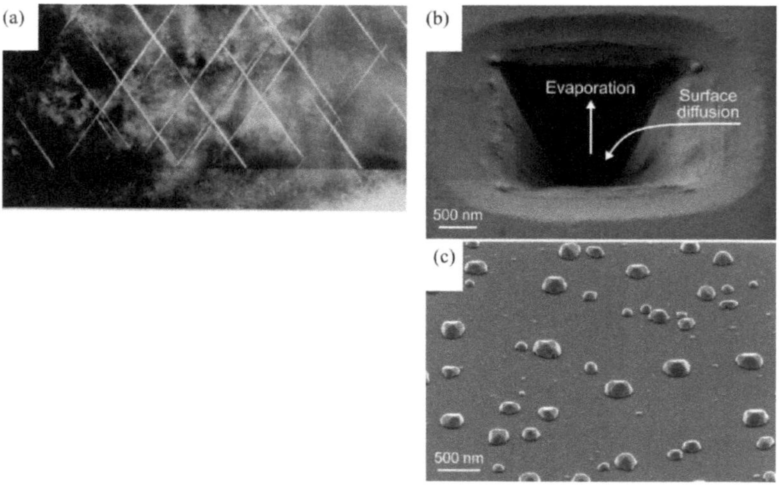

Figure 1.12 TEM dark field cross-section images of ZnCdSe grown on InP substrates, which shows a high density of microtwins in epilayer as indicated by arrows (a); SEM image of an etch on Ge (001) showing a pyramid shape in the center and circular depression on the edge; and (c) SEM image of Ge crystallites deposited on carbon-contaminated Ge(001).

At conventional temperature, electrons in the shallow donor states could be excited into the conduction band and holes in the shallow acceptor states excited to the valence band, (i.e., electrons are captured by the acceptor states from the valence band). Figures 1.12a and 1.12b present the transmission electron microscope (TEM) image of ZnCdSe expitaxially grown on InP substrate (cross-sectional view) and scanning electron microscope (SEM) image of Ge etch pits [24, 25].

1.4 Built-in Electric Field

1.4.1 *p-n* Junction

Modern semiconductor devices rely more and more on junctions formed by contacting between one semiconductor and another. The two semiconductors could either be the same semiconductors (homojunction) with different types of doping or different semiconductors (heterojunctions). Junctions could also be constructed with a semiconductor and an insulator, as well as with a semiconductor and a metal. Limited by narrow room for material selection and few technologies for material preparation, ideal junctions are extremely difficult to obtain. There are always defects, impurities, contaminations, as well as atomic interdiffusion in the junction region. For better understanding the fundamental properties of a junction, one could assume ideal contacts, meaning that the two materials are in intimate contact on atomic scale with no defects, no absorbed foreign atoms, and no atomic interdiffusion.

When a *n*-type Si and a *p*-type Si are brought into contact with the interface being atomic smooth (technically, this is realized by using a *n*-type wafer with one side doped with *p*, or vice versa), as shown in Fig. 1.13a, there is now a system in which there are high density of free electrons on the left side of the junction and high density of holes on the right side. The free electrons (holes) on the *n*-type (*p*-type) region will start to diffuse into the *p*-type (*n*-type) region of low electron (hole) concentration (see Fig. 1.13b). The diffusion is driven by the gradient of electron and hole concentrations in the system. The diffusion of electrons and holes breaks the original charge neutrality of the materials so that *p*-type region become negatively charged and *n*-type region become positively charged. Thus, band edges in positively net charged *n*-region moves down and those in negatively net charged *p*-region move up referring to electron energy, forming a junction potential, as shown in Fig. 1.13c. The net charge imbalance results in a macroscopic electric field pointing rightward, the opposite direction as that of electron diffusion and the same direction as that of hole diffusion. Now, there occurs electromagnetic force that drives the drift motion of electrons and holes, in addition to the thermodynamics force that drives the diffusion. As shown in Fig. 1.13c, electrons (holes) are swept by the macroscopic junction electric field to the *n*-region (*p*-region). Thus, the drift direction of electrons (holes) is opposite to the diffusion direction of electrons (holes). Finally, a net charge transfer stops and a dynamic balance between electron (hole) diffusion and drift is established. The *p-n* junction with a certain value of junction electric field is then formed. Theory of a *p-n* junction has been described in a large number of literature and books.

Charges participating in diffusion are those located in the vicinity of the junction. At the initiation of thermal diffusion, free electrons in the *n*-region side (nearby the junction) diffuse into the *p*-region and recombine with majority carriers—holes, leaving behind with positively charged fixed ions since free carriers are depleted. Similarly, in the *p*-region nearby the junction charges are dominated by negatively charged fixed ions. Thus, the built-in electric field is mainly due to the fixed charges, as shown by Fig. 1.14a. The junction region in which an electric field is built is called the space charge region, which is also called depleted region since free carriers are nearly depleted. The distribution of carrier concentrations for electrons and holes is schematically depicted in Fig. 1.14b. Note at the junction, the carrier concentrations for both electrons and holes are essentially zero. The width of the depletion layer depends on the semiconductor properties, specially doping concentrations of both *p*- and *n*-regions. Figure 1.14b–e present schematical distributions of carrier density, electric field, voltage, as well as charge quantity.

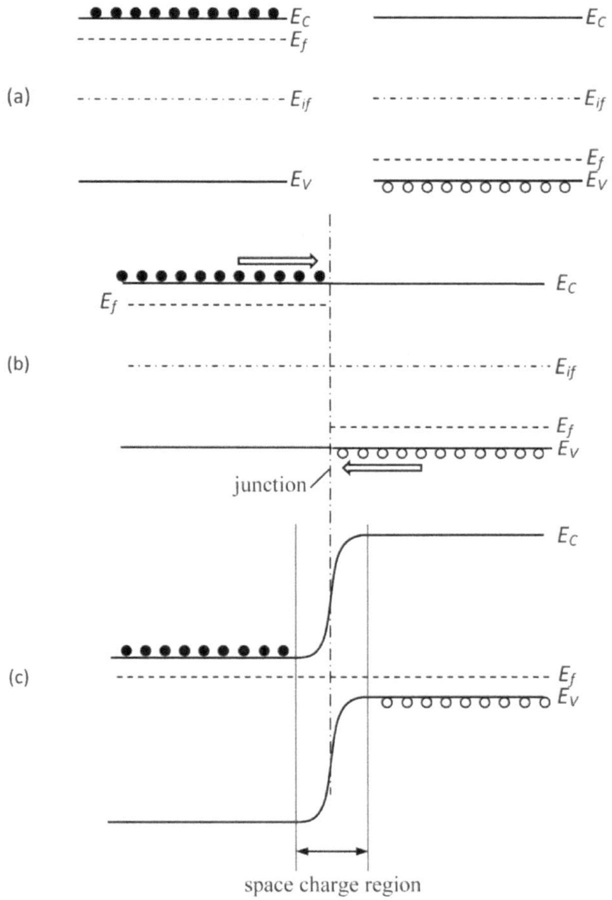

Figure 1.13 A *p–n* junction and a built-in electric field.

A *p–n* junction plays the key role in an opto-electronic device. In wafer-based Si solar cells, for instance, it is the built-in electric field in the *p–n* junction that splits photoexcited electron–hole pairs (excitons), which is the critical part of the photovoltaic process. A strong built-in electric field is of most importance for achieving high conversion efficiency of photovoltaic devices. For c-Si solar cells, high-energy photons could be absorbed by the surface layer before they could reach to the junction region, since the penetration depth in Si is photon-energy dependent. On the other hand, the low energy photons might penetrate through the junction if the junction is very shallow, though shallow junction is good for high-energy photons. Photon-exited electron/hole pairs (excitons) in junction region are likely to be split by the built-in electric field, while the excitons generated outside of the *p–n* junction are likely to recombine radiatively or non-radiatively so that they make no contribution to photovoltaic effect.

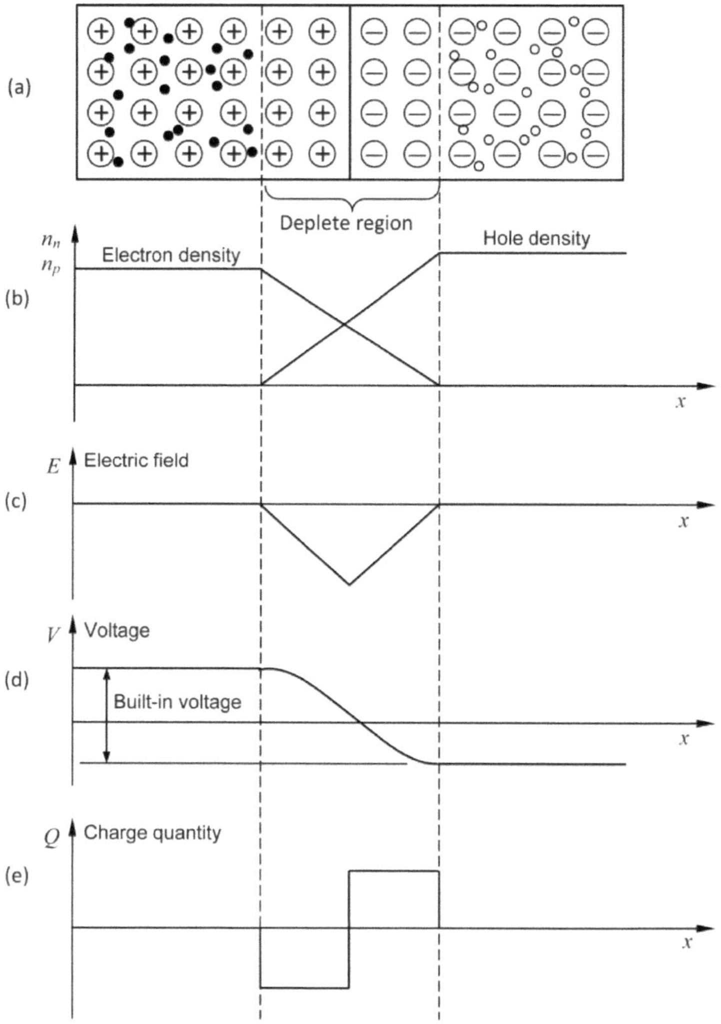

Figure 1.14 A *p–n* junction with carrier depletion (a) and carrier density (b), electric field (c), voltage (d), as well as charge quantity (e) distributions.

1.4.2 Heterojunction

A heterojunction can be constructed by bringing two different materials into contact with atomic smooth and atomic bonding interface. For most semiconductor heterojunctions, the two materials are single crystals. The required conditions for the formation of heterojunction are rigorous and typical heterojunctions are usually prepared by vacuum thin-film growth technologies like molecular beam epitaxy (MBE) and metal organic chemical vapor deposition (MOCVD) [26, 27]. Those heterojunctions usually have large interface area in square centimeters or square inches. The heterojunction could also be formed by nanomaterials with their interface areas in the sizes of nanometer squares. Nano heterojunctions are usually prepared using chemical ways and the junction quality is not as good as those grown by MBE.

A typical configuration of heterojunction consists of p-doped AlGaAs and n-doped GaAs. An exchange of electrons and holes, between the two differently doped materials, occurs due to thermal diffusion. As shown in Fig. 1.15, very important parameters for the characterization of the heterojunction are the band discontinuities ΔE_C and ΔE_V in the conduction band and the valence band at the junction, respectively, which are the partitions of bandgap difference ΔE_g in the conduction and the valence bands so that $\Delta E_C = \Delta E_g - \Delta E_V$.

It is, however, rather difficult to measure the parameters precisely. Theoretical prediction on the energy band discontinuities is also challenging even for a nearly ideal junction interface. Assuming a perfect interface, the energy positions of bulk energy bands of two semiconductors forming the heterostructure are affected mutually, due to the charge exchange between the differently doped semiconductors. Consider a heterojunction consisting of p-type doped AlGaAs and n-type doped GaAs, with the band edges and Fermi energy levels in GaAs and AlGaAs are aligned in the way shown in Fig. 1.15a before they are brought into contact. Note that the Fermi energies of the two materials are not aligned when they are separated. Upon being brought into contact, charge migration occurs in the way that holes migrate from the p-type AlGaAs toward the n-type GaAs, while electrons the *n*-type GaAs toward the *p*-type AlGaAs, driven by thermodynamic driving force (concentration gradients of electrons and holes on both sides of the heterojunction). The electron and hole migration won't stop until the Fermi levels on both semiconductors line up, giving rise to the band alignment of the unified system as shown in Fig. 1.15b.

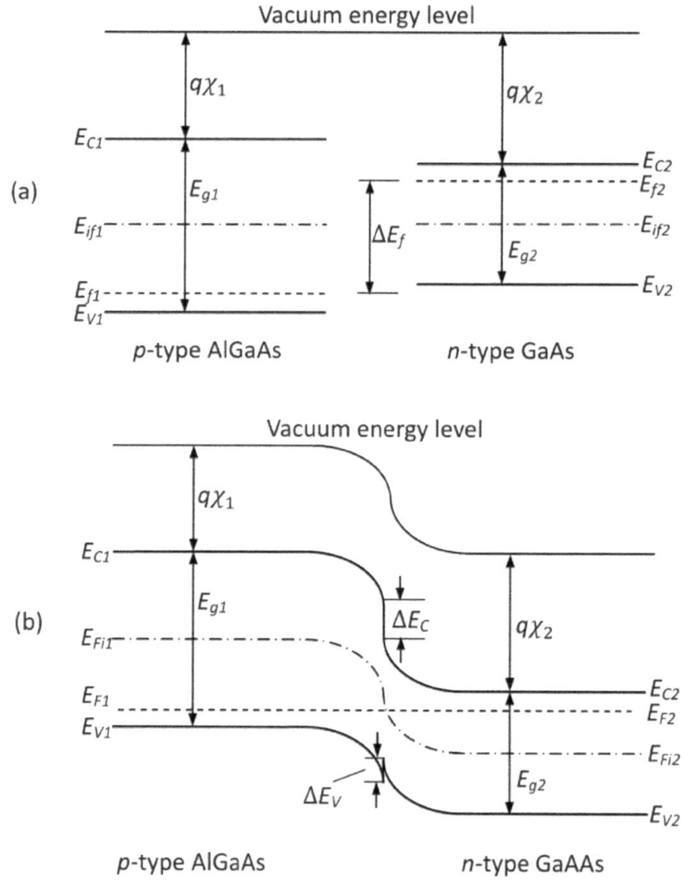

Figure 1.15 Schematic diagrams of energy band profiles of the *p*-type AlGaAs with relatively large bandgap and *n*-type GaAs, prior to (a) and after (b) contacting.

Assuming there is no surface band bending prior to contact, the energy bands in the junction region have to shift for a measure to accommodate aligning of the Fermi levels of the two semiconductors. Thus, the built-in potential across the junction is approximately equal to difference in Fermi levels, i.e.,

$$qV_b = E_{F1} - E_{F2} \tag{1.12}$$

where V_b is the built-in voltage. qV_b is mainly determined by p-doping and n-doping levels, so that above equation can also be written as

$$qV_b = k_B T \ln\left(\frac{N_a N_d}{n_i^2}\right) \tag{1.13}$$

where N_a and N_d are the densities of acceptors and donors, respectively, assuming complete ionization of both kinds of impurities. Apparently, large built-in potential, beneficial for splitting the electron and the hole of an exciton in a photovoltaic process, can be obtained on semiconductors with large bandgaps. The junction properties can be tuned by an electric field externally applied cross the junction through electric contacts. The applied electric field either adds on or reduces the built-in potential. In either case, there is a charge redistribution induced by the applied electric field.

When two different nanoparticles (or differently doped) are brought into contact at atomic level (form atomic bonding), an energy level discontinuity occurs. Followed by the self-consistent process of charges, an electric potential is then established at the contact region where photogenerated electrons and holes are driven to move toward opposite directions. However, the contacting-induced electric field is so far in poor quality, as far as photovoltaic effect is concerned. The major problems lie in the following:

(1) Contact is not perfect in the sense that nanoparticles are often coated with organic ligand that is difficult to remove completely. As a result, good electric contact is hard to achieve.
(2) The whisker contact between nanoparticles is often point, instead of face contact. It is thus difficult for the contact to carry large current required by solar cells.
(3) The stability of the contact is poor.

Due to above-mentioned problems, together with existing high density of defects and surface states, the photo conversion efficiency of nanomaterial-based solar cells is currently very low. This is shown very clearly by poor short-circuit current typically in the order of $\mu A/cm^2$. A number of works were reported to improve the performance of nanomaterial solar cells. One sees the future for breaking through, though more works need to be done right now.

Nano solar cells are largely based on heterojunctions for charge separation. The hetero-interfaces are, however, usually far from atomic smooth and defect free. A typical example are nano solar cells using quantum dots for sensitization, where electrons are separated from holes at TiO_2/QD junction [28]. The quantum dots are often coated with organic ligands that are extremely difficult to clean up completely. The coated layer acts as a barrier resisting electron transfer, which is part of the reason why the quantum dot solar cells have low short-circuit current compared to wafer-based Si ones. High density of defects on surface of quantum dots is often a more serious problem. The defects often act as trapping centers causing charge trapping effect and inducing electron–hole recombination. Since the quantum dots are of nano-sizes, so are the areas of heterojunction interfaces.

1.4.3 Semiconductor–Metal Contact

Metals are reservoirs of electrons. When a semiconductor, with limited number of electrons and holes, and a metal are brought into contact forming an atomic smooth interface, a self-consistent process occurs where charges are exchanged between the metal and the semiconductor, which modifies the electric potential both at the contact boundary and in the depth. When the Fermi level in an *n*-type semiconductor is higher than that in the metal, electrons in the semiconductor move into the metal upon contacting, resulting in a Schottky potential barrier in the semiconductor adjacent to the contact interface. A Schottky barrier blocks electron injection from the semiconductor into the metal. For a *p*-type semiconductor with its Fermi level below that of a metal, a Schottky barrier in the valence band forms due to hole migration from the semiconductor into the metal. A voltage could be applied cross the junction to tune the Schottky barrier [29]. On the other hand, a semiconductor–metal junction could also exhibit an Ohmic behavior where *I* is linearly dependent on *V* in the cases that the Fermi level in an *n*-type semiconductor is lower than that in a metal or the Fermi level in a *p*-type semiconductor is higher. Ohmic contacts are the fundamental requirement for devices where a current needs to be injected into the devices through an external circuit.

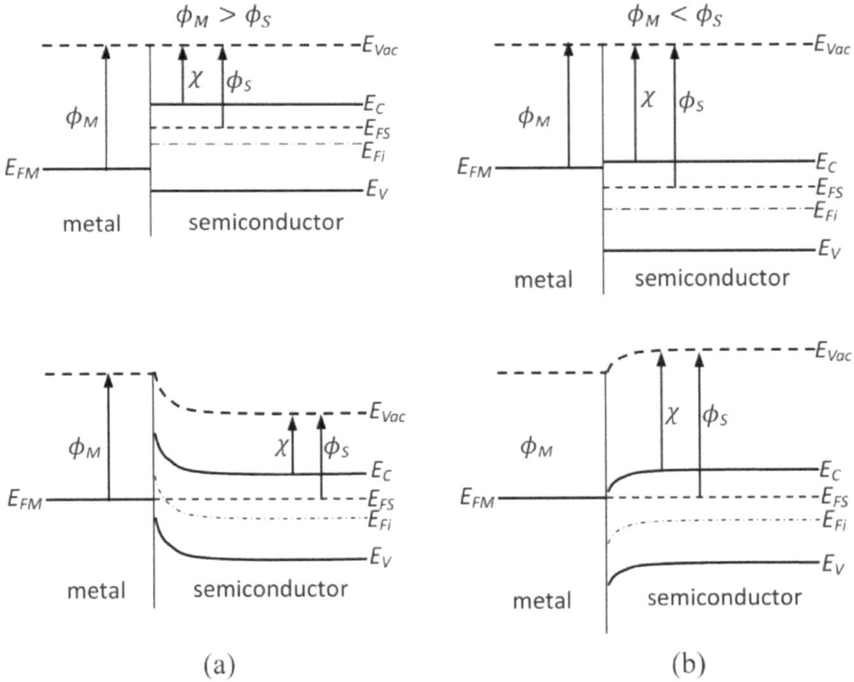

Figure 1.16 (a) A junction formed by an *n*-type semiconductor and a metal for $\phi_M > \phi_S$ prior to (up panel) and after self-consistent process (bottom panel) and (b) a junction for $\phi_M < \phi_S$.

The way of band bending in the junction region depends on the work functions of the metal and the semiconductor. If the metal work function ϕ_M is larger than the work function ϕ_S of an n-type semiconductor, electrons are extracted from the semiconductor to the metal. The band in the semiconductor then bend upward at the junction from in-depth due to its loss of electrons, as shown in Fig. 1.16a. The surface layer is then enriched with holes. When ϕ_M is smaller than ϕ_S, electrons flow from the metal to the semiconductor in the interfacial region, causing the band in semiconductors bend downward and the surface is enriched with electrons, as depicted in Fig.1.16b.

If electrode contacts are made on both sides of the semiconductor and the metal, an I–V curve (current-voltage relationship) can be measured and the Schottky junction displays a rectifying property. Thus, solar cells can be fabricated based on the Schottky barrier that is the driving force for exciton splitting and the property of the potential barrier is critical for the performance of the solar cells. Semiconductor–metal junctions of various materials exhibit interesting physical properties useful for many applications.

1.4.4 Semiconductor–Insulator Interface

A semiconductor–insulator junction is another fundamental structure for many opto-electronic devices [30]. When contacting is made between a semiconductor and an insulator, the insulator performs as an electric barrier but not necessary an electric isolator. In a metal oxide semiconductor field effect transistor (MOSFET), the oxide insulator layer is used for applying a gate electric voltage to allow or terminate electric current through the source and drain. In the case of a quantum tunneling junction, a voltage is applied cross the insulator layer as a barrier but it allows for electrons of certain energies to penetrate through the barrier.

Nanomaterials have very large and irregular surface area, which could be troublesome for achieving high device performance and stability. Passivation becomes pre-requirement and insulators are often chosen to perform the function. Bare quantum dots, as high efficient light absorbers (as a result, they are often referred as light antenna), are not stable, so that they are coated with insulating layers to form core/shell structures, as shown in Fig. 1.17a. The core/shell quantum dots can be grown by, for instance, the hydro-thermal method with a nearly perfect core–

shell interface. The core/shell structure stabilizes the quantum dots, as demonstrated in Figs. 1.17b and c, showing that the fluorescence spectrum of CdSe quantum dots with ZnSe shell hardly changes with time. Like planar interface grown using epitaxial techniques, type-I and type-II interface band profiles can be constructed, as shown in Fig. 1.17a. When the quantum dots are located in an electric field like space charge region in a *p-n* junction, electrons and holes are separable in both the type-I and type-II quantum dots, if the shell layers are thin enough.

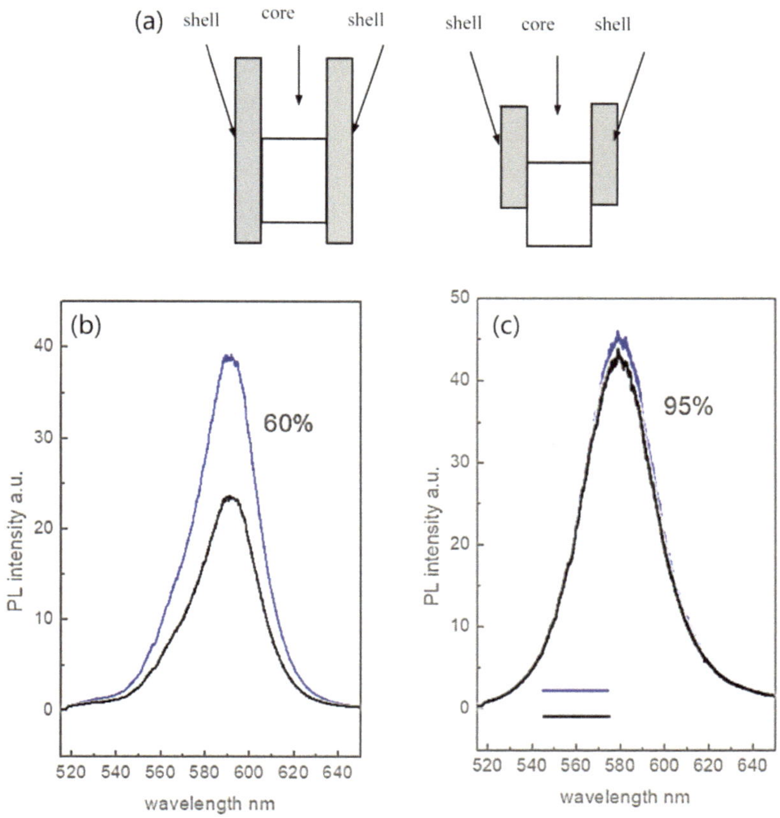

Figure 1.17 Schematic energy band profile of core/shell QDs with type-I (left) and type-II (right) band alignments (a). Photoluminescence spectra of bare CdSe QDs (b) and CdSe/ZnSe core/shell QDs (c) all suspended in solution, where the blue curves are measured just after growth and the Black curves after exposed to sunlight for two hours.

1.5 Nanomaterials

Nanomaterials refer to materials that have sizes in 1–100 nm range at least in one dimension. To be more accurate, nanomaterials should exhibit size effect—some fundamental material parameters, such as electronic states, diffusion length, carrier mean free path, etc., are size-dependent. The size-dependent effect is called nano effect since, in most cases, it occurs when the size of the materials is in the nanometer range, although the size-dependent effect does not have to show up in nano-size materials. In a GaAs two-dimensional electron gas (2DEG) system formed at the heterojunction between intrinsic GaAs and *n*-type AlGaAs, the electron mobility in the *xy* plane (in plane) can be as high as 3×10^7 $cm^2/(V \cdot s)$, corresponding to an electron mean free path of 300 μm. In another words, electrons moving in the *xy* plane could feel the boundary if the size of the 2DEG system is close to 1 mm in the *x* or *y* direction. Thus, the best GaAs 2DEG system displays size-effect at the size of 1 mm and it is called "nano effect"! Nano effect, which does not require that a material has a nano size, really means size effect.

The formation of energy bands in a bulk crystalline semiconductor is due to the strong inter-atom coupling among valence electrons. Electrons excited into the conduction band and hole into the valence band are delocalized, so that they are free to migrate in the conduction and the valence band, respectively. Within a semiconductor QD, electrons and holes still experience the periodic potential but they are also confined by the boundary. In macroscopic nano semiconductor thin films (or bulk) shown in Fig. 1.18, semiconductor QDs (often termed as nanoparticles) are embedded in an insulting matrix, making electron migration in the nano semiconductors become quite difficult due to the scattering of boundaries. Electron and hole transport through the materials relies on tunneling or hopping. For a given electron energy, tunneling is sensitively dependent on barrier height and barrier width, which are difficult to control in nanomaterials. As a result, the conductivity in a nano thin film is low. However, some characteristic parameters, such as an average electron (hole) velocity, drift length, and diffusion length can still be defined in the case of nano thin films. Similarly, since a nano thin film is not a material with homogenous phase, defect density and doping density are different inside the nanoparticle and in the matrix region and some measurement can be performed to describe such non-uniformity. In fact, an averaged doping density and defect density are often used for a macroscopic area or volume. The justification for the effectiveness of the "average" parameters is that diffusion length, drift lengths, and depletion region width in the film needs to be much larger than the size of the nanograins.

Figure 1.18 SEM image of nano CdSe thin film, with the average CdSe particle size of 5.3 nanometer.

Nanomaterials (nanostructures) have very large surface area-to-volume ratio, which depends on the size and shape of nanoparticles. If a 1 m^3 cube, for example, is cut into smaller and smaller cubes, the total surface area of the cubes increase significantly. As listed in Table 1.1, when a cube of 1 m side size is cut into cubes of 1 nm side with the same total volume, the surface area increases from 6 m^2 to 6×10^9 m^2, a factor of 10^9! Thus, nanomaterials with small particles have relatively large surface area. In some cases, large surface area allows for the enhancement of material properties for diverse applications and, more importantly, the material properties are tunable via varying the size and the shape. On the other hand, large surface area could sometimes be troublesome for device performance.

Table 1.1 The surface area change as a 1 m^3 cube is cut into smaller and smaller cubes

Number of cubes	Size of cube side	Surface area
1	1 m	6 m^2
10^3	0.1 m	6×10 m^2
10^6	0.01 m	6×10^2 m^2
10^9	1 mm	6×10^3 m^2
10^{12}	0.1 mm	6×10^4 m^2
10^{15}	0.01 mm	6×10^5 m^2
10^{18}	1 μm	6×10^6 m^2
10^{21}	0.1 μm	6×10^7 m^2
10^{24}	0.01 μm	6×10^8 m^2
10^{27}	1 nm	6×10^9 m^2

1.5.1 Quantum Confinement and Quantum Confinement States

In textbooks of condensed matter physics, ideal materials are regarded as infinitively large. In dealing with electronic states in semiconductors, boundary conditions are introduced. Two forms of boundary conditions, namely, fixed boundary (standing wave) and periodic boundary conditions are usually used. Since conventional materials are big enough, the number of the surface atoms is much less than that of bulk atoms. The materials can still be regarded as infinitively large, even if they do have boundaries. As a result, stacking the macroscopic materials together could not change the electronic states inside the materials.

1.5.2 Moving Electrons in Quantum Confined System

In nanomaterials, the numbers of surface atoms and atoms below surface are comparable. This gives rise to dependence of the electronic state properties on material sizes and surface conditions including surface absorption and surface potential. Nanomaterials include two-dimensional (2D) quantum wells and superlattices, while quantum wires, nanotubes such as carbon nanotubes, as well as nanorods (nano pins) are one-dimensional materials (1D). Quantum dots and clusters are regarded as zero-dimensional (0D) materials. 2D, 1D, and 0D materials receive constraint from one, two, and all three directions, respectively. Thus, at least in one degree of freedom, the size of the materials is comparable to the feature parameters such as electron mean free path, Bohr radius of excitons, etc. The electron motion is then constrained in the corresponding direction(s), leading to the variation of electronic state. Quantum confinement effect gives rise to significant modification on electron density of states that is essential to the performance of opto-electronic devices. Figure 1.19 schematically depicts the bulk, quantum well, quantum wire and quantum dot (Fig. 1.19a) and their corresponding density of states (Fig. 1.19b).

Nanostructures formed by heterojunctions are the most fascinating architectures that display wide spectrum of opto-electronic characteristics, extremely useful for device purposes. To a great extent, the opto-electronic behaviors are largely dependent on the band alignment of the heterojunction. Figure 1.20 schematically shows the band line-up of semi-conductor heterostructures, where a narrow (wide)-bandgap nano semiconductor is sandwiched between another semiconductor with wide (narrow) bandgap.

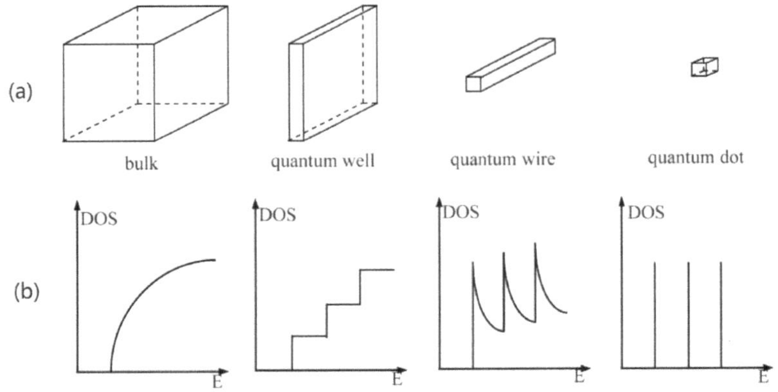

Figure 1.19 Schematic diagrams of bulk, quantum well, quantum wire, and quantum dot (a) and their corresponding density of states (b).

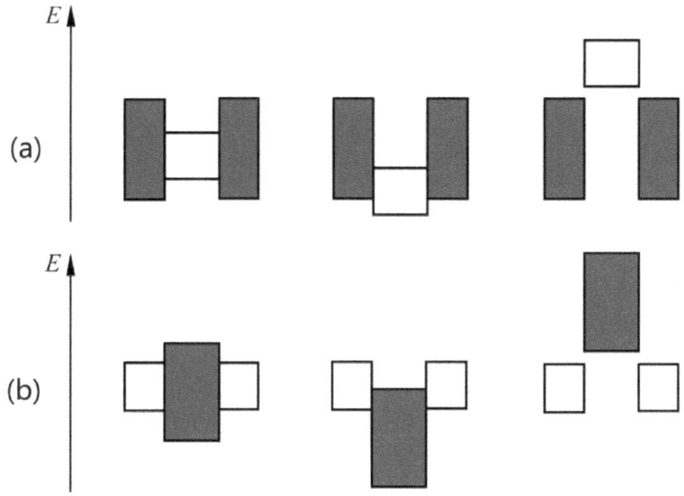

Figure 1.20 Possible band alignment of heterojunctions formed by a narrow bandgap nano semiconductor (the open areas) sandwiched between two same semiconductors of wide band gap (the shaded areas) (a) and by a wide-bandgap semiconductor between narrow–bandgap semiconductors (b).

In bulk semiconductors, electron motion is described by the Schrödinger equation in single-electron approximation

$$\left[-\left(\frac{\hbar^2}{2m_x}\frac{\partial^2}{\partial x^2}+\frac{\hbar^2}{2m_y}\frac{\partial^2}{\partial y^2}+\frac{\hbar^2}{2m_z}\frac{\partial^2}{\partial z^2}\right)+V(x,y,z)\right]\psi(x,y,z)=E\psi(x,y,z), \tag{1.14}$$

where m_x, m_y, and m_z are electron masses in the x, y, and z directions, respectively; $V(x,y,z)$ is the crystal potential, $\psi(x,y,z)$ the electron wavefunction, and E the electron eigen energy. Now we consider the cases of 2D, 1D, and 0D materials and find for electronic eigenstates of electrons.

Two-dimensional systems

For 2D materials as quantum wells, electrons move freely in the xy plane, while there is a potential discontinuity in the z direction as shown in Fig. 1.21. Very often, the 2D thin layer acts as a potential well for electrons confined by the potential barriers.

In effective mass approximation, the crystal potential in (1.14) is characterized by effective masses. By making $m_x \to m_x^*$, $m_y \to m_y^*$, and $m_z \to m_z^*$, $V(x, y, z)$ is removed in (1.14). There are potential discontinuities on the boundaries between the potential well and the potential barrier. Introducing a potential $V_b(Z)$ depending only on z coordinate, the potential discontinuities can be counted and (1.14) then take the form of

$$\left[-\left(\frac{\hbar^2}{2m_x^*}\frac{\partial^2}{\partial x^2}+\frac{\hbar^2}{2m_y^*}\frac{\partial^2}{\partial y^2}+\frac{\hbar^2}{2m_z^*}\frac{\partial^2}{\partial z^2}\right)+V_b(z)\right]\psi(x,y,z)=E\psi(x,y,z). \quad (1.15)$$

For an approximation good for many cases, electron's in-plane and z direction motions are decoupled, implying that the electron wavefunction can be written as $\psi(x,y,z) = \varphi(x,y)\phi(z)$. Now we have two separate Schrödinger equations for the in-plane and the z direction motions of electrons:

$$\left[-\left(\frac{\hbar^2}{2m_x^*}\frac{\partial^2}{\partial x^2}+\frac{\hbar^2}{2m_y^*}\frac{\partial^2}{\partial y^2}\right)\right]\varphi(x,y)=E_{\text{in-plane}}\varphi(x,y) \quad (1.16a)$$

$$\left(-\frac{\hbar^2}{2m_z^*}\frac{\partial^2}{\partial z^2}+V_b(z)\right)\phi(z)=E_n\phi(z) \quad (1.16b)$$

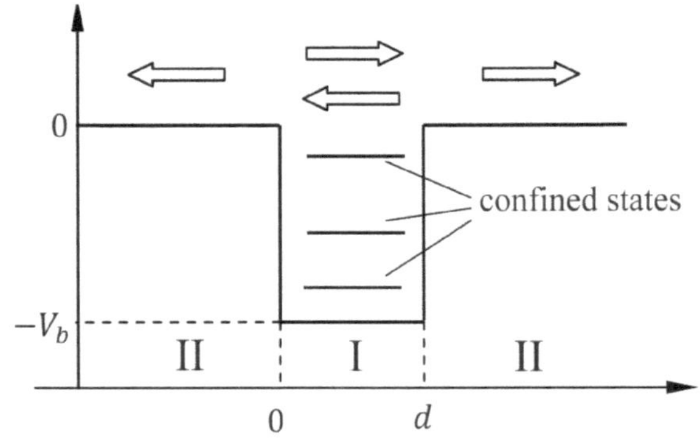

Figure 1.21 Schematic diagrams of energy band profile of a narrow-bandgap semiconductor I sandwiched between a relatively wide-bandgap semiconductor II (only the conduction band profile is shown), where "I" is the well region and "II" is the barrier region. The thin lines in the quantum well region represent the eigen states confined in the well region and the arrows designate the directions of electron waves.

From wave mechanics picture, the electron motion in a confinement system is characterized by interference of in-coming and reflective waves on a boundary, as schematically shown by the arrows in Fig. 1.21, while outside the well region (in the potential barriers), electron wavefunction should be decaying wave. From Fig. 1.21, the energy of the confined states (as labeled in the figure) in the well is $E_n = \frac{\hbar^2 k_w^2}{2m_z^*} - V_b$, corresponding momentum k_w is then

$$k_w = \sqrt{\frac{2m_{I,z}^*}{\hbar^2}(E_n + V_b)} \tag{1.17a}$$

In the barrier, however, the electron momentum is imaginary. A "real momentum"

$$k_b = \sqrt{-\frac{2m_{II,z}^*}{\hbar^2}E_n} \tag{1.17b}$$

can be introduced for describing a decaying electron wavefunction. In (1.17), $m_{I,z}^*$ and $m_{II,z}^*$ specify the effective masses in region I and region II, respectively, which are different for different semiconductors. For an asymmetric quantum well, the materials in left side barrier and the right side barrier are different materials. Using (1.17) and taking the simplest plane wave approximation, the electron wavefunctions in the regions I, II, and III (see Fig. 1.21) take the forms of

$$\text{I: } \phi_b(z) = A\exp(k_b z) + B\exp(-k_b z)\,,\, z \le 0 \tag{1.18a}$$

$$\text{II: } \phi_w(z) = C\exp\left[ik_w\left(z - \frac{d}{2}\right)\right] + D\exp\left[-ik_w\left(z - \frac{d}{2}\right)\right], -0 < z < d \tag{1.18b}$$

$$\text{III: } \phi_b(z) = E\exp\left[k_b(z-d)\right] + F\exp\left[-k_b(z-d)\right], z \ge d. \tag{1.18c}$$

In (1.18a) and (1.18c), the coefficients B and E should both be zero for convergence of wavefunctions at z approaching $-\infty$ and $+\infty$, respectively. Furthermore, $C = D$ or $C = -D$ should hold from the symmetry of the quantum well in Fig. 1.21, then we have

$$\text{I: } \phi_b(z) = A\exp(k_b z),\ z \le 0 \tag{1.19a}$$

$$\text{II: } \phi_w(z) = C\sin\left[k_w\left(z - \frac{d}{2}\right)\right],\ 0 < z < d. \tag{1.19b}$$

$$\text{III: } \phi_b(z) = F\exp[-k_b(z-d)],\ z \ge d. \tag{1.19c}$$

Now, using continuity conditions for both $\phi(z)$ and $(1/m_z^*)$ $d\phi(z)/dz$ on the boundaries at $z = 0$,

$$\phi_b(0) = \phi_w(0), \tag{1.20a}$$

$$\frac{1}{m_{I,z}^*}\frac{d\phi_b(z)}{dz}\bigg|_{z=0} = \frac{1}{m_{II,z}^*}\frac{d\phi_w(z)}{dz}\bigg|_{z=0}, \tag{1.20b}$$

and at $z = d$

$$\phi_w(d) = \phi_b(d), \tag{1.20c}$$

$$\frac{1}{m_{II,z}^*}\frac{d\phi_w(z)}{dz}\bigg|_{z=b} = \frac{1}{m_{I,z}^*}\frac{d\phi_b(z)}{dz}\bigg|_{z=b}, \tag{1.20d}$$

we then have

$$\frac{k_w}{m_{II,z}^*}\tan\left(\frac{d}{2}k_w\right) = \frac{k_b}{m_{I,z}^*} \text{ for even state} \tag{1.21a}$$

$$\frac{k_w}{m_{II,z}^*}\text{cotan}\left(\frac{d}{2}k_w\right) = -\frac{k_b}{m_{I,z}^*} \text{ for odd state} \tag{1.21b}$$

Equations (1.21a) and (1.21b) can be solved numerically. The eigen energy should line up in the well and in the barriers, Using (1.17a) and (1.17b) and assuming $m_{I,z}^* = m_{II,z}^* = m_z^*$ for simplicity, we thus obtain

$$k_w^2 = \frac{2m_z^* V_b}{\hbar^2} - k_b^2. \tag{1.22}$$

Substituting (1.22) into (1.21a) and (1.21b), we obtain

$$\cos\left(\frac{d}{2}k_w\right) = \sqrt{\frac{\hbar^2}{2m_z^* V_b}} k_w, \tag{1.23a}$$

$$\sin\left(\frac{d}{2}k_w\right) = \sqrt{\frac{\hbar^2}{2m_z^* V_b}} k_w. \tag{1.23b}$$

Equations (1.23a) and (1.23b) are plotted in Fig. 1.22, which gives solutions graphically where allowed k values are labeled by arrows. Apparently, the allowed k values are discrete and so do the eigen energies (see Fig. 1.21).

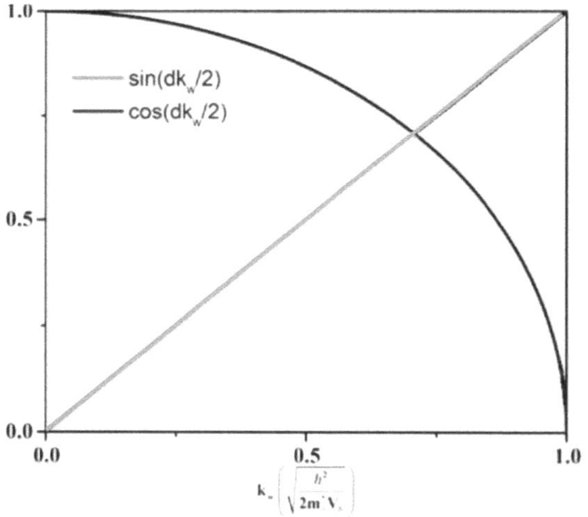

Figure 1.22 sin $\left(\frac{d}{2}k_w\right)$ and cos $\left(\frac{d}{2}k_w\right)$ vs k_w, which gives allowed k as labeled by arrows.

Equation (1.16a) describes in-plane free electron motion corresponding to the eigen energy of

$$E(k_x, k_y) = \frac{\hbar^2}{2}\left(\frac{k_x^2}{m_x^*} + \frac{k_y^2}{m_y^*}\right) \tag{1.24}$$

while (1.16b) gives confined motion in the z direction that leads to discrete energy spectrum. For infinitively high barrier of the quantum well,

$$V_b(z)=\begin{cases}0, -\frac{d}{2}\le z\le\frac{d}{2} \\ \infty, -\frac{d}{2}\ge z\ge\frac{d}{2}\end{cases}, \tag{1.25}$$

with $\phi_w(0) = \phi_w(d) = 0$ for fixed boundary,

$$\phi_w(z)=\begin{cases}\sqrt{\frac{2}{d}}\cos\left(\frac{n_z\pi z}{d}\right), \text{for } n_z=1,3,5..... \\ \sqrt{\frac{2}{d}}\sin\left(\frac{n_z\pi z}{d}\right), \text{for } n_z=2,4,6.....\end{cases} \tag{1.26}$$

Inserting (1.26) into (1.16b), we have $E_{n_z}=\left(\frac{\hbar^2}{2m_z^*}\right)\left(\frac{n_z\pi}{d}\right)^2$ that has a discrete energy spectrum in the z direction. Thus the total 2D eigen state is

$$E\left(k_x,k_y,n_z\right)=\frac{\hbar^2}{2}\left(\frac{k_x^2}{m_x^*}+\frac{k_y^2}{m_y^*}\right)+\left(\frac{\hbar^2}{2m_z^*}\right)\left(\frac{n_z\pi}{d}\right)^2 \tag{1.27}$$

Superlattices are multiple "quantum wells" in which the barrier layers are thin enough to allow for electron wavefunctions in the neighboring quantum wells to overlap, minibands thus forming in superlattices. Figure 1.23 presents the TEM images of AlGaAs/GaAs quantum wells and AlGaAs/GaAs superlattice.

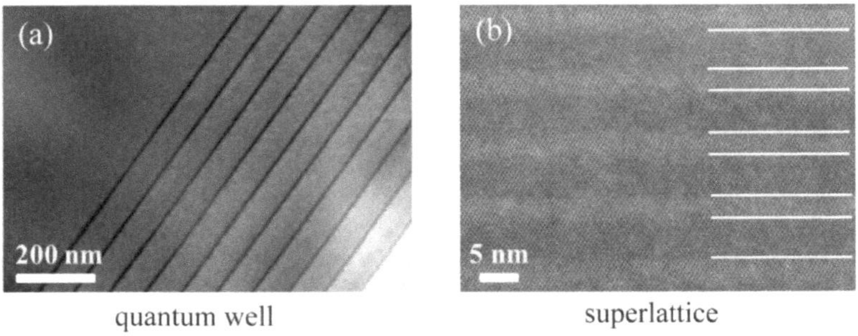

Figure 1.23 TEM images of quantum well (a) and superlattice (b) prepared by MBE. The AlGaAs/GaAs superlattice layers are highlighted by white lines.

One-dimensional systems

One-dimensional (1D) materials are systems in which electrons are free to move in one direction while they receive constrain in the other two directions. Schrödinger equation for 1D systems takes the form of

$$\left[-\left(\frac{\hbar^2}{2m_x^*}\frac{\partial^2}{\partial x^2}+\frac{\hbar^2}{2m_y^*}\frac{\partial^2}{\partial y^2}+\frac{\hbar^2}{2m_z^*}\frac{\partial^2}{\partial z^2}\right)+V_b(y,z)\right]\psi(x,y,z)=E\psi(x,y,z),$$

(1.28)

for free electron motion in the x direction. Again, the yz plane and the x direction are treated independently, the wavefunction of electrons is

$$\psi(x,y,z) = \varphi(x)\phi(y,z), \quad (1.29)$$

which leads to the following Schrödinger equation in the yz plane similar to (1.16b),

$$\left(-\frac{\hbar^2}{2m_y^*}\frac{\partial^2}{\partial y^2}-\frac{\hbar^2}{2m_z^*}\frac{\partial^2}{\partial z^2}+V_b(y,z)\right)\phi(y,z)=E_n\phi(y,z) \quad (1.30)$$

where $V_b(y, z)$ is the barrier potential. Simply assuming infinitive barrier of

$$V_b(y,z)=\begin{cases}0,0\le y\le d_y;\,0\le z\le d_z\\ \infty,0\ge y\ge d_y;\,0\ge z\ge d_z\end{cases}, \quad (1.31)$$

the wavefunction takes the standing wave form of

$$\phi(z,y)=\frac{2}{\sqrt{d_yd_z}}\sin(k_yy)\sin(k_zz) \quad (1.32)$$

that vanishes in the barrier. Here $k_y=\frac{n_y\pi}{d_y}$, $k_z=\frac{n_z\pi}{d_z}$ (n_y and n_z are positive integers) are the momentum in well region. The eigen energy, obtained by inserting (1.32) into (1.30) in the well region, is

$$E(n_y,n_z)=\frac{\hbar^2\pi^2}{2}\left(\frac{n_y^2}{m_y^*d_y^2}+\frac{n_z^2}{m_z^*d_z^2}\right) \quad (1.33)$$

in the yz plane. Adding on the continuous energy spectrum $E_x = \frac{\hbar^2 k_x^2}{2m_x^*}$ in the x direction gives the total eigen energy of

$$E(k_x, n_y, n_z) = \frac{\hbar^2 k_x^2}{2m_x^*} + \frac{\hbar^2 \pi^2}{2}\left(\frac{n_y^2}{m_y^* d_y^2} + \frac{n_z^2}{m_z^* d_z^2}\right) \quad (1.34)$$

Quantum wires are 1D semiconductor and can be prepared by several routes. Short quantum wires are called nanorods. Figure 1.24 presents the TEM images of nanorods (Fig. 1.24a) and a nanowire (Fig. 1.24b), with the zoom-in images of a nanorods and a nanowire are shown in (a) and (b), respectively, with their crystalline planes marked. Both nanowire SCs and nanorods SCs have been intensively investigated due to their unique advantages including light trapping capability.

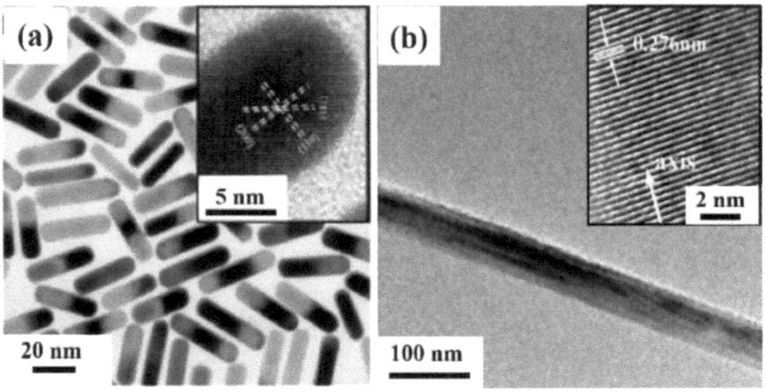

Figure 1.24 TEM images of nanorods and a nanowire, where the insets in (a) and (b) present the zoom-in images of a nanorod and a nanowire with marked crystalline planes, respectively.

Zero-dimensional systems

As shown in Fig. 1.19, electrons are confined in all three directions in zero-dimensional (0D) materials. A typical 0D system is semiconductor quantum dots (QDs). Assuming decoupled electron motion in the x, y, z directions, i.e., $\psi(x, y, z) = \varphi(x)\phi(y)\chi(z)$, Schrödinger equation for 0D systems takes the form of

$$\left(-\frac{\hbar^2}{2m_x^*}\frac{\partial^2}{\partial x^2} - \frac{\hbar^2}{2m_y^*}\frac{\partial^2}{\partial y^2} - \frac{\hbar^2}{2m_z^*}\frac{\partial^2}{\partial z^2} + V_b(x, y, z)\right)\varphi(x)\phi(y)\chi(z)$$
$$= E_n\phi(x)\phi(y)\chi(z). \quad (1.35)$$

For the simplest 0D material of a quantum box with infinitive barrier height and the size of $d_x \times d_y \times d_z$, $\psi(x, y, z) = \varphi_{n_x}(x)\phi_{n_y}(y)\chi_{n_z}(z)$. Equation 1.40 can then be solved for fixed boundary condition of $\varphi_{n_x}(0)=\varphi_{n_x}(d_x)=\phi_{n_y}(0)=\phi_{n_y}(d_y)=\chi_{n_z}(0)=\chi_{n_z}(d_z)=0$ to give standing wave solution of

$$\varphi_{n_x}(x)\phi_{n_y}(y)\chi_{n_z}(z)=2\sqrt{\frac{2}{d_x d_y d_z}}\sin(k_x x)\sin(k_y y)\sin(k_z z), \qquad (1.36)$$

with $k_x=\frac{n_x\pi}{d_x}$, $k_y=\frac{n_y\pi}{d_y}$, and $k_z=\frac{n_z\pi}{d_z}$ (n_x, n_y, n_z are positive integers). The total energy is then

$$E(n_x,n_y,n_z)=\frac{\hbar^2\pi^2}{2}\left(\frac{n_x^2}{m_x^* d_x^2}+\frac{n_y^2}{m_y^* d_y^2}+\frac{n_z^2}{m_z^* d_z^2}\right) \qquad (1.37)$$

that has no continuous part.

Many routes have been demonstrated to prepare nanoparticles of various shapes. Figure 1.25 shows the TEM images of for Si quantum dots (nanoparticles) (a) and GaAs quantum plates (b). Quantum confinement is usually realized through constructing quantum heterostructure in which a nano-sized low band edge semiconductor is surrounded by relatively high band edge semiconductor. In the case that both electrons and holes are confined as shown in Fig. 1.26, the discrete eigen states for both electrons and holes are formed and confined in the well. In the figure, $E_{e1}(E_{h1})$ is the ground state for electrons (holes), while $E_{e2}(E_{h2})$ and $E_{e3}(E_{h3})$ are the first and second excited states, respectively.

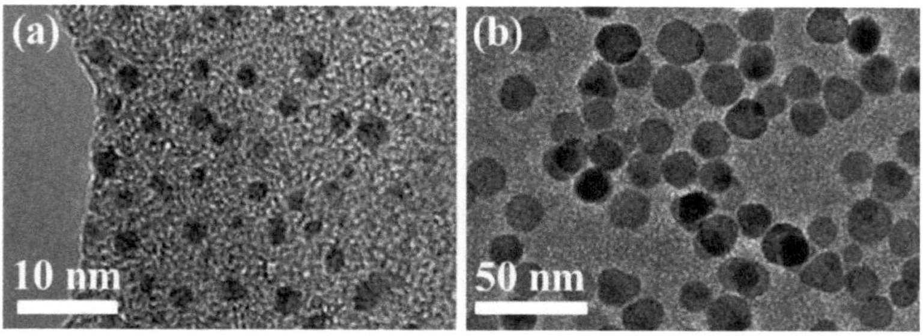

Figure 1.25 TEM images of quantum dots (a), quantum plates (b) in which electrons receive constrain in all three directions.

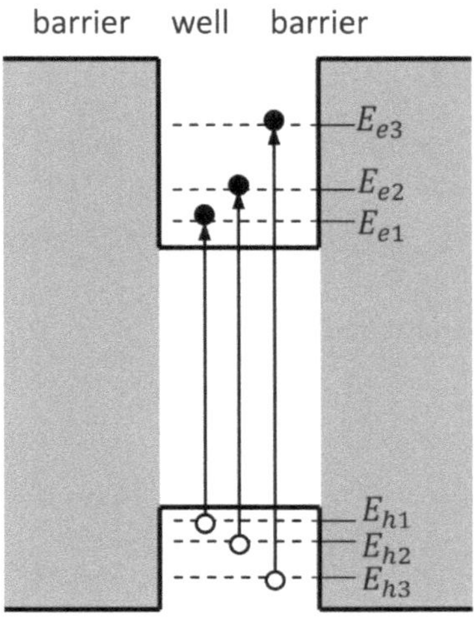

Figure 1.26 Schematic diagram of energy band alignment, indicating that due to quantum confinement the continuous conduction and valence bands become discreet energy levels, as shown by E_{e1}, E_{e2}, and E_{e3} for electrons and E_{h1}, E_{h2}, and E_{h2} for holes.

1.5.3 Density of States of Quantum Confined System

Density of states (DOS) describes how electronic states distribute in materials and it determines many important features of materials, especially optoelectronic properties. At nanoscopic scale, variations in size, shape, and surface adatoms modify material's DOS and, as a result, lead to change in material's optoelectronic properties. Nanotechnology offers routes to tune material parameters through varying the size, shape, and surface condition aiming at specific requirements for optoelectronic applications.

Now, starting from a simple 3D box with side length L, let us find out how DOS changes as the 3D box becomes a 2D plane, 1D wire, and 0D dot. In the framework of an electron Fermi gas system, the Schrödinger equation in single-electron framework is

$$-\left(\frac{\hbar^2}{2m_x}\frac{\partial^2}{\partial x^2}+\frac{\hbar^2}{2m_y}\frac{\partial^2}{\partial y^2}+\frac{\hbar^2}{2m_z}\frac{\partial^2}{\partial z^2}\right)\psi(x,y,z)=E\psi(x,y,z) \quad (1.38)$$

Note the absence of potential $V(x, y, z)$ in the case of free electrons. The corresponding solution of eigen function is then

$$\psi(x,y,z)=\frac{1}{\sqrt{L^3}}\exp\left[-i(k_x x+k_y y+k_z z)\right]. \tag{1.39}$$

The periodic boundary condition requires

$$\psi(x,y,z)=\psi(x+L,y,z)=\psi(x,y+L,z)=\psi(x,y,z+L), \tag{1.40}$$

which can only be satisfied when

$$k_x=\frac{2\pi n_x}{L} \tag{1.41a}$$

$$k_y=\frac{2\pi n_y}{L} \tag{1.41b}$$

$$k_z=\frac{2\pi n_z}{L} \tag{1.41c}$$

where n_x, n_y, n_z = ±1, ±2, ±3, ... Using $k_i=\frac{2\pi n_i}{L}$ ($i = x, y, z$) for (1.39) and inserting (1.39) into (1.38), we obtain

$$E=\frac{\hbar^2}{2m}(k_x^2+k_y^2+k_z^2)=\frac{\hbar^2}{2m}\left[\left(\frac{2\pi n_x}{L}\right)^2+\left(\frac{2\pi n_y}{L}\right)^2+\left(\frac{2\pi n_z}{L}\right)^2\right] \tag{1.42}$$

Due to very large L, E spectrum is almost continuous with k_i. In $\boldsymbol{k}$-space, the allowed $\boldsymbol{k}$-states described by (1.41) distribute uniformly, with each allowed $\boldsymbol{k}$-state occupying a volume of $\left(2\pi/L\right)^3$. Thus, the number of allowed $\boldsymbol{k}$-states per unit volume in $\boldsymbol{k}$-space is $\left(L/2\pi\right)^3$. At zero temperature (T = 0), a Fermi gas system of N electrons is in its ground state which, described in $\boldsymbol{k}$-space, is characterized by N electrons filling all the allowed $\boldsymbol{k}$-states below the spherical Fermi surface with the spherical radius $\boldsymbol{k}_F$ since $E\propto k_x^2+k_y^2+k_z^2$. Allowed states outside the Fermi ball are not occupied by electrons. Using density of states $\left(L/2\pi\right)^3$ in $\boldsymbol{k}$-space, the total number of allowed $\boldsymbol{k}$-states in the Fermi ball with the volume of $\left(4/3\right)\pi k_F^3$ is

$$N_{\boldsymbol{k}\text{-states}}=\left(\frac{L}{2\pi}\right)^3\cdot\frac{4}{3}\pi k_F^3=\frac{V}{6\pi^2}k_F^3, \tag{1.43}$$

where $V = L^3$ is the volume of the material. Pauli exclusion principle allows each $\boldsymbol{k}$-states to accommodate two electrons with opposite spin orientations. Thus, the total number of electrons occupying all the $\boldsymbol{k}$-states within Fermi ball (below Fermi surface $\boldsymbol{k}_F$) is

$$N = \frac{V}{3\pi^2} k_F^3, \text{ or } n = \frac{1}{3\pi^2} k_F^3, \tag{1.44}$$

where $n = N/V$ is the density of electrons in the Fermi gas system. DOS is defined as the number of eigen states at E per unit energy interval so that DOS can be expressed by

$$D(E) = \frac{dN(E)}{dE}. \tag{1.45a}$$

From (1.42) or $k = \sqrt{2mE}/\hbar$, we have for 3D materials

$$D(E) = \frac{Vm^{3/2}}{\hbar^3 \pi^2} \sqrt{2E}. \tag{1.45b}$$

Thus, $D(E) \propto E^{1/2}$ in 3D case. Figure 1.27 presents the calculated energy band and DOS for bulk Si [31].

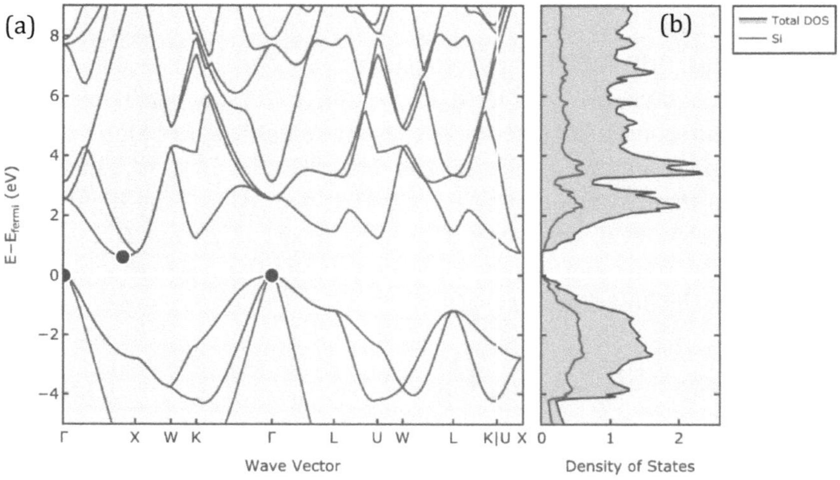

Figure 1.27 Energy band of bulk Si (a) and the density of states vs. energy (b).

Assuming that a thin film has macroscopic size L in the x and y directions and nanometer size in the z direction. Electron motion is then confined in the z direction (usually the quantum well growth direction) in the quantum well, reinforcing the periodic boundary condition for the xy plane gives allowed in-plane $\boldsymbol{k}$-states of $k_x = 2\pi n_x/L$ and $k_y = 2\pi n_y/L$. Each allowed $\boldsymbol{k}$-state occupies an area of $\left(2\pi/L\right)^2$ so that the density of allowed $\boldsymbol{k}$-states in $\boldsymbol{k}$-space is $\left(L/2\pi\right)^2$. At $T = 0$, all the $\boldsymbol{k}$-states within the Fermi "disc" of radius k_F is occupied and outside the Fermi "disc" is empty. Thus, the total number of electrons within the Fermi "disc" equals to the density of allowed $\boldsymbol{k}$-states $\left(L/2\pi\right)^2$ multiplied by the area of the Fermi "disc", i.e.,

$$N = 2\cdot\left(\frac{L}{2\pi}\right)^2\cdot\pi k_F^2 = \frac{A}{2\pi}k_F^2, \tag{1.46}$$

Here the factor 2 is due to electron spin degeneracy and A is the area of the 2D material. From $E = \frac{\hbar^2k^2}{2m}$, (1.46) can be written as $N = \frac{AmE}{\pi\hbar^2}$. DOS in the 2D case is

$$D(E) = \frac{dN(E)}{dE} = \frac{Am}{\pi\hbar^2}. \tag{1.47}$$

Thus, DOS in 2D is thus a constant.

As shown by Eq. (1.16) in Section 1.5.2, the eigen energy spectrum consists of discrete energy levels in the z direction and is continuous in the xy plane. The fact that 2D DOS is constant implies that at each energy level DOS is a constant. The higher is the energy level, the larger the DOS. In fact, $D(E)$ in 2D case is step-like following more rigorous derivation, which can be empirically described by [32]

$$H(E) \sim \frac{1}{\pi}\left\{\frac{\pi}{2} + \mathrm{Arctan}\left[\frac{(E - E_n)}{\Gamma_n}\right]\right\}, \tag{1.48}$$

where E_n is eigen energy and Γ_n the broadening parameter due to crystal imperfection including dislocations. With slight modification on (1.27) for a free electron Fermi gas system confined in the z direction and with a periodic boundary in the x and y directions, the energy spectrum for a quantum well system is, for infinite barrier height,

$$E(k_x,k_y,n_z)=\frac{\hbar^2}{2m}(k_x^2+k_y^2)+\frac{\pi^2\hbar^2}{2md^2}n_z^2, \tag{1.49}$$

where $k_x = \frac{2\pi n_x}{L}$ and $k_y = \frac{2\pi n_y}{L}$. Note that in (1.49), we keep the notation of k_x and k_y to show that they are nearly continuous in order to differ from E_{n_z} so that the in-plane energy spectrum can still be treated as continuous as in (1.27). Figure 1.28 presents calculated in-plane band dispersion relation ($E \sim k_{\text{in-plane}}$) and 2D DOS in a quantum well [33].

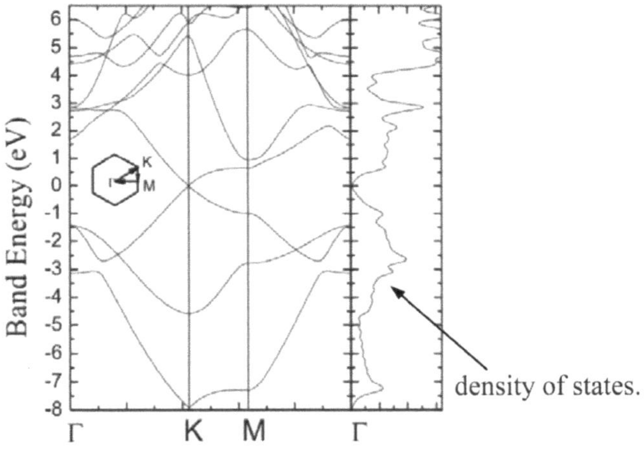

Figure 1.28 Energy band of 2D Si and the density of states. Reproduced with permission from [33].

For a 1D nanowire in Fig. 1.24 with a macroscopic length L in the x direction and nanometer size l in both y and z directions, the Fermi "surface" is at two ends $\pm\boldsymbol{k}_F$ of a Fermi "rod". The allowed $\boldsymbol{k}$-states are $2\pi n/L$ that occupies a length of $2\pi/L$ in $\boldsymbol{k}$-space. The density of $\boldsymbol{k}$-states in $\boldsymbol{k}$-space is then $L/2\pi$. The number of electrons within the Fermi "rod" is then

$$N=2\cdot\frac{L}{2\pi}\cdot 2k_{\mathrm{F}}=\frac{2L}{\pi}k_{\mathrm{F}} \tag{1.50}$$

when spin degeneracy is taken into consideration. This leads to DOS of

$$D(E) = \frac{dN(E)}{dE} = \frac{L\sqrt{m}}{\pi\hbar}\sqrt{\frac{2}{E}}. \tag{1.51}$$

Again with slight modification on (1.34), the eigen energy spectrum of the 1D nanowire is

$$E(k_x, n_y, n_z) = \frac{\hbar^2}{2m}k_x^2 + \frac{\pi^2\hbar^2}{2md^2}(n_y^2 + n_z^2). \tag{1.52}$$

Note that in the x direction periodic boundary condition was used, while in the y and z directions the fixed boundary condition was adopted, so that $n_x = \pm1, \pm2, \ldots$ and n_y, $n_z = 1,2,3, \ldots$. Figure 1.29 presents the energy band structure and DOS of Si nanowires [34].

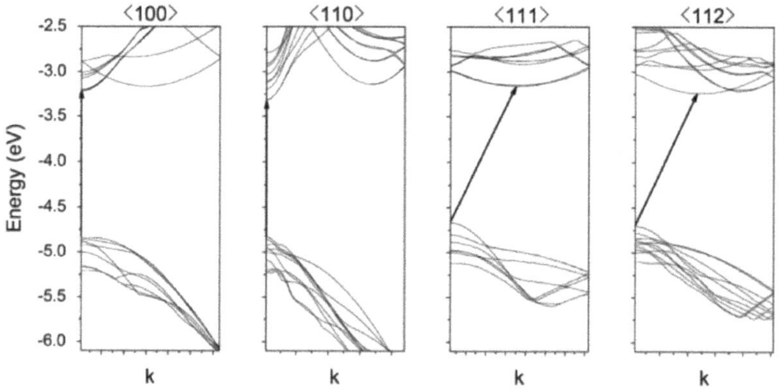

Figure 1.29 Energy band of 1D Si in several crystalline orientations. Reproduced with permission from [34].

Now consider a nano cubic box of side length d, the eigen energy of the confined electron Fermi gas is, similar to (1.37),

$$E(n_x, n_y, n_z) = \frac{\pi^2\hbar^2}{2md^2}(n_x^2 + n_y^2 + n_z^2), \tag{1.53}$$

where n_x, n_y, n_z are all positive integers, so that the lowest energy is $E = \frac{3\pi^2\hbar^2}{2md^2}$. Apparently, there is no continuous part of the energy spectrum. We thus expect a δ-function type of DOS for this nano quantum box with the form of

$$D(E) = 2\sum_{n_x, n_y, n_z} \delta(E - E(n_x, n_y, n_z)). \tag{1.54}$$

The 0D nanomaterials have a typical band dispersion relationship and DOS as shown in Fig. 1.30. The δ-like DOS implies that high density of electrons piles up at the same energy level, which is extremely beneficial for opto-electronic devices.

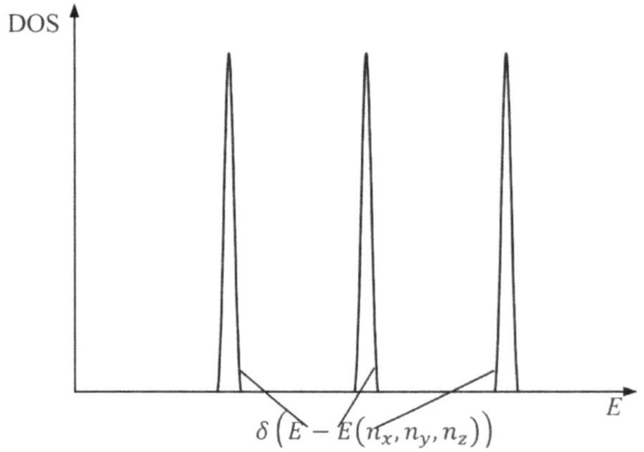

Figure 1.30 Schematic δ-like density of states for 0D quantum dots. Note that broadening is introduced to avoid divergence of δ-function at $E = E(n_x, n_y, n_z)$.

1.5.4 Surface Modification of Nanomaterials

Semiconductor quantum dots have a large surface-to-volume ratio relative to bulk materials, beneficial for many applications such as opto-electronic detectors, lasers, as well as high sensitivity bio-chemical sensing. On the other hand, the large surface could also be troublesome for optoelectronic devices due to unsaturated atomic bonds on surfaces that often act as defects for electron trapping or channels for non-radiative recombination of electrons and holes. There are usually a huge number of surface atoms in nanomaterials! Thus, proper surface passivation is thus necessary for practical uses. Typical surface passivation technique is coating the surface of nanoparticles with layers of organic, inorganic molecules to saturate the dangling bonds.

Surface ligand exchange

Free surfaces of nanoparticles are usually not stable. Under practical circumstances, surface atoms on nanoparticles are reconstructed or absorb ex-atoms to decrease the surface energy.

The "unintentional passivation" is helpful for the reduction of the density of surface states but far from satisfaction since it is uncontrollable. Among many ways of intentional passivation, organic materials are good candidates for the controllable passivation on nanoparticle surfaces which leads to decrease of density of surface states and modify the physical properties of the nanoparticles at the same time. Chemically prepared CdSe quantum dots, namely, colloidal quantum dots are dispersed in liquid and coated with a layer of trioctylphosphine oxide (TOPO) or hexadecylamine (HDA) ligands, for instance, that acts as passivation layers [35]. The TOPO or HDA layer can be replaced by alternative ligands to achieve better passivation. Changing the surfactants modify the surface states distribution. Mostly, the passivation ligands for QDs are long chain (8–20 carbons) organic materials and the capping layers produce insulating barriers that block the charge transport. Very often, the organic ligands are unable to completely eliminate the deep trap states on the QD surfaces that act as charge trapping centers. Improvements were made to use inorganic ligands. The use of inorganic metal chalcogenide ligands has shown promising improvement on photoelectric conversion efficiency for QD sensitized solar cells [36]. The improvement stems from the low density of shallow surface states and decreased inter-particle distance that enables good electron transport. Thiol-connection is often adopted to functionalize the quantum dots for biology use. Figure 1.31 depicts the QDs passivated by surface ligands.

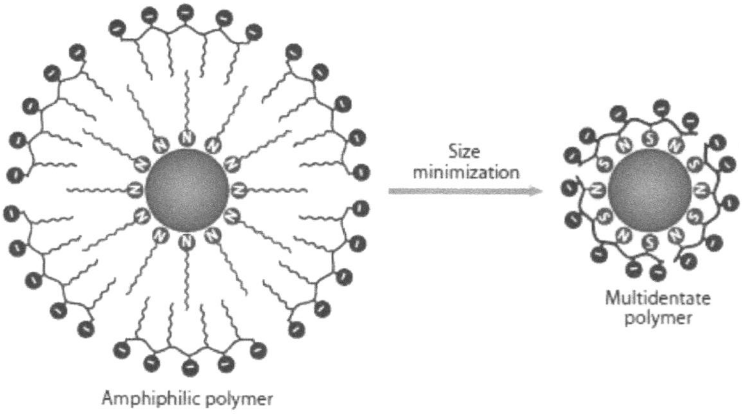

Figure 1.31 Multidentate ligands wrap around the QD, leading to the minimization of QD size. The ligands contain a mixture of amine and thiol functional groups [37].

Core/shell structures

The best passivation route to QD surface is the growth of core–shell structure—a thin layer of another semiconductor is grown on the surface of the core QDs. In this case, the surface dangling bonds are well-passivated, giving rise to very low surface defect density, due to the use of closely lattice-matched shell layer, as compared to the core. Figure 1.32 presents the schematic diagram ZnSe/CdSe core/shell quantum dots and the band alignments. The CdSe shell layer has narrower bandgap than that of ZnSe. When the core/shell QD is surrounded by a material with wide bandgap (air, for instance), CdSe is the well layer for both electrons and holes. Increasing CdSe thickness leads to successive capture of the electrons and holes by the CdSe layer.

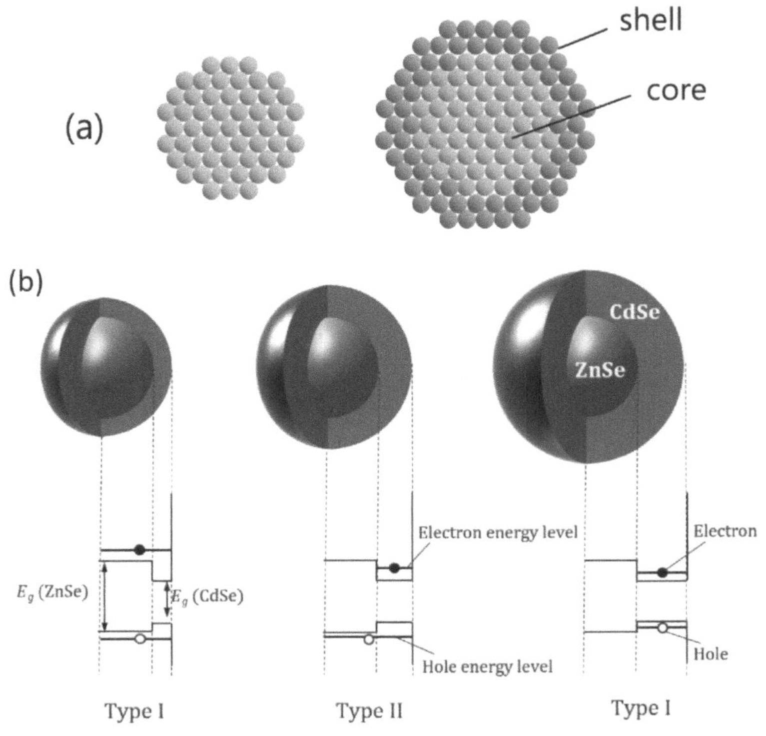

Figure 1.32 A bare quantum dot, core/shell quantum dot (a) and their band alignment of ZnSe/CdSe core/shell quantum dots with different shell layer thickness (b). Note that, in this case, the shell layer is the well layer for both electrons and holes. With increasing CdSe thickness, the electrons and holes become localized in CdSe layer successively.

Embedded in matrix

Another effective way to passivate nanoparticles is to grow the particles in matrix network materials such as crystalline or amorphous semiconductors or other organic and inorganic materials. For instance, enhanced luminescence has been demonstrated with PbSe QDs embedded in silicon random photonic crystal microcavities [38]. The colloidal QDs show improved wavelength-tunable photoluminescence emission with efficiency as high as 80%. Silicon nanoparticles of variable diameters embedded in amorphous silicon-nitride matrix (nc-Si/a-SiN_x:H) could be prepared by radio frequency inductively coupled plasma chemical vapor deposition (RF-ICPCVD [39]. Unlike bulk crystalline silicon and amorphous silicon, the nc-Si/a-SiN_x:H shows intense PL emission in the visible wavelength. Dangling bonds on silicon are the most influential defects on electronic and optical properties. Due to the hydrogen atoms, the dangling bonds in a-SiN_x are passivated and the density of dangling bonds is significantly reduced, which favors the PL emission.

1.5.5 Preparation of Nanomaterials

There are many routes to prepare nanoparticles of various shapes, which can be classified into top-down and bottom-up methods. In the top-down route, the preparation starts from bulk materials and ends up with nanoparticles. The nanoparticles are prepared either via mechanical ball milling/attrition that involves a great number of mechanical cycles or photolithography that involves wet chemical/dry ion etching. Mechanical ball milling/attrition is probably the simplest approach for the fabrication of nanoparticles. In most cases, ball milling/attrition yields a rather broad size distributions (10–1000 nm) and varied shapes that is hard to control, and gives rise to contamination due to impurities from milling or attrition process. In addition, the mechanical procedures are likely to damage the materials and introduce a great deal of defects. Photolithography is generally used for the fabrication of two-dimensional patterns formed by nanoparticles of various shapes, starting with bulk materials that are often quite expensive. The left-over nanomaterials after wet chemical/dry ion etching are often a small portion of bulk materials, most of the materials being wasted. Although the yield is low and the methods involve relatively expensive wet or dry etching, it has the advantage that both the size and the shape of the nanoparticles can be well-controlled.

The bottom-up route starts from small building blocks such as atoms and molecules and terminates at the formation of nanoparticles each containing 100–10000 atoms or molecules. Many bottom-up routes of fabricating the nanoparticles have been proposed and some of them are now widely used via the processes of solvothermal reaction, sol-gel, pyrolysis, inert gas condensation, etc. Structured media are sometimes used as templates for nanomaterial growth. A bottom-up route usually has low yield and both the size and the geometry of nanoparticles are not well-controlled. The bottom-up routes includes vapor-phase and liquid-phase processes with the preparation steps as shown in Fig. 1.33a,b, respectively. The vapor-phase fabrication starts with source materials in the forms of solid, liquid, or gas. The source materials are heated to form a vapor phase precursor. In this fabrication procedure, an element or material vapor is formed by heating a precursor through thermal heating, sputtering, pulsed laser ablation, spark discharge, electrospraying, etc. Subsequently nucleation occurs due to supersaturation of the material vapor, through vapor cooling, expansion, and chemical reaction which can sometimes be induced by adding a reactive gas or heating the oven. Very often, supersaturation is achieved by vaporizing material into a background gas, followed by cooling the gas. The nucleation could either occur on a solid substrate or in liquid.

Once the nucleation is initiated, the supersaturation is relieved by further reaction of vapor-phase molecules on formed particles, leading to the growth of nanomaterials. The crystals grow as long as the solution remains supersaturated upon the formation of nuclei. To obtain nanoparticles with narrow size distribution, new nucleation should be stopped as soon as growth starts. The particle growth can be terminated after nucleation by rapid quenching through lowering the temperature down or removing the source of supersaturation. Vapor-phase processes are suitable for the growth of many nanoparticles. Compared to liquid-phase processes, vapor-phase fabrications have the advantages of continuous growth, better material purity, surfactant-free, and sometimes high yield. However, the apparatus required by vapor-phase fabrication is much more complicated and expensive than that by liquid-phase fabrication. Liquid-phase processes can synthesize many kinds of nanoparticles and their fabrication is cheap and requires no large apparatus. Since a liquid-based synthesis (hot-injection or hydrothermal route, for instance) is carried out in a hot pot and its growth rate is often very fast, the difficulty in achieving temperature uniformity and good time control could lead to poor control over the particle size. Additionally, the batch form growth often gives rise to batch-dependent product.

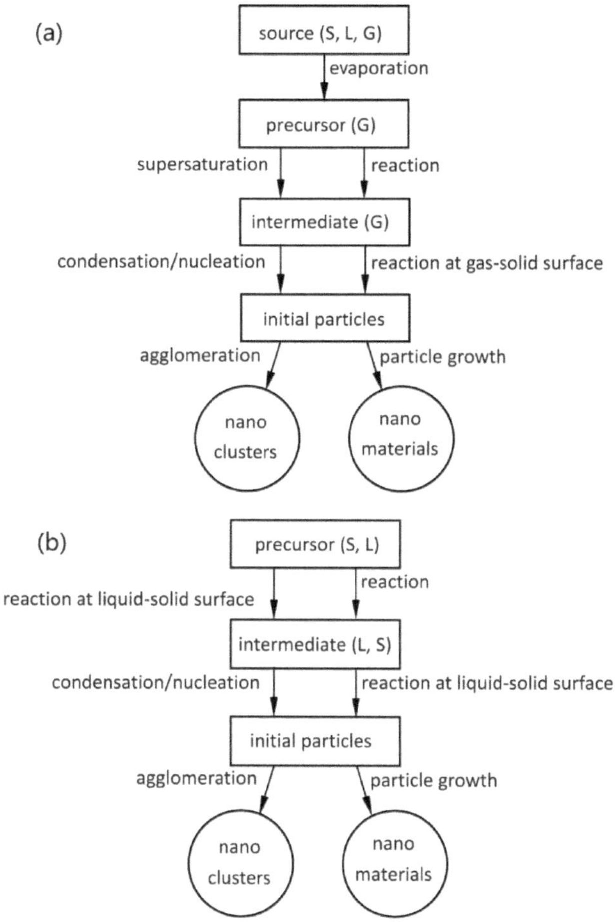

Figure 1.33 Preparation steps of nanomaterials based on vapor phase (a) and liquid phase (b), where letters S, L, and G represent solid, liquid, and gas forms of materials, respectively.

The vapor-phase fabrication includes spray pyrolysis, inert gas condensation, etc. In a spray pyrolysis process shown in Fig. 1.34, the precursor is atomized and thin films are grown by spraying the precursor on a substrate heated to a given temperature. In inert gas condensation shown in Fig. 1.34, nanoparticles are grown in vacuum by sputtering atoms from a source target and then agglomerate into clusters in an inert gas flow, followed by deposition on a substrate. This yields monodisperse nanoparticles with selectable size by varying the growth parameters.

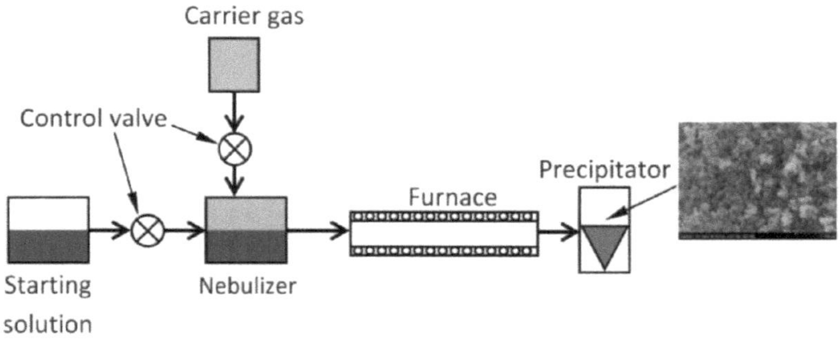

Figure 1.34 A setup and process of spray pyrolysis growth.

Solvothermal and hydrothermal synthesis are methods for the growth of chemical compounds. Solvothermal and hydrothermal are similar nanoparticle growth routes. Solvothermal method refers to nanoparticle synthesis route through chemical reactions in non-aqueous solutions while, by the name, in hydrothermal synthesis chemical reaction occurs in water. Both solvothermal and hydrothermal synthesis allows for the growth of nanoparticles with good crystallinity and narrow distributions of sizes and shapes determined by reaction temperature, reaction time, solvent type, surfactant type, and precursor type. Solvothermal/hydrothermal synthesis was used to prepare nanostructured TiO_2 (titanium dioxide), graphene, and some other materials. Solvothermal/hydrothermal preparation is particularly useful for fabrication of various colloidal semiconductor nanoparticles that are called quantum dots by physicists. The quantum dots are monodispersed in water or organic solvent with controllable sizes and shapes. The surface of the quantum dots can be functionized by attaching functional surfactants for biological uses. The route for liquid preparation is shown as follows.

Pyrolysis refers to the thermal decomposition of materials at elevated temperatures. In the fabrication of nano semiconductors, pyrolysis usually occurs in inert environments such as a vacuum, inert gas, or nitrogen gas. In a spray pyrolysis process, aerosol is created by spray and the droplets are heated to form solid particles.

Sol-gel preparation is a chemical synthesis technique for preparing. A sol is a dispersion of colloidal particles in amorphous or crystalline phases in a solvent. An aerosol is those particles in a gas. A gel is a three-dimensional continuous material network and a colloidal gel is the network formed by agglomeration of colloidal particles. A gel could be constructed by many materials including linked polymer chains and nanomaterials of various structures. Having the advantages of simple and easy fabrication and low cost, the sol-gel technique is widely used for the growth of oxide nanoparticles. It, however, has some major drawbacks. The grown materials are limited to oxides, i.e., non-oxide semiconductor nanoparticles cannot be grown by sol-gel. The reaction rate is too quick so that the nanoparticles grow very fast during nucleation, resulting in poor control of the nanoparticle growth. The fact that the growth temperature is below boiling point of reagents such as water and alcohols often leads to low crystallinity. Low growth temperature is good for the preparation of glasses (for instance), but bad for the growth of semiconductor nanoparticles in crystalline forms. The growth process of the sol-gel preparation is shown schematically in Fig. 1.35. During the growth procedure, the precursor (sol) evolves into a di-phase gel composed of a liquid phase solvent and solid phase material with morphologies ranging from nanoparticles to polymer networks. Sequentially, the solution is often spin-coated or dip coated on a substrate with controllable thickness. The thickness of the film is determined by the viscosity of the precursor and the spinning rate of the spinner. A thermal annealing is then usually carried out to remove the organic or aqueous solvent and, at the same time, crystallize the nanoparticles or porous network. The ultimate microstructure of the final nanomaterial is determined by details of the precursor and preparation conditions including annealing temperature. Grain growth is completed during annealing and the thermal treatment is often a key step for obtaining nanomaterials with optimized electric and mechanical properties.

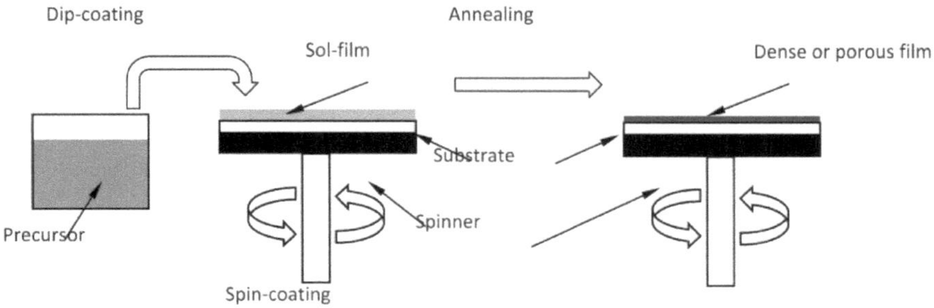

Figure 1.35 A setup and process of sol-gel growth.

The sol–gel technique allows the control of the size and chemical composition of the nanomaterials. The sol-gel growth is also very cost-effective and it can be carried out at low-temperature. In addition, doping in the sol-gel-grown nanomaterials is possible and convenient since dopants can be incorporated into the sol. In optimized condition, dopants disperse uniformly in the fabricated dense or porous film.

Microemulsions are mixtures of water, oil, and surfactants, which are thermodynamically stable and isotropic liquid. Microemulsions are ternary systems consisting of two immiscible phases—water and oil separated by surfactant molecules, the surfactants forming a monolayer at the oil and water interface. The hydrophobic chains of the surfactants are dissolved in the oil phase and the hydrophilic head groups in the aqueous phase. The microemulsion process is one of the most widely used techniques for the synthesis of monodispersed nanoparticles. In microemulsion synthesis, two microemulsions, the one with aqueous phase often containing salts and/or other ingredients, and another one with the complex mixture of different hydrocarbons and olefins as the precipitating agent, are mixed and both reactants react to form nanoparticles. The reaction is due to the droplets collisions and coalescence with the precipitate being in the microemulsion droplets. Compared to other methods, the microemulsion synthesis is able to fabricate a great variety of nanoparticles with controlled quality, size and composition using very simple and cheap growth system. There are many reports on the synthesis of nanoparticles with crystalline structure.

1.6 Interaction between Light and Matter

An optoelectronic device is based on the theory of light interacting with matter. In other words, the mechanism and the process of the light–matter interaction are the fundamental part of optoelectronic effect. In other words, an optoelectronic effect consists of the mechanism and the process of light–matter interactions. The light–matter interaction leads to optical phenomena such as diffraction, reflection, refraction, transmission, scattering, polarization, photoluminescence, and various photoelectric effects. In the case of nanoscale matter, the interaction could lead to the confinement of light on nanoscale or microscale sizes comparable to or smaller than the wavelength of light. A photoelectric process is thus confined in a nanoscale or microscale region.

1.6.1 Some Facts about Solar Radiation

As an approximate 5800°C black body, the sun radiates light in the wavelength ranging from near ultraviolet to far infrared with most radiation energy in visible and near infrared. Out of the earth's atmosphere, the solar illuminance is [40]

$$E_{\text{out}} = E_{\text{sl}} \cdot \left[1 + 0.022412 \cdot \cos\left(2\pi \cdot \frac{N_{dn} - 3}{365} \right) \right]. \tag{1.55}$$

where E_{sr} is the solar illuminance constant (=128000), N_{dn} the day number of the year (N_{dn} = 1, and 356 on January 1 and the December 31, respectively, except for the leap year). In (1.55), $\cos\left(2\pi \cdot \frac{N_{dn} - 3}{365} \right)$ is the correction for the elliptic orbit. The sun radiates a power of 1366 W/m^2 at the top of Earth's atmosphere and it decreases to about 1100 W/m^2 at sea level, depending on the locations. The solar radiation spectra at the top of the atmosphere and at the earth sea level are presented in Fig. 1.36, covering a wavelength range from 200 to 3200 nm.

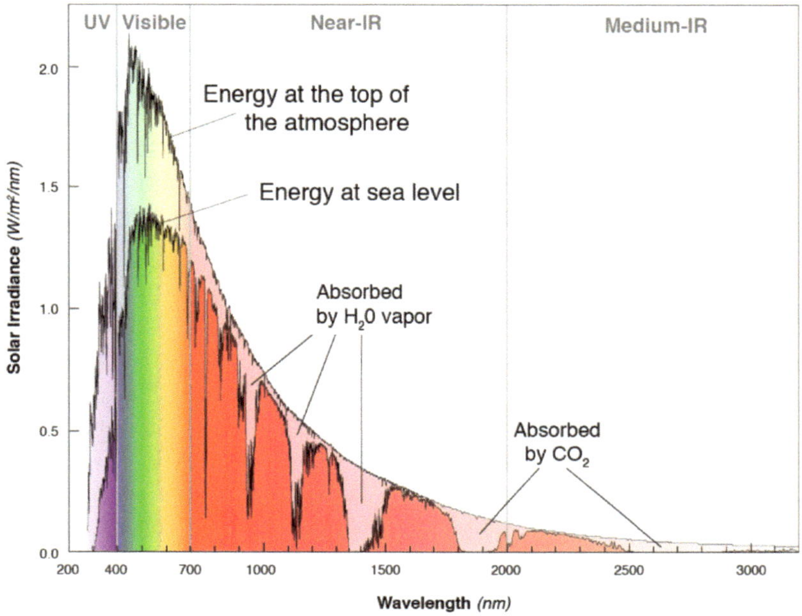

Figure 1.36 Solar radiation spectra at the top of the atmosphere and the sea level in wavelength range from 200 to 3200 nm. The difference of the two spectra indicates the atmospheric absorption.

Solar radiation is attenuated due to scattering and absorption, as sunlight passes through the atmosphere. The closer it is to the sea level, the greater the attenuation. On the earth at the sea level, the solar radiation power varies from place to place depending on solar zenith angle. At different zenith angles, the light passes through different atmosphere thickness. Air mass 1.5 (AM 1.5) is a standard terrestrial solar spectral irradiance distribution, at a mid-latitudes with a solar zenith angle of 48.2°. AM1.5 is a value characterizing the overall yearly average solar radiation received in mid-latitude region. The difference of the solar radiation spectra at the top of the earth's atmosphere and at the sea level is due to atmospheric absorption, mainly the CO_2 gas and H_2O vapor absorption, as labeled in Fig. 1.36. Strictly speaking, the solar radiation spans a wavelength range from 100 to about 10^6 nm. The whole solar radiation band is usually divided into five wavelength ranges.

The spectrum from 100 to 280 nm is called the ultraviolet C (UVC) range, with the radiation frequency higher than that of violet light. Due to atmospheric absorption, very few UVC photons can arrive at the Earth's sea level, so that it is also termed as solar blind ultraviolet. The spectrum from 280 to 315 nm is the ultraviolet B (UVB) range. Photons in this wavelength range are also significantly absorbed by the atmosphere. Ultraviolet A (UVA) range covers the spectrum from 315 to 400 nm. The visible solar radiation spans 380 to 780 nm. This wavelength range is visible to the human eyes as self-explanatory by the name. As shown in Fig. 1.36, visible light dominates the solar radiation in terms of photon flux. The wide spectrum covering 780 to 10^6 nm are the infrared range that is non-visible to human eyes. The infrared range takes about 50% of the total solar radiation energy. The whole infrared spectrum is divided into three ranges: 700 to 1400 nm (Infrared-A), 1400 to 3000 nm (Infrared-B), and 3000 to 10^6 nm (Infrared-C). In terms of total solar radiation energy, the ultraviolet light, visible light and infrared light account for 10%, 40% and 50% at the top of atmosphere, respectively. With the sun at the zenith and at the sea level, the energy fractions become 3% (ultraviolet), 44% (visible), and 53% (infrared), corresponding to the power fractions of 32, 445, and 527 W/m^2, respectively. Photon flux in extreme ultraviolet and X-ray range makes up a very small portion of the solar radiation.

1.6.2 Optical Transition Probability

Semiconductors are characterized by their energy bands which determine their major optical and electronic properties. A semiconductor may absorb photons with energies equal to and larger than its bandgap and a photovoltaic device made from the semiconductor transfers the photons into electric power. The major steps for the completion of the photovoltaic process include radiation photon absorption, electron–hole pair generation, electron–hole separation, and charge collection on electrodes.

Even in the single-electron framework, the photoelectric process might include several particles like an electron, a photon, a phonon, as well as impurities and defects. In the simplest case, the event is described by a process in which an incident photon is absorbed by an electron in valence band and the electron is then excited to the conduction band or above the conduction band becoming a hot electron. The "simplest" means no phonon, impurity, and defect takes a role in the process. A photoelectric process can be described quantum-mechanically. In dipole approximation, the interaction Hamiltonian takes the form of

$$H_I = -\frac{e}{m}\boldsymbol{A}\cdot\boldsymbol{p} = \frac{e\hbar i}{m}\boldsymbol{A}\cdot\nabla, \tag{1.56}$$

where $\boldsymbol{A}$ is the vector potential of electromagnetic field (incident light) which, in plane wave approximation, is described by

$$A(\boldsymbol{q},\boldsymbol{r},t) = A_0\hat{x}\left\{\exp\left[i(\omega t - \boldsymbol{q}\cdot\boldsymbol{r})\right] + \exp\left[-i(\omega t - \boldsymbol{q}\cdot\boldsymbol{r})\right]\right\} \tag{1.57}$$

Here $\hat{x}$ is the unit vector of the direction of vector potential $\boldsymbol{A}$, and A_0 is the amplitude. Thus, the interaction Hamiltonian H_I can be expressed by

$$\begin{aligned} H_I &= -\frac{e}{m}A_0\exp\left[i(\omega t - q\cdot r)\right]\hat{\boldsymbol{x}}\cdot\boldsymbol{p} - \frac{e}{m}A_0\exp\left[-i(\omega t - q\cdot r)\right]\hat{\boldsymbol{x}}\cdot\boldsymbol{p} \\ &= H_I^{+}\exp(i\omega t) + H_I^{-}\exp(-i\omega t) \end{aligned} \tag{1.58}$$

where H_I^{+} and H_I^{-} are the space-dependent (time-independent) parts in the interaction Hamiltonian. In Schrodinger wave mechanics, an electron in crystal is described by Bloch wave—a multiplication of a plane wave $\exp[i(\omega t - \boldsymbol{k}\cdot\boldsymbol{r})]$ and a wave $u(\boldsymbol{k}, \boldsymbol{r})$ with the periodicity of given crystal. The initial and the final states of the electron are then

$$\phi_i(\boldsymbol{k},\boldsymbol{r},t) = \exp[-i(\omega_i t - \boldsymbol{k}\cdot\boldsymbol{r})]u_i(\boldsymbol{k},\boldsymbol{r}) \tag{1.59a}$$

and

$$\phi_f(\boldsymbol{k},\boldsymbol{r},t) = \exp[-i(\omega_f t - \boldsymbol{k}\cdot\boldsymbol{r})]u_f(\boldsymbol{k},\boldsymbol{r}). \tag{1.59b}$$

respectively. Here $\omega_i = \frac{1}{\hbar}E_i$ and $\omega_f = \frac{1}{\hbar}E_f$, E_i and E_f being the energies of the initial and the final eigen states, respectively. Thus, the mathematical form of Fermi Golden Rule for electron transition is described by

$$\Theta(\boldsymbol{k},\boldsymbol{k}',\omega,t)=\left|\int_0^t \mathrm{d}t'\int_r \phi_f(\boldsymbol{k}',\boldsymbol{r},t)H_I\phi_i^*(\boldsymbol{k},\boldsymbol{r},t)dr\right|^2 \tag{1.60}$$

Or, using (1.59a) and (1.59b), it is written as

$$\Theta(\boldsymbol{k},\boldsymbol{k}',\omega,t)=\left|\int_0^t \exp[(\omega_f-\omega_i\pm\omega)\hbar t']dt'\int\phi_f(\boldsymbol{k}',\boldsymbol{r})H_I\phi_i^*(\boldsymbol{k},\boldsymbol{r})dr\right|^2, \tag{1.61}$$

where $\phi_f(\boldsymbol{k}', \boldsymbol{r})$ and $\phi_i(\boldsymbol{k}, \boldsymbol{r})$ are the time independent parts of the initial and the final states of electrons, respectively. Photons in the visible spectrum have momenta much smaller than those of electrons. Visible photons have large energy of ~2 eV and small momenta of ~1.3×10^{-27} kg·m/s. Thus, no matter absorbing a photon or emitting a photon, the electron transition is approximately vertical, $\boldsymbol{k}' \sim \boldsymbol{k}$ as a direct transition. The time integral part of (1.61) yields a δ-function, which can then be written as

$$\Theta(\boldsymbol{k},\boldsymbol{k}',\omega)=\frac{2\pi}{\hbar}\left|\int\phi_f(\boldsymbol{k}',\boldsymbol{r})H_I^{\mp}\phi_i^*(\boldsymbol{k},\boldsymbol{r})dr\right|^2\delta(\hbar\omega_f-\hbar\omega_i\pm\hbar\omega) \tag{1.62a}$$

or

$$\Theta(\boldsymbol{k},\boldsymbol{k}',\omega)=\frac{2\pi}{\hbar}\left|\langle f|H_I^{\mp}|i\rangle\right|^2\delta(E_f-E_i\pm\hbar\omega) \tag{1.62b}$$

using simple Kronic–Delta notation. Here "–" corresponds to a process of absorbing a photon and "+" to that of emitting a photon. Accounting for the distribution of the initial and the final states, a summation over all initial and final states gives rise to the total transition probability of

$$\begin{aligned}\Theta_{\text{Total}}(\boldsymbol{k},\boldsymbol{k}',\omega)&=\sum_{i,f}\Theta_{i,f}(\boldsymbol{k},\boldsymbol{k}',\omega,t)\\&=\frac{2\pi}{\hbar}\sum_{i,f}\left|\langle f|H_I^{\mp}|i\rangle\right|^2\delta(E_f-E_i\pm\hbar\omega).\end{aligned} \tag{1.63}$$

$\Theta_{\text{Total}}(\boldsymbol{k},\boldsymbol{k}',\omega)$ includes both the photon emission and the photon absorption processes, which can be separated if the low and the high energy states are explicitly used. Let $|l\rangle$ and $|h\rangle$ denote the low- and the high-energy states, respectively. The probability for photon absorption and emission then takes the form of

$$\Theta_{ab}(\boldsymbol{k},\boldsymbol{k}',\omega)=\frac{2\pi}{\hbar}\sum_{l,h}\left|\langle h|H_I^-|l\rangle\right|^2\delta(E_h-E_l-\hbar\omega) \tag{1.64a}$$

and

$$\Theta_{em}(\boldsymbol{k},\boldsymbol{k}',\omega)=\frac{2\pi}{\hbar}\sum_{l,h}\left|\langle l|H_I^+|h\rangle\right|^2\delta(E_l-E_h+\hbar\omega), \tag{1.64b}$$

respectively. δ-functions in (1.64), $\delta(E_h-E_l-\hbar\omega)$ and $\delta(E_l - E_h + \hbar\omega)$, are an indication of energy conservation. Since photon energy $\hbar\omega$ is positive, the variables in the above δ-functions describes electron transition from the low energy state $|l\rangle$ to the excited energy state $|h\rangle$ companied by absorbing a photon (Eq. 1.64a) or electron transition from the high-energy state $|h\rangle$ to the low-energy state $|l\rangle$ while emitting a photon (Eq. 1.64b). $\delta(E_h-E_l-\hbar\omega)$ and $\delta(E_l - E_h + \hbar\omega)$ represent energy conservation by $E_h - E_l = \hbar\omega$ and $E_l - E_h = \hbar\omega$, guaranteed both in a photon absorption and a photon emission processed, respectively.

The photon absorption and photon emission described in (1.64) are due to the light–matter interaction through the interaction Hamiltonian H_I. The absorption and emission in (1.64) due to the interaction Hamiltonian between electrons and photons are stimulated absorption and stimulated emission, since both Θ_{ab} and Θ_{em} are zero in the absence of light field. A spontaneous emission that is not stimulated by a light field is possible to occur. In the spontaneous emission, an electron transits from a high-energy to a low-energy state, giving away energy by emitting a photon.

1.6.3 Optical Transition in Low-Dimensional Semiconductors

The fundamental process of photovoltaic effect includes exciting an electron from the ground state (the top of the valence band) into the first excited state (the bottom of the conduction band), left with a positively charged hole, via photon absorption. As a good approximation, the electron can be described in single-electron scheme, though in reality both electrons and holes are charged particles and they interact with each other due to Coulomb interaction. The photoexcited electron and hole are still bound by the Coulomb potential, forming a hydrogen-like system called an exciton. Similar to a hydrogen atom, the exciton is described by a Bohr radius a_B^* and Rydberg energy Ry^* of an exciton. The exciton Bohr radius can be expressed as

$$a_B^* = \varepsilon_r \frac{m_e}{m_r} a_B, \tag{1.65}$$

where $a_B = \frac{4\pi\varepsilon_0\hbar^2}{m_e e^2}$ (= 0.053 nm) is the Bohr radius of the ground state electron in hydrogen atom, ε_r the relative permittivity, m_e the electron mass at rest, $m_r = \frac{m_e m_h}{m_e + m_h}$ (m_h is the mass of the hole) the reduced mass of the electron–hole pair, and ε_r is the relative permittivity of the material. For common semiconductors, a_B^* is in the interval of 10–200 Å. The energy level of an electron in a hydrogen atom takes the form of

$$E_n = -\frac{m_e e^4}{32\pi^2\varepsilon_0^2\hbar^2}\frac{1}{n^2} = -\frac{Ry}{n^2} \tag{1.66}$$

where n is the principle quantum number. One of the most important outcomes for the quantum confinement is that it leads to discreet energy levels in both the conduction (for electrons) and the valence (for holes) bands that are originally continuous. $R_y = \frac{m_e e^4}{32\pi^2\varepsilon_0^2\hbar^2}$ is the Rydberg energy of a hydrogen atom. A Rydberg energy of an exciton can be simply defined by

$$R_y^* = \frac{m_r}{\varepsilon_r^2 m_e} R_y. \tag{1.67}$$

In semiconductors, the exciton Rydberg energy R_y^* takes the values in the range of 1–150 meV. Using hydrogen model, the exciton energy spectrum can be expressed by

$$E_n(k) = E_g + \frac{\hbar^2 k^2}{2m_r} - \frac{Ry^*}{n^2}, \quad (1.68)$$

where E_g is bandgap, k the wavevector. Rydberg energy Ry^* of an exciton is related to that of a hydrogen atom by a factor of $\frac{m_r}{\varepsilon_r^2 m_e}$, as shown by (1.67). In fact, the wavevector of a photon is extremely small and the transition occurs vertically. At the center of Brillion zone ($k = 0$), one can write

$$E_n = E_g - \frac{Ry^*}{n^2}. \quad (1.69)$$

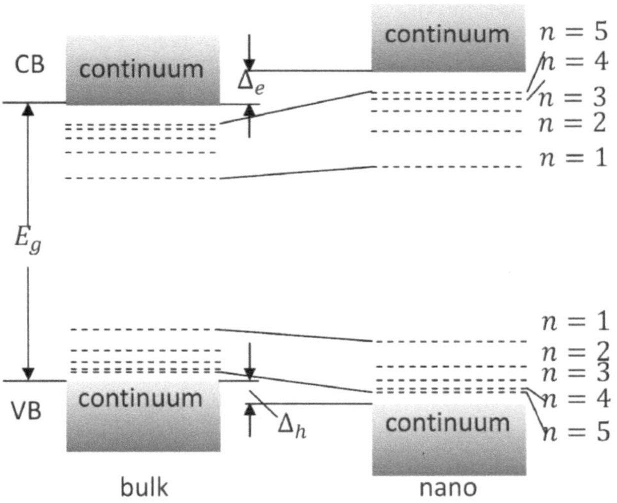

Figure 1.37 Schematic diagram of band edge continuum and excitonic states with n = 1, 2, 3, 4, 5 for both electrons and holes. The quantum confinement effect induces the band edge shift Δe and Δh in CB and VB, respectively.

Thus, the exciton energy is lower than the bandgap due to attractive Coulomb interaction between an electron and a hole, as schematically shown in Fig. 1.37. Strictly speaking, electrons in the conduction band are all bounded to holes in the valence band in semiconductors, since electrons and holes are generated simultaneously. The so-called free electrons and free holes are ionized excitons in the energy spectrum of exciton continuum (see Fig. 1.37). Figure 1.37 depicts the lowest five discrete exciton states with n = 1, 2, 3, 4, and 5 for both electrons and holes that compose excitons. At $n \to \infty$, the exciton state approaches to the band edges of the conduction band (CB) and the valence band (VB). Quantum confinement induces shift of both the band edges and the exciton states. Since the exciton states are relatively more localized, they are less affected with respect to the band edge states.

In nano-sized materials, both the electron wavefunctions and eigen state energies are different from those in bulk case. Thus, the optical transition energy, transition probability, and transition selection rule are all modified. Excitons are squeezed in a semiconductor with size comparable to or smaller than that of exciton's Bohr radius, due to quantum confinement. The quantum confinement leads to shift of energy levels of electrons and holes. The energy level shift is well-described by a theory in which the energies of the eigen states depend on the size of the nanocrystal. The confinement energies for electrons and holes confined in a box with the size a are approximately written as

$$E_{\text{conf},e} = \frac{\pi^2\hbar^2}{2a^2 m_e} \tag{1.70a}$$

$$E_{\text{conf},h} = \frac{\pi^2\hbar^2}{2a^2 m_h}, \tag{1.70b}$$

respectively. Thus the total confinement energy is

$$E_{\text{conf}} = E_{\text{conf},e} + E_{\text{conf},h} = \frac{\pi^2\hbar^2}{2a^2}\left(\frac{1}{m_e} + \frac{1}{m_h}\right) = \frac{\pi^2\hbar^2}{2a^2 m_r} \tag{1.71}$$

Apparently, quantum confinement tends to increase transition energies for the type-I band alignment in opposite to the exciton (Coulomb) effect. Thus the optical transition energy is, including exciton effect,

$$E_n = E_g - \frac{Ry^*}{n^2} + \frac{\pi^2\hbar^2}{2a^2 m_r} \tag{1.72}$$

The resultant effect of exciton and quantum confinement is shown in the right panel of Fig. 1.37. In fact, the quantum confinement acts to enhance the exciton effect since, due to the confinement the electrons and holes are likely to stay closer. The third term on the left side is positive and could be quite large for small a. Note that, in the case of a broad size distribution of the QD size a, $E_n(k)$ has a distribution in certain energy range rather than a single energy level, corresponding to a broad distribution of light absorption. Using quantum dots of different sizes, photons in the whole solar radiation spectrum can be absorbed by photovoltaic devices designed properly.

References

1. P. Y. Yu and M. Cardona, *Fundamentals of Semiconductors: Physics and Materials Properties*, 4th ed., Springer (2010).
2. C. Kittel, *Introduction to Solid State Physics*, 8th ed., John Wiley & Sons, Inc (2005).
3. K. Y. Tsao and C. T. Sah, Temperature dependence of resistivity and hole conductivity mobility in p-type silicon, *Solid-State Electronics*, **19**, 949–953 (1976).
4. Y. P. Timalsina, A. Horning, R. F. Spivey, K. M. Lewis, T.-S. Kuan, G.-C. Wang, and T.-M. Lu, Effects of nanoscale surface roughness on the resistivity of ultrathin epitaxial copper films, *Nanotechnology*, **26**, 075704 (2015).
5. M. D. Sturge, Optical absorption of Gallium Arsenide between 0.6 and 2.75 eV, *Phys. Rev.*, **127**(3), 768–773 (1962).
6. D. D. Sell, S. E. Stokowski, R. Dingle, and J. V. DiLorenzo, Polariton reflectance and photoluminescence in high-purity GaAs, *Phys. Rev.* **7**(10), 4568–4586 (1973).
7. Ed. by P. Butcher, N. H. March, M. P. Tosi, *Physics of Low-Dimensional Semiconductor Structures* (Plenum Press, New York and London) (1993).
8. P. Granitzer and K. Rumpf (eds.), *Nanostructured Semiconductors: From Basic Research to Applications* (Pan Stanford Publishing Pte. Ltd.) (2014).
9. M. Y. Ge, Y. Yue, Y. F. Liu, J. Wu, Y. Sun, X. Chen, and N. Dai, Facile capping CdS and ZnS shells by thermolysis of ethylxanthate

precursors for CdSe/CdS/ZnS nanocrystals, *Colloids Surface A*, **384**, 574 (2011).

10. S. M. Sze and Kwok K. Ng, *Physics of Semiconductor Devices*, 3rd ed. (John Wiley & Sons, Inc., Hoboken, New Jersey) (2007).
11. O. Madelung, *Introduction to Solid-State Theory*, Springer Series in Solid-State Sciences (M. Cardona, P. Fulde, and H. J. Queisser, series eds.), 3rd printing (Springer-Verlag Berlin Heidelberg New York) (1996).
12. W. A. Harrison, *Solid State Theory*, (Dover Publications Inc. New York) (2011).
13. J. C. Slater and G. F. Koster, Simplified LCAO method for the periodic potential problem, *Phys. Rev.*, **94**, 1498 (1954).
14. W. A. Harrison, *Electronic Structure and the Properties of Solids: The Physics of the Chemical Bond* (W. H. Freeman, San Fransisco) (1980).
15. W. A. Harrison, *Pseudopotentials in the Theory of Metals* (W.A. Benjamin, New York) (1966).
16. J. R. Chelikowsky and M. L. Cohen, Nonlocal pseudopotential calculations for the electronic structure of eleven diamond and zinc-blende semiconductors, *Phys. Rev. B*, **14**, 556 (1976).
17. B. I. Shklovskii and A. L. Efros, *Electronic Properties of Doped Semiconductors* (Springer-Verlag, Berlin Heidelberg New York Tokyo) (1984).
18. J. C. Ho, R. Yerushalmi, Z. A. Jacobson, Z. Fan, R. L. Alley, and A. Javey, Controlled nanoscale doping of semiconductors via molecular monolayers, *Nat. Mater.*, **7**, 62 (2008).
19. K. Takei, R. Kapadia, Y. Li, E. Plis, S. Krishna, and A. Javey, Surface charge transfer doping of III–V nanostructures, *J. Phys. Chem. C*, **117**, 17845–17849 (2013).
20. J. Wallentin and M. T. Borgström, Doping of semiconductor nanowires, *J. Mater. Res.*, **26**(17), 2142–2156 (2007).
21. D. Mocatta, G. Cohen, J. Schattner, O. Millo, E. Rabani, and U. Banin, Heavily doped semiconductor nanocrystal quantum dots, Science, **332**(6025), 77–81 (2011).
22. R. C. Newman, Defects in silicon, *Rep. Prog. Phys.*, **45**, 1163 (1982).
23. W. Bergholz and D. Gilles, Impact of research on defects in silicon on the microelectronic industry, *Phys. Status Solidi B*, **222**, 5 (2000).

24. N. Dai, A. Cavus, R. Dzakpasu, and M. C. Tamargo, F. Semendy and N. Bambha, D. M. Hwang, and C. Y. Chen, Molecular beam epitaxial growth of high quality $Zn_{1-x}Cd_xSe$ on InP substrates, *Appl. Phys. Lett.*, **66**(20), 2742–2744 (1995).

25. A. Becker, C. Wenger, and J. Dabrowski, Control of etch pit formation for epitaxial growth of graphene on germanium, *J. Appl. Phys.*, **126**, 085306 (2019).

26. J. Y. Tsao, *Materials Fundamentals of Molecular Beam Epitaxy* (Academic Press, Inc., Harcourt Brace Jovanovich, Publishers, Boston, San Diego, New York, London, Sydney, Tokyo Toronto) (1993).

27. P. D. Dapkus and J. J. Coleman, Metalorganic chemical vapor deposition, in *III–V Semiconductor Materials and Devices*, R. J. Malik, ed. (Amsterdam: North-Holland) (1989).

28. S. Emin, S. P. Singh, L. Han, N. Satoh, A. Islam, Colloidal quantum dot solar cells, *Solar Energy*, **85**, 1264–1282 (2011).

29. W. Mönch, Electronic structure of metal-semiconductor contacts, *in Perspectives in Condensed Matter Physics* (Springer, Dordrecht) (1990).

30. G. Margaritondo, Electronic structure of semiconductor hetero-junctions, *in Perspectives in Condensed Matter Physics* (Springer, Dordrecht) (1988).

31. G. Bastard, *Wave Mechanics Applied to Semiconductor Heterostructures* (Halsted Press, New York) (1988).

32. J. R. I. Lee, H. D. Whitley, R. W. Meulenberg, A. Wolcott, J. Z. Zhang, D. Prendergast, D. D. Lovingood, G. F. Strouse, T. Ogitsu, E. Schwegler, L. J. Terminello, and T. van Buuren, Ligand-mediated modification of the electronic structure of CdSe quantum dots, *Nano Lett.*, **12**, 2763–2767 (2012).

33. J. Tang, K. W. Kemp, S. Hoogland, K. S. Jeong, H. Liu, L. Levina, M. Furukawa, X. Wang, R. Debnath, D. Cha, K. Wei Chou, A. Fischer, A. Amassian, J. B. Asbury, and E. H. Sargent, Colloidal-quantum-dot photovoltaics using atomic-ligand passivation, *Nat. Mater.*, **10**, 765–771 (2011).

34. B. A. Kairdolf, A. M. Smith, T. H. Stokes, M. D. Wang, A. N. Young, and S. Nie, Semiconductor quantum dots for bioimaging and biodiagnostic applications, *Annu. Rev. Anal. Chem.*, **6**, 143–162 (2013).

35. J. Yang, J. Heo, T. Zhu, J. Xu, J. Topolancik, F. Vollmer, R. Ilic, and P. Bhattacharya, Enhanced photoluminescence from embedded

PbSe colloidal quantum dots in silicon-based random photonic crystal microcavities, *Appl. Phys. Lett.*, **92**, 261110 (2008).

36. B. Sain and D. Das, Development of nc-Si/a-SiN_x:H thin films for photovoltaic and light-emitting applications, *Sci. Adv. Mater.*, **5**(2), 188–198 (2013).

37. C. Kandilli and Koray Ulgen, *Solar Illumination and Estimating Daylight Availability of Global Solar Irradiance, Energy Sources, Part A: Recovery, Utilization, and Environmental Effects*, **30**(12), 1127–1140 (2008).

Chapter 2

How Solar Cells Work

Photoelectronic devices, which convert solar radiation energy into electric power, are called solar cells or photovoltaic devices. The most important parameter to characterize the performance of solar cells (SCs) is photoelectric conversion efficiency (PCE), defined as

$$\eta = \frac{\text{Output electric power by the solar cell}}{\text{Solar radiation power on the surface of the solar cell}}. \tag{2.1}$$

Photovoltaic process includes photon absorption, charge separation, charge transportation, and charge collection. Solar cells should be properly designed to perform these functions to complete the photovoltaic process with high efficiency. The materials performing the photovoltaic functions include semiconductors, insulators, and metals. Semiconductors are the key materials that are used as the active layers to absorb light and split excitons. Insulators are used as passivation layers for electric isolating, and the metals as contact electrodes for charge collecting.

2.1 Photovoltaic Effect

The physical processes by which photons are converted into electrical potential and current are well explained by fundamental

Introduction to Nano Solar Cells
Ning Dai

ISBN 978-981-4877-49-7 (Hardcover), 978-1-003-13198-4 (eBook)
www.jennystanford.com

physical theory. Theoretical studies and simulations on solar cells with various structures have guiding significance. Theoretical works are essential to understand details in device working processes, helping to predict fundamental losses and device efficiencies in solar cells. The theory offers guidance on device design. Although nano solar cells (NSCs) are volatile and complicated in device architectures, the photon-converted-to-electrical-current (-potential) processes follow the same physical principle as the traditional wafer-based Si SCs.

2.1.1 The Physical Process of Photovoltaic Effect

Using suitable devices—solar cells, photons are converted to electrical potential through the following steps:

- Photon incidence on the solar cell surface. For the photons incident on the surface of a solar cell, some of them are reflected and diffracted by the surface of the device. On polished semiconductor surface, roughly about 35% of the incident photons are reflected.
- Photon absorption by semiconductors such as silicon. Photons with energy equal to and greater than bandgap can be absorbed by electrons in semiconductors. The photons with energies smaller than the bandgap will penetrate through the semiconductor.
- Generation of an electron/hole pair. After a photon is absorbed, an electron/hole pair (an exciton) is generated, i.e, an electron in the valence band is excited into the conduction band and a hole is left in the valence band. Via such a photoelectric conversion process, the photon energy is converted into electric energy.
- Photon-excited electrons and holes flowing apart toward opposite directions. By means electric field (buildt-in electric field of a *p–n* junction, for instance) or non-electrostatic force (due to a heterojunction), the photoexcited electron–hole pair can be split into a free electron and a free hole.
- Electrons and holes collected by separate electrodes. Photogenerated electrons and holes transport in opposite directions and collected by their respective electrodes.

Thus, a photovoltaic process consists of five major sub-processes: light absorption, photoexcited generation of excitons—electron–hole pairs—due to light absorption, separation of electrons and holes, carrier transport, and carrier collection on electrodes. To achieve high conversion efficiency, all the sub-processes need to be optimized. However, each sub-process is always companied by loss that leads to the reduction of photoelectric conversion efficiency. Optically, for instance, some of the incident photons are reflected from the device surface. Electrical losses include but are not limited to recombination of photoexcited electrons and holes. Device architecture needs to be carefully designed and constructed. However, the device parameters for some optimized sub-processes are mutually restricted. For instance, materials should be thick enough in order to gain sufficient light absorption, while thin materials are beneficial for high carrier collection efficiency. The collection probability is high if the material thickness is comparable or less than the diffusion length of minority carriers.

2.1.2 The Photoelectric Conversion Efficiency

In a real solar cell, each of the above five steps can be completed in limited probability less than 100%. In another words, the overall conversion efficiency equals to

$$P_{\text{Total}} = P_1 \times P_2 \times P_3 \times P_4 \times P_5. \tag{2.2}$$

Since P_i (i = 1, 2, 3, 4, 5), the probability of completion of each of the above five steps, is less than 100%, P_{Total} is definitely below 100%. In fact, the conversion efficiency of the current wafer-based Si solar cells is typically around 20%, i.e., 80% incident solar energy cannot be converted into electric power by Si solar cells.

As shown in Fig. 1.36, solar radiation covers a large photon energy range from far infrared to ultraviolet. When a photon hits the surface of a solar cell, three events could happen: the photon is reflected or diffracted off the surface Si (assuming a Si solar cell), the photon penetrates through the solar cell, which happens likely for photons with energy lower than the bandgap of Si, or the photon is absorbed by an electron in Si. In the first and the second evens, the photon cannot excite an electro/hole pair, so that it makes no contribution to a photovoltaic effect. In the third event, which requires that the photon has energy equal to or larger than the bandgap, an electron/hole pair is generated. Photon energy is then given to the electron in the valence band where the electron is bounded with neighboring atoms in covalent bonding, the electron being excited into the conduction band. If the energy that the photon gives to the electron is exactly equal to the bandgap energy of Si, the electron becomes nearly free electron in the bottom of the conduction band. In the valence band, a "hole" is generated by the "missing" electron. A bonded electron in neighboring atom could jump in to fill into the hole and generate another hole on the neighboring atom, as if the photon-generated hole can move freely in the crystal. The photon-excited electron and hole are still loosely bounded to form an exciton due to Coulomb interaction.

Very often, an absorbed photon has energy larger than the bandgap, the electron is then excited to an energy state above the bottom of the conduction band, becoming a hot electron. The hot electron relaxes to the bottom of the conduction band via non-radiative channels through which it gives away its extra energy to the crystal usually through interaction with the crystalline lattice by emitting phonons. The extra energy given to the crystal makes no contribution to photovoltaic process. On the other hand, Si could absorb two or more low energy photons with a sum of their energy larger than its bandgap, so that low energy photons can be used for photovoltaic process via two- or multi-photon absorption. However, such a non-linear process for the two- or multi-photon absorption requires a high intensity light source, the efficiency being very low for the light intensity of solar radiation.

Note that the photon can also be absorbed by phonons and, in this case, heat, instead of an exciton, is generated, which is unfavorable for a photovoltaic process. In addition, photon-excited electrons and holes have a certain probability to recombine to give away their energy gained through photon emission or other processes. Thus, long carrier lifetime (diffusion length) is required for the electrons and holes to be collected by metal contacts before they could recombine.

2.1.3 Modeling a Conventional Semiconductor Solar Cell

Containing a large number of charges (electrons and holes), a piece of semiconductor is still charge neutral because the negative charges are all balanced by positive ones. An utmost important issue for design of photovoltaic devices is highly efficient separation of the negative charges (electrons) from the positive ones (holes). There are two macroscopic motions of charged particles in a photovoltaic device, namely, drift and diffusion. Carrier drift refers to the motion of charged particles driven by an electric field in devices. Carrier diffusion is caused by the carrier density gradient in a device, the driving force for diffusion being random thermal motion. In a wafer-based Si solar cell, a *p–n* junction is constructed to form a built-in electric potential that separates photon-generated electrons and holes (excitons). As a result, the carrier densities of both electrons and holes are then not equal in the *p*- and the *n*-regions of the device, i.e., there is an intentional gradient of electron density and hole density toward opposite directions. Thus, electrons and holes move in solar cells in both the drift and the diffusion modes. If photons are absorbed in region out of the space charge region (the built-in electric field is located within the space charge region), motion and separation of electrons and holes have to rely on diffusion. As a result, lifetime of minority carriers needs to be long enough to allow the carriers to travel through the device before they recombine with majority carriers. For thin devices such as thin-film amorphous Si (a-Si) solar cells, scattering due to high density of defects and dislocation gives rise to very short minority carrier lifetime. In nano SCs, efficient charge separation requires strong electric field similar to the space charge region at the junction region in a wafer-based solar cell. In a Si thin-film cell, the built-in electric field extends the entire thickness of devices.

A Si solar cell is characterized by a strong built-in electrostatic field, as shown in Fig. 1.13, functioning to split the excitons and drive electrons and holes to migrate in opposite directions. Figure 2.1 schematically depicts device structure of typical Si wafer-based SCs.

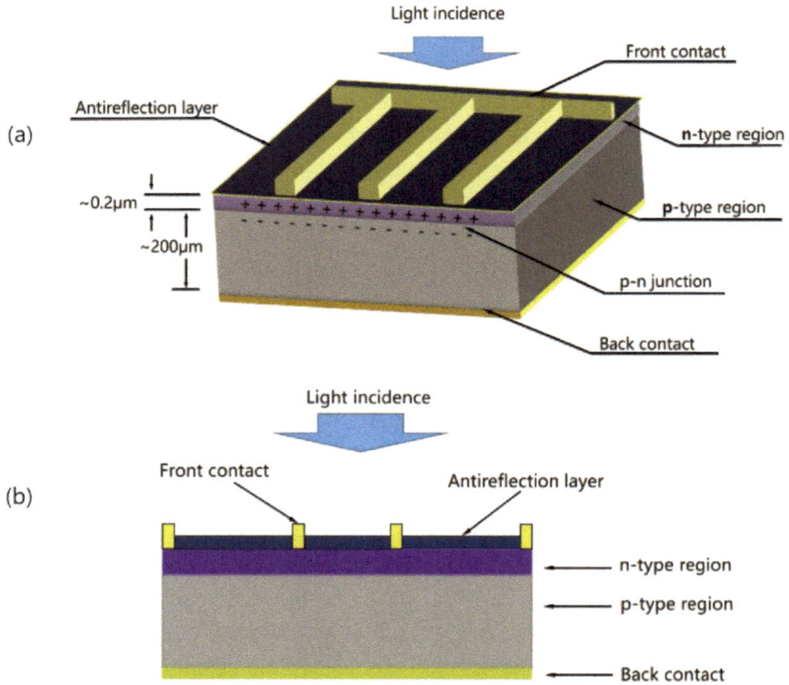

Figure 2.1 Schematic top view (a) and side view (b) of wafer-based Si solar cells.

In order to gain output electric power from solar cells, electric contact needs to be made to the two-terminal devices. Good contacts are critical for high efficiency power transfer to an external load. Electrons are collected by the contact electrode connected to the *n*-side of the device and holes are collected by the contact at the *p*-side (see Fig. 2.1). The electric power is transferred to the external load after the electrons passes through the load and collected by the terminal at *p*-side of the device. The electrons then recombine with holes in *p*-side. This process can be described using an equivalent circuit model as shown in Fig. 2.2a. This is an ideal solar cell model consisting of a current source in parallel with a diode. According to the theory of a p–n junction, the ideal model of equivalent circuit is described by

$$I_{\text{out}} = I_{\text{ph}} - I_D \tag{2.3}$$

$$I_D = I_0\left[\exp\left(\frac{qV_d}{nk_BT}\right) - 1\right] \tag{2.4}$$

where I is the solar cell output current, I_{ph} the photon-generated current, I_D diode current, q the elementary charge, V_d the voltage on the diode, n the ideality factor that is a measurement of the junction quality, and I_0 the reverse saturation current which is a measure of leakage carriers across the p–n junction in reverse bias.

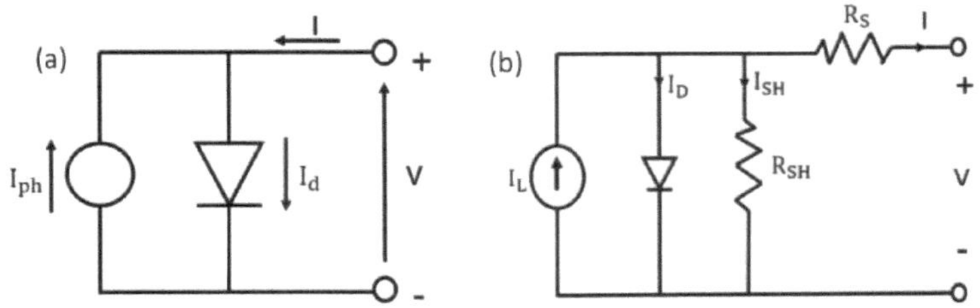

Figure 2.2 (a) Ideal equivalent circuit and (b) real equivalent circuit where both the parasitic series resistance and the shunt resistance are taken into account.

Practical solar cells are not ideal. As an improved model, a realistic and simple one is shown in Fig. 2.2b in which a series resistance and a shunt resistance component are added to the ideal model. In Fig. 2.2b, R_s is the parasitic serial resistance, the sum of structural resistances depending on the junction depth, metal contacts, the impurity concentration of the p- and the n-type regions, and connection wires. The shunt resistance R_{sh} connects parallel with the diode and accounts for the leakage current of the p–n junction. Thus, the shunt resistance is caused by poor module installation and the technological defects in solar cell fabrication [1]. An ideal solar cell has zero R_s and infinitively large R_{sh}. Low shunt resistance leads to significant power loss since it bypasses a current through a path other than load. At low level of solar radiation, the effect of a shunt resistance can become very significant, owing to the relatively low photon-generated current. Many factors affect the series resistance R_s. Series current I_s passing through R_s and shunt current I_{sh} running through R_{sh} are also marked in Fig. 2.2. For the model including R_s and R_{sh}, Eq. 2.3 can be rewritten as

$$I_{out} = I_{ph} - I_D - I_{sh}. \tag{2.5}$$

In other words, the current offered by a solar cell is equal to photogenerated current of the solar cell subtracted by the current flowing through the diode and the current flowing through the shunt resistor. Correspondingly, the voltage across the diode (also the shunt resistor) is

$$V_d = V_{out} + I_{out} R_s, \tag{2.6}$$

where V_{out} is the output voltage applied on a load, V_d the voltage across both diode and shunt resistor, and I the output current. Clearly, a series resistor gives rise to decrease of output voltage. Thus, the shunt current is

$$I_{sh} = \frac{V_d}{R_{sh}}. \tag{2.7}$$

Using (2.6) and substituting (2.4) and (2.7) into (2.5) yields the characteristic equation of a solar cell,

$$I_{out} = I_{ph} - I_0 \left[\exp\left(\frac{qV_{out} + qI_{out} R_s}{nk_B T} \right) - 1 \right] - \frac{V_d}{R_{sh}}. \tag{2.8}$$

This equation directly connects solar cell parameters with the device output current and voltage, which is useful for calculations. In Eq. 2.8, I_{out} and V_{out} are simply measurable quantities and, as a result, some device parameters can be calculated using Eq. 2.8, such as I_0, R_{sh}, and R_s. n is the ideality factor, that will be discussed later. Note that the equation is in a transcendental form since both sides contain I_{out}, so that the calculation needs to be numerical in which non-linear regression is adopted to extract the parameters that occur with combined effect.

In above equation, the parameters I_{out}, I_0, R_{sh}, R_s, I_{ph}, etc., are all extensive quantities. In comparing two identical cells with different area, the cell with a larger surface area have a larger I_{out}. It is sometimes more meaningful to use intensive quantities to describe the relationship among the material and device parameters. Instead of I_{out}, I_0, R_{sh}, R_s, and I_{ph}, one could use J_{out} (current density in A·cm^{-2}), J_0 (reverse saturation current density in A·cm^{-2}, r_{sh} (specific shunt resistance in Ω·cm^2), r_s (specific series resistance in Ω·cm^2), as well as J_{ph} (photoexcited current density A·cm^{-2}) to rewrite Eq. 2.8 as

$$J_{\text{out}} = J_{\text{ph}} - J_0\left[\exp\left(\frac{qV_{\text{out}} + qJ_{\text{out}}r_s}{nk_BT}\right) - 1\right] - \frac{V_d}{r_{\text{sh}}}. \tag{2.9}$$

Due to the use of the intensive quantities of J_{out}, J_{ph}, r_{sh}, etc., one does not have to refer to the size of a solar cell, which offer easiness for research in which device parameters of solar cells of very different sizes are compared. In the cases of nano solar cells, one needs to note that those intensive quantities are just an average over area, since there are a large number of interface and material discontinuities, i.e., those intensive quantities (J_{ph}, for instance) are sensitively location dependent, even in microscopic sense. For wafer-based single crystal Si solar cells, one just need to note the difference of any of the quantities at the edge and the center and, for wafers of large sizes, the edge effect is ignorable. This is, however, not the case for nano solar cells. Due to very large interface area, J_0 could be quite large and r_{sh} could be small, which is detrimental for a solar cell.

For a solar cell, there are several important parameters that are widely used for the characterization of device performance, i.e., open-circuit voltage V_{oc}, short-circuit current density J_{sc}, filling factor *FF*, and power conversion efficiency η. V_{oc} is the output voltage when J_{out} is zero, i.e., the voltage measured on an illuminated photovoltaic device without external load. Short-circuit current J_{sc} is defined, under solar illumination, to be the current flowing through a photovoltaic device when its electrodes are shortly connected. Assuming very large shunt resistance, the open-circuit voltage V_{oc} can be calculated by setting $J_{out} = 0$ and using the characteristic equation (the transcendental equation of above Eq. 2.9) which takes a very simple form of

$$V_{oc} = \frac{k_B T}{q} \ln\left(\frac{J_{ph}}{J_0} + 1\right). \tag{2.10}$$

Similarly, by setting $V_{out} = 0$, the short-circuit current J_{sc} takes the form of

$$J_{sc} \approx J_{ph} \tag{2.11}$$

for a solar cell with very low J_0 and R_s and very high R_{sh}.

The basic structure of solar cell includes p–n junction formed by p-type and n-type doping, and metal electrodes, which displays energy band profiles as shown in Fig. 2.3. The band profile and electron occupation (shaded areas in Fig. 2.3) determine important device parameters J_{sc} and V_{oc}, the maximum output current and maximum output voltage, respectively, that are obtainable on a solar cell. However, J_{sc} and V_{oc} cannot be obtained simultaneously, i.e., a solar cell is unable to deliver a power of $J_{sc} \times V_{oc}$. In fact, no power can be extracted from a solar cell if the device is operated at either open-circuit or short-circuit situations. Thus, on an J–V curve of a solar cell, there must be a point corresponding to maximum out power P_m that can be written as

$$P_m = J_m \times V_m, \tag{2.12}$$

where J_m and V_m are the current and the voltage values that give maximum output power, as shown in Fig. 2.4. Obviously, J_m must be less than J_{sc} and V_m less than V_{oc}. Filling factor FF is defined as the ratio of the maximum actual power that one can get from a solar cell to the product of its short-circuit current and its open-circuit voltage, i.e.,

$$FF = \frac{J_m \times V_m}{J_{sc} \times V_{oc}}, \tag{2.13}$$

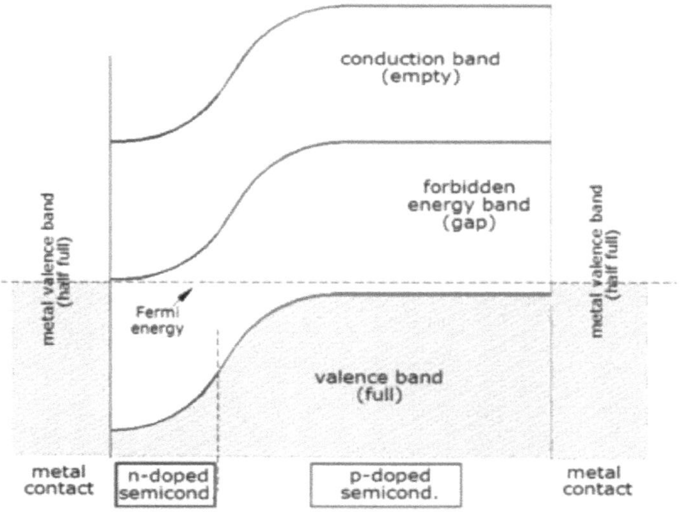

Figure 2.3 Energy band profile of a *p–n* junction, where shaded are states with electron occupation.

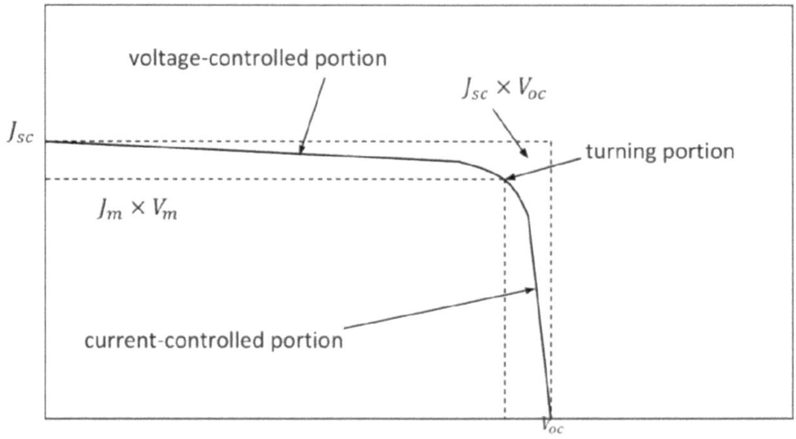

Figure 2.4 *J–V* curve of a wafer-based Si solar cell. The *J–V* curve is labeled as three portions: voltage-controlled portion, current-controlled portion, and turning portion.

Figure 2.5 presents a typical *J–V* curve of a wafer-based Si solar cell made by J. A. Solar. $J_m \times V_m$ is the area of the small rectangle and $J_{sc} \times V_{oc}$ the area of the large rectangle. The current–voltage curve reflects the performance of a solar cell. *FF* is the ratio of the small and the large rectangle. FF is an important parameter that characterizes the solar cell quality and wafer-based Si solar cells have a typical $FF \sim 80\%$.

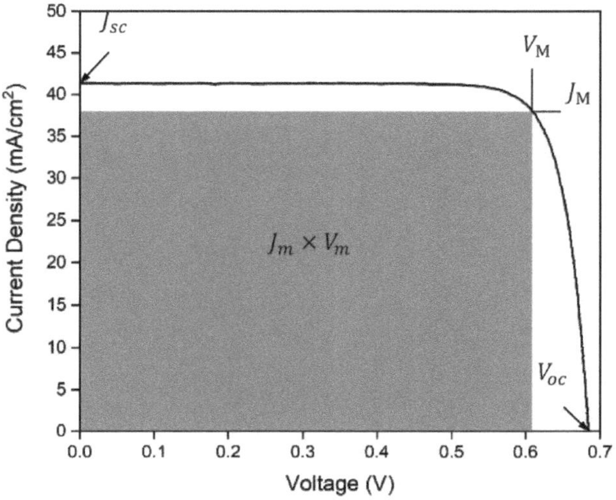

Figure 2.5 The J–V curve of a crystalline solar cell made by J.A. Solar Inc, with parameters J_{sc}, V_{oc}, J_m and V_m labeled.

For an ideal solar cell, η is equal to 100%, meaning that the J–V curve is a rectangle (the small rectangle and the large rectangle are then completely overlapped). Any deviation from ideal situation, non-zero R_s and finite J_0 and R_{sh}, for instance, results in sagging of an J–V curve toward the origin. The idea of series resistance was first proposed by Prince [2]. As shown in Fig. 2.2b, output current passes through the series resistor. This implies that part of the electric power transformed from solar radiation will be consumed on the resistor. As increase of resistance of the resistor R_s, more and more power ($=J_{out}R_s^2$) will be consumed by the resistor. In addition, there is a voltage drop on R_s, which is equal to $J_{out}\,R_s$. Thus, output voltage of a solar cell is going to be significantly reduced for a series resistor with large resistance. On a J–V curve, this results in deviation of J–V curve away from ideal situation. Nano solar cells often have high values of R_s, which is paid by a relatively poor J–V curves. For devices having very large R_s, their J–V property could be dominated by the behavior of resistors, as shown by the J–V curve for a nano solar cell with large R_S (see Fig. 2.6).

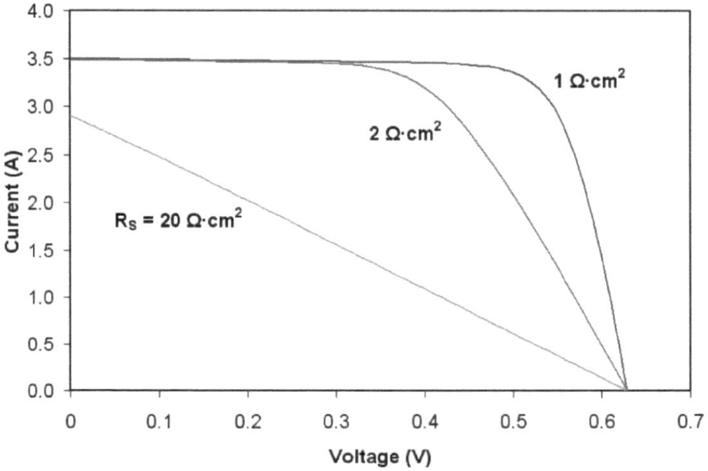

Figure 2.6 J–V curve calculated for R_S = 1, 2, and 20 $\Omega\cdot cm^2$.

Shunt resistance

Caused by manufacturing incompleteness and poor module installation procedure, shunt resistance is connected to an ideal solar cell in a parallel way, as shown in Fig. 2.2b. For large shunt resistance, the resistor acts as a bypass channel to decrease the output current and its effect on J–V curve is to make the voltage-controlled portion decline toward voltage axis. Shunt resistance results in a significant reduction in the output current J_{out} while the output voltage remains essentially unchanged. The larger the voltage, the more it declines. For low shunt resistance caused by any reason, both the voltage-controlled portion and the current-controlled portion are strongly affected, as shown in Fig. 2.7, including significant reduction in output voltage. For very low shunt resistance, a device could show J–V characteristics similar to that of a resistor.

Temperature

According Eq. 2.8, the characteristic equation is temperature dependent, as shown in the exponential term. With increase of temperature, the value of $\exp\left(\frac{qV_{out}+qJ_{out}R_s}{k_BT}\right)$ decreases. But, the decrease is not significant in the temperature range usually considered, as shown in Fig. 2.8a for the value of $\exp\left(\frac{qV_{out}+qJ_{out}R_s}{k_BT}\right)$ at 293, 303, 313, and 323 K. In fact, J_0 increases exponentially with temperature, a fundamental property of a diode with a *p–n* junction. This in turn reduces the open-circuit voltage V_{oc} of a solar cell. Approximately, V_{oc} decreases linearly with the increase of temperature. But for solar cells with higher V_{oc} values, their open-circuit voltages reduce slower than those of cells with lower V_{oc}. For wafer-based Si solar cells, the reduction rate is typically below −0.5%/°C. For nano solar cells, the situation varies significantly among solar cells with different materials and structures. Strictly speaking, all the material and device parameters are temperature dependent, but temperature dependence of device parameters strongly dependent on device architectures.

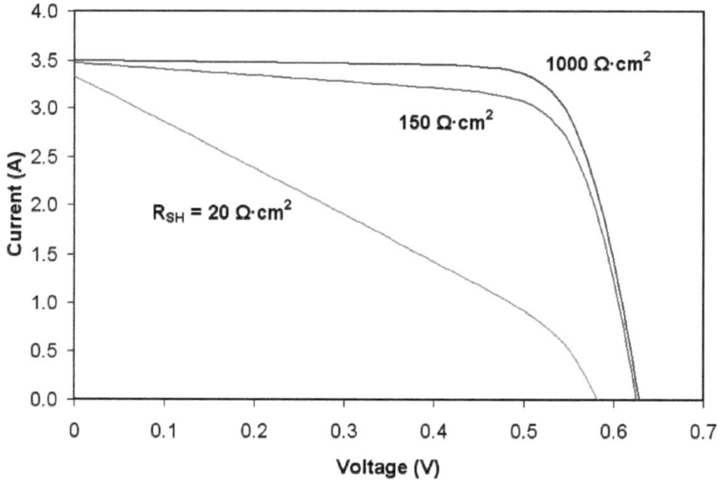

Figure 2.7 *J–V* curve calculated for R_{sh} = 1000, 150, and 20 Ω·cm².

Calculation on temperature-dependent power conversion efficiency is rather complicated, since one need to consider the temperature behavior of all those parameters in the characteristic equation. Roughly, temperature dominantly affects voltage, rather than current. Thus, the cell efficiency will be affected by temperature the same way that the open-circuit voltage is affected. In the case of wafer-based Si solar cell, a cell efficiency decreases ~5% in efficiency for 10°C increase in temperature, while amorphous cells decreases ~2% for the same value of temperature increase. Figure 2.8b presents the calculated *J–V* curves of a crystalline silicon solar cell at 20, 30, 40, and 50°C.

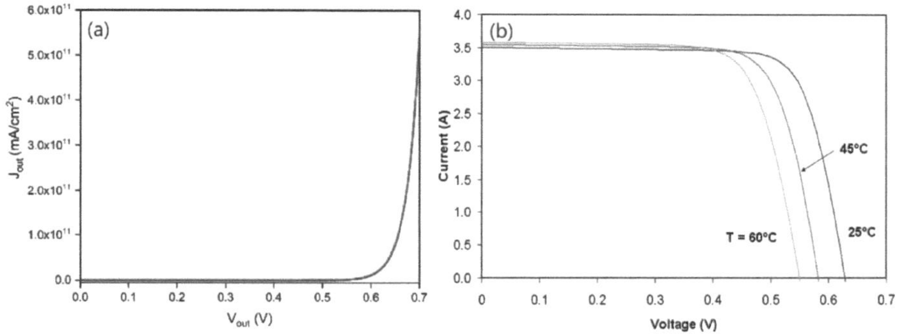

Figure 2.8 (a) Change of the value of $\exp\left(\frac{qV_{out} + qJ_{out}R_s}{k_BT}\right)$ and (b) *J–V* characteristics at 20, 30, 40, and 50°C calculated using Eq. 2.8.

Reverse saturation current J_0

Increase in J_0, which could be due to poor device technology or increasing tremperature, gives rise to reduction in V_{oc} according to Eq. 2.10, assuming infinitively high shunt resistance. The reverse saturation current is actually a measure of the "leakage" of carriers across the *p–n* junction in reverse bias, due to the carrier recombination in the neutral parts of *p*- and *n*-region. The reverse saturation current J_0 is related to the properties of semiconductors and takes the form of

$$I_0 = qAn_i^2\left\{\frac{D_nF_p}{L_nN_A} + \frac{D_pF_n}{L_pN_D}\right\}, \tag{2.14}$$

where D_n (D_p) is the diffusion coefficient, L_n (L_p) the diffusion length, and F_n (F_p) the recombination rate at the surfaces, of electrons (holes). n_i is the intrinsic carrier concentration and A the area of the solar cell. N_D and N_A are the densities of donors and acceptors, respectively. F_n and F_p are sensitively dependent on doping situation and properties of both front and back surfaces of the solar cell, such as surface state densities and distributions. D_n and D_p are mainly decided by carrier mobility in materials.

Figure 2.9 presents the calculated results of J–V characteristics for $J_0 = 10^{-10}$, 10^{-11}, and 10^{-12} A·cm^{-2}, which clearly shows that, as the increase of reverse saturation current, open-circuit voltage decreases quickly.

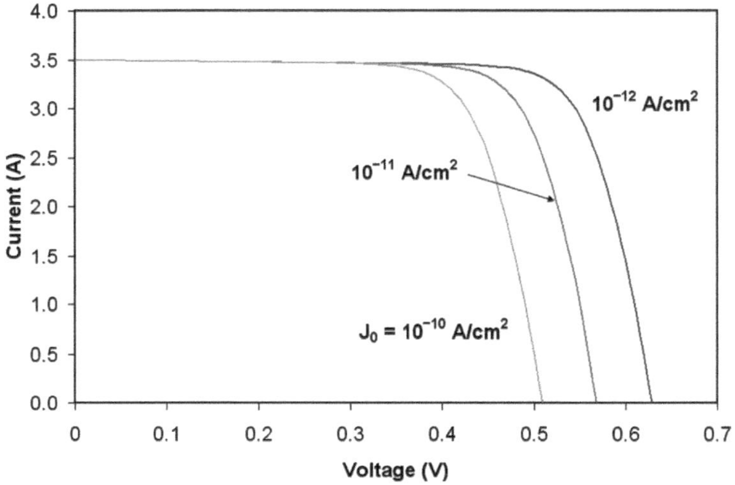

Figure 2.9 Calculated J–V characteristic of a wafer-based Si solar cell for several reverse saturation current values of 10^{-10}, 10^{-11}, and 10^{-12} A·cm^{-2}.

Ideality factor

In some literature, the ideality factor is also called the emissivity factor. The ideality factor is a characterization of the conformity degree between a theoretical and real solar cells: $n = 1$ for a perfect conformity, no recombination in the space-charge region. However, the ideality factor could reaches to $n = 2$, if recombination in the space-charge region dominates, accompanied by an increase in J_0. Figure 2.10 presents the calculated J–V for $n = 0.9$, 1.0, and 1.1 from Eq. 2.9, assuming all other parameters remain unchanged. Thus, the increase of n leads to the enhancement of solar cell output voltage. However, this is counter-balanced by the increase of J_0, which tends to decrease the output voltage. Unfortunately, J_0 is usually a dominant factor so that an increasing n gives rise to reduction of output voltage in combined effects.

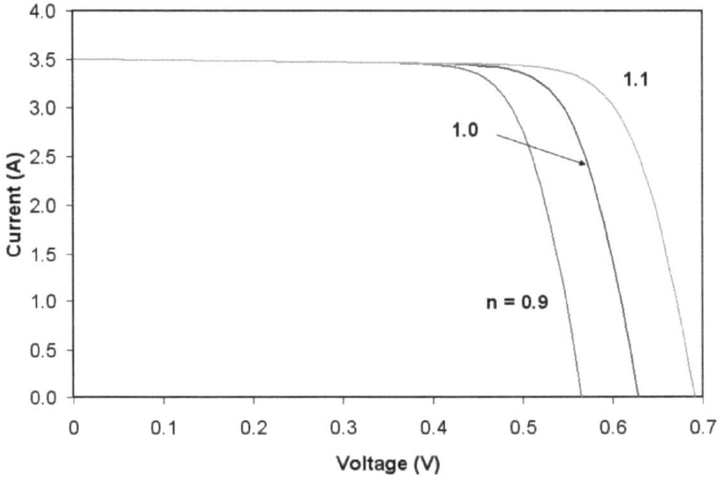

Figure 2.10 Calculated J–V characteristic of a wafer-based Si solar cell for the ideality factor values (n) of 0.9, 1.0, and 1.1.

2.1.4 Charge Separation in Solar Cells

A photovoltaic process in which solar radiation is converted to electricity involves the consecutive five steps: light absorption, exciton generation, electron–hole separation, carrier transport and carrier collection. A photon of given energy incidents on a solar cell as shown in Fig. 2.11. It goes through the cell surface into the region where it excites an electron in the valence band into the conduction band, left a hole in the valence band. Due to the strong built-in electric field, the electron is driven to the n-region while the hole to the p-region. After leaving the built-in electric field region, the electron (the hole) diffuses to the anode (cathode) to complete the whole photovoltaic process. Photon-generated carriers (electrons and holes) might recombine or trapped by impurities and defects, so that the photovoltaic process could be hindered by electron–hole recombination, various defects, relaxations, and other effects. The semiconductors need to be grown with high quality and appropriate device structure should be designed to eliminate those effects.

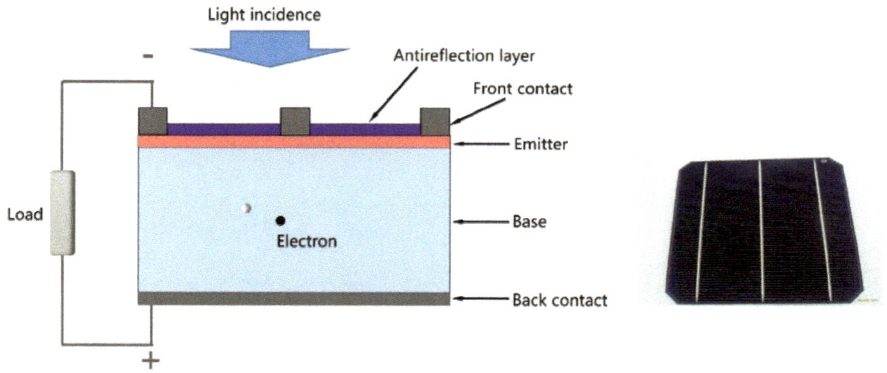

Figure 2.11 The device structure of a solar cell showing a photovoltaic process. The right side presents a photo picture of a Si solar cell.

Charge Separation in the built-in electric field

Motion of charged particles is driven by electrostatic force in *p*–*n* junction. The magnitude of this force is proportional to the Fermi level offset of *p*- and *n*-doped regions and the steepness of the potential in the space charge area (or inversely proportional to the width of the depletion region). The electrostatic force in space charge region drives the electrons and holes to move in opposite directions, leading to separation of photoexcited electrons and holes in excitons. Very importantly, electrons and holes have the tendency to recombine, which is one of the important cause of reduced conversion efficiency for photovoltaic devices.

The *p*–*n* junction of wafer-based crystalline Si solar cells usually has a planar geometric shape. For nano solar cell, the *p*–*n* junction could take any shapes, depending on the kinds of nanomaterials, structures, and shapes. For nanowire solar cells, for instance, their *p*–*n* junctions could have a radial or an axial geometry. The unique shapes of the *p*–*n* junctions give tunable and enhanced functions of charge separation, light absorption, and charge transportation. A photoelectrochemical cell (PEC) offers an alternative way to construct a *p*–*n* junction (a liquid/

solution junction) at the interface between a semiconductor and electrolyte. The different Fermi levels of the semiconductor and the electrolyte are responsible for the formation of a space charge region through a self-consistent process in which charge redistribution occurs through diffusion and drift. In fact, the liquid/solution junction can be made with very high quality with no need for doping. For dye-sensitized solar cells, such a junction plays the key function for the enhancement of charge separation and transportation.

Charge separation in heterostructures

A heterojunction could be constructed between two different semiconductors. An abrupt junction corresponds to a band edge discontinuity at the junction. For ternary compounds, on the other hand, heterojunctions can also be constructed using two semiconductors of different material compositions. The band edge discontinuity results in quasi-electric field for carriers [3]. In fact, the formation of a heterojunction is often followed by a self-consistent process since there exist charged carriers in real semiconductors. An electric potential then forms in the heterojunction region. This potential is similar to a *p–n* junction in behavior, acting to separate photoexcited electrons and holes. Thus, a photovoltaic effect could take action for a type II heterojunction where electrons and holes are driven in opposite directions. For nano solar cells, the separation of photoexcited electrons and holes are mostly executed by non-electrostatic force—heterojunctions. Solar cells with the device structure of nanowire geometry, for instance, offers enhanced capabilities for charge separation and charge transportation. The utilization of heterojunction enables the fabrication of solar cells without need for doping, which is one of the important advantages of nano solar cells.

2.2 The Major Losses in Solar Cells

Not all solar radiation power that hits solar cells could be transferred to electric power, since many loss mechanisms will take effect to diminish the photovoltaic process.

2.2.1 The Loss Mechanisms

Sequentially, the loss mechanisms in solar cells include

(1) reflection and diffraction of light on the surface of solar cells;
(2) absorption of light before reaching to the built-in electric field region;
(3) below E_g loss;
(4) lattice thermalization loss;
(5) recombination loss;
(6) trapping loss of photo *n*-excited electrons (holes) when diffusing towards anode (cathode);
(7) junction voltage loss;
(8) contact voltage loss;
(9) defect absorption;
(10) Joule heat loss;
(11) thermodynamic loss.

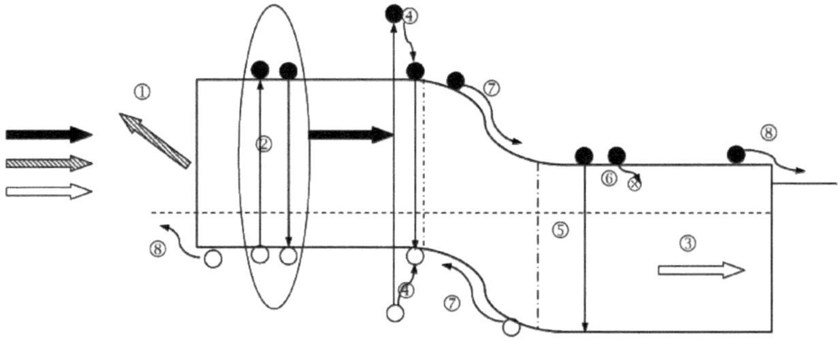

Figure 2.12 The loss process when a solar cell works.

As shown in Fig. 2.12, the above listed losses of (1) → (8) are schematically labeled by ① → ⑧, respectively. We now discuss those losses one by one.

2.2.2 Optical Losses

Reflection and diffraction of light on the surface of device

Optical reflection and diffraction occurs on any solar cell surface characterized by discontinuity in dielectric constant, and optical diffraction happens on any surface (interface) that is not perfectly smooth, as shown by the process ① marked in Fig. 2.12. The optical reflection is mirror reflection and the optical diffraction is scattered. At normal light incidence, the reflectivity takes the simple form of

$$R = \frac{n_1 - n_0}{n_1 + n_0}, \tag{2.15}$$

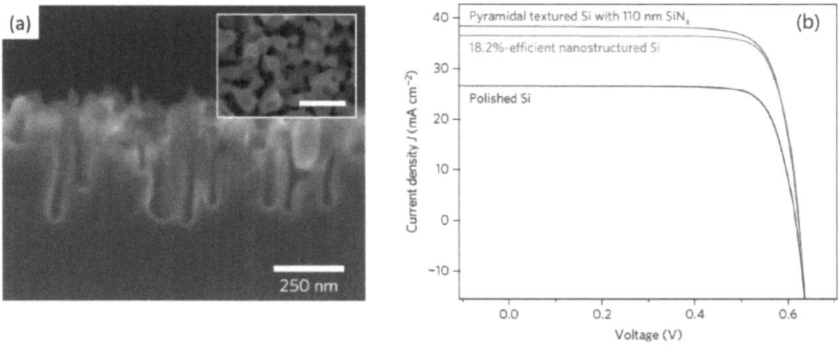

Figure 2.13 (a) Cross-sectional and top-view (inset) SEM images of silicon nanostructure prepared using metal-assisted etching. (b) *J–V* curves of solar cells: black silicon, polished silicon and pyramid-textured silicon with a SiN_x antireflection coating. The black silicon solar cell displays a PCE of 18.2% under AM 1.5G illumination. Reproduced with permission from [4-6].

where n_1 and n_0 are the refractive indices of the solar cell surface material and the air, respectively. Thus, the greater the refractive index difference, the stronger the surface reflection. Without any measurement, a smooth semiconductor surface reflects about 35% of light. Multi-layer coating, surface texture, or other light trapping methods are generally adopted to reduce the surface reflection. The surface of Si wafer-based solar cells is usually textured, through chemical etching, to suppress light reflection. Oh and co-workers fabricated solar cells based on black Si technology, with nano-sized cone-like surface as shown in Fig. 2.13a. The solar cell is extremely effective for light trapping, which leads to very high power conversion efficiency (PCE) exceeding 18% [4]. Figure 2.13b *J–V* curves of the Si solar cells with the surfaces of polished silicon, black silicon, and pyramid-textured silicon with a SiN_x coating. A pattern of nanopillars (nanocones) simulates the continuous transition of the average refractive index from air to solar cell surface. The effectiveness of those measurements have been demonstrated in applications [5, 6].

Absorption of light in locations far away from the space charge region

Since the built-in electric field region is underneath the surface, a photon could be absorbed before reach to the space charge region (② in Fig. 2.12). The photon could create an electron/hole pair there. However, without a strong electric field as in the space charge region, the electron and hole are unlikely to be split effectively so that they could quickly recombine to give away their energy to the lattice or defects non-radiatively or radiatively. Even if the photoexcited carriers are not recombined, they have to migrate to electrodes through diffusion, which has very low efficiency. Thus, the photoexcitation of electron/hole pairs far away from the space charge region should be avoided due to low probability of those electrons and holes being collected by electrodes.

Below bandgap loss

The sun is roughly a 5800°C blackbody that radiates continuum light ranging from 300 to 2500 nm and about 50% of the radiation energy is in infrared region. If a solar cell is based on a single semiconductor material with a given bandgap E_g, the solar cell is transparent to the part of light with photon energies below E_g, as shown by the process ③ in Fig. 2.12. A silicon wafer solar cell, for instance, can only use photons with energy exceeding 1.12 eV (the bandgap of silicon). Thus, a big portion, close to 40% of the total solar energy output cannot be used, as shown in Fig. 2.14.

2.2.3 Electric Losses

Above bandgap loss

An electron in the valence band absorbing a photon with energy larger than bandgap becomes a hot electron and the hot electron will give up some energy to relax to the bottom of the conduction band through colliding with lattice and heating up the lattice in carrier-phonon scattering events, as described by ④ in Fig. 2.12. Since a large number of the solar radiation photons have energies above 1.12 eV, as seen in Fig. 2.13, the "thermalization loss" takes an important part of the total loss. In fact, it alone limits the efficiency of the optimum bandgap solar cell to 44%.

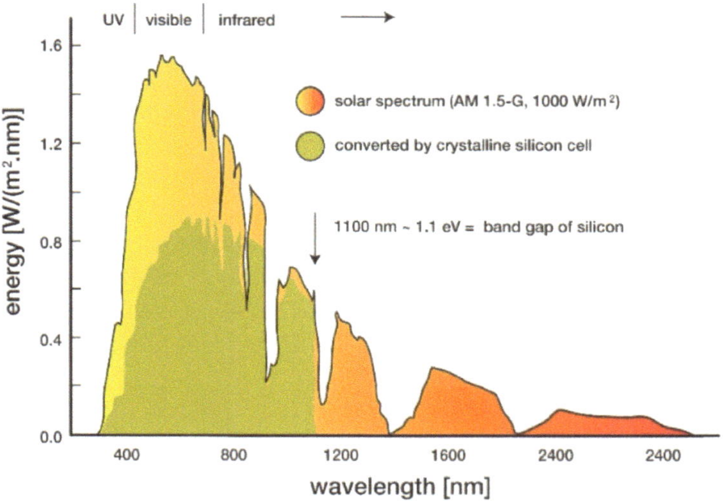

Figure 2.14 AM 1.5G solar spectrum and the portion unconvertible by Si solar cells.

Recombination loss

Recombination loss is due to the electron–hole recombination after their being photo-excited. The recombination might occur in junction area, or on the way of diffusion to the electrodes (see ⑤ in Fig. 2.12). Recombined electron/hole makes no contribution to the photovoltaic effect. This is the intrinsic loss that is difficult to suppress. Fortunately, the recombination loss usually takes only an ignorable portion of the total loss.

Trapping loss of the photo-excited electron (hole)

Photoexcited electrons (holes) might be trapped on their way travelling towards anode (cathode), as shown by ⑥ in Fig. 2.12. The trapping is usually due to the dislocations, defects, and impurities in the materials. Semiconductors used for solar cells require a certain qualities. In the case of Si, for instance, it requires

6N for the purity of the material and well-crystallized structure, in order to suppress this part of loss.

Junction voltage loss

The process ⑦ in Fig. 2.12 corresponds to a junction resistance in the junction region, which causes a drop of voltage cross the region that can be described by a junction resistor in series with a load resistor. As the increase of the junction resistance (the increase of series resistance), the whole J–V curve moves toward the origin, leading to a decrease in the output voltage V_{out} and the output current J_{out}. The short-circuit current J_{sc} decreases slightly, although the open-circuit voltage remains unchanged, as shown in Fig. 2.6. At high values of R_s, J_{sc} decrease significantly and the J–V curve resembles that of a resistor (see the $R_s = 20\ \Omega\cdot\text{cm}^2$ curve in Fig. 2.6). The series resistance would consume a power of I^2R_s, (I is the total current passing R_s) which become significant for large R_s.

Contact voltage loss

No electrode contact is perfect. Therefore, the contact resistance leads to a voltage drop cross the metal/semiconductor interface, which causes a power loss.

Defect absorption

There are a number of dislocation, defects, and impurities that act as trapping centers for photoexcited electrons and holes. The trapped electrons and holes make no contribution to the photo current. Usually, nanomaterials have much higher defect densities than crystalline materials. Therefore, measurements such as passivation need to be adopted to suppress the defect trapping effect in nanomaterials.

2.2.4 Thermal Losses

Joule heat loss

There exists resistance throughout the device so that, when carriers pass through, it generates heat. Joule heat could be generated in the materials, junction region, contact region, etc.

Thermodynamic loss [7]

Not all the energy of an electron can be extracted as photovoltaic energy. Free energy of a system is defined as the maximum energy available for use. At non-zero temperature, thermal excitation leads to distribution of electrons in the conduction band. According to thermodynamics, an electron at finite temperature has and kinetic energy of $\frac{3}{2}k_BT$, in 3D case. That part of energy cannot be extracted. There are other causes for power conversion efficiency loss due to temperature. As an example, raising temperature gives rise to increase of entropy, which in turn leads to voltage drop. Another thermodynamic effect is called etendue loss, which is a direct result of increasing entropy. At output voltage of V_{oc} (without a load), light absorption power is equal to emission power. The photoexcited electrons give maximum chemical energy. Light hits the solar cell as parallel beam with a solid angle = 0. If the light beam is focused (the solid angle increases) using mirrors or lenses, light absorption of a cell is larger than light emission. V_{oc} then increases. This effect occurs at a temperature $\neq 0$.

Currently, the conversion efficiency of a conventional crystalline Si is about 22%. Among the 78% loss, roughly, 56% is due to color mismatch, 8% due to reflection and transmission, 9% fundamental recombination, 5% excess recombination, resistance, etc.

2.3 Photovoltaic Devices: Solar Cells

Semiconductors are suitable candidates for photovoltaic devices due to their unique energy band structures and doping performance. Among all semiconductors, Si takes an un-substitutional position in solar cell technology due to the advantages such as earth-abundance, relative low cost, suitable bandgap, mature material growth and purification technology, well controllable doping, and tunable morphology. Si are the most intensively studied semiconductor material for various forms of solar cells, such as wafer-based, thin-film, as well as many architectures of nano devices.

2.3.1 Wafer-Based Si Solar Cells

To utilize the solar energy conveniently, it is required to transfer the solar energy, in the form of light, into electric energy. A device that converts solar energy into electric energy is the key component in the use of solar radiation and the photoelectric conversion devices are designed to achieve the best conversion efficiency with the lowest cost. The conversion of sunlight energy into electric energy is controlled by physics of photoelectric interaction that obeys the energy conservation law. Now, a variety of technologies have been developed for the fabrication of solar cells, photoelectric conversion devices, to convert solar energy to electric power efficiently and cost-effectively.

The first generation solar cells, which dominate the solar cell market, are based on crystalline silicon (c-Si) wafers and the photovoltaic devices are well-matured in terms of their technology. While the conversion efficiency of the poly-crystalline Si (pc-Si) SCs has hit 20%, that of single crystalline Si (sc-Si) SCs has reached to ~23% in industry. A large portion of the total cost of c-Si SCs is from solar grade Si crystal. Research on crystalline-based solar cells is now focused on reducing the cost through minimizing the material cost and maximizing the solar conversion efficiency. As an indirect bandgap semiconductor, silicon has very small absorption coefficient. Thus, a thick absorption layer is needed for Si solar cells.

The active region of a solar cell is usually made of light-sensitive semiconductor materials such as silicon and germanium (elementary semiconductor materials) as well as gallium arsenide and cadmium telluride (compound semiconductors). Now, some organic materials and nanomaterials are also used in making solar cells. Figure 2.15 shows the structure of a typical conventional silicon wafer solar cell, consisting of a silicon *p*-type region, *n*-type region, a *p–n* junction, top and bottom electrode. To suppress the surface light reflection, the top surface facing the sunlight incidence is textured and/or coated with anti-reflection coating (see Fig. 2.15). When solar cells absorb photons with energies equal to or larger than bandgap E_g, electrons and holes are excited in the junction region where the photoexcited electrons and holes experience a strong electric field that drives the electrons and holes in opposite directions. Electrons move upward and holes move downward as shown in Fig. 2.15. Upon leaving the space charge region where electric field is essentially absent, the motion of the photoexcited electrons and holes are govern by diffusion process. With continuous sunlight incidence, more and more electrons and holes are generated and are driven to the photoanode and photocathode, respectively, leading to the collection of the photogenerated charge carriers at the junction terminals. The accumulation of charge carriers on the electrodes gives rise to macroscopic photovoltaic effect. When external circuit is connected to the top and the bottom electrodes, there will be a continuous photogenerated current flowing through the circuit.

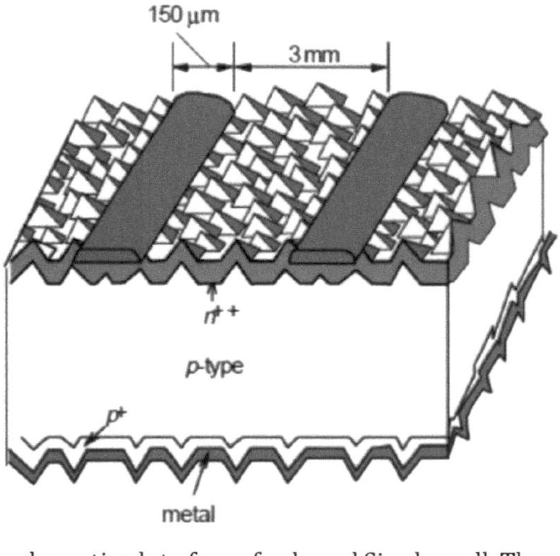

Figure 2.15 A schematic plot of a wafer-based Si solar cell. The antireflection layer and the surface texture are for the reduction of light reflection. The device uses metal Ag/Pb/Ti fingers as the top contact on *n*-type Si and aluminum layer as the bottom contact. The heavily doped *n*-region (n^+) and *p*-region (p^+) are for the reduction of contact resistance in *n*- and *p*-region, respectively. Under solar illumination, photoexcited electrons and holes are driven by the build-in potential towards opposite directions.

2.3.2 Important Parameters of Solar Cells

Although the materials and configurations may change dramatically from solar cells to solar cells, their working principle is essentially the same—all the solar cells work based on the photovoltaic effect. Microscopically, photovoltaic process is governed by quantum mechanics concerning multiple particle interaction while macroscopically one can measure a sunlight-generated voltage cross the device.

The key part of the above solar cell is the *p*–*n* junction that acts to separate the photoexcited electrons and holes. A good *p*–*n* junction should be strong enough to separate the electrons and holes very efficiently, so that the electron/hole separation happens in a time much shorter than the time constant of electron–hole recombination. The abruptness of a *p*–*n* junction is determined by doping and the bandgap of semiconductor materials. The use of large bandgap semiconductors is helpful for the construction of large potential drop. However, large bandgap implies that few photons absorption in the solar spectrum. A balance should be found to give the best device performance. Other measurements could also be taken to give best efficiency to separate the electrons and hole from each other. An appropriate doping density distribution in the device, for instance, might be beneficial to block the electrons to move down-ward and the hole to move up-ward (see Fig. 2.15). The doping density distribution in Fig. 2.15 corresponds to a potential distribution for the carriers. The potential variation in the interface of n–n^+ is beneficial for electrons to move up-ward, while the same potential change tends to block the up-ward motion of holes.

In fact, the best way of minimizing the injection of electrons from the absorber into the *p*-type region and holes into the *n*-type region is to introduce a potential barrier in the conduction and the valence bands. Figure 2.16 presents the energy band alignment of an ideal solar cell under sun illumination in open circuit (left) and short-circuit (right) situations. For an ideal solar cell, the potential jump ΔE_c from the absorber to the *p*-type region in the conduction band and the potential drop ΔE_v from the absorber to the *n*-type region in the valence band serve to block electron injection into the *p*-type region and hole injection into the *n*-type region, respectively. Blocking electron injection into *p*-region can be done by using a *p*-type semiconductor with larger bandgap than that of the absorber layer and dominant band offset in the conduction band. Similarly, blocking hole injection into the *n*-region requires a larger *n*-type semiconductor bandgap (than that of the absorber). The potential barriers are very effective in suppress the carrier injection in wrong direction. However, as shown in Fig. 2.16, achieving those potential barriers requires growth of heterojunctions, so that one has to use very expensive epitaxial semiconductor growth techniques such as molecular beam epitaxy (MBE) and metal organic chemical vapor deposition (MOCVD). In conventional crystalline Si solar cells, blocking electron and hole injection is achieved by growing layers of different doping densities (see Fig. 2.15). Hereby it should be emphasized again that keeping electrons and holes away from each other is of utmost importance to suppress the electron–hole recombination effectively.

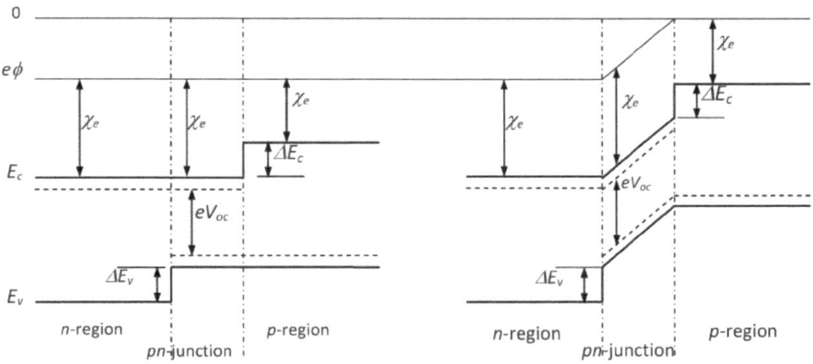

Figure 2.16 Energy band alignment of an ideal solar cell under sun illumination in open circuit (left) and short circuit (right) situations.

In darkness, the Fermi levels for electrons and holes in solar cells are lined up. Solar illumination breaks the original thermal dynamic balance by generating photoexcited electrons and holes in the conduction and the valence bands, respectively. For the description of solar cells under sun illumination, the concept of quasi-Fermi levels for electrons and holes can be introduced and they are labeled with dashed lines, as illustrated in Fig. 2.16. The energy spacing between the quasi-Fermi levels for electrons and holes is a measure of sun's radiation energy converted into electrochemical energy. In the open-circuit case, there is no current flowing through the solar cell and external circuit, i.e., the current is zero. But one can measure a voltage, called open-circuit voltage between two terminals of the solar cell. A crystalline Si solar cell has a typical open-circuit voltage $V_{oc} \sim 0.6$–0.7 V. The open-circuit voltage depends on the properties of materials, including the bandgaps and doping properties that make up the solar cell, as well as the device configuration. V_{oc} is one of the most important parameters for a solar cell that determine the device performance.

In the short-circuit case, the two terminals of the solar cell are connected to an external circuit with very low resistance. A current flows through the external circuit and the device, called the short-circuit current—another very important parameter for the characterization of solar cell performance. Except for the materials properties and device configurations, the short-circuit current density denoted by J_{sc} depends strongly on the technology that the solar cell is fabricated. The J–V characteristics gives general performance of a solar cell and, on a measured J–V curve, V_{oc} and J_{sc} represent the maximum voltage and current one can obtain from the solar cell. A so-called solar cell filling factor equals the area of the largest rectangle enclosed by the J–V curve and the horizontal and vertical axes, normalized by $V_{oc} \times J_{sc}$.

Additionally, the geometric configuration of the solar cells should also be optimized. The thickness of the n-region and p-region in devices, for instance, is also critical for the performance of solar cells. Dislocations, defects, and impurities exist in semiconductors and for any material the mean free path and diffusion length of electrons and holes are limited. As a result, the thickness of the n-type and the p-type region should be less than the mean free path of the electrons and hole, respectively.

2.3.3 Thin-Film Solar Cells

Up to now, more than 80% of the solar cells are based on the crystalline silicon wafer (multi or single crystalline) and the thickness of the wafers is around 180 μm. The cost of the material is nearly 30% of the total cost of the devices. A Si thin-film solar cell uses only below 1% Si material of a wafer Si solar cell, with a thickness less than 0.5 μm, a significant reduction in the cost for materials. The use of much less material is benefited by the direct bandgap of the thin-film Si in amorphous form, which has much higher light absorption efficiency than crystalline Si. The amorphous Si (a-Si) shows improved light absorption property in the visible range except that light absorption in the red part is still rather weak. The major cost for crystalline Si is in the process in which Si is purified and prepared into crystal, though Si technology is mature and relatively cost-effective to be made high quality. Si element itself is abundant on earth. However, the power conversion efficiency of Si thin-film solar cells is typically 1/3~1/2 that of the Si wafer solar cell. The diffusion length of minority carrier is typically less than 300 nm and the minority carrier lifetime is very short, due to high density of defects and dangling bonds in amorphous Si. The dangling bonds in a-Si are passivated with hydrogen atoms (a-Si:H), which appears to be quite effective. One of the disadvantages of the thin-film solar cells lies in the difficulty in light trapping measurements. Since the thin-film thickness is less than 0.5 μm, light trapping effect cannot be enhanced using surface texture schemes as used in the case of wafer-based Si solar cells.

Crystalline Si is an indirect bandgap semiconductor. A thick material layer is thus required due to low optical absorption coefficient. It needs a layer of 20 micron in thickness for 95% light absorption of above-gap energy photons. Thin films for SCs are usually direct bandgap semiconductors so that only very small thickness is required due to high absorption. However, it is extremely difficult to prepare thin films in crystalline forms and the non-crystalline feature of the thin films leads small carrier diffusion length and high recombination rate. Nevertheless, thin-film SCs have the advantages of low material cost and low material quantity cost, simple and multiple preparation techniques, various forms of material available (including amorphous, nanocrystal, microcrystal, polycrystal, etc.). Thin-film SCs have more room to play for device purposes due to the tailorability on many opto-electronic properties of materials, such as bandgap, work function in the preparation processes.

Other than those, it is extremely flexible in thin-film SC techniques to design and prepare various types of junctions, such as homo, hetero, Schottky, photoelectrochemical cells, etc. Similarly, it is quite easy to realize multiple junction and tandem cells. For thin films, it is relatively easy to achieve large area, a requirement critical to cut down cost and realize integration for modules. In addition, thin-film techniques are compatible to solar thermal devices. The most widely investigated and manufactured thin-film solar cells are based on amorphous Si (a-Si), cadmium telluride (CdTe), and (CIGS). Figure 2.17 presents the structures of the three popular thin-film solar cells.

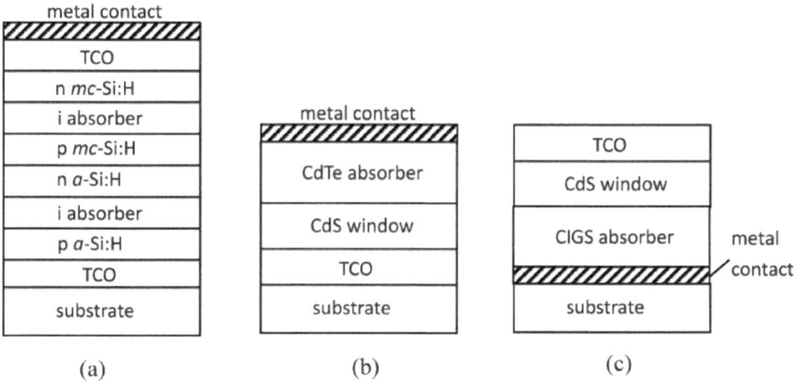

Figure 2.17 The fundamental structures of an a-Si:H (a), CdTe (b), and CIGS (c) solar cells.

Amorphous silicon can be deposited at low temperatures (e.g., as low as 75 degrees Celsius) onto a variety of substrates, allowing for the film to grow not only on "hard" substrates like glass, but also on flexible ones such as plastic as well, suitable for a roll-to-roll preparation. Unlike crystalline Si, amorphous Si has a direct bandgap of 1.7 eV, close to the peak gap of 1.5 eV at which the best conversion efficiency is expected. Amorphous silicon has a very high above-gap photon absorption coefficient ($>10^5$ cm^{-1}), 100 time larger than that of c-Si, so that only a thickness of 200–500 nm is needed for the solar cells. More importantly, a-Si can be doped to obtain both *n*-type and *p*-type materials, essential for opto-electronic devices.

In crystalline Si, a silicon atom is fourfold coordinated and tetrahedrally bonded to its four nearest neighbors, and the crystal structure keeps translational invariant over a long range. In an amorphous form, however, the material does not have such a long range order anymore, though it does have a short range order, and a large number of Si atoms are no longer fourfold coordinated.

As a result, there exists high density of dangling bonds that behave as defects in the random network, harmful to optoelectronic properties of the material. Fortunately, a-Si can be passivated by hydrogen. In fact, plasma enhanced chemical vapor deposition (PECVD), the technique suitable for deposition of large area a-Si thin films, uses silicon containing gas SiH_4 mixed with H_2. As SiH_4 and H_2 flow into the growth chamber, a plasma will occur between two electrode plates where an RF power is applied. The gas was excited and decomposed. The deposited silicon films on substrates mounted on the electrode plates is then hydrogenated. The hydrogenated amorphous silicon, a-Si:H, films contain a significant percentage of hydrogen atoms bonded into the amorphous silicon structure and passivated the films, leading to significant improvement of the electronic properties of the material by reducing the dangling bond density by several order of magnitudes.

The first a-Si SC was made in 1976 by Carlson and Wronski [8]. The device consists of a single *p*–*i*–*n* junction (the active layer, where "i" means "intrinsic") and transparent conductive oxide (TCO) front electrode and aluminum back contact, with 2.4% conversion efficiency. Later on, with improvement in material preparation and device design, the conversion efficiency has reached to 15% in laboratory and 7% in large-scale production. The major problems for single layer a-Si thin-film SC are low conversion efficiency and degradation. A number of strategies for improvement have been proposed, including graded gap, multi-junctions, light trapping that are effective.

Figure 2.17a presents a double material layer a-Si SC. The active part of the cell consists of a *p*-type a-Si layer and an *n*-type mc-Si layer (mc-Si means microcrystal Si), both the a-Si layer and the mc-Si layer being *p*–*i*–*n* configured, where the *i* layer is the absorber for photon absorption. The active layers are sandwiched by two TCO contacts. The double junction SC has a conversion efficiency of about 12%. The conversion efficiency can be improved further by adding more junctions. However, the complicated device structure of multi-junction solar cells often brings up material cost due to the requirement of expensive and complex film deposition systems. Other measurements for achieving cheaper and more efficient cells include the use of better light trapping surfaces and low cost substrates.

Cadmium telluride (CdTe), a chalcogenide binary compound semiconductor, has a zinc blende crystal structure with a direct energy gap of 1.5 eV, optimum for solar spectrum. Its direct bandgap yields high optical absorption coefficient; only 1–3 micron thin layers of CdTe/CdS are needed for sunlight harvesting. CdTe can be grown with both *p*-type and *n*-type. The study on CdTe solar cells is dated since 1950, when the first *p–n* homojunction gave a poor conversion efficiency of only 3%. It took about 20 years to double the efficiency to 6%. Currently, the highest efficiency record is 19% set up by First Solar Inc.

The basic structure of CdTe SCs is shown in Fig. 2.17b, where the active part of the device consists of a CdTe layer and a CdS layer as window, the *p–n* junction being on CdTe/CdS interface. There is about ~10% lattice mismatch between CdTe and CdS and they have different crystal structures, CdTe being zinc blende and CdS being wurtzite. A good heterojunction should be made with lattice-matched semiconductors to minimize the strain and strain-induced defect density. The 10% mismatch between CdS and CdTe is large enough for the formation of high-density interface defect states. Fortunately, two factors tend to mitigate the seriousness of the problems. The first one is the micro-crystal characteristies of CdTe and CdS. The CdTe and CdS layers are formed by large number of crystal grains imbedded in amorphous network. The amorphous phase is more tolerant to strain than crystal phase of the materials. The second one is the intermixing between CdTe and CdS layers that facilitates a smooth and continuous transition between the two sets of crystal lattices, resulting in strain relaxation. The intermixing between CdTe and CdS can be tuned in certain extent by varying the growth temperatures of the two materials.

CdTe can be grown by close space sublimation, which gives the best material for solar cells though a high growth temperature up to 650°C is required. Other growth techniques include physical vapor deposition (PVD), vapor transport deposition (VTD), metal organic chemical vapor deposition (MOCVD), and electrodeposition that can be performed at very low temperature of 90°C. The fact that CdTe and CdS can be made with a variety of deposition techniques is really helpful for lowering the cost of the cells.

The purpose of the CdS layer in the cell structure is to form a window for photon incidence and the *n*-region of a *p*–*n* junction. The bandgap of CdS is 2.42 eV, almost 1 eV higher than that of CdTe; optical absorption of CdS is very low in most of the solar spectrum, since, in addition, the thickness of CdS is often chosen to be very thin (typically around 50 nm). The CdTe layer is the absorber where photoelectric conversion occurs. In addition to the direct bandgap of around 1.5 eV in the optimal range for maximum photovoltaic conversion, the minority carrier diffusion length is a few micron, allowing for a majority of photoexcited carriers to be collected at contacts. The diffusion length of several micron is a surprise. In fact, the micro-crystalline CdTe cells show better performance than crystalline CdTe cells. This can be explained by microscopically exploring the energy band distribution in the micro grains and grain boundaries. The grain boundary acts as a potential barrier for electrons and accumulation for the holes. This is beneficial for photovoltaic process since such a band distribution favors the separation of photoexcited electrons and holes, reducing the probability of electron recombination.

As a heterojunction solar cells, the optical and electrical characteristics of CdTe and CdS must be compatible, one of the most important characteristics being the energy band alignment on the junction, which affects the performance of cells significantly. The energy band distribution of the CdTe/CdS solar cells is schematically presented in Fig. 2.18. In the case of a heterojuction, the *p*–*n* junction (built-in electric field) is induced by both different types of doping and the interfacial energy discontinuity. The *p*-type doping density on CdTe is difficult to control due to native defects and unintentionally introduced impurities during film growth process. Some of the impurities act as deep acceptors and deep donors, making high density of *p*-type doping extremely difficult. On the other hand, the intermixing of CdTe and CdS, which leads to flat *p*–*n* junction, appears to be a problem. All of these lead to poor performance of the *p*–*n* junction in CdTe/CdS solar cells. It should be pointed out that elemental cadmium is highly toxic. However, CdTe and

CdS have extremely low water solubility and low vapor pressure. Both the Cd–Te and the Cd–S bonds are very stable. CdTe and CdS are thus not so dangerous.

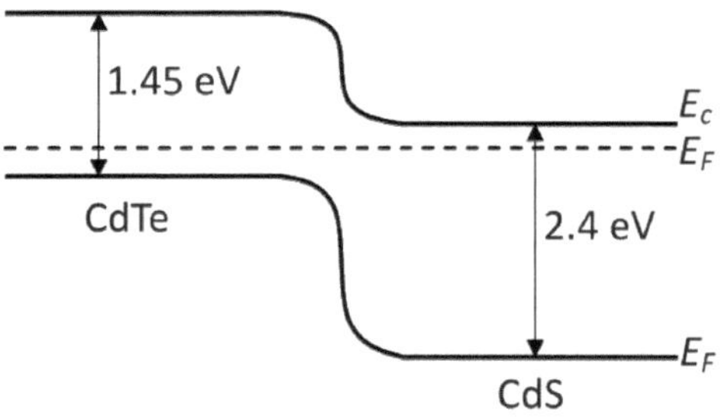

Figure 2.18 Schematic band alignment of the CdTe/CdS solar cells.

Copper indium gallium diselenide, $CuIn_xGa_{1-x}Se_2$ (the pseudo-binary alloys of $CuInSe_2$ (CIS) and $CuGaSe_2$ (CGS)), is an I–III–VI compound semiconductor having a tetragonal chalcopyrite crystal structure as shown in Fig. 2.19. Involved with four elements that belong to group I (Cu), III (In and Ga), and VI (Se), the material has four relevant phases, namely, α-, β-, δ- and Cu_2Se-phase, depending on the preparation conditions. At room temperature (RT), the α-phase is formed in the range of 24–24.5 at% Cu, while the optimal range for high efficient $CuIn_xGa_{1-x}$ Se_2 (CIGS) solar cells is 22–24 at % Cu. From its phase diagram, the α-phase occurs at the growth temperature of 500–550°C. At RT, the material is in the α-phase + β-phase and the β-phase ($CuIn_3Se_5$) corresponds to a defect phase formed by ordered array of defect pairs. δ-phase, the high-temperature phase is due to the Cu (In) sub-lattice disordering. The Cu_2Se phase can be viewed as a chalcopyrite structure with Cu interstitials and Cu–In anti sites, similar to the β-phase. The co-existence of the multiple phases in CIGS makes the growth of the material for optimal cell efficiency extremely difficult. The narrow α-phase range can be widened by the introduction of other elements. It can be done by around 25% In replacement by Ga. It is also demonstrated that incorporation of 0.1 at % Na into the material leads to better film morphology and passivation of grain-boundaries. The improved thin film also has reduced defects concentration and higher p-type conductivity, critical for the performance of solar cells.

CIGS has a direct bandgap in 1.04–1.68 eV, as shown in Fig. 2.20 for the optical absorption spectrum. In high-quality CIGS material, electron diffusion length is several micron, electron mobility in the order of 1000 $cm^2 \cdot V^{-1} \cdot s^{-1}$, and the minority-carrier lifetime around several nanoseconds. Those great parameters make the material an excellent candidate for the fabrication of thin-film solar cells, with a fundamental structure shown in Fig. 2.17c. Chirilă and co-workers fabricated CIGS solar cells on flexible polyimide substrate, which showed a strong absorption in the visible and near infrared region matching well to the solar radiation spectrum, as shown in Fig. 2.21a for the overall solar spectrum absorption length > 1 micron [9]. As a result, the thickness for the CIGS active layer is only around a few microns. They achieved a power conversion efficiency of 20.4%, corresponding to an open-circuit voltage of 736 mV, a short-circuit current density of 35.1 $mA \cdot cm^{-2}$, and a fill factor of 78.9.

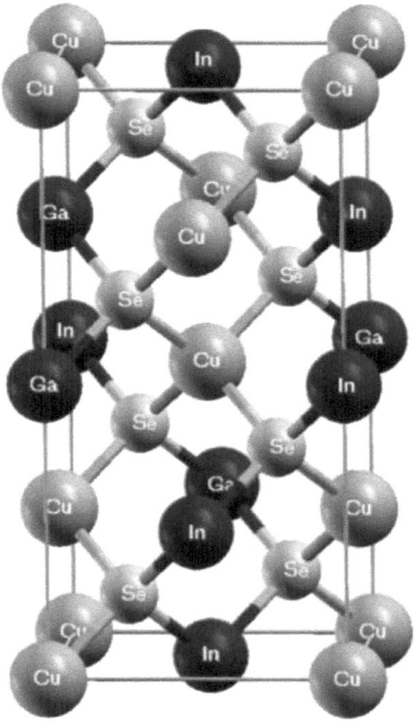

Figure 2.19 The tetrahedrally coordinated unit cell of CIGS crystal.

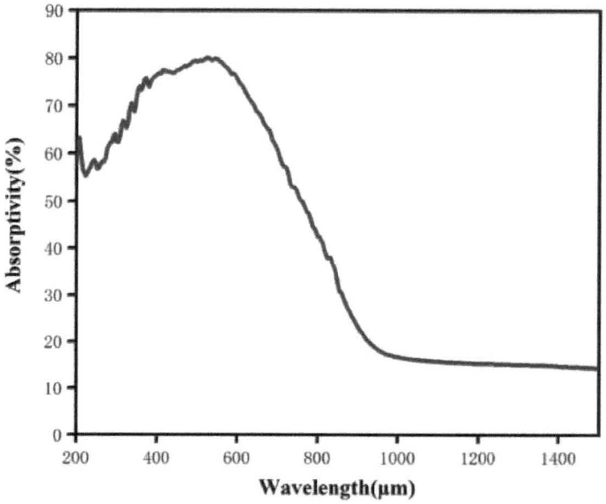

Figure 2.20 Absorption spectrum of CIGS in visible and near infrared.

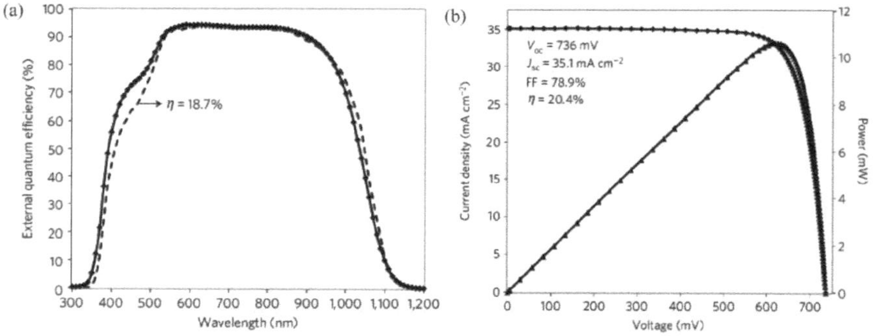

Figure 2.21 Measured EQE (a) and *J–V* (diamonds) and *P–V* (triangles) curves (b) of a CIGS solar cell with CdS/CIGS junction prepared on flexible polyimide substrate (solid line). The EQE of an earlier device is also presented for comparison (dashed line in (a)). Reproduced with permission from [9].

CIGS is now at the heart of research and manufacture of thin-film photovoltaic devices. A chalcopyrite-based solar cell was first proposed in 1974 with a conversion efficiency ~5% [10]. The cell structure of $CuInSe_2$/CdS was prepared in the vacuum, where the *p*-type $CuInSe_2$ was used as an absorber and *n*-type CdS as window layer. Although other alternative structures were suggested later, such a combination of *p*-type chalcopyrite absorber and *n*-type window remains to be the fundamental idea. In 2000, Satoh et al. proposed CIGS solar cells on flexible stainless steel substrates with a similar active layers of CuInGaSe/CdS (see Fig. 2.21b) [11]. Note that here the $CuInSe_2$ layer was replaced by CuInGaSe, which was helpful for higher open-circuit voltage V_{OC}. Now, the lab conversion efficient of CIGS cells on glass substrates has exceeded 21%, which is highest among all thin-film-based solar cells. The theoretical conversion efficiency is ~28–30%. There is still room for CIGS solar cells to improve! The reported manufacturing conversion efficiency is 18% on flexible metal substrates, comparable to c-Si solar cells, and 15% on polyimide foils.

CIGS films can be prepared using a variety of methods, including layered vacuum deposition followed by selenization with Se or H_2Se, spray deposition, screen printing, electroplating, sputter deposition, as well as co-evaporation and homogenization, etc. Among those, co-evaporation and homogenization yield the best cell quality. Recently, CIGS film preparation using chemical ways has received more and more attention since it is promising for reducing the cost significantly.

The energy band alignment in CIGS solar cells is schematically shown in Fig. 2.22. The *p*–*n* junction is located between the CdS window layer and CIGS. Note that there are high potential barriers for photogenerated holes between the ZnO contact and the CdS window and between the CdS window and CIGS layer which effectively stop the hole injection into the ZnO layer. The solar radiation photon with energy higher than 1.2 eV excites electrons from the valence band into the conduction band, leaving holes in the valence band. The photogenerated electrons and holes are then driven by the built-in electric field in the junction area to ZnO contact and the Mo contact at the right side of CIGS layer, forming photovoltaic process.

In 2017, Solar Frontier achieved CIGS solar cell efficiency of 22.9%. The CIGS solar cells have three major advantages. First, they are thin-film SCs that have the efficiency and stability of crystalline Si SCs. Second, they have very high specific power up to nearly 1000 W/Kg, which is the highest among all solar cells, including crystalline-based and thin-film solar cells [12]. Third, CIGS SCs are very robust against the cosmic radiation, rendering them the usefulness in space technology. Except for those advantages, some problems exist for CIGS solar cells. As a quarternary semiconductor with four elements, the CIGS material has multiple binary phases that lead to polymorphism, as well as structural and electronic disorder, making sophisticate control of the growth rather difficult. Another serious issue is the availability of In and Ga elements that are not very abundant on our earth, which limits the wide application of CIGS solar cells. Alternative candidates for absorber materials includes Cu_2ZnSnS_4 (CZTS) and $Cu_2ZnSn(S,Se)_4$ (CZTSSe) that contain only earth abundant elements. CZTS and CZTSSe are relatively low-cost, so that they are suitable for large-scale application. Xu and co-workers reported the fabrication of $Cu_2ZnSn(S,Se)_4$ nanoparticle ink solar cells on flexible molybdenum foil substrates [13].

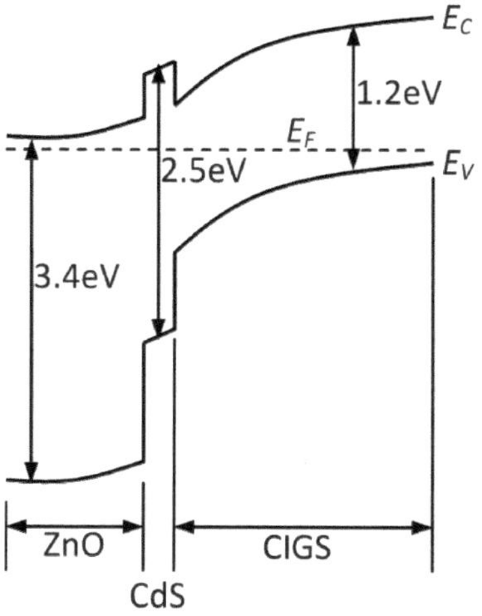

Figure 2.22 Energy band alignment in CIGS solar cells.

2.3.4 Nanomaterial-Based Solar Cells

Not only the thin-film Si, CdTe, and CIGS solar cells, other thin-film solar cells include, for instance, organic material solar cells, dye sensitized solar cells, and multi-junction tandem cells one can find in a great number of articles. Thin-film solar cells can also be categorized into nanomaterial/nanostructure solar cells, since in those solar cells, either the materials have nanosized structures (e.g., the grain size of amorpous Si thin films) or they are in nano sizes (e.g., the thickness of the CdS window layer in CdTe/CdS and CIGS solar cells is in the nanometer range). On the other hand, collection of solar spectrum requires devices with large areas; nano solar cells need to be prepared in the form of thin-film materials. It is thus impossible to draw a clear-cut distinction between thin-film solar cells and nanomaterial solar cells.

In literature, nano solar cells are referred to as the photovoltaic devices made up of nanomaterials (at least some of the materials), such as semiconductor quantum dots, quantum wires, nano-rods, and nano-pipes. The preparation routes of those nanomaterials are now quite mature, except that large-scale production and fine control of the material parameter are still under investigation. Nanomaterials and nanostructures are potentially cheap for fabrication, since many of them can be prepared using self-assembled technologies. Apparently, the major problem for thin-film solar cells and nano solar cells is low conversion efficiency. However, those solar cells have a large scope for the improvement of cell efficiency by fine tailoring material properties, optimizing device structures and, more importantly, using new physical mechanisms. It is more hopeful for nano solar cells to have their conversion efficiencies closer to theoretically predicted values. CIGS solar cells, for instance, have achieved the efficiency as high as 26.4% with the structure of CIGS/WS_2 [14]. The device has an open-circuit voltage V_{oc} of 1.026 V, a short-circuit current density J_{sc} of 29.57 mA·cm^{-2} and a fill factor FF of 86.96%.

Despite the big difference in device structures, a nano solar cell and a conventional solar cell work according to the same mechanism. In a solar cell, an absorber absorbs a photon to generate a photon-excited electron/hole pair. Both the electron and the hole are in their excited states, electron in the conduction band and the hole in the valence band. The photon-excited electron and hole need to be separated and transported to two separate contacts. A nano solar cell should be designed to execute such a process with high efficiency. In a nano solar cell, a photon absorber is often a unique nanomaterial called quantum dot with tunable bandgap so that it is able to act as a photon absorber for a range of photon energies. The quantum dot functions to absorb a photon to create a photon-excited electron–hole pair inside of the QD. Next, splitting the electron–hole pair requires something similar to the *p–n* junction in a conventional crystal Si solar cell. The "*p–n* junction" in a nano solar cell is often the hetero-interface between the quantum dot and another semiconductor with wide bandgap to allow photon to penetrate through, as shown in Fig. 2.23. Such a step-like band edge arrangement looks quite similar to a *p–n* junction in a conventional solar cell. Upon absorbing a photon to generate an electron in the conduction band and a hole in the valence band, the interface potential will push the electron into the material on the left side of the heterojunction, while the photon-excited hole will stay in the QD, leading to the special separation of the electron and the hole.

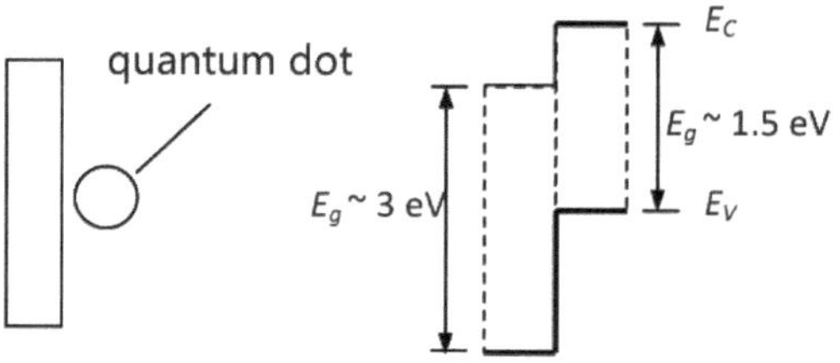

Figure 2.23 The contact between a quantum dot and a material that takes the photoexcited electrons, SEM image (a) and schematic band alignment (b). The conduction band and the valence band edges are arranged quite similar to a *p–n* junction in a conventional Si solar cell.

The key for the formation of such a hetero-interface in Fig. 2.23 is apparently, to find the right materials, a wide-bandgap (~2.7 eV) semiconductor transparent to most part of the solar spectrum and a semiconductor of suitable bandgap (~1.5 eV) absorbing most solar radiation photons. There are many candidates to choose from in big nanomaterial family, but unfortunately, having the right materials is often not enough for the device purpose. The right materials often do not have a band edge arrangement as ideal as that in Fig. 2.23. When two semiconductors form contact next to each other, they could form different band edge alignments.

In type I band alignment, electrons and holes are confined in the same material so that they are not separated in space. In type II arrangement, however, electrons and holes are located in different materials in equilibrium. In other words, the potential well for electrons is in one material and that for holes is in another material. Figure 2.24 schematically presents the type I and type II band alignments on hetero-interface formed by a quantum dot with bandgap ~1.5 eV and a semiconductor with a wide bandgap ~3 eV. An incident photon with energy equal to or larger than that of the QD bandgap will be absorbed by the QD and an electron is excited into the conduction band and a hole be excited into the valence band. In the case of type I arrangement (see Fig. 2.24a), the photon-excited electron and the holes are all confined in the QD. Very often, the electron and the hole will recombine by emitting a photon with the bandgap energy of the QD (see Fig. 2.24b) or giving away their energy through interaction with lattice and defects. Thus, such a system makes no contribution to the photovoltaic process in which the electron and the hole should be separated further. When the band edges form a type II alignment, the photoexcited electron and hole feel a interfacial potential that tends to drive the electron out of the QD (Fig. 2.24c–d) or the hole out of the QD (Fig. 2.24e–f), depending on the band arrangement.

Nano solar cells with a type II core/shell heterojunction have the advantage of broad solar absorption spectrum utilizing different bandgaps of core and shell materials and interfacial transitions crossing the interface. Another advantage is the enhancement of the carrier lifetime due to specially separated electrons and holes in type II heterojunction which slows down electron–hole recombination. In addition, since the electrons and holes transport in separate materials after the quick splitting of photoexcited electrons and holes, it provides easiness to adjust the material thickness of the core and shell and optimize the transport effectiveness of electrons and holes, so as to maximize the device conversion efficiency.

Solar cells based on nanomaterials could provide substantial reductions in material consumption and in production costs. In addition, the growth of nanomaterials relaxes the severe requirement of lattice-matching for epitaxial growth of heterostructure, allowing the III–V- or II–VI-based nano solar cells to be prepared on low-cost substrates such as silicon. Currently, nanomaterial solar cells still face big challenges mainly with respect to raising cell efficiency, which is the focus of global research.

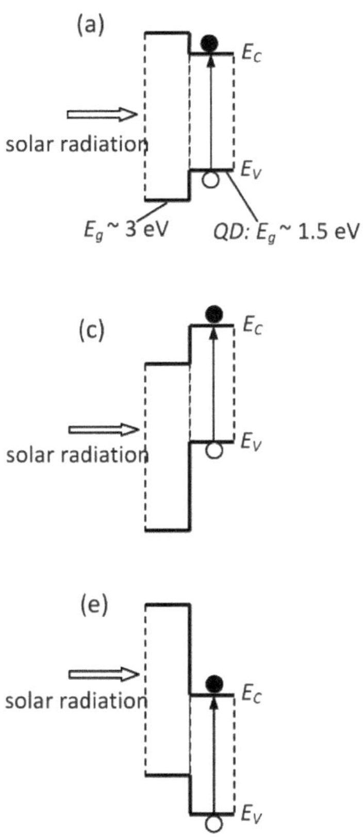

Figure 2.24 Possible band edge arrangements in nano solar cells. (a) The QD absorbs a photon and an electron (a hole) is excited into the conduction (valence) band. (b) The electron and the hole cannot be separated in the type I band alignment system where the electron and the hole are confined in the QD; the electron and the hole could recombine by emitting a photon or relax with lattice to give away energy. (c) A photon is absorbed and a photon excited electron and hole are generated. (d) The type-II band alignment leads to a hetero-interface potential that separates the electron and the hole. (e) and (f) are the case similar to (c) and (d), respectively, but with photon-excited hole being driven out of the QD.

2.4 Factors That Affect the Efficiency of Nano Solar Cell

Photovoltaic semiconductors should have high light absorption coefficient (> 10^5 cm^{-1}) in order to convert as many photons as possible into electric power. Thus, direct-bandgap semiconductors are more suitable and the bandgap is preferably around 1.5 eV for achieving best matching to solar spectrum. Others include low electron–hole recombination rate, high minority lifetime, long carrier diffusion lengths for both electrons and holes, as well as high quantum efficiency. It is also required that PV materials be grown and processed easily. For PV devices, since a high-performance electric potential is critical for splitting the electron–hole pairs, the nanomaterials should form atomic smooth interface and suitable band discontinuity at the heterojunction, or the materials can be doped controllably with both *n*-type and *p*-type dopants for the formation of a good *p*–*n* junction. More importantly, PV devices should have high power conversion efficiency to shorten the energy payback period below 2 years. Manufacture of PV devices should be simple and inexpensive.

2.4.1 Special Characteristics of Nano Photovoltaic Materials and Devices

The use of nanostructures and nanomaterials for photovoltaic devices offers great room and flexibility to tune the material properties and to construct device architectures to meet the parameters for achieving high photoelectric power conversion efficiency. It also reduces the manufacturing costs since nanomaterials can often be prepared at relatively low temperatures and the manufacturing of devices is based on low-cost technology such as printing instead of the high-temperature vacuum deposition that is the conventional way to produce conventional solar cells based on crystalline semiconductors. In addition, nano solar cells make it possible for the incorporation of advanced ideas for high efficiency devices, such as intermediate band, multiple electron–hole pair generation, etc. In recent studies, some of the ideas have been prototypically realized in nanostructures such as quantum dots and quantum wires.

Compared with traditional crystalline silicon solar cells, nano photovoltaic materials and devices show some characteristics different from those mentioned previously such as the size effect and the surface effect.

(1) Optical antenna effect

At some specific wavelengths, incident photons beyond their physical cross section are absorbed, a phenomenon called the "optical antenna effect."

(2) Enhanced light trapping and photocarrier collection

Nanomaterials and nanostructures offer large room for designing various device configurations for maximizing light trapping and photocarrier collection, benefited by the great diversities in the choice of materials, nanomaterial sizes and shapes, as well as device structures.

(3) Low cost and easy to fabricate

Technologies for fabricating nanomaterials have become quite mature. A large number of nanomaterials for solar cell use can be fabricated in cost-effective ways and on a large scale now. Thus, cost-effective production of photovoltaics is now possible.

In Section 2.2, the loss processes in solar cells are discussed, which determine the solar cell efficiency. Achieving high cell efficiency relies on the suppression of the lost processes. Thus, a perfect design of the solar cell device and a well-controlled technology for the device fabrication are essential. In fact, one knows how to design a solar cell with maximized cell efficiency in most cases, but we cannot find a suitable fabrication technology to make the device cost-effectively. Nanomaterials provide a way to reach the destination since the fabrication of nanomaterials requires no dedicated instruments, the growth of nanomaterials being popular and affordable technology. The family of nanomaterials is huge, i.e., there are a large number of nanomaterials one can choose for solar cell fabrication. Additionally, the physical and chemical properties of nanomaterials can be easily tuned, so that it is relatively easy to make up the gap between experimental efficiency and theoretical efficiency. Thin-film solar cells aim to reduce the cost rather than to achieve high efficiency, but the nanomaterial solar cells should take both.

2.4.2 Existing Problems That Affect Solar Cell Efficiency

Essential requirements for photovoltaic materials are low cost, simple, stable, and abundance, while those for photovoltaic devices include—simple and cheap manufacturing process, high power conversion efficiency, stable (with a lifetime of 25–30 years), and environmentally friendly manufacture technology scalable to large-scale manufacturing. For wafer-based Si solar cells, those are still existing problems that have not been completely resolved except that the device is quite stable. There are much more existing problems for solar cells based on nanomaterials. Among them, low power conversion efficiency and poor stability are the main manifestations. There are quite a few existing problems that affect nano solar cell efficiency.

(1) Unmatched optical and electrical performances

Nano-sized materials such as quantum dots have excellent optical properties, while their electric performance is rather poor, given the fact that quantum dot thin films do not have excellent current-carrying ability. The excellent properties of a single nano-sized material are not relevant when the nanomaterial is integrated into collective materials like a thin film.

Nano-sized materials such as quantum dots have excellent optical properties, while their electric performance is rather poor, evidenced by the fact that quantum dot thin films do not have excellent current-carrying ability. The quantum confinement in quantum dots gives rise to excellent photon capture and bandgap tuning abilities. However, the confinement also captures carriers, limiting the carriers' move that is necessary for the photovoltaic effect. In a solar cell, the transport properties of carriers are an essential issue. The interface between nanomaterials often has poor performance so that carrier transport is often blocked there. Nanostructured materials are not single crystal, and there are a large number of defects, boundaries and interfaces, and potential variation since the materials consist of many different phases. Carriers in nanomaterials have low diffusion length and short minority carrier lifetime, which is a more serious problem for the carrier collecting devices than carrier injection ones (like LEDs and LDs).

(2) Low volume of nanostructure materials

Low volume of nanostructure materials leads to low light absorption, which could be a serious problem for nanostructured solar cells. Limited by rather short carrier diffusion length, increasing the thickness of the absorption layer is usually not a solution. For enhancing the device performance, it is necessary to keep the thickness of the devices comparable to the carrier diffusion length so that photogenerated carriers can be collected by contact electrodes efficiently. The diffusion length in nanomaterials is very short, usually a few to a few hundred nanometers. An appropriate thickness should be chosen to balance light absorption and the carrier diffusion.

(3) Poor interface properties

The interface between nanomaterials often has poor performance so that photo-carriers could be captured very likely by high density of interfacial defects. Strictly speaking, it is the materials and structures used that determine the power conversion efficiency, including materials themselves and their combination. Among various problems, the interface between different materials is an essential problem that is not solved in many cases. In nano solar cells, there are a great number of wanted and unwanted interfaces. A large number of defects is another big problem to be solved. The quantum dots that prepared in hydrothermal process are covered with organic molecules that are not conductive. Even if there is a strong potential jump between a photon absorber (a quantum dot) and a photoanode, a photon-excited electron might not be able to transport to the photoanode due to their poor connection.

(4) Lack of high-performance "built-in electric field"

In conventional solar cells, built-in electric field is formed by constructing a *p*–*n* junction through *p*-type and *n*-type doping. In the nanomaterial solar cells, "built-in potential" is often formed by natural interfaces between two nanomaterials. In PbS QD sensitized solar cells, for instance, the potential is located on the connecting region ("interface") between a PbS QD and TiO_2 photoanode. It is difficult in the nanomaterial solar cells to construct a built-in electric field like the *p–n* junction in crystal silicon solar cells, whereas the band discontinuity at the heterointerface is usually inefficient for separating electrons and holes. Thus, the separation of photoexcited electrons and holes has to rely on such a naturally formed interfacial potential discontinuity. At present, this is still a problem to be solved, since the potential discontinuity is insufficient for the separation of electrons and holes. These are the challenges but, in the case of nanomaterials, there exist a large number of approaches for overcoming the challenges.

(5) Design rules for device structures are not clear

The conventional crystal Si solar cells might have different device configurations. However, they have the same basic structure consisting of an anti-reflection coating surface to reduce light reflection for maximum light absorption, a front contact to collect carriers, a *p*-type and an *n*-type Si layers that form a *p–n* junction, and a back contact for carriers collection. The all back-contact solar cells are just a variation of the conventional Si cells, where all contact electrodes are moved to the back. In nanomaterial solar cells, it is hard to design a cell according to the nanomaterial properties and fundamental parameters. In other words, it is difficult to find a design rule for nano cells, as seen from the literature on nanomaterial solar cells that looks so much different in device configurations and materials used.

(6) Performance degradation of nano materials when incorporated into composite materials or made into devices

This may be reflected in many aspects. Quantum dots (QDs) are called photon antenna due to their strong ability in collecting photons. When quantum dots are imbedded in other material matrix, strong interface light reflection often leads to poor device performance, diminishing the advantage of quantum dots. In the case of nano solar cells, it is rather difficult to incorporate with surface texture or anti-reflection coating that are adopted strategies for wafer-based Si solar cells to minimize surface reflection. Limited by the preparation process, on the other hand, a thin film might have low density of quantum dots, resulting in insufficient light absorption by QDs. A typical example are thin-film TiO_2:QDs, where QDs are grown on nanostructured TiO_2. The excellent properties of QDs are thus not relevant when they are integrated into composite materials such as thin films.

Difficulties in modeling and simulation: The important part of the photovoltaic effect is associated with the light–matter interaction. Theoretically, this is described by either Lorentz–Mie theory [15–17], or electromagnetic simulations [18, 19], with the typical examples of NWs. The finite difference time domain (FDTD) technique is a numerical full-field simulation which has become popular recently. It is powerful and relatively straight-forward with efficient formulation. FDTD is especially suitable for modeling devices, such as solar cells, and material system with various structures. FDTD employs volume elements to accurately mesh the computational space, and the volume elements are calculated on a non-uniform rectilinear mesh on which geometric and material parameters adjusted. Stability of calculation is guaranteed, with choosing a time step corresponding to the smallest mesh cell.

Nano solar cells are structurally extremely complicated, though their fabrication processes are simple. The complexity is particularly reflected in the high density and different shapes of the interface, leading to complex electromagnetic field distribution and even various cavity modes. The cavity modes in nanomaterials have localized spatial profiles, including Fabry–Pérot resonance, whispering gallery modes, as well as higher-order complex modes at certain wavelengths. A combination of resonant modes and antenna effects in nanostructures gives rise to enhanced light trapping and absorption effects that differ significantly from bulk cases.

2.4.3 Some Fundamental Rules in Designing a Solar Cell

Although it is rather difficult to give details in designing nano solar cells with maximized power conversion efficiencies, there are some fundamental rules that should be followed.

(1) Using materials with sizes so that the major absorption peak is around the photon energy of 1.5 eV or, in the case of a tandem solar cell, the absorption spectrum matches the solar radiation spectrum.

(2) Minimizing the electric losses by using nanomaterials with good crystalline quality and low interface (grain boundary) defect densities.

(3) Minimizing the surface and interface recombination rates of photoexcited carriers by using suitable passivation techniques.

(4) Minimizing the optical losses by reducing the light reflectance and diffraction and maximizing the light absorption.

(5) Optimizing design of device structure with close-to-zero series resistance and close-to-infinite shunt resistance in the cell.

A device with best conversion efficiency may be obtained if the above conditions are all simultaneous realized. Currently, the best reported wafer-based Si solar cell has the conversion efficiency = 26.3% under AM1.5 global illumination. Figure 2.25 presents the efficiencies of the various best research cells.

The ultimate goal of solar cells is the realization of high power conversion efficiency and low manufacturing cost, so that electric power can be generated in cost-effective ways. The low-cost and large-scale manufacturing of solar cells means that nanomaterials and nanostructures can be efficiently manufactured through low-cost routes such as self-assembly, including effective material use and simple process technologies. The solution-based colloidal approach, for instance, has been used in a large number of device structures, especially those with colloidal quantum dots as absorber materials in dye-sensitized solar cells.

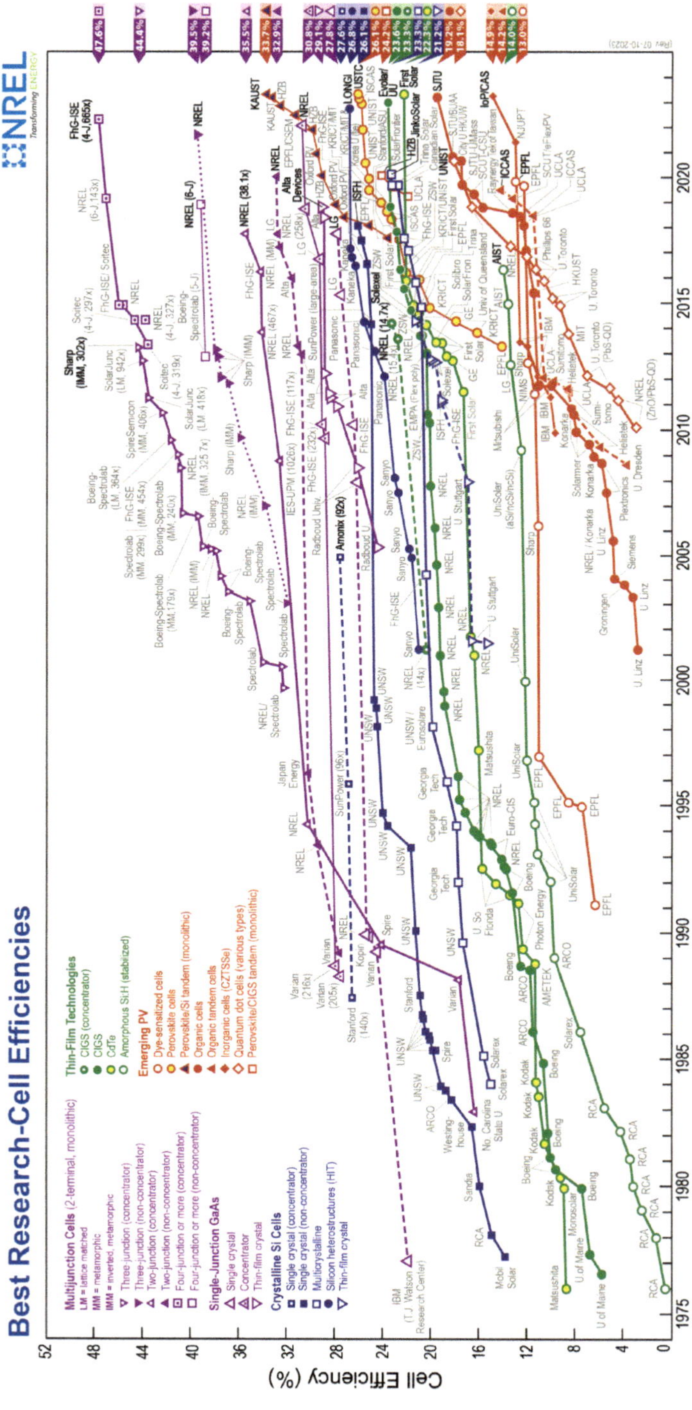

Figure 2.25 Best efficiencies for various research cells.

2.5 Existing Problems in Solar Cells Used for Power Generation

Essential requirements for photovoltaic materials are low cost, simple, stable, and abundance, while those for photovoltaic devices include—low cost, stable (with a lifetime of 25–30 years), simple and environmental friendly manufacture technology, and scalable to large-scale manufacturing. Photovoltaic semiconductors should have high light absorption coefficient (>10^5 cm^{-1}) in order to convert as many photons as possible into electric power. Thus, direct-bandgap semiconductors are more suitable and the bandgap is preferably around 1.5 eV for achieving best matching to solar spectrum. Others include low electron–hole recombination rate, high minority lifetime, long carrier diffusion lengths for both electrons and holes, as well as high quantum efficiency. It is also required that PV materials can be grown and processed easily. For PV devices, since a high-performance electric potential is critical for splitting the electron–hole pairs, the nanomaterials should form atomic smooth interface and suitable band discontinuity at the heterojunction, or the materials can be doped controllably with both *n*-type and *p*-type dopants for the formation of a good *p–n* junction. More importantly, PV devices should have high power conversion efficiency to shorten the energy payback period below 2 years. Manufacture of PV devices should be simple and inexpensive.

The use of nanostructures and nanomaterials for photovoltaic devices offers great room and flexibility to tune the material properties and to construct device architectures to meet the parameters for achieving high photoelectric power conversion efficiency. On one hand, the use of nanomaterials reduces the manufacturing costs since nanomaterials can often be prepared at relatively low temperatures and the manufacturing of devices is often based on low-cost technology such as printing instead of the high-temperature vacuum deposition that is the conventional way to produce conventional solar cells based on crystalline semiconductors. In addition, nano solar cells make it possible for the incorporation of advanced ideas for high efficiency devices, such as intermediate band, multiple electron–hole pair generation, etc. In recent studies, some of the ideas have been prototypically realized in nanostructures such as quantum dots and quantum wires.

The ultimate goal of solar cells is the realization of high power conversion efficiency and low manufacturing cost, so that electric power can be generated in cost-effective ways. Solar cell manufacturing, at low cost and large scale, means that nanomaterials and nanostructures may be fabricated efficiently including efficient material usage and simple technology, with low-cost routes such as self-assembled nanostructures. The solution-based colloidal approach, for instance, has been used in a large number of device structures, especially those with colloidal quantum dots as absorber materials in dye-sensitized solar cells.

References

1. U. Stutenbaeumer and B. Mesfin, Equivalent model of monocrystalline, polycrystalline and amorphous silicon solar cells, *Renewable Energy,* **18**(4), 501–512 (1999).
2. M. B. Prince, Silicon solar energy converters, *J. Appl. Phys.,* **26**, 534–540 (1955).
3. F. Capasso, Compositionally graded semiconductors and their device applications, *Annu. Rev. Mater. Sci.*, **16**, 263–291 (1986).
4. J. Oh, H. C. Yuan, H. M. Branz, An 18.2%-efficient black-silicon solar cell achieved through control of carrier recombination in nanostructures, *Nat. Nanotech.,* **7**(11), 743–748 (2012).
5. P. Repo, J. Benick, V. Vähänissi, J. Schön, G. von Gastrow, B. Steinhauser, M. C. Schubert, M. Hermle, and H. Savin, *n*-type black silicon solar cells, *Energy Procedia,* **38**, 866–871 (2013).
6. H. Savin, P. Repo, G. von Gastrow, P. Ortega, Eric Calle, Moises Garín and Ramon Alcubilla, Black silicon solar cells with interdigitated back-contacts achieve 22.1% efficiency, *Nat. Nanotech.* **10**, 624–629 (2015).
7. T. Markvart, Thermodynamics of losses in photovoltaic conversion, *Appl. Phys. Lett.,* **91**, 064102 (2007).
8. D. E. Carlson and C. R. Wronski, Amorphous silicon solar cell, *Appl. Phys. Lett.*, **28**, 671–673 (1976).
9. A. Chirilă, P. Reinhard, F. Pianezzi, P. Bloesch, A. R. Uhl, C. Fella, L. Kranz, D. Keller, C. Gretener, H. Hagendorfer, D. Jaeger, R. Erni, S. Nishiwaki, S. Buecheler, and A. N. Tiwari, Potassium-induced

surface modification of Cu(In,Ga)Se_2 thin films for high-efficiency solar cells, *Nat. Matter.*, **12**, 1107 (2013).

10. S. Wagner, J. L. Shay, P. Migliorato, and H. M. Kasper, $CuInSe_2$/CdS heterojunction photovoltaic detectors, *Appl. Phys. Lett.*, **25**(8), 434–435 (1974).
11. T. Satoh, Y. Hashimoto, S. Shimakawa, S. Hayashi, and T. Negami, CIGS solar cells on flexible stainless steel substrates, in *Proceedings of the Conference Record of the 28th IEEE Photovoltaic Specialists Conference*, p. 567 (2000).
12. S. Ishizuka, A. Yamada, P. Fons, and S. Niki, Flexible Cu(In,Ga)Se_2 solar cells fabricated using alkali-silicate glass thin layers as an alkali source material, *J.* Renew. *Sustain. Ener.*, **1**, 013102 (2009).
13. X. Xu, Y. Qu, V. Barrioz, G. Zoppi, and N. S. Beattie, Reducing series resistance in $Cu_2ZnSn(S,Se)_4$ nanoparticle ink solar cells on flexible molybdenum foil substrates, *RSC Adv.*, **8**, 3470–3476 (2018).
14. K. Sobayel, M. Shahinuzzaman, N. Amin, M. R. Karim M. A. Dar, R. Gul, M. A. Alghoul, K. Sopian, A. K. M. Hasan, Md. Akhtaruzzaman, Efficiency enhancement of CIGS solar cell by WS_2 as window layer through numerical modelling tool, *Solar Energy*, **207**, 479–485 (2020).
15. L. Cao, P. Fan, A. P. Vasudev, J. S. White, Z. Yu, W. Cai, J. A. Schuller, S. Fan, and M. L. Brongersma, Semiconductor nanowire optical antenna solar absorbers, *Nano Lett.*, **10**, 439–445 (2010).
16. G. Brönstrup, N. Jahr, C. Leiterer, A. Csáki, W. Fritzsche, and S. Christiansen, Optical properties of individual silicon nanowires for photonic devices, *ACS Nano*, **4**, 7113–7122 (2010).
17. W. F. Liu, J. I. Oh, and W. Z. Shen, Light absorption mechanism in single c-Si (core)/a-Si (shell) coaxial nanowires, *Nanotechnology*, **22**, 125705–125708 (2011).
18. E. S. Barnard, R. A. Pala, and M. L. Brongersma, Photocurrent mapping of near-field optical antenna resonances, *Nat. Nanotechnol.*, **6**, 588–593 (2011).
19. T. J. Kempa, J. F. Cahoon, S.-K. Kim, R. W. Day, D. C. Bell, H.-G. Park, and C. M. Lieber, Coaxial multishell nanowires with high-quality electronic interfaces and tunable optical cavities for ultrathin photovoltaics, *Proc. Natl. Acad. Sci. U. S. A.*, **109**, 1407–1412 (2012).

Chapter 3

Nanomaterials and Structures for Photon Trapping

According to Maxwell's equations, light is a smooth and continuous wave which is described in terms of its wave-like properties. In 1905, Albert Einstein found that light shows a different character—it behaves as though it is composed of many individual particles that are now called *photons*, when light interacts with electrons. In fact, it is more appropriate to treat light as particles, rather than waves, when light interacts with a matter. The interaction picture is quite clear when the particles, as discrete packets of momentum and energy, or *quanta*, interact with electrons also as particles. Quantum field theory treats the light wave as quantized packets of energy, i.e., the light only exists in discrete quantities—photons.

3.1 Wave-Particle Duality of Light and Light Reflection, Diffraction, and Refraction

Light is a kind of matter that shows typical wave and particle duality [1]. The wave characteristics of light are manifested by the phenomena of interference, reflection, refraction, diffraction, coherence, etc. In the wave nature, light is electromagnetic wave consisting of two fluctuating fields—an electric field and a

Introduction to Nano Solar Cells
Ning Dai

ISBN 978-981-4877-49-7 (Hardcover), 978-1-003-13198-4 (eBook)
www.jennystanford.com

magnetic field that propagate at the speed of light, *c*. In homogenous free space, the two fields are in phase and orthogonal to each other, and they are both perpendicular to the traveling direction of the light. Figure 3.1 schematically illustrates the geometric relationship of the electric field, magnetic field, and the wave propagation directions of the electromagnetic wave (light), together with their phases and wavelength. Both the electric component and the magnetic component are characterized by amplitude, wavelength (frequency), polarization, and phase. As shown schematically in Fig. 3.1, for light propagating in the *z* direction, the electric and the magnetic components can be written as the wave equations of

$$\mathbf{E}_x = \mathbf{E}_0 \exp[i(\omega t - \mathbf{k}_z \cdot \mathbf{z})] \tag{3.1a}$$

and

$$\mathbf{B}_y = \mathbf{B}_0 \exp[i(\omega t - \mathbf{k}_z \cdot \mathbf{z})], \tag{3.1b}$$

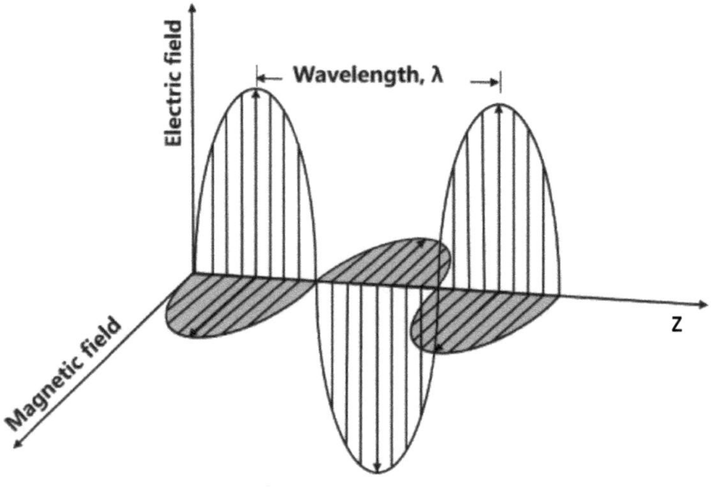

Figure 3.1 Schematic electric and magnetic fields of light (electromagnetic wave) in free or homogenous space. It is shown that the electric component is perpendicular to magnetic one and both electric and magnetic components are orthogonal to the light propagation direction. z is spatial coordinate of light propagation direction.

where $\mathbf{k}_z = 2\pi/\lambda$ is the wave number, $\omega = 2\pi\nu$ the angular frequency and ν the frequency. $\omega - \mathbf{k}_z \cdot \mathbf{z}$ is the phase of the wave and the subscripts x and y indicate the electric field and magnetic field are polarized in the x and y directions, respectively. The Sun radiates light with frequencies from ultraviolet to far infrared.

Thus, light can be characterized by wavelength λ, period T, frequency ν, and propagation velocity c and these parameters satisfy

$$\nu = c/\lambda = 1/T. \tag{3.2}$$

The particle properties are shown by the photoelectric effect and photoelectric scattering.

In the particle nature, light consists of photons. A photon strikes a matter and is absorbed by an electron in the matter. The electron is excited to a higher energy state, or the electron gets enough energy to leave the matter (see Fig. 3.2) if the energy received exceeds the work function of the material. In semiconductors, an electron can be excited from the valence band to the conduction bands if the electron absorbs a photon with energy equal to or larger than the bandgap. As a particle, a photon has a defined energy E that is proportional to the frequency of radiation ν with the form of

$$E = \hbar\nu = h\omega, \tag{3.3}$$

where h is the Plank constant and $\hbar = h/2\pi$. Thus, the wave nature (3.1) and the particle nature (3.2) of an photon is connected by Plank constant, as written by

$$E = h\frac{C}{\lambda}. \tag{3.4}$$

In solar cells, the wave-particle duality of light is shown vividly, where the wave nature is demonstrated by surface reflection, diffraction, interference, light trapping effects, etc., and the particle nature by photoelectric effect, i.e., electrons are excited into the conduction band by absorbing photons.

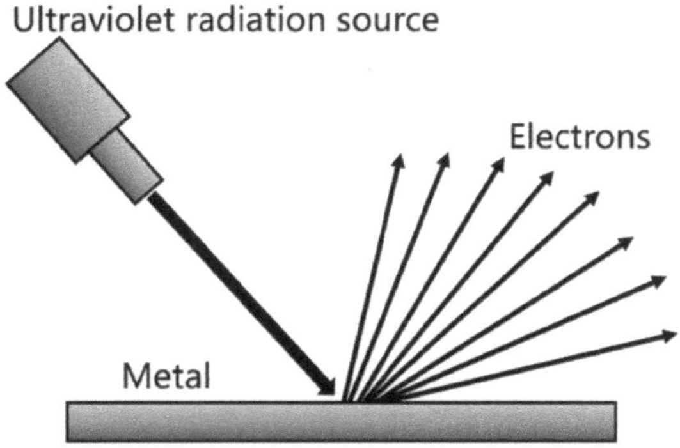

Figure 3.2 A schematic diagram showing that incident photons causes the emission of electrons from the matter surface due to the absorption of the photons by the electrons. Note that a minimum photon energy is needed before the electron is ejected.

3.1.1 Light Reflection and Diffraction at Interface between Two Dielectrics

Reflection, diffraction, refraction, and transmission occur when light propagates in inhomogeneous media. Consider a simple case that a material slab with index of refraction n_1 is in surrounding medium with index of refraction n_0, as shown in Fig. 3.3. The incident angle (the angle between incident light beam and surface normal) equals to the reflection angle (the angle between reflected light beam and the surface normal). The incident angle (θ_0) and the refraction angle (θ_1) (180° - the angle between refraction light beam and the surface normal) has the relation of

$$n_0 \sin\theta_0 = n_1 \sin\theta_1, \tag{3.5}$$

which is called refraction law. Assuming normal incidence and zero diffraction, the reflectivity is

$$R = \frac{n_2 - n_1}{n_1 + n_2} \tag{3.6a}$$

for the amplitude of light wave, or

$$R_p = \left(\frac{n_1 - n_0}{n_0 + n_1}\right)^2 \tag{3.6b}$$

for the power of light. The refractive index of Si is around 4 in the visible range. For light incidence on smooth Si surface from the air (the refractive index of the air is approximately 1.0), $R_P \cong 35\%$. Thus, about 35% of the solar radiation power is reflected. For visible light travelling from air into a piece of glass ($n_1 \cong 1.5$), R_P is calculated to be only 4% on a single reflection. Generally, if a smooth Si surface is coated with a thin-film layer with lower index of refraction, reflection could be reduced apparently. In fact, optimal anti-reflection effect can be achieved by choosing an anti-reflection coating with the index of refraction $n' = \sqrt{n_0 n_1}$. Anti-reflection coating has been widely used in wafer-based solar cells.

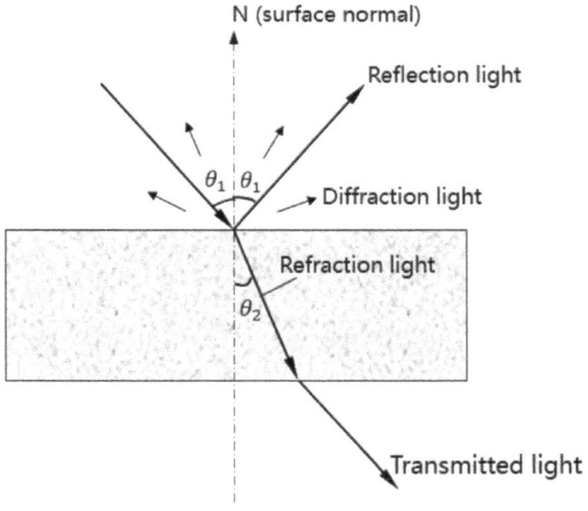

Figure 3.3 Geometric relationships of incident light, reflected light, diffraction light (small arrows), refraction light and transmitted light. For mirror-like Si surface, diffraction can be ignored.

3.1.2 Rayleigh Scattering and Mie Scattering

Rayleigh scattering and Mie scattering are both bulk scattering due to the particles distributed in a medium [2]. Raleigh scattering refers to scattering to light caused by particles with sizes much smaller than the wavelength of the incident light. Rayleigh scattering is strongly wavelength dependent. In Rayleigh scattering approximation, a small particle is considered as an individual dipole and the intensity of scattered light the form of

$$I = I_0 \frac{8\pi^4 N\alpha^2}{\lambda^4 r^2}(1+\cos^2\theta) \sim \frac{1}{\lambda^4}, \tag{3.7}$$

where I_0 is the intensity of incident light, N the number of particles, α the polarizability of the particle, λ the wavelength of the light, θ the angle between the surface normal and the observer, and r the distance from the particles to the observer. Thus, the scattering is proportional to $1/\lambda^4$, so that shorter wavelength light receives stronger scattering, which explains why the sky is blue. Note that the forward scatter equals the back scatter and at θ = 90° the scattering intensity is half of that of the forward scattered light (or backscattered light), as shown in Fig. 3.4a.

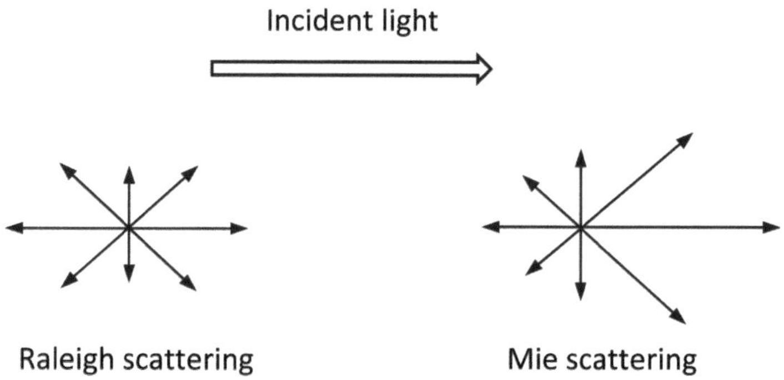

Figure 3.4 The Rayleigh scattering (a) and the Mie scattering (b) patterns, with the incident light direction as shown on the top.

Mie scattering, named after Gustav Mie occurs when the dimension of the homogeneous spheres is much larger than the wavelength of the incident light. Thus, Mie scattering predominates for particle sizes much larger than a wavelength. The Mie scattering has a pattern with relatively more intense forward scattering, as shown in Fig. 3.4b. The Mie scattering is more complicated than Rayleigh scattering when a mathematical formula is tried to be derived. The Mie scattering, which accounted for the sizes of the particles (spheres), can be expressed by an infinite series of multi-pole expansion of polarization. The corresponding electromagnetic wave of such polarization then also takes the form of an infinite series. Each of the terms is a representation of the contribution of a multi-pole polarization component. So far, a classical Mie scattering theory and several numerical calculations including finite difference time domain (FDTD) method, finite element method (FEM), discrete dipole approximation (DDA), etc., have been developed.

In nano solar cells, the optical absorption layer is often a 3D heterojunction one. Therefore, either the Rayleigh scattering or the Mie scattering plays a certain role and, sometimes, the effects affect the cell performance to a large extent.

3.1.3 Lambertian Scattering

Assuming an ideal diffuse radiator, Lambert's law states that luminous intensity from a surface element dA is proportional to the cosine of the angle θ between the surface normal and the direction of the surface element to the observer [3], as shown in Fig. 3.5,

$$\mathrm{d}I = I_0 \cos\theta \mathrm{d}\Omega \, \mathrm{d}A, \tag{3.8}$$

where dA is the area element, I_0 the light intensity (in the unit of photons/(s·m^2·sr)) observed per unit area and per unit solid angle, dΩ the solid angle subtended from the emission area element dA on the surface to the observer. The surface, when its scattering to light obeys Lambert's cosine law, is called "Lambertian."

When sunlight strikes the surface of a semiconductor, some of the ray is reflected at the interface between the semiconductor and the air. The remainder passes through the interface into the material, where it is scattered through absorption and reemission in a new direction. The absorption and reemission could randomize the direction of the light path, so that when light is re-emitted into the air with no preferred orientation. Of course, the surface of a solar cell is not perfectly Lambertian. One always observes relatively stronger reflection in mirror reflection direction.

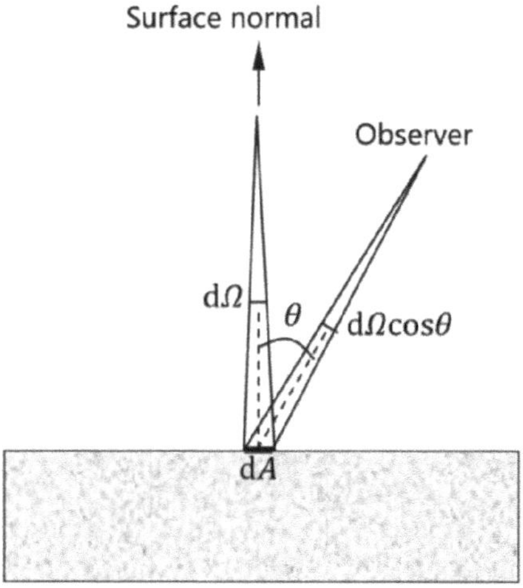

Figure 3.5 Light intensity for a normal and off-normal observer at θ.

3.2 Light Trapping Methods

In materials where there is a large mismatch between the minority-carrier diffusion length and light absorption depth, the solar cell efficiency is usually low, which is the case for thin-film and various nano SCs [4]. When photons incident on the surface of a solar cell, part of them are reflected and part of them transmit into the solar cell and are converted into free electrons and holes. Light trapping is for the enhancement of light absorption by solar cells and the suppression of photon reflection, photon management thus being essential for high-efficiency photon energy conversion. For SCs, especially thin-film SCs, photon management aims at enhancing the light trapping. The purpose of enhancing the light trapping is two folds: one is increasing the photon path in solar cells so that photons have high chance to be absorbed by the active layer, which is especially critical for thin-film SCs; another one is to make the photon absorbable by the active layer in SCs. Since the device requires to convert photons with wavelength covering a wide band, the trapping effect needs to be efficient to the whole solar spectrum. The use of trapping technology tolerates the low-purity Si that often has a small minority carrier diffusion length. One can then use relatively low-quality Si materials since the device can be structured to have a short minority carrier collection path.

Currently, some measures have been developed to trap light in solar cells, with different technologies being suitable for different photovoltaic devices [5]. Anti-reflection coating is adopted for wafer-based and, sometimes, for thin-film solar cells. Surface texture is only suitable for wafer-based solar cells.

3.2.1 Light Trapping in Conventional Crystalline Si Solar Cells

Light trapping through surface texturing

Another effective way for light trapping is through surface texturing. A light beam might undergoes multiple reflection on a surface that is covered with rough texture, as shown in Fig. 3.6a where incident beam is reflected back and forth between the inclined surfaces. Thus, the light beam goes through two reflections before it is bounced back to air, leading increasing probability of photon absorption. Figure 3.6b presents an SEM image of textured (100) silicon surface formed by etching with 40% KOH:IPA:DIH_2O with a volume ratio of 1:3:46 for 15 min at 80°C [6].

With the pyramid structures of 2 to 10 μm in size etched into the surface of wafer-based SCs for light trapping, the surface texturing methods are, however, not feasible for thin-film SCs with a typical thickness of 1 μm, especially for nano SCs with even much smaller characteristic structures. The textured surface helps to minimize the reflection loss by allowing multiple times of light reflection before escaping and by refracting light into large angles that leads to increasing in optical path. The increase in optical path length is helpful for the matching of optical path length and electron (hole) diffusing length. Other methods for the formation of surface texture are given in Ref. [7].

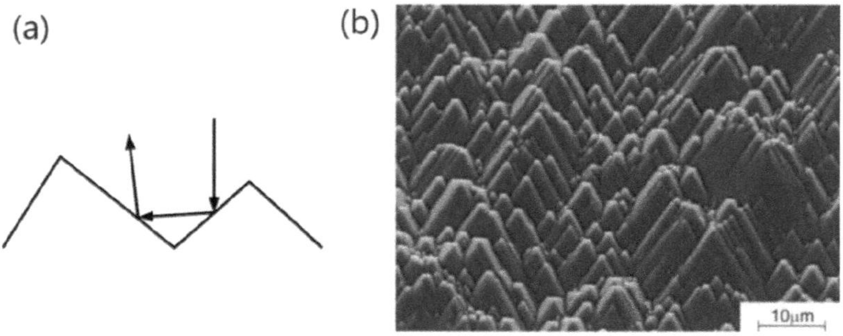

Figure 3.6 Schematic process in which Incident light experiences two reflections before it leaves the cell surface (a) and SEM image of texture (100) silicon surface formed by etching with 40% KOH:IPA:DIH_2O with a volume ratio of 1:3:46 for 15 min at 80°C (b). Part (b) reproduced with permission from [6].

Being a semiconductor with an indirect bandgap, crystalline silicon (c-Si) has a light absorption length over 5 mm for photon energy at its near infrared bandgap (1.12 eV) and a bit over 10 μm for photon energy around 1.55 eV [8]. Thus, 35% of the solar photons with energy above the Si bandgap are in this weak absorption region [9]. The commercial c-Si solar cells with a typical thickness of 180 μm is even not sufficient for the effective absorption of photons in this energy range. The analyzing on light trapping effect might base on wave optics or on geometrical optics. Wave optics has been proven to outperform geometrical optics approaches through taking light wavelength into active considerations.

Aiming at the increase of light path length in the active layer in solar cells, the fundamental measurement consists of anti-reflection layer(s) on the front surface and a reflecting back layer, as shown in Fig. 3.7. In conventional c-Si solar cells, the front anti-reflection measurement is the combination of surface texture and SiN_x layer both for anti-reflection and for surface passivation. The back reflecting layer uses metal Al that reflects the light back into the Si layer. For perfect surface texture (no light loss), the light path increases a factor of $4n^2$, which is roughly 50 for Si and 30 for TiO_2 [10]. Together with the back metal reflecting layer, this could lead to significant increase of the conversion efficient, due to the cutting-down of light loss.

The effect becomes more prominent for thin-film solar cells due to their low baseline of conversion efficiency.

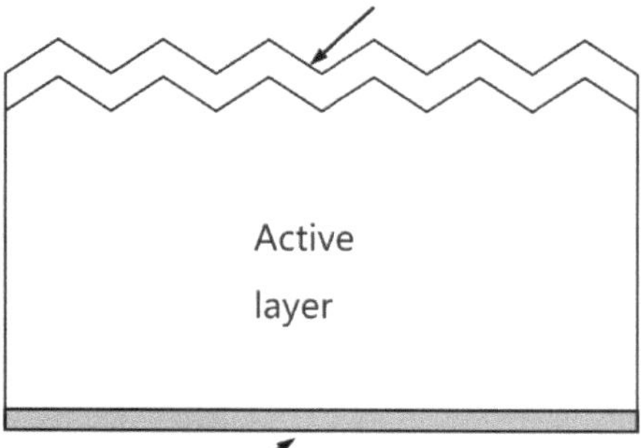

Figure 3.7 Schematic diagram of a solar cell with light trapping measurements: frond anti-reflection coating, surface texture, and back-reflector.

Wafer-based Si solar cells need about 100 μm Si layer thickness to absorb sufficient incident light due to the material's indirect bandgap, while for direct-bandgap semiconductors, nearly 100% incident light is absorbed in 1 μm thickness. As a result, if light can be trapped in 1 μm thickness, 99% of Si material, which takes 30% of the total cost of the solar cells, can be saved. For thin-film solar cells, the thickness of the active layers is in the order of 1 to 3 micron. This is to say that light must be trapped within the thickness for the best utilization of incoming solar energy for photoelectric conversion. In fact, there has been a growing interest in developing Si-based thin-film solar cells for lowing material costs, although the conversion efficiency of thin-film Si SCs is still much lower (about 30%) than their wafer-based counterpart.

A polished semiconductor surface reflects about 30–40% incident light with which a solar cell cannot use for photoelectronic conversion. An ideally rough surface randomly scatters light of various wavelengths such that in the limit of weak light absorption, optical path length enhances by a factor of $4n^2$ (n is the refractive index of the material)—a Lambertian light trapping limit that is the thermodynamic limit of maximum absorption based on geometric optics framework. The Lambertian light trapping limit can be surpassed by directing the light to travel in an acceptance angle 2θ only using certain nano "elements," instead of allowing random scattering of light. The Lambertian factor then become $4n^2/\sin\theta$ for non-isotropically incident radiation. Thus, for small θ the optical path of light can be further enhanced significantly. For wafer-based Si solar cells, total thickness of 100 to 200 μm is usually required for complete light trapping (absorption). With proper surface light trapping measurements using nanomaterial architectures, thickness of 1 μm would be enough to absorb all the light, a reduction of the crystalline Si wafer thickness by 2 orders of magnitude. Note that since Si is a dominant material for solar energy harvesting, many examples are presented with the Si material and the Si nanostructure. The solar cell design principles, however, are applicable to other kinds of materials. Efficient photon management design is essential for nano SCs and thin-film SCs but rather challenging, primarily due to the broadband solar spectrum that covers the wavelength range from 400 to 1500 nm range and the wide range use of structures and materials of the SCs.

3.2.2 Light Trapping in Nano Solar Cells

It is required that the general idea of photon management design for nano SCs should work for a broad range of the materials and structures, as well as for broadband solar spectrum. Fortunately, nanomaterials offer wide easiness in tailoring of morphologies for optimized photon management design, in terms of their large variety of structures shapes, and feature sizes in the wavelength and sub-wavelength regime that are controllable in a wide range and achievable in a large scale. Nanomaterials and their corresponding structures, enable the control to photons via transmission, reflection, scattering at the boundaries of nanomaterials. In fact, a variety of designs for photon management based on nanostructure have been proposed and investigated both theoretically and experimentally, the architectures including anti-reflection coating, resonant scattering, photonic crystals, dielectric nanostructures, metamaterials, and plasmonics.

The design principle of the photon management is twofold: suppressing the reflectance and enhancing the absorption in the solar spectrum range.

Currently, a large number of technologies have been developed to engineer various semiconductors at nanoscale. Unique and easily variable geometries of nanowires (NWs) offer the systematic and wide potential for light absorption enhancement. The easiness in NW synthesis, alloying, and heterojunction formation with various geometries such as axial and radial junctions allow for great control in fabricating materials and nanostructures, which offers ways to tailor absorption spectra for enhancing the efficiency of solar power conversion [11].

3.2.3 Light Trapping Using NW Arrays

Aiming at maximizing the absorption of solar radiation energy, an antireflection coating is designed to inhibit the reflection of light incident on the battery surface. Thus, intensifying light absorption by suppressing light reflection is first step toward the achievement of high power conversion efficiency for solar cells. Significant effort has focused on how to enhance light absorption properties of the surface of solar cell devices. Very often, like the textured surface of Si-wafer solar cells, a suitably roughed surface is highly beneficial for the suppression of light reflection and refraction. With geometry of created roughing on solar cells surface, nanowires (NWs) are excellent material configuration for light trapping. The effectiveness of light trapping with NWs depends on material, size, shape, and density of NWs. A strongly enhanced absorption can be realized with synthetic control over the parameters, rendering the NWs with absorption characteristics far superior to those of wafer-based Si or thin-film solar cells. The major advantages of light trapping using NWs are broadband anti-reflection and tolerance in wide light incident angle. Excellent light trapping has been realized in both horizontally and vertically oriented NW arrays with appropriate pitch sizes.

A number of studies were carried out for modeling the light absorption of NW arrays, including those for Si, InAs/InP, and ZnO NWs [12–14]. The simulations were able to give light field distributions along NWs and in the surrounding media. Most of the studies are based on a 2D wave-guiding model. As shown in Fig. 3.9, the power absorption along the NWs is apparently frequency dependent and the anti-reflection layer is effective in suppressing the light reflection. More specifically, the light absorption is strongly dependent on the special locations of the field distribution of the light modes that are determined by geometry of the NWs/substrate structures rather than material composition of NWs. One of the important geometrical parameters of NW arrays on substrates is the NW filling ratio. Strictly, an optimized filling ratio depends on geometrical shape and the kind of materials, as well as substrate material, the material filled between NWs, and wavelength of light. For instance, reducing filling ratio leads to decrease in absorption for light at long wavelengths and increase in absorption for light at shorter wavelengths. Figure 3.8 presents normalized power absorption spectra of 2 μm-long InAs NWs with and without oxide layer, indicating that the absorption sensitively depends on width of the NWs.

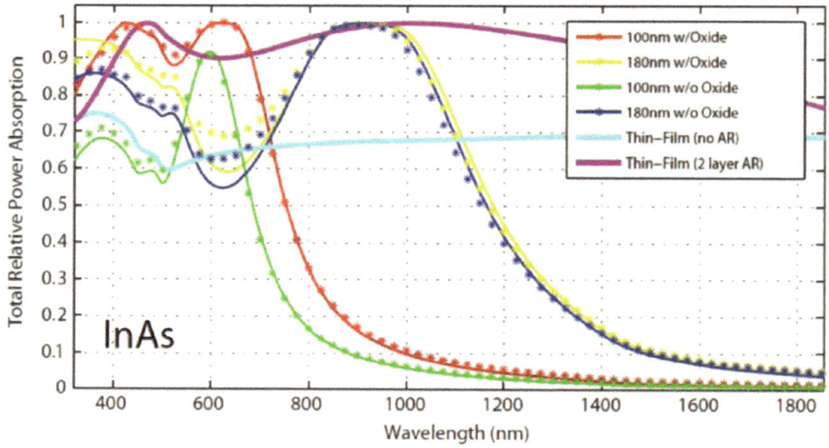

Figure 3.8 Normalized spectral power absorption of 2 μm-long InAs NWs with and without oxide layer. The NWs are 100 or 180 nm in diameter and the spectra are calculated by 2D waveguide model including impedance matching and Beer-Lambert law (solid lines) and 3D calculation (asterisk), together with the thin-film results showing the effect of anti-reflective properties. Reproduced with permission from [13].

NW arrays are not a completely disordered system. In fact, light absorption also depends on relative orientation between polarized light vectors and the direction along NWs. As a result, light absorption is elevation- and azimuth-angle dependent. The effect of absorption anisotropy weakens with the reduction of NW size and increase in NW array density, since the disorder effect in arrays increases with density and thinning of NWs. Furthermore, the use of high-refractive-index media to fill the space between NWs is beneficial for the reduction of light scattering. With a mechanism of multiple diffuse light scattering [15], disordered NW arrays show excellent properties for light reflection suppression. The ease of lower cost in the fabrication offers disordered NW arrays a bright future in the use for enhancing light absorption in opto-electronic devices like solar cells.

Optical properties of disordered NW arrays were investigated both theoretically and experimentally [16, 17]. Muskens and co-workers studied experimentally the optical properties of InP, Si, and GaP nanowires for the use in solar cells [18]. They proposed a design principle for the suppression of reflective losses. There are a large number of interesting works that demonstrate the effectiveness of using nanomaterials and nanostructures to trap light, including refs. [19–23].

3.2.4 Suppression of Light Reflection Using Tapered Two-Dimensional Gratings

Han and Chen studied the suppression of light reflection by periodically patterning 2D nanostructures with tapered two-dimensional gratings [24]. The system investigated is shown in the inset of Fig. 3.9. The pyramids are coated with a 90 nm Si_3N_4 layer for anti-reflection purpose. Figure 3.9 clearly demonstrates that the absorptance is close, but below the Lambertian curve in almost the whole spectrum range studied even in normal light incident condition and a periodicity of 80 nm (optimized condition). The absorptance can be enhanced by using asymmetric two-dimensional gratings. In fact, the use of skewed pyramid structure leads to optical absorptance very close to the Lambertian trapping limit at normal light incidence.

Furthermore, the use of asymmetric two-dimensional gratings consisting of rod arrays with asymmetric tapered tops gives rise to enhanced absorptance close to the Lambertian limit even in the case that all directions of incidence are averaged. Thus, nano Si thin films might have comparable absorption as thick crystalline Si wafers [24].

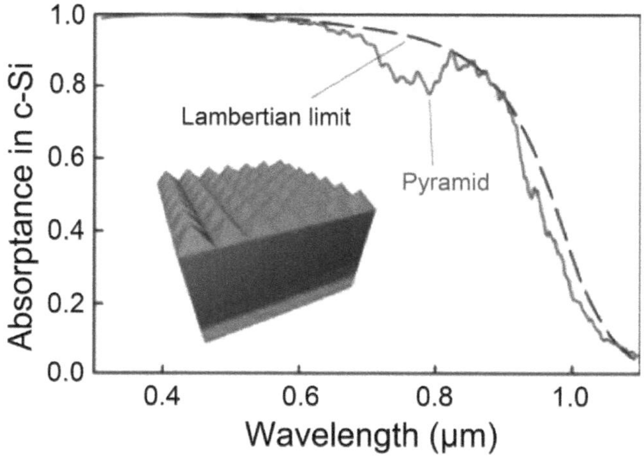

Figure 3.9 Optical absorptance as a function of light wavelength for the pyramid-layer/Si/SiO_2/Ag structure at normal incidence. The solid line is the absorptance of the pyramid structure and the dashed line the Lambertian limit of a crystal Si thin film of the same thickness. The nano pyramid has the height of 566 nm and the base length of 800 nm and the pyramids are coated with a 90 nm Si_3N_4 layer. The thickness of the Si thin film is 2.012 μm and that of the SiO_2 layer is 717 nm. The spectrum was averaged over $\Delta\lambda/\lambda^2 \sim 0.05$ μm^{-1} in order to smooth out the arrow peaks. Reproduced with permission from [24].

Efficient light trapping scheme, particularly anti-reflection measurement, is indispensable in thin-film solar cells. Light trapping is, however, a technical barricade for thin-film solar cells, since those mature technologies such as surface texturing is not suitable to adopt. Fortunately, the nano properties of the NSCs give more room for the manipulation of light travelling for both light absorption enhancement and light reflection reduction. Experimentally testing the effectiveness of light trapping for a particular material system and device structure is often time consuming and expensive. Modeling for effectiveness is often extremely efficient in the case of nano SCs and, as a result, theoretical investigations on nanostructured SCs have gained tremendous attention, aiming at improving conversion efficiency through enhancement of light trapping and carrier collection efficiency. Many works devoted to light absorption of nanostructure arrays have been reported for efficient solar cells [25, 26].

3.2.5 Enhancement of Light Absorption Reflection by Periodic Arrays of Nanowires

Photon trapping properties of nanowire arrays can be simulated and a number of software packages are available now. In Chen and co-workers' work, for instance, the transfer matrix method was applied to analyze how the length, diameter, and filling ratio affect light trapping, the nanostructures being nanowires and nanoholes [27, 28]. They demonstrated that with proper selection on the nano parameters, the surface of solar cells might trap 99% of the incoming light for a given range of light wavelength. The theoretical simulation is greatly helpful for the understanding of photoelectric conversion processes in the complicated nano solar cells and serves as guidance of optimized PV device structures. Detailed analysis of the optical absorption of nanowire arrays is helpful for the structural design and performance optimization of nanowire solar cells. Except for the material, optical absorption of nanowire arrays depends on the periodicity, wire diameter, and wire length. Hu and Chen reported theoretical analysis on the optical properties of periodic silicon nanowire arrays, including NW arrays, nanohole arrays and nano-pyramid arrays based on numerical modeling [27]. The calculation is performed by solving a full-wave vector Maxwell equation and the periodicity of the nanowires is accounted for using the transfer matrix method. The transfer matrix method is a very powerful tool in dealing with a periodic structure. It was also found that nanowire arrays have an intrinsic anti-reflection effect that causes increasing absorption in short wavelength range.

Figure 3.10a schematically shows a periodic nanowire (nanorod) array of the length L, period a, and diameter d. The coordinates are set such that the wires are along the z direction and the 2-D array is in x–y plane. The calculated absorption spectrum is presented in Fig. 3.10b, where a broadband absorption shows how the incident light interacts with the periodic nanowire array. It is interesting to see that the absorption edge depends on the length of the nanowires, the absorption edge shifting to lower energy as the increase of the nanowire length, indicating that it is thus possible to change the nanowire length to absorb photons of various energies for photoelectric conversion.

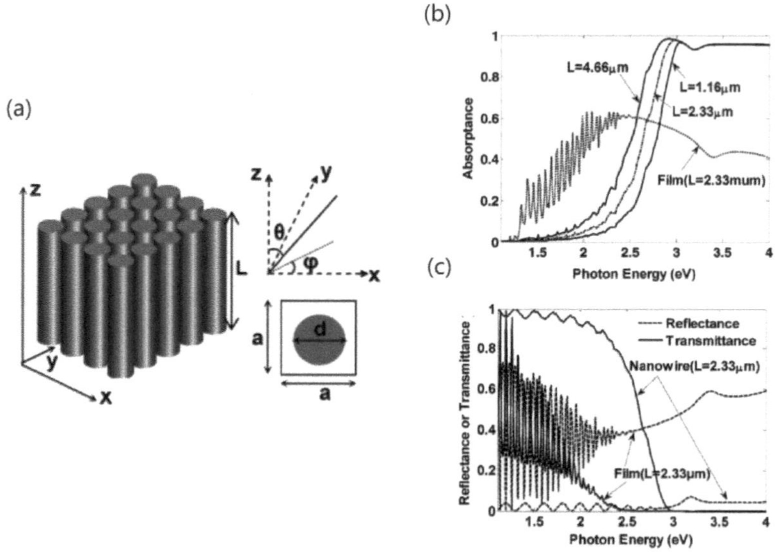

Figure 3.10 Light absorptance (a) and reflectance and transmittance (b) as a function of photon energy for nanowires with L = 1.1, 2.33, and 4.66 μm and fixed d = 50 nm and a = 100 nm. The absorption of a Si thin film is also presented as a reference. The inset in the figure shows schematically the structure of the periodic nanowires array and the geometric relationship of the light incidence onto the sample surface. Light is normally incident on top. Reproduced with permission from [27].

Hu and Chen's calculation shows enhanced intrinsic anti-reflection effect of the nanowire arrays in comparison to the Si thin film with a thickness of 2.33 μm, as shown by Fig. 3.10. Light absorption, reflectance, and transmittance are presented vs. photon energy for nanowires with their length L = 1.1, 2.33, and 4.66 μm and fixed wire diameter d = 50 nm and the period a = 100 nm, as compared to those from the Si thin film. It shows that at the photon energies below the indirect bandgap of

1.12 eV, the absorption is extremely weak and, at the photon energy above 1.12 but below 2.5 eV that is below the direct gap of Si, Si is more or less transparent as indicated by the strong oscillation due to Fabry–Perot interference measured on the 2.33 μm Si thin film. With further increase of photon energy, the light absorption enhances until a plateau of nearly 100% absorptance is reached. Apparently, in the plateau region, the absorptance of all the nanowire arrays is higher than that of the Si thin film. The result is a bit astonishing, given the fact that the 2.33 μm Si thin film is optically denser than all the nanowire arrays with d = 50 nm and a = 100 nm. The reflectance and the transmittance of the nanowires show corresponding behavior matched to the absorptance, as shown in Fig. 3.10c, where only the results from the nanowires array of L = 2.33 μm is presented. The 2.33 μm nanowire array shows very low reflectance in the whole measured wavelength range and varied transmittance, the high transmittance signifying low absorption in long-wavelength region. This indicates that the nanowire array is effective for the suppression of reflection in the wide wavelength range, while effective absorption only occurs when photon energies are above the bandgap of the nanowires. In the case of photon trapping using nanowires, nanowire arrays are often expected to enhance transmittance through reflectance suppression so that photons can be absorbed as much as possible by the solar cells underneath the nanowires. Hence, nanowires of wide-bandgap materials should be used to avoid light absorption by the wires.

From Fig. 3.10, one could note that the absorption-enhancing effect is significant in the range of short wavelength, which is due to the multiple scattering. With the increase of light wavelength, the nanowire array behaves more and more like effective (average) medium of nanowire/air with less and less medium discontinuity. The Si thin film absorbs more effectively than nanowire arrays due to low nanowire filling effect and small extinction coefficient.

The calculations show superior light trapping property of the NW array, in comparison to that of a common thin film, due to the fact that in the NW array the light undergoes multiple scattering-absorption by the walls of the NWs, while in the thin film the photons experiencing the first scattering on the top surface will have no chance to be absorbed again. The light trapping effect is also dependent on the polarization, since the NWs are themselves directional. Figure 3.11 presents the calculated light absorptance for TE and TM polarized light in the photon energy range between 1.0 to 4.0 eV, for the NW array with a geometry of a = 100 nm, d = 80 nm, and L = 2.33 μm at several θ and φ angles. At normal incidence (θ = 0°), their calculation showed isotropic TE absorption in x–y plane, as indicated by the negligible change of the absorption spectra at different φ (the figure gives the results of φ = 0°, 20°, and 40°). At oblique incidence of light (θ = 30°, for instance), the TE polarized light receives stronger absorption than the TM polarized light, due to an electric field component in TE light along the NWs (z direction in Fig. 3.10a) that favors the absorption.

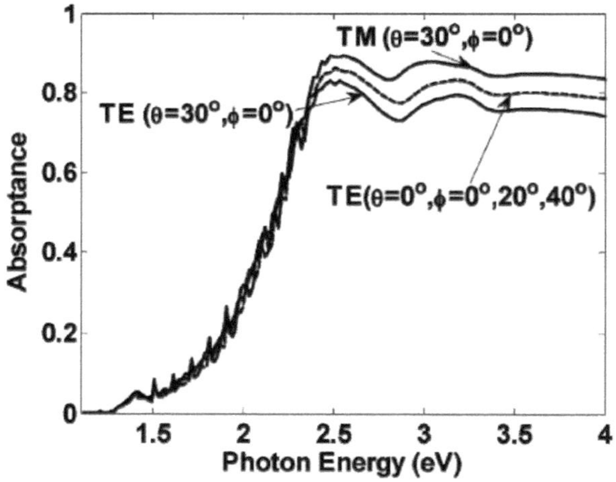

Figure 3.11 Calculated angular dependent absorption as a function of photon energy for TE and TM polarized lights at various θ and φ. Reproduced with permission from [27].

The use of nanomaterials makes it possible for broadband suppression of both specular and diffusive reflections over a wide spectral and a wide incident angular range. A solar cell should absorb light as much as possible, i.e., a wide absorption band with nearly perfect absorption covering the major part of the solar spectrum for obtaining PV devices of high conversion efficiency. In the case of NWs, the nearly perfect absorption in the solar radiation range is pursued through optimizing the selection of NW parameters of periodicity *a*, wire diameter *d*, as well as wire length *L*. However, the optimization of those parameters, sometimes, cannot give perfect photon trapping in wide solar radiation wavelength range. There is still room for the improvement of above NW array to trap more light.

Intuitively, arrays with thin and long nanowires should have good absorption of light due to the smooth and flat refractive index gradient from air to substrate. Multiple and diffuse reflection/scattering could lead to enhanced non-photoelectric processes that make no contribution to photoelectric conversion. The multiple scattering effect leads to a strong coupling mismatch with incident light, which is responsible for the enhanced light absorption. This occurs when the wavelength of light is comparable to the geometrical dimensions of the NWs. This occurs speaking, long and thin NWs are more efficient for light trapping. However, they suffer from the issues of high surface defect density and short carrier lifetime. In addition, long and thin NWs are flexible and van der Waals forces could even cause collapsing into disordered clusters, leading to decrease in light trapping efficiency and increase in carrier recombination.

Note that the reduced reflection due to nanowire arrays is insensitive to the polarization and the incident angle of light over a broad range of wavelengths. The improved light trapping in the nanowire arrays is primarily due both to the gradual or step-like reduction of effective refractive index and internal scattering of the nanowires. Not only Si, nanowire arrays of other materials are widely used in solar cells, even including semiconductors with wide bandgaps. ZnO, a wide-bandgap semiconductor (E_g = 3.37 eV) for instance, is also used for light trapping [28, 29].

It is always efficient to screen out nano configurations that trap light efficiently, from a large variety of nanotechnology-based methods. The simulation results are extremely instructive to the device design for optimized light management. Additionally, although the above calculation focuses on nanowire arrays for particular materials, the results obtained are also of interest to photovoltaic devices based on other materials.

3.2.6 Enhancement of Light Absorption by Periodic Arrays of Nanoholes

Nanoholes are "anti-nanowires." While nanowires offer good light trapping effect, nanoholes may present the similar function, though preparation techniques are different due to their mutual "anti-geometry." Han and Chen studied a theoretical light trapping effect on periodic nanohole arrays and made comparison to previous nanowire results [30]. Figure 3.12a presents the optical absorptance vs. filling fraction at the wavelength of 670 nm for a nanowire and nanohole arrays with the same structure parameters. As clearly demonstrated, the absorption of the nanohole array is stronger than that of the NW array in the whole range of filling fraction studied. The calculated absorptance as a function of light wavelength for the NW and nanohole arrays is depicted in Fig. 3.12b for the thickness (wire and hole lengths) d = 2.33 μm and d = 1.193 mm. The better overall absorption of the nanohole array is clearly displayed.

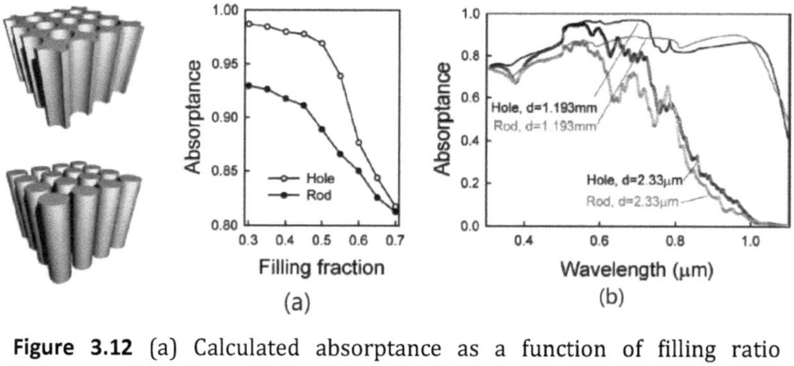

Figure 3.12 (a) Calculated absorptance as a function of filling ratio for nanorod and nanohole arrays at the wavelength of 670 nm (the nanohole and nanowire arrays are shown schematically on the right); (b) calculated absorptance spectra for the nanohole and the nanorod arrays having the thickness d = 2.33 μm and d = 1.193 mm. Both arrays have the period a = 500 nm and the material is crystal Si with a filling fraction 50%. Reproduced with permission from [30].

From their simulation, Han and Chen demonstrated that with decreasing filling fraction, the less denser optical density results in increasing light absorption at the wavelength of 670 nm for both the nanowire and nanohole arrays. Still, the nanohole array are comparatively better in optical absorption property than its nanowire counterpart in the studied filling fraction range (0.4–0.6), due to mainly the good waveguide property of the nanoholes.

3.2.7 Light Trapping Using Step-Like Nanocone Array

As early as 1879, Lord Rayleigh proposed the idea of suppressing light reflection using a transparent thin film with gradually changed refractive index. Such a gradual transition in the refractive index eliminates the interface between the thin film and the substrate, i.e., there is essentially no distinct optical interface. Lord Rayleigh modeled the continuously changed refractive index system as the stacking of a sequence of thin films with small refractive index difference between neighboring films for the convenience of calculation. In fact a nanocone array, instead of nanowire, has a continuously changing average dielectric constant distribution from the cone tips to cone bottoms, if the wavelength of light is much larger than the characteristic size of nanostructures according to the change of average dense of the nanocone array. In this case, the nanocone array is viewed as a continuum media, the continuous distribution of dielectric constant being determined by the shape and the size of the nanocones. Jeong and co-workers demonstrated that nanocone arrays, with the radial size of the nanocone changing gradually, correspond to a smooth transition of effective refractive index and, as a result, optical reflectance is significantly reduced [31]. The nanocone array is particularly suitable for thin film solar cells where the traditional surface texture strategy cannot be adopted [32].

Figure 3.13 presents SEM image of PEDOT (poly(3,4-ethylenedioxythiophene)) nanocone array, together with the reflective spectra for planar PEDOT thin film, ZnO on planar PEDOT thin film, PEDOT nanocone array, and ZnO-coated PEDOT nanocone array, showing that the optical reflectivity is significantly suppressed on the nanocone surface [33].

An efficiency suppression of light reflection can also be completed by using multi-diameter nanopillars, which can be grown with either a top-down or a bottom-up process.

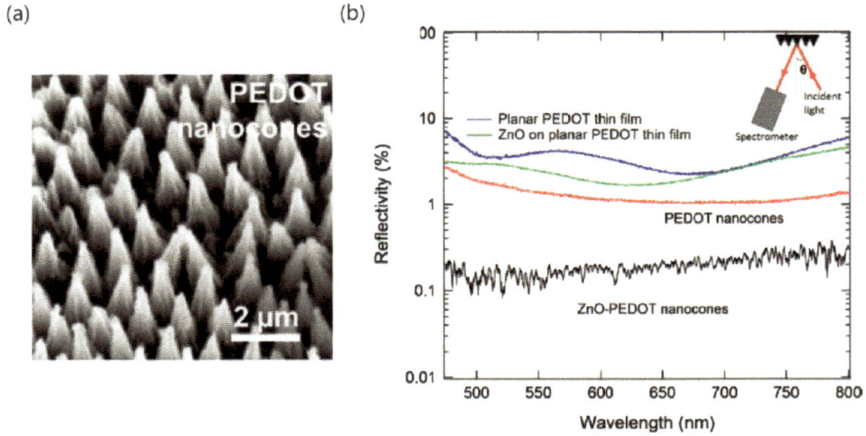

Figure 3.13 (a) Schematic illustration and SEM image of Si nanocone arrays; (b) optical reflective spectra from 450 to 800 nm, measured on planar PEDOT thin film (black curve), ZnO-covered planar PEDOT thin film (green), PEDOT nanocone array (red), and ZnO-coated PEDOT nanocone array (blue). Reproduced with permission from [33, 34].

Fan and co-workers used Ge to construct the 1D NW arrays of multi-diameter nanopillars for the suppression of reflection to a broadband solar spectrum [34]. In Fig. 3.14a–d, 1-diameter, 2-diameter, 3-diameter, and continuously-varying-diameter (nanocone) nanopillar arrays are illustrated, respectively. The simulation indicates that light absorption of the nanopillar array relies on both the combination of material properties and the geometries, such as material filling ratio. The light absorption also depends on the wavelength and incident angle of the incident light. At an optimized design, a multi-diameter nanopillar array could achieve efficient reflection suppression on a broad wavelength range, similar to a nanocone array. As shown in Fig. 3.15, when the segment number of multi-diameter Ge nanopillar arrays is equal to 7, calculated broadband-integrated absorption of the nanopillar array approached that of the nanocone array (continuously-varying-diameter nanopillar array). Fan and co-workers also demonstrated suppression of light reflection based on dual-diameter Ge NWs [35]. As a result, light absorption is significantly enhanced due to the gradient effect that is modeled by NWs with thinner parts on top and thicker parts at the bottom. The multi-diameter nanopillars can be grown with either a top-down or a bottom-up approach.

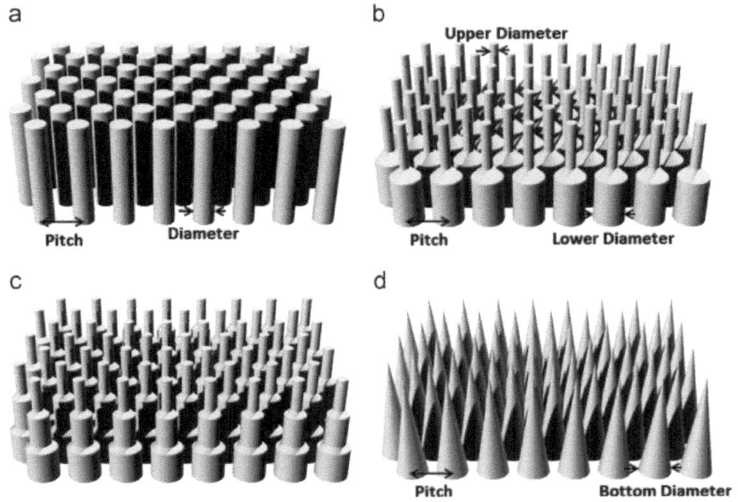

Figure 3.14 Schematics of multi-diameter nanopillar arrays: single-diameter nanopillar array (a), dual-diameter nanopillar array (b), multi-diameter nanopillar array (c), and continuous-diameter nanopillar (nanocone) array (d). The parameters of pitch, diameter, lower diameter, and bottom diameters are labeled. Reproduced with permission from [34].

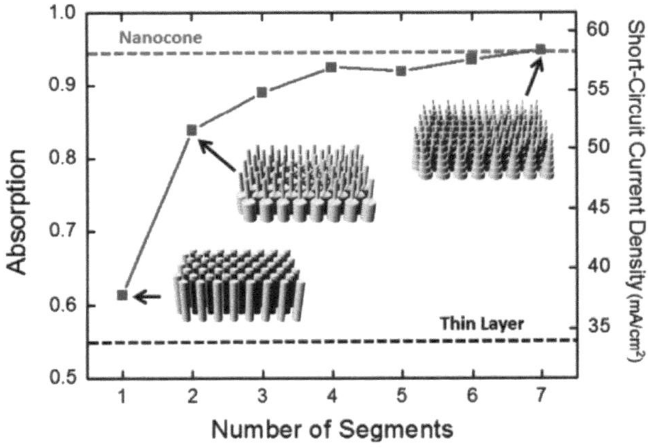

Figure 3.15 Calculated broad band-integrated absorption of multi-diameter Ge nanopillar arrays, as a function of the segment number (the segment number = 1, 2, 3, and ∞, for single-diameter, dual-diameter, and three-diameter nanopillar arrays, and continuously-varying-diameter, respectively). Red dashed line is the broadband-integrated absorption of nanocone array and black dashed line that of 2000 nm thick Ge thin layer. The pitch is 1000 nm used in the calculation. Reproduced with permission from [34].

A so-called black Si surface traps light of a wide wavelength band based on its macroscopically graded change of dielectric constant from air to Si. As an anti-reflection coating, black silicon shows excellent light management property by offering nearly 100% light trapping. Its electrical properties, however, are often unsatisfactory due to the enlarged surface area and poor interface with Si which gives rise to increased carrier recombination. Figure 3.16a presents the SEM image of black Si that shows morphology quite similar to that of the nanocone array shown in Fig. 3.13 [36]. Black Si has the advantages of simple preparation, high efficiency, and low cost, compared to other technologies of nanopatterning. On a black Si surface, very low reflectance of 1% can be achieved in the visible range, as shown in Fig. 3.16b. When a 20 nm Al_2O_3 anti-reflection thin layer is coated on the black Si surface, the reflectance is further reduced to less than 1% in the spectral range from 300 to 1000 nm (see also the inset in Fig. 3.16b. The black Si surface offers excellent light trapping function, as compared to the reference surface of random pyramid coated with 90 nm-Al_2O_3 layer.

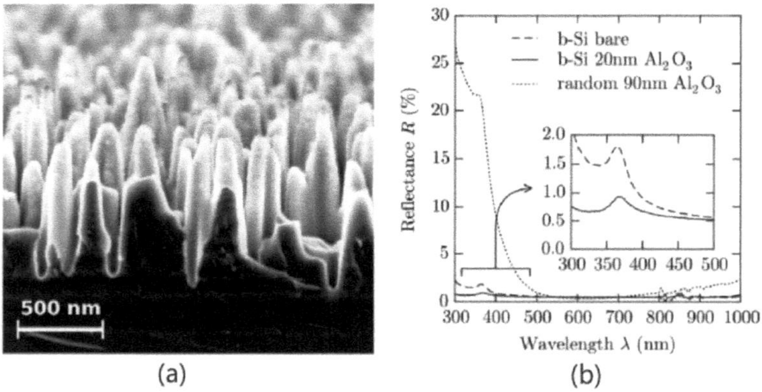

Figure 3.16 (a) Cross-sectional SEM image of the black Si surface. The average height and bottom diameter of the Si nanopillars are roughly 800 and 200 nm. The Si nanopillars are coated with the 20 nm Al_2O_3 layer shown as the brighter layer. (b) Reflectance spectra measured on the bare black Si (the dashed line), 20 nm-Al_2O_3-coated black Si (the black solid line), and 90 nm-Al_2O_3-coated random pyramids (Si) coated (the dotted line) as a reference. The inset shows the curves of the bare black Si and 20 nm Al_2O_3 coated black Si in the zoomed range (300–500 nm). Reproduced with permission from [36].

3.3 Light Trapping Using Plasmonic Technique

3.3.1 Localized Surface Plasmons

There is high density of free electrons in metal. In certain condition such as external excitations (light excitation, for instance), the high-density electrons oscillate collectively. The quantized oscillation is called plasmons. In nanomaterials, electrons are confined. The properties of plasmons, including frequencies, magnitudes, and energy distributions, depend on sizes, shapes, material properties, and boundary conditions. Thus, nano-sized metal particles offer room to manipulate plasmon properties through their sizes and shapes. There are two kinds of plasmonic modes, namely the localized modes and the propagating modes. The localized mode can be confined in a particular region extending into a dielectric layer, while the propagating plasmon modes are also confined but located at the metal-dielectric interface where the negative part of the real component of the dielectric constant of the metal is larger than the corresponding value of the dielectric layer. In order to increase light absorption particularly for thin-film SCs, light trapping could be enhanced through incorporation of metal nanoparticles onto SC surfaces via excitation of surface plasmon resonance [37–40]. Due to the enhancement in solar spectrum absorption in Si solar cells, photocurrent response in Si *p–n* junction increases over the wavelength ranges corresponding to the nanoparticle plasmon resonance wavelengths [41, 42]. The light trapping effect due to pasmonic resonance of nano-sized metal particles was also applied on a-Si:H SCs, which is more important due to the incomplete absorption in the thin film [43, 44]. In most cases, noble metal particles such as Au, Ag, and Cu are used for absorption enhancement with their resonance wavelength close to the bandgap edge of semiconductor active layer. The use of nanoparticles of noble metals, Ag and Au, for instance, leads to the enhancement of device performance.

For the use of plasmonic resonance for light trapping, metal particles are usually incorporated on top of the device surfaces. As a result, some short wavelength light at the resonant frequency is scattered away or absorbed by metal. Another important thing to worry about is that the incorporation of metal particles does not affect the surface passivation layer that needs to remain the same to avoid surface recombination losses. Fortunately, the incorporation of metal particles can be introduced individually and subsequently (after the completeness of device fabrication processes).

Consider a simple system in which nano metal particles sit on a semiconductor. When the system is under the irradiation of light, light might be scattered and absorbed by metal particles. Electrons are driven by light, generating plasmonic oscillation in nanoparticles, as shown in Fig. 3.17. The semiconductor substrate with larger dielectric constant than air could then trap most light scattered by the metal particles, due to dense optical modes in the substrate. Proper configuration of the substrate can be designed to optimize the light trapping effect by introducing waveguide and periodic back for the substrate. Under optimized design, a dominant amount of light can be trapped in desired regions.

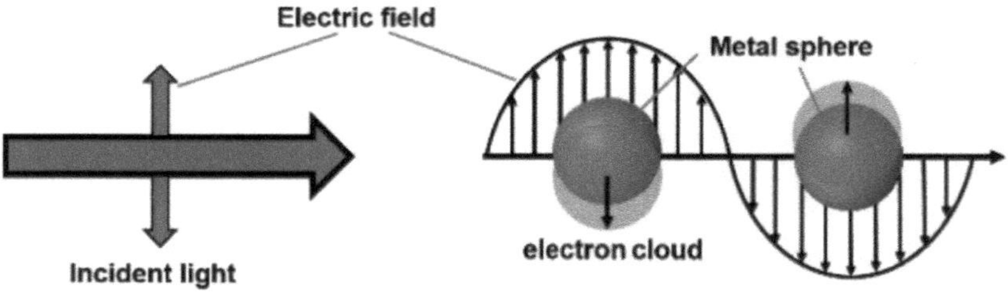

Figure 3.17 Plasmon induced by incident light.

Thus, light trapping through plasmonic effect is based on the above-mentioned three factors, light scattering, localized surface plasmons in metal particles, and in-plane surface plasmons at metal/semiconductor interface. The resonant behavior of those modes relies on material and structural paramenters, including geometry sizes and structures. When the wavelength of incident light is much larger than the size of metal nanoparticles, plasmonic scattering cross section and absorption cross section can be expressed by [45].

$$C_{sc} = \frac{k^4}{6\pi}|\alpha|^2 = \frac{8\pi}{3}k^4V^2\left|\frac{\varepsilon - \varepsilon_m}{\varepsilon + 2\varepsilon_m}\right|^2 \tag{3.9a}$$

and

$$C_{ab} = k\,\mathrm{Im}[\alpha] = 4\pi kV\,\mathrm{Im}\left[\frac{\varepsilon - \varepsilon_m}{\varepsilon + 2\varepsilon_m}\right], \tag{3.9b}$$

respectively, at dipole approximation. Here $k = 2\pi/\lambda$ is the wavevector of incident light, ε the dielectric constant of metal particles, ε_m the dielectric function of the medium in which metal particles are embedded, V the volume, and α the polarizability of metal particles that takes the form of

$$\alpha = 3a^3\left(\frac{\varepsilon - \varepsilon_m}{\varepsilon + 2\varepsilon_m}\right). \tag{3.10}$$

Clearly, polarizability become very large at $\varepsilon = -2\varepsilon_m$, a case known as surface plasmon resonance at which the scattering cross section could be much larger than the geometrical size of the nanoparticles. Thus, nanoparticles act as photon antenna, leading to the strong absorption of light. Note that the dipole approximation is good for small size particles. For large particle sizes, higher-order excitation modes, such as quadruple and octupole modes, should be included. Another effect for particle sizes is the dynamic depolarization, which needs to be taken into account since the conduction band electrons driven by the electric field of light no longer oscillate in phase in the case of large size particles. The depolarized field generated by surrounding medium decreases, which leads to a reduction of the restoring force and the red shift of the resonance wavelength of the metal particles. Thus, although metal particles with large geometrical size have large scattering cross section, they tend to broaden the plasmon resonance spectrum. In addition, the radiation damping, caused by the radiation of electrically polarized metal particles, might also play a non-negligible part [46]. The radiation damping acts to broaden the surface plasma resonance peak, which is helpful for achieving effective light trapping in the broad solar spectrum range. Increasing the size of the nanoparticle leads to increase in scattering cross section. However, the ration of cross-section/geometrical-area is reduced.

Assuming good approximation of Drude model for damped free electrons in metal, the dielectric response of the metal particle exited by incident light of frequency ω can be expressed by

$$\varepsilon = 1 - \frac{\omega_p^2}{\omega^2 + i\Gamma\omega} \tag{3.11}$$

for ignorable interband absorption. Here Γ is the damping broadening parameter, ω_p the bulk plasma frequency and $\omega_p = Ne^2/m\omega_0$ for metals, where e is electron charge, N the density of free electrons, m the effective mass of the electrons, and ε_0 the dielectric constant in vacuum. Combining (3.10) and (3.11), we obtain the polarizability for metal particles in free space with the form of

$$\alpha = 3V \frac{\omega_p^2}{\omega_p^2 - 3\omega^2 - 3i\Gamma\omega}. \tag{3.12}$$

This is the equation for the calculation of metal particle polarizability and the surface plasmon resonance occurs at the frequency of $\omega_{sp} = \sqrt{3}\omega_p$, at which light–electron coupling becomes very strong.

3.3.2 Light Trapping Effect Using Plasmons

Plasmonic effect offers routes for the enhancement of performance for various opto-electronic devices through light trapping, which can be classified into three geometries as shown in Fig. 3.18, based on different light trapping mechanisms [39]. In Fig. 3.18a, light is trapped through scattering by plasmons in metal nanoparticles in selective direction (right panel), where light reflection in backward direction is effectively suppressed. Another route for light trapping is through local field enhancement, where light is concentrated and localized in the vicinities of metal nanoparticles, as shown in Fig. 3.18b. In this case, the metal particles act as photon antennas. In third geometry for plasmonic light trapping, light is coupled and converted into surface plasmon polaritons that are quasiparticles resulting from strong coupling of light with an electric or magnetic dipole excitation. The surface plasmon polaritons travel along the interface (see Fig. 3.18c).

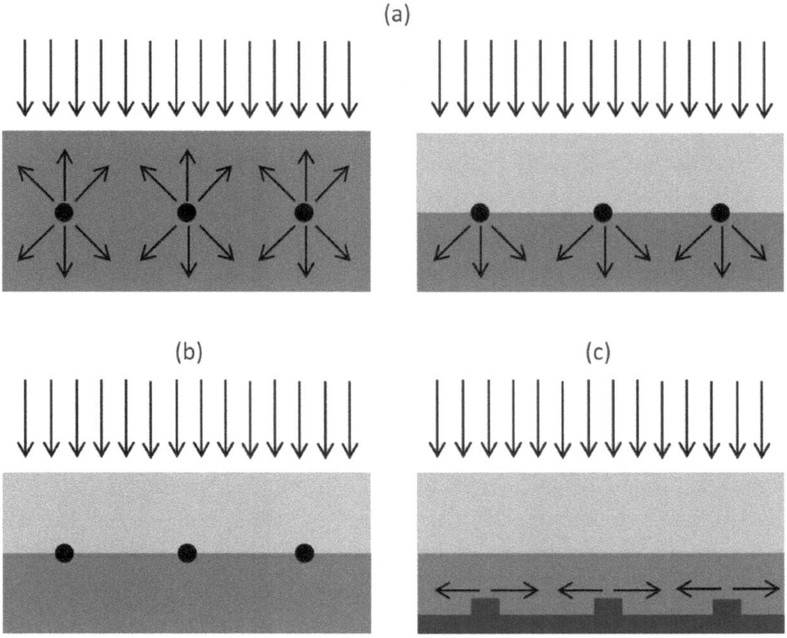

Figure 3.18 Processes of plasmonic light trapping. (a) metal nanoparticles imbedded in homogenous medium shown symmetric forward and backward scattering (left panel) or enhanced forward scattering by metal nanoparticles sandwiched by media of different dielectric constants (right panel); (b) excitation of localized surface plasmons near the particle surfaces; and (c) in-plane surface plasmon modes in the vicinity of corrugated metal surface.

For nano metal particles imbedded in homogenous matrix, they scatter light nearly symmetrically with respect to the forward and reverse directions, as shown in left panel of Fig. 3.18a. If the particles are sandwiched between two media with different dielectric constants, however, a forward scattering anisotropy can be created at the expense of reverse scattering, if the geometric and material parameters of the dielectric media and nanoparticles are carefully chosen (see right panel of Fig. 3.18a). Forward scattered light has an angular distribution in medium and for a scattered light at an angle larger than the critical angle (16° for air/Si interface), it is trapped in medium due to total reflection at the interface. Note that the scattering is due to the surface plasmon of nano-sized metal particles. Stuart and Hall pioneered discovery of enhance light coupling into semiconductor by plasmon-induced scattering. In their work, they found a 20-fold increase of infrared current on their detector by the use of dense metal nanoparticles [47, 48].

The trapping effect can be chosen and tuned by geometry and selection of proper materials. For instance, light trapping is extremely sensitive to the geometrical shape of the metal nanoparticles. Catchpole and Polman studied the dependence of the light trapping effect on metal nanoparticle sizes and shapes, found that the small particles could couple 96% of light into a Si substrate when the effective dipole moments are located near semiconductor, as show in Fig. 3.19 [37, 42]. However, very small metal particles could bring with significant Ohmic losses that should be avoided. Thus, the resonance frequency and the resonance spectrum can be tuned by the choice of materials for metal particles and dielectric layer, size/shape of metal nanoparticles, dielectric medium around the particle, location of the metal nanoparticles, thickness of the dielectric layer, as well as back surface structural design. Note that those material and structural parameters would be selected depending on the kinds of solar cells. Another very important thing is the surface passivation for the reduction of surface recombination of carriers. Very often, there could be a tradeoff between SP coupling and surface passivation.

Light scattering by metal nanoparticles due to excitation of plasmons is a promising technique for trapping light inside the active volume of a solar cell. Plasmonic nanoparticles have been shown to give substantial photocurrent enhancements on thick substrates and on thin waveguides.

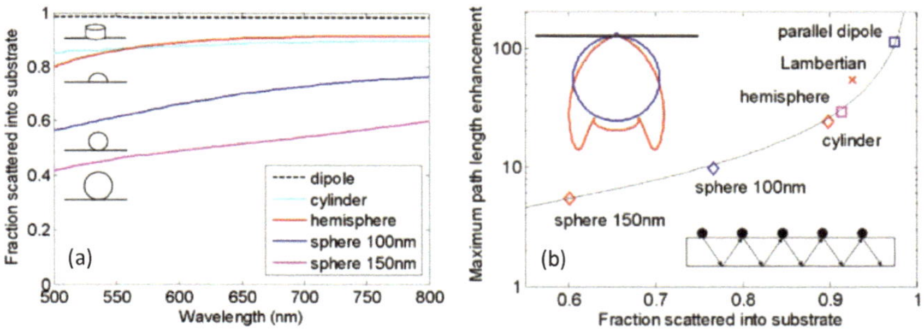

Figure 3.19 Fraction of light scattered into Si substrate for different sizes and shapes of Ag on Si, normalized by total scattering power (a) and maximum path-length enhancement for the geometries in (a) at the wavelength of 800 nm (b). Reproduced with permission from [42].

3.3.3 Nanoparticle/Dielectric/Metal Structures for Light Trapping

Figure 3.19 shows that for Ag diameter above 150 nm nearly all light is scattered, while 50% light is absorbed for Ag diameter ~50 nm. The incident light can also be trapped by exciting localized surface plasmon modes on the surfaces of the nanoparticles so that energy is concentrated and localized in the structure. Thus, the metal nanoparticles act as photon antenna. This works well specially for small size particles due to their low albedo (ratio of reflected radiation from the surface to incident radiation). The concentrated and localized near-field light could excite electron–hole pairs with high efficiency to make contribution for the photovoltaic process. Very importantly, the controllable distribution of the concentrated light is a gift that is extremely useful for the enhancement of device performance. As a fact, the light field or, more specifically, a polariton field can be concentrated in the active region (i.e., *p–n* junction region) in a device to enhance the opto-electronic conversion efficiency. In thin-film and nano solar cells, this is particularly beneficial since there exist high density of defects and large area of surfaces and interfaces that cause small carrier diffusion lengths.

Zhang and co-workers prepared a structure of nanoparticle/dielectric/metal, a close analogy of a typical nano solar cell architecture of nanoparticle/dielectric/semiconductor [49]. The

ZnO dielectric layer was deposited on the Ag-coated Si substrate by atomic layer deposition (ALD). The thickness of the Ag layer was around tens of nanometers. A 3 nm-thick Au layer was then thermally deposited on the ZnO dielectric followed by thermal annealing that led to the formation of Au particles randomly distributed with a wide size distribution. The formed Au particles were present on the ZnO oxide layer with a flat contact interface, with a total surface coverage of 22%. Figure 3.20 presents calculated distribution of the plasmon modes in Au-nanoparticle/ZnO-layer/Ag-layer structure at λ = 487 nm (a) and λ = 677 nm (b) which are two maxima on absorption curve. Apparently, the mode intensities are distributed in the vicinities of interfaces either between Au particles and the dielectric ZnO layer or between the ZnO layer and the Ag metal. The mode distribution depends on the material parameters of the dielectric and metals and the structure parameters such as the dielectric layer thickness. Thus, the nano-metal/dielectric/metal structure intensifies the coupling between light and electron system and the coupling is tunable.

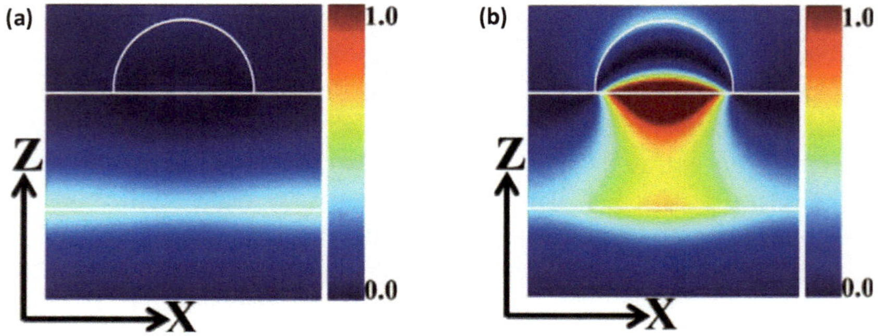

Figure 3.20 Calculated localized and concentrated surface plasmon modes in metal-nanoparticle/ZnO layer/metal structure at two absorption maxima of λ = 487 nm (a) and λ = 677 nm (b). Reproduced with permission from [49].

Measured reflectance spectra at nearly normal incidence are shown in Fig. 3.21, where the black solid line is the absorption spectrum measured on ZnO-layer/Ag-layer without Au nanoparticles and Au-nanoparticles/ZnO-layer/Ag-layer. The incorporation of Au nanoparticles gives rise to significant reduction of optical reflectance in a very broad spectral range. The study aims at suppressing solar reflectance in the whole visible and near-infrared range for raising power conversion efficiency, which has been proven to be effective by the overall much weak reflection compared to the system without Au-particle coating. The lowest part of the reflectance is even below 1%.

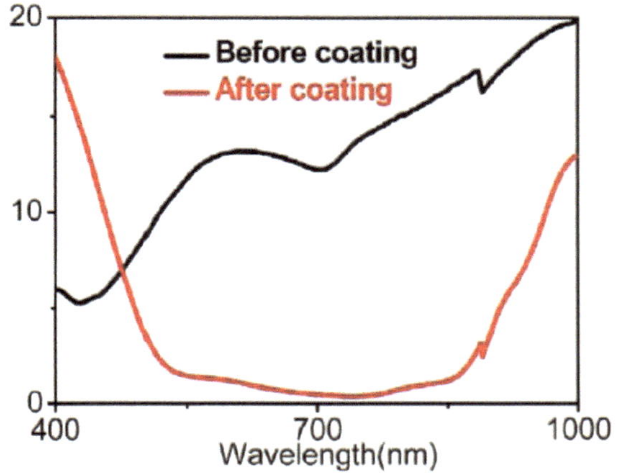

Figure 3.21 The reflectance spectra measured at nearly normal incidence on ZnO-layer/Ag-layer structures without (black curve) and with (red curve) coated Au nanoparticles. Reproduced with permission from [49].

3.3.4 Plasmonic Light Trapping Used for Solar Cells

Nanomaterial arrays can be prepared by many top-down and bottom-up methods. A well-adopted way is masked thermal deposition through a template. Figure 3.22 presents the SEM images of Ag nanomaterial arrays deposited through anodic aluminum oxide templates [50]. The anodic aluminum oxide templates were fabricated following a two-step anodization process and the density of the nanopores is controlled by anodization conditions. In preparing the template with an array density of 3.3×10^9 cm^{-2} (1.8×10^9 cm^{-2}), the anodization was carried out at 80 V (120 V) in a mixture of 0.3 mol l^{-1} oxalic acid and 0.3 mol l^{-1} malonic acid solutions (see Ref. [50] for details). The heights of the nanomaterials were controlled by the evaporating time.

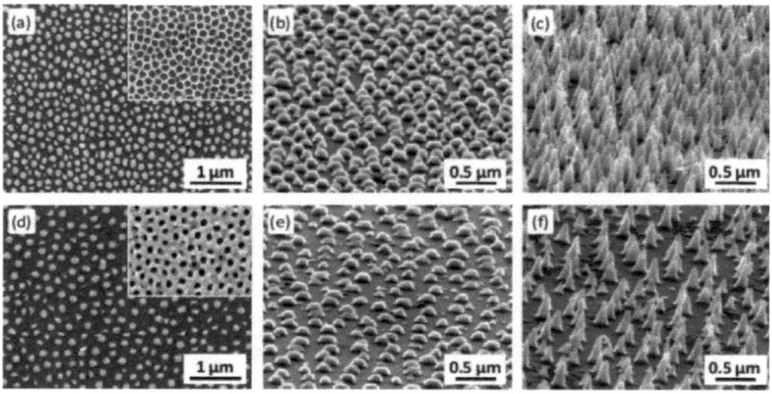

Figure 3.22 SEM images of Ag nanparticles of 110 nm diameter, but different average heights: (a) and (d) very low, (b) and (e) 55 nm, (c) and (f) 220 nm. All the images were taken at the angle of 75°. The insets of (a) and (d) show the images of anodic aluminum oxide templates fabricated with a anodization voltages at 80 and 120 V, respectively. The densities of the nanomaterials are (a–c) 3.3×10^9 and (d–f) 1.8×10^9 cm^{-2}. Reproduced with permission from [50].

The Ag nanoparticles were coated on top of the studied GaAs solar cells, in order to utilize the plasmonic effect to trap light. Table 3.1 lists measured photovoltaic parameters, including short-circuit current J_{sc}, open-circuit voltage V_{oc}, filling factor *FF*, and power conversion efficiency η at two Ag nanoparticle densities of 3.3×10^9 and 1.8×10^9 cm^{-2}, and two different nanoparticle heights of 80 and 120 nm, together with those of a reference GaAs cell without Ag nanoparticle decoration. Apparently, the performance of the photovoltaic cells is improved by using the particles. Among all the photovoltaic parameters, the increase of the filling factor is distinct. The study shows that the improvement is mainly due to the enhancement of photoresponse in the long-wavelength region of the solar radiation spectrum. This is clearly demonstrated in Fig. 3.23, where the normalized external quantum efficiencies (EQE) of the Ag-nanoparticle-decorated cells are above 1 from 600 to 900 nm. The 9% to 18% increase in *FF* (with respect to the reference cell) makes the major contribution to the overall enhancement of the cell performance. Part of the increase in *FF* is caused by reduction in the surface sheet resistance due to the conductive

Ag nanoparticle. The #3 cell with Ag-nanoparticle parameters of 220 in height and 3.3 × 10^9 in density has the overall superior performance in J_{sc}, V_{oc}, FF, and η, showing an 25% increase in conversion efficiency. The enhancement of the EQE is due to the plasmonic effect, which in turn leads to the increasing absorption of the incident light. The nanoparticle coverage on the solar cell surface is thus an effective way to enhance the performance of the solar cells through plasmonic effect. Note that the GaAs cell, with a conversion efficiency of only 4.7%, is not optimized. Otherwise, the gain due to the use of Ag-nanoparticle array should be more significant.

Table 3.1 Structural parameters of Au nanoparticles and corresponding photovoltaic parameters of the GaAs solar cells, measured under one-sun illumination

Structure parameters of the Ag nanoparticles							
Ag array	Diameter (nm)	Height (nm)	Density (cm^{-2})	J_{sc} (mA cm^{-2})	V_{oc} (V)	FF	η (%)
#1	110	55	3.3 × 10^9	10.9	76	0.62	5.1
#2	110	55	1.8 × 10^9	11.4	75	0.60	5.1
#3	110	220	3.3 × 10^9	11.9	76	0.65	5.9
#4	110	220	1.8 × 10^9	10.8	77	0.64	5.3
#5				11.0	73	0.55	4.7

Note: #5 is the reference cell without Au nanoparticle decoration.

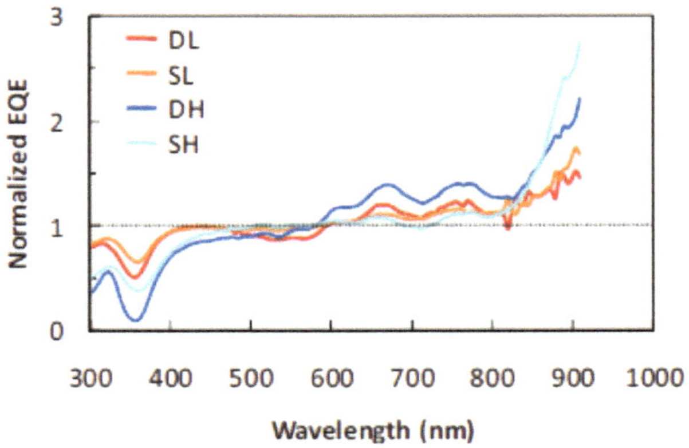

Figure 3.23 Normalized external quantum efficiency (EQE) by the spectral response of the reference cell. Reproduced with permission from [50].

Figure 3.24a shows an SEM image of randomly distributed Ag nanoparticle array. The array was fabricated via a two-step process in which a 14 nm-thick Ag film was thermally deposited on a thermally oxidized Si wafer with 10 nm-thick SiO_2 followed by annealing at 200°C for 60 min in a N_2 ambient [39]. Agglomeration of the Ag nanoparticles is induced by the effect of surface tension. The formed Ag nanoparticle array has a 135 nm average particle diameter and a 26% surface coverage. Figure 3.24b present the hexagonal Ag nanoparticle array prepared by substrate-conformal imprint lithography [39]. Even with a geometrical areal coverage of about 10%, a calculation shows that the scattering cross section is nearly 10 times the geometrical area of the metal particles.

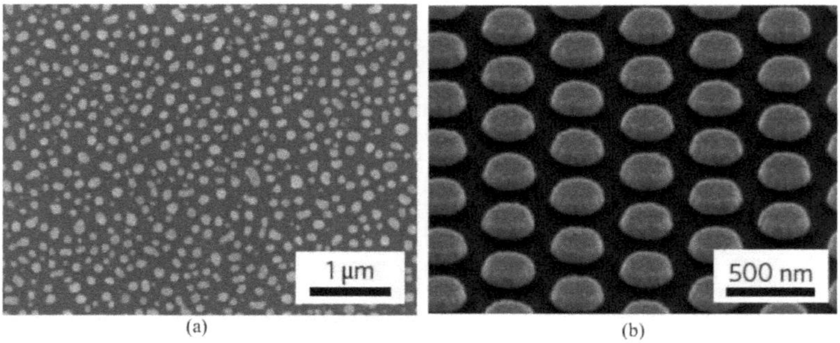

Figure 3.24 (a) Random Ag nanoparticle array on thermally oxidized Si substrate (the thickness of the formed SiO_2 layer is around 10 nm). The average diameter of the Ag nanoparticle is 135 nm and the surface coverage is 26%; (b) hexagonal Ag nanoparticle array deposited using substrate-conformal imprint lithography using the substrate conformal imprint lithography technique. The particle diameter is 300 nm. Reproduced with permission from [39]. Reproduced with permission from [39].

Plasmonics are useful for both thin-film solar cells and wafer-based solar cells as an alternative to texturing. Aiming at enhancing scattering and coupling efficiency of the plasmonic layer, the design of plasmonic solar cells is based on optimization of various factors like size, shape, location, passivation layer thickness, and dielectric medium, etc. In order to raise the efficiency of light coupling, it is very important to avoid possible absorption in the metal particles that is wavelength dependent and other losses. Mitigating absorption in the metal particles and on metal particle surfaces is always something needed to be considered carefully.

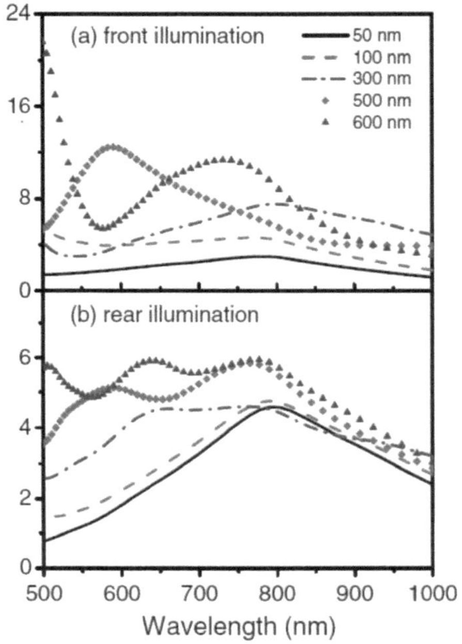

Figure 3.25 Calculated scattering cross-sections normalized to the geometrical area of the nanoparticles, with several nanoparticle heights, as a function of wavelength. (a) Ag nanoparticles on the front (illuminated) side and (b) Ag nanoparticles on the rear side of the substrate. The Ag nanoparticles have a cylindrical shape with a 200 nm base diameter and the heights of 50, 100, 300, 500, and 600 nm, as marked. Reproduced with permission from [40].

The light trapping efficiency is closely dependent on the shapes of metal particles. As an example, only the variation of the height of nanoparticles is considered. Figure 3.25 shows calculated scattering cross-sections with several nanoparticle heights, versus wavelength. Note that the presented cross-section has been normalized to the area of the nanoparticles. The taller nanoparticles tend to enhance the scattering cross-section via plasmonic standing waves resonant excitation and they exhibit higher coupling efficiency if they are placed on the illuminated surface of the SCs than on the rear determined by the forward scattering nature. The coupling effect depends on the phase shift of the surface plasmons along the nanoparticle (nanowires) while they undergo reflection from the Ag/Si (in fact Ag/SiO_2 due to the native oxide layer on Si) or the Ag/air interface. Mokkapati and co-workers demonstrated that at the incident wavelength of 700 and with a cylindrical nanoparticle of height 500 nm and diameter 200 nm, the effective travel path (in a single pass through the solar cell) might increase by a factor of 16 with respect to the thickness of the SC for optimized device structure [40].

3.3.5 Light Trapping Using High-Index Nanostructures

The incoming light from free space onto a planar high-index materials tend to be strongly reflected. However, nanostructures of high index materials offer possibility to trap light by enabling the formation of a variety of resonant modes. Thus, nanostructures of high-index materials, including nanoparticles, nanowires, and other nano forms of materials, offer great room to manipulate light with wavelength at subwavelength scales through local optical scatterings and resonances. As a result, light absorption is enhanced.

Interaction between light and matters is enhanced for materials with high index of refraction in the form of slowing down the travel of light and offering strong confinement to light. Light undergoes reflection and diffraction on the boundaries of two different materials and, when travelling from optically dense material to optically dilute material might be totally reflected at certain condition, which gives rise to the phenomena such as waveguide. When the materials have sizes comparable to the wavelength of light, i.e., in the range of a few hundred nanometers, the interaction causes strong optical resonances. The resonances are made extremely significant especially when the nanomaterials are properly sized and shaped, which offers unique ways to manipulate light travelling and trapping through nano-sizing and nano-shaping. The resonance is due to the reflection, transmission, and diffraction on the boundaries that leads to instructive and destructive interferences among the light waves, while the optical absorption plays minor role since the materials are usually non-absorptive. The resonant optical field is quantized and can be described with various resonant modes.

As shown in Fig. 3.26 that schematically depicts a simplified nano SC consisting of a thin-film active layer and on-top nanoparticles for light trapping. There are four possible outcomes of light interacting with the system, namely, resonance in the nanoparticles, light raveling in the surface plane, the confined in the thin film (waveguide modes), as well as transmission light [51]. Note that in Fig. 3.26, the reflection has been ignored and the transmission is suppressed due to the use of the back reflective coating. Correspondingly, calculated optical field distributions are presented in Fig. 3.27, as a function of incident angle and wavelength. The optical field distribution shows a number of features corresponding to optical resonance maxima with marked four resonance mechanisms, depending on the incident angle and wavelength, as shown in Fig. 3.27. It is thus possible that for a given incident angle, optical fields of different wavelength be trapped at desired regions (different resonances), which is beneficial for nano SCs to achieve high conversion efficiency.

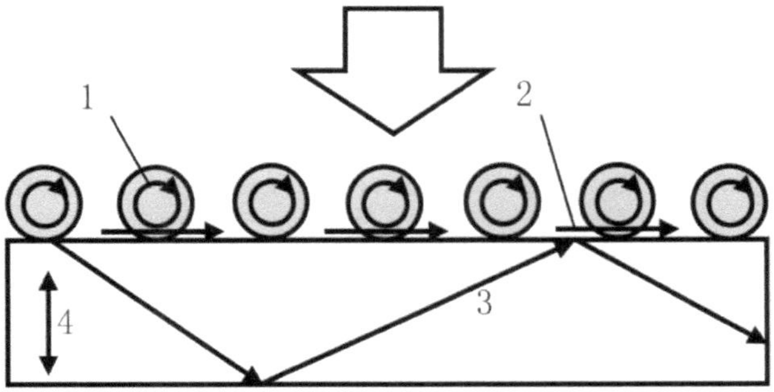

Figure 3.26 Four possible resonance light trapping mechanisms: (1) light trapped in nanowires (the nanowires are pointed in and out of the paper), (2) diffractive resonance in the light trapping layer, (3) light confined in the thin-film waveguide, and (4) standing-wave resonance.

The underlying physics for these interesting features is the resonant coupling of sunlight with wide spectrum range to multiple modes allowed and controlled by the cell structures. Multi-parameter (the wavelength and the incident angle of sunlight, as well as the size, shape, distribution, etc.) choices make the resonant coupling easily tunable, a given resonance corresponding to a given electro-magnetic field distributions. In addition, the field distribution can be easily calculated, which is helpful for the understanding of the types of resonant coupling. Brongersma and co-workers simulated the optical absorption as a function of light incident angle and wavelength, as shown in Fig. 3.27, where the labels 1–4 specify the angle and the wavelength at which the four possible resonances schematically illustrated in Fig. 3.26 can be excited [51].

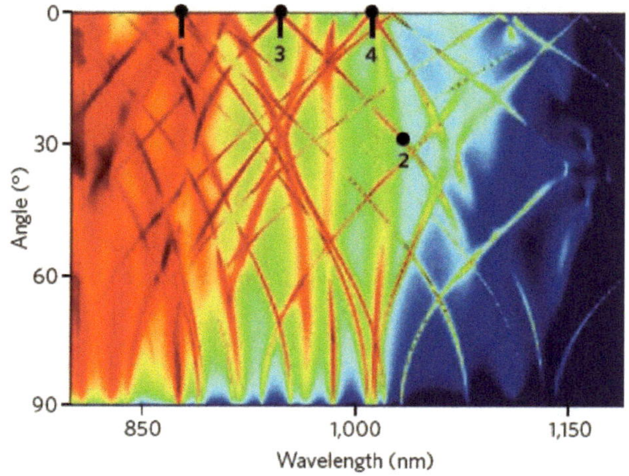

Figure 3.27 Angle vs. wavelength of resonance distributions. Marked points are for the four possible resonance light trapping mechanisms. Reproduced with permission from [51].

One could take a close look at the nano SCs to see optical field distributed. Figure 3.28a–d presents calculated distribution over the nano SCs (Fig. 3.26) for the four resonance mechanisms [51]. For the resonance trapped in the nanowires, Fig. 3.28a shows that with a nanowire grating layer on top of the nano SCs, optical resonance of hexapolar symmetry can be excited, with the optical field modes being confined in the nanowires. Figure 3.28b presents the modes inside the thin film due to the hybridization with the waveguide resonance within the thin film. The diffractive resonance in the light trapping layer is excited by light interaction with the periodic nanowires. The modes are trapped on the top surface of the thin film and also extend into the underneath thin film due to hybridization. The waveguide resonance are the travelling resonance wave in the thin film confined by the front surface and the back reflector of the thin film, characterized by uniformly distributed resonance modes in the thin film shown by Fig. 3.28c. The standing wave resonance is Fabry–Perot-like, caused by interference of the light reflected by the front and the back surface of the thin film. A layered distribution of resonance optical field is clearly demonstrated (Fig. 3.28d). Using the model calculation, it is possible to separate the contribution of the different mechanisms, which is instructive for the optimized effective design of high efficiency nano SCs of proper structure.

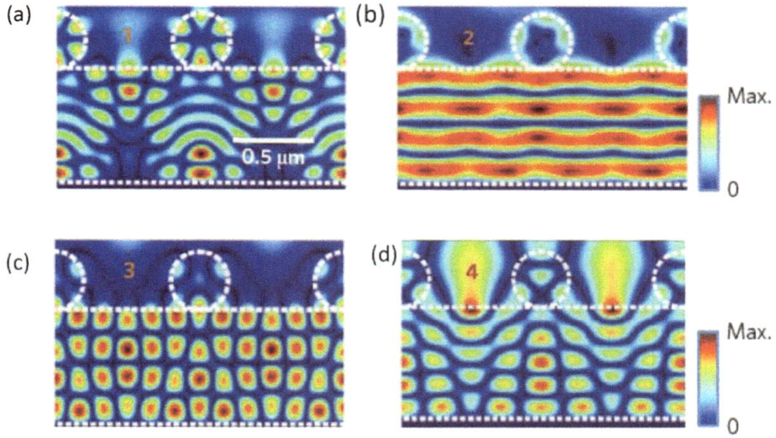

Figure 3.28 Optical field distributions for the four possible resonance light trapping mechanisms. (a) resonance trapped in nanomaterials, (b) diffractive resonance in the light trapping layer, (c) light confined in the thin-film waveguide, and (d)standing-wave resonance. Reproduced with permission from [51].

The nano-sized structure is not limited to the grating form of nanowires. A 2D array of nano-cubes (or quantum dots) and vertically aligned nanowire array are also good for light trapping.

3.4 Photon Trapping Using Photonic Crystals

Photonic crystals are periodically structured artificial media, possessing photonic bandgaps. In the frequency range of the photonic bandgaps propagating light is forbidden in the crystal.

In comparison to homogenous media, photonic crystals changes the propagating path of light and, in the case of resonance, light can be trapped.

In dealing with light trapping using dielectric materials, there are two frameworks in which light trapping effect can be described in distinct ways: geometrical optics and wave optics. In geometrical optics framework, light travel is changed at interfaces depending on the dielectric constants of the materials and the incident angle. In wave optics framework, on the other hand, the light is treated as wave with different wavelength, which is especially powerful in dealing with light propagating in complex dielectric configurations. In conventional c-Si solar cells, the surface texture is a typical example that light trapping is described by geometrical optics. In using photonic crystal for light trapping, wave optics approach has to be adopted.

3.4.1 Photonic Crystals for the Manipulation of Light Propagation

Photonic crystal structure based on well-defined nanostructure arrays are extremely effective to achieve easy manipulation on light reflection and absorption [52, 53]. Photonic crystals are artificial materials whose dielectric constants vary periodically in one, two or three dimensions, as shown in Fig. 3.29. With properly chosen materials and structural parameters (dielectric constants and sizes), photonic crystals could act to enhance light trapping by controlled constructive/destructive interface through light reflection, refraction and diffraction in dielectric interfaces inside photonic crystals. Photonic crystals exhibit photonic bandgaps tunable by varying those parameters. The major feature of photonic crystal is their bandgaps—electromagnetic waves of certain frequency range cannot propagate inside. The propagation of electromagnetic waves inside photonic crystal structures is described by macroscopic Maxwell's equations:

$$\nabla \cdot \mathbf{D} = \rho \tag{3.13}$$

$$\nabla \cdot \mathbf{B} = 0 \tag{3.14}$$

$$\nabla \times \mathbf{E} + \frac{\partial \mathbf{B}}{\partial t} = 0 \qquad (3.15)$$

$$\nabla \times \mathbf{H} - \frac{\partial \boldsymbol{D}}{\partial t} = \mathbf{J}, \qquad (3.16)$$

where **E** and **B** are the macroscopic electric and magnetic field quantities, **D** and **H** are electric displacement and magnetic field intensity, respectively. ρ is charge density and **J** is current density. In free space, ρ and **J** are zero. **E** and **D**, as well as **B** and **H**, are related in linear isotropic materials by

$$\mathbf{D} = \varepsilon_0 \mathbf{E} + \mathbf{P} \qquad (3.17a)$$

$$\mathbf{B} = \mu_0 \mathbf{H} + \mathbf{M}, \qquad (3.17b)$$

where **P** is the polarization field, **M** the magnetization field, ε_0 the dielectric function in vacuum, and μ_0 magnetic permeability in vacuum. In linear isotropic materials, the above two equations can be rewritten as

$$\mathbf{D}(r, \omega) = \varepsilon(r, \omega)\, \mathbf{E}(r, \omega) \qquad (3.18a)$$

$$\mathbf{B}(r, \omega) = \mu(r, \omega)\, \mathbf{H}(r, \omega) \qquad (3.18b)$$

Here $\varepsilon(r, \omega)$ is frequency dependent complex dielectric function and it is also a periodic function of position due to the periodic distribution of dielectric constants in photonic crystals. The dielectric function varies appreciably within the frequency range of solar spectrum. $\mu(r, \omega)$, magnetic permeability, is essentially a constant for non-magnetic materials.

There is no free charge in dielectrics, so that $\nabla \cdot \mathbf{D} = \rho = 0$ and $\mathbf{J} = 0$ since conductivity $\sigma = 0$. We then have

$$\nabla^2 \mathbf{E} - \mu_0 \varepsilon_r \varepsilon_0 \frac{\partial^2 \mathbf{E}}{\partial t^2} = 0 \qquad (3.19a)$$

$$\nabla^2 \mathbf{H} - \mu_0 \varepsilon_r \varepsilon_0 \frac{\partial^2 \mathbf{H}}{\partial t^2} = 0. \qquad (3.19b)$$

Apparently, (3.19a) and (3.19b) are wave equations, which have the solutions of

$$E_x = E_0 \exp[i(\omega t - kz)] = E_0 \exp\left[i\omega\left(t - \frac{z}{v}\right)\right] \tag{3.20a}$$

$$H_y = H_0 \exp[i(\omega t - kz)] = H_0 \exp\left[i\omega\left(t - \frac{z}{v}\right)\right] \tag{3.20b}$$

assuming that the electromagnetic wave propagates along the z direction and electric and magnetic polarizations are along the x and y directions, respectively. Here E_0 and H_0 are the amplitudes of the electric and the magnetic components of the electromagnetic wave (light). Thus, both the electric and the magnetic components are plane waves.

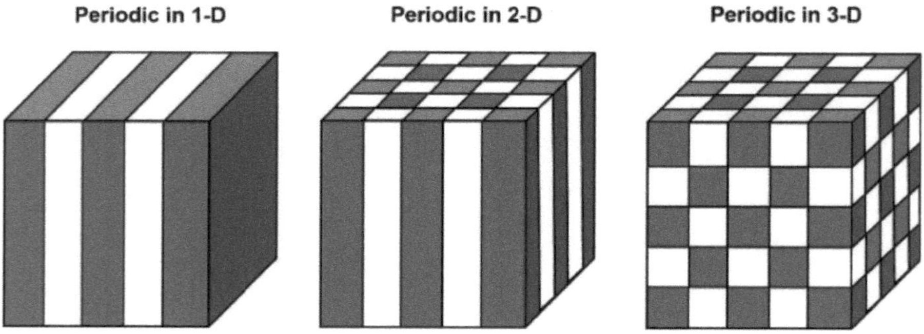

Figure 3.29 Schematic illustration of 1D, 2D, and 3D photonic crystals formed by two materials with different refractive indices.

In a semiconductor (including artificial structures like quantum wells and superlattices), travelling electrons experience a periodically distributed potential due to an atomic lattice and man-made structures. The Bragg scattering to the electron waves gives rise to constructive or destructive interference between the waves, leading to electron wavefunction of Bloch form that is quite different from the free electron wavefunction in vacuum. A complete description of electronic states in a semiconductor is given by the material's energy band structure, i.e., $\boldsymbol{k} \sim E$ dispersion relation in first Brillioun zone for all branches.

The major feature of the energy band structure is the bandgap caused by Bragg scattering of destructive interference. Electrons with energy falling into the bandgap are not allowed to propagate in the semiconductor.

A similar analogy applies for photons or electromagnetic waves, since photons are also waves described by (3.20). In this case, the corresponding potential is created by periodic modulation of the refractive index of materials, as shown in Fig. 3.29. Similar to the semiconductor bandgap, light waves of various wavelengths travelling in a periodic medium, waves are bounced back and forth with their ways depending on wavelengths governed by Bloch's theorem. In semiconductors and various man-made structures such as quantum wells and superlattices, it is electron waves that undergo such back and forth bouncing. In photonic crystals where dielectric constant varies periodically, the light waves is modulated by the periodic dielectric "potential," the energy distributions of the light waves are no longer uniform, characterized by energy accumulation in some regions and energy depletion in other regions. The energy can be accumulated either in high-refractive-index dielectric or in low refractive index dielectric, corresponding to two allowed energy states. The two allowed energy states are separated by the photonic bandgap, where photons with energy within the gap are not allowed to travel—they will be reflected if the light is incident from a medium to the photonic crystal.

An artificial crystal for photons, or a photonic crystal, exhibits a characteristic dispersion relation between the propagation constant and the photon energy (corresponding to frequency) called a photonic band structure. Commonly, photonic crystals are fabricated by using two materials with different refractive index, e.g., silicon and air. While an electronic bandgap can generally be found in any type of semiconductor, formation of a photonic bandgap requires a strong refractive index contrast between the two materials and a rather specific geometry. The ratio of the two refractive indices has to be larger than 2 for a formation of a small bandgap and a ratio greater than 3 is generally necessary for a large bandgap. Furthermore, among various types of the three-dimensional (3D) lattices, only the diamond or diamond-like structures have been found to exhibit a large photonic bandgap in 3D. This difference between electronic and photonic bandgaps is a result of the difference between the scalar wave equation of electrons and the vectorial wave equation of photons.

Photonic crystals are periodically structured artificial ones formed by at least two materials, as shown schematically in Fig. 3.29. It possesses a photonic structure. Pavarini and co-workers calculated the photonic band structure of an fcc lattice composed of close-packed opals in air and the results are shown in Fig. 3.30 [54]. The photonic bands are shown in dimensionless units $\omega a/2\pi c$ is used for vertical axis, where ω is circular frequency, a the lattice constant, and c the speed of light. The photonic band structures are presented along the symmetry lines of the fcc Brillouin zone (Fig. 3.30a) and along symmetry lines on the (111) face (Fig. 3.30b), with the Symmetry points in the Brillouin zone as shown in Fig. 3.30c. One could see clearly a photonic bandgap, for instance, along the $\Gamma - \mathrm{U}$ line in Fig. 3.30a.

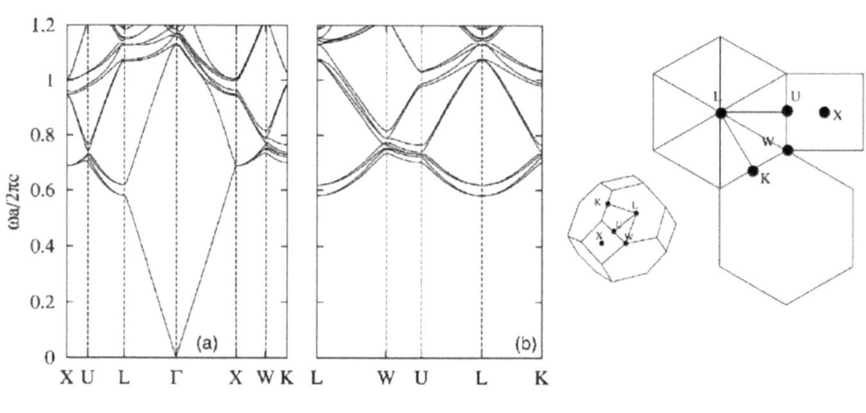

Figure 3.30 (a) calculated photonic band structure for close-packed opals with (ε_1 = 2.35) in air (ε_2 = 1), assuming an fcc lattice and (b) symmetry points in the Brillouin zone. Reproduced with permission from [54].

In the frequency range of the photonic bandgap, light cannot propagate in the photonic crystal. Photonic crystals manipulate how light travels and how photons are absorbed or reflected. Photonic crystals are usually made of dielectric nanostructures that allow light with certain wavelengths to pass through while blocking light with other wavelengths to travel. Photonic crystals enhance solar cell performance by changing light travel directions and coupling light with important wavelengths into devices. The travel of light can be manipulated by photonic crystals that change the travel direction and select the wavelength. Except for the underline physics, the usefulness of the photonic crystals also relies on the scalability and feasibility of techniques to be incorporated with the real structures of solar cells.

3.4.2 One-Dimensional Photonic Crystals

The reflectivity of bare silicon surface is about 35%, i.e., 35% of the solar energy is wasted if no measure for light trapping is adopted. For wafer-based Si SCs, the adopted methods to reduce the light reflection include anti-reflection coating and surface texturing. In anti-reflection scheme shown in Fig. 3.31, the front surface of a SC is coated with anti-reflection thin film. Assuming a perfect dielectric anti-reflection without absorption (extinction coefficient k = 0), the reflection of the three-layer system can be written as

$$R=\frac{(n_0-n_2)^2\cos^2\frac{\delta}{2}+\left(\frac{n_0n_2}{n_1}-n_1\right)^2\sin^2\frac{\delta}{2}}{(n_0+n_2)^2\cos^2\frac{\delta}{2}+\left(\frac{n_0n_2}{n_1}+n_1\right)^2\sin^2\frac{\delta}{2}}, \tag{3.21}$$

with

$$\delta=\frac{4\pi}{\lambda}n_1 d\cos\theta \tag{3.22}$$

where δ is the optical path difference between neighboring reflection light beams, θ the diffraction angle of light beam in the anti-reflection layer, n_0, n_1, n_2 the refractive indices of air, the anti-reflection layer, and Si, respectively, d the thickness of the anti-reflection layer, and λ the wavelength of incident light, as shown in Fig. 3.31. At normal incidence (θ - 0) and a given wavelength λ_0, for n_1d - $\lambda_0/4$ (the optical thickness is quarter of incident wavelength), the light reflection becomes

$$R=\frac{\left(\dfrac{n_0 n_2}{n_1}-n_1\right)^2}{\left(\dfrac{n_0 n_2}{n_1}+n_1\right)^2}. \tag{3.23}$$

Obviously, zero reflection of $R = 0$ can be obtained if $n_1 = \sqrt{n_0 n_2}$. Unfortunately, $R = 0$ occurs only at a particular wavelength, while solar radiation covers a wide spectrum range from ultraviolet to infrared. Obviously, anti-reflection coating of a proper material (n_1) and layer thickness needs to be optimized for lowest surface reflection for the whole solar spectrum range. The use of multi-layer tends is helpful for the improvement of the anti-reflection effect in a wide spectrum. However, cost and technological feasibility are often the issues of limitation.

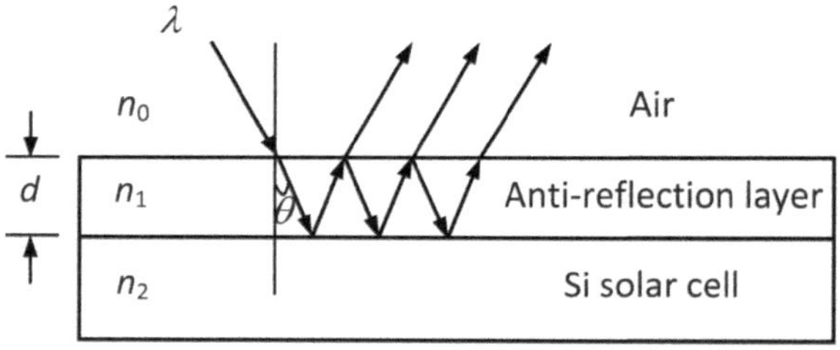

Figure 3.31 Reflection of the incident light.

It is rather difficult to achieve excellent anti-reflection effect on SCs using a layer of anti-reflection coating, since good anti-reflection effect can only be achieved for a single wavelength. In other words, the quarter wavelength condition for efficient anti-reflection effect works only within a very narrow range of wavelengths. On high efficient SCs, multilayer anti-reflection coating is usually adopted. Multilayer anti-reflection coating could have good anti-reflection effect in a wide solar radiation spectrum depending on the dielectric constants of the layers in coating. Anti-reflection coating is extremely important measure especially for thin-film SCs, since, in this case, other measures such as surface texturing cannot be used. Beye and co-workers performed a study on double-layer anti-reflection coating for thin-film Si SCs on glass through numerical simulation and comparing the result with the single layer case, as shown in Figs. 3.32a,b for normal incidence of light [55]. The reflectance curve has a "V" shape for single-layer coating while it has a "W" shape. This indicates that there is only one reflectance minimum of the "V" shape curve, corresponding to only one specific wavelength. The "W" shaped reflection curve has two minima; thus double-layer anti-reflection coating has an anti-reflecting effect in a wider spectrum range than single layer coating. In addition, double-layer coating offers quite a few degrees of freedoms to tune the wavelengths of reflection minima and spectrum width of the low reflection band. In fact, wide low reflection band can be achieved using relatively large dielectric constant difference for the two dielectric layers in the reflection coating.

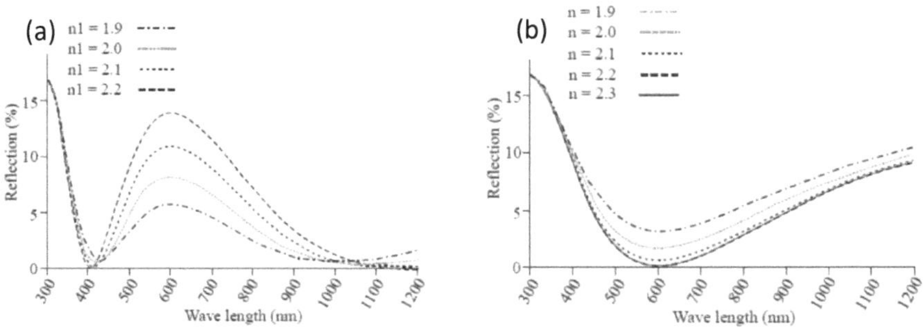

Figure 3.32 (a) Reflectance spectra for SiN_x double layer anti-reflection coating under normal incidence for different dielectric constants for the first layer (the refractive index for the second layer is 2.3). (b) Reflectance spectra for a single SiN_x anti-reflection layer of different refractive indexes at normal incidence. Reproduced with permission from [55].

The graded-refractive index can also be realized using nanomaterials. In Fig. 3.33, the graded-refractive index is achieved by growth of nanowire layers of different materials and different densities. The stacking of several SiO_2 and TiO_2 layers forms a sequence that gives rise to an approximate graded-refractive index distribution as shown in the inset of Fig. 3.33 [56]. The surface with the graded-refractive index traps 99.5% light in a broad wavelength and wide incident angle ranges.

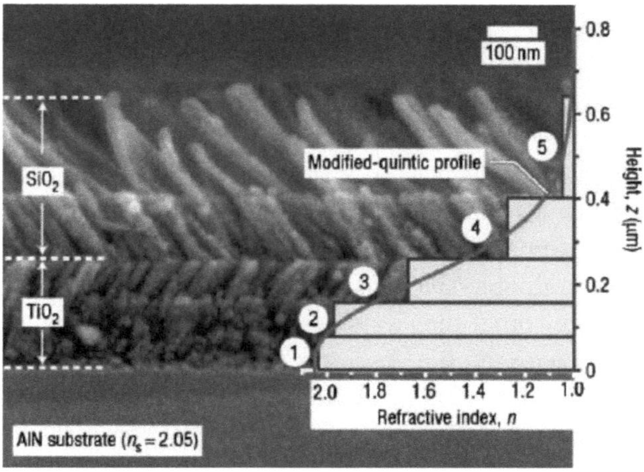

Figure 3.33 The stacking of several SiO_2 and TiO_2 layers, forming an graded-refractive index distribution with the step-like refractive index as shown by the inset. Reproduced with permission from [56].

Nano SCs are usually thin, so that light could easily penetrate the cells. A device architecture is usually adopted as shown in Fig. 3.34a, where the back electrode is also a perfect metallic reflector that reflects light back into the SC. Thus, light passes through the device twice before it goes out the cell, as the light beam denoted by *r* in the figure. The SC efficiency is then improved. However, if a photonic grating is used instead of the perfect back mirror, the photonic crystal could generate strong diffracted light that has long light path, as shown in Fig. 3.34b, which is beneficial for higher conversion efficiency.

Photonic crystals are a class of materials with their refractive index modulated periodically in the scale of optical wavelength. The most important characteristic is the formation of a photonic bandgap which blocks photons of selective frequencies. Photonic crystals are device for photon control and manipulation. They are used to enhance light absorption by trapping photons and changing the directions of photon propagation and polarization. Optical waveguide, resonator, multiplexer, etc., are typical photonic crystal devices.

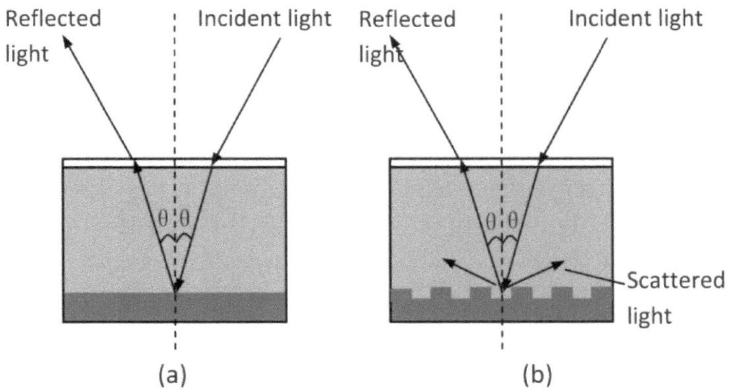

Figure 3.34 Schematics of two solar cell architectures: (a) with a perfect reflecting back electrode where there is only mirror reflection and (b) with a photonic grating on the back electrode so that not only the mirror reflection but also diffraction occurs. The front electrode is transparent.

In order to couple light into oblique angles, a grating, instead of a reflection mirror, could be used to increase the optical path in the cell. The absorption spectrum presented in Fig. 3.35 are calculated for a 1D grating (red curve), in comparison to that of a perfect planar reflector [57]. Apparently, the cell with 1D grating back electrode exhibits stronger overall light absorption than that with a perfect back reflector. As shown in Fig. 3.34b, the enhanced absorption is due to the diffractive light trapping effect (see r_D in the figure). The fact that part of the light is coupled into large angle with respect to surface normal through diffraction increases the length of light path, leading to the further improvement of the device efficiency. The absorption curve was calculated by a simple model that can be found in textbooks.

The grating does not have to be metal. Chutinan and co-workers simulated the typical solar cells with dielectric grating with the structures as shown in Fig. 3.36 [58]. The device architecture consists of a top anti-reflective coating, a Si active layer, a bottom grating layer. The grating on a bottom reflector, a perfect mirror, is either rectangular or a triangular. In their simulation, the structure parameters of the cell include t_{AR}, (thickness of the top coating), d (depth of the grating), p (periodicity of the grating), and d_C (duty cycle of the grating). Figure 3.36 presents solar cell structures with two different gratings: a rectangular grating (a) or a triangular one (b). For t_{AR} = 70 nm, d = 180 nm, p = 620 nm, and d_C = 0.5, the optimized cell efficiency of 16.76% is obtained for 2 μm cell thickness. For the cell thickness of 10 μm, the optimized cell efficiency of 22.24% is obtained at t_{AR} = 76 nm, d = 720 nm, p = 720 nm, and d_C = 0.5. For the similar cell structures but having a triangle-shaped grating, with t_{AR} = 72 (77) nm, d = 260 (310) nm, and p = 750 (860) nm, the optimized cell efficiency of 18.82% (23.24%) is achieved for the cell of 2 (10) μm in thickness.

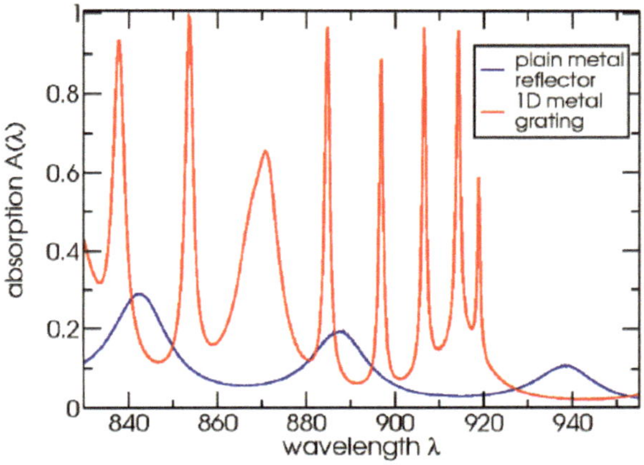

Figure 3.35 Calculated absorption spectra on the cell in Fig. 3.34a (blue curve) and on the cell in Fig. 3.34b (red curve). The c-Si thin film is 2 μm thick and the metal grating has the period 255 nm and etch depth 67 nm. Reproduced with permission from [57].

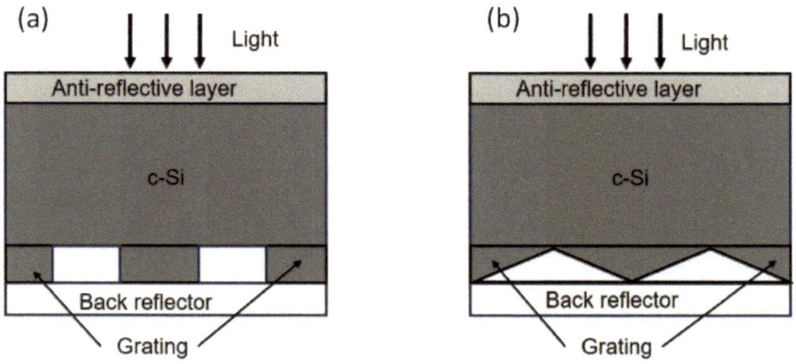

Figure 3.36 (a) Simple cell structures with an antireflective top layer, a c-Si layer, a rectangular grating layer, and a back reflector. (b) The similar cell structures as in (a) having a triangle-shaped grating.

Strictly speaking, multi-layer anti-reflection coating on wafer-based Si solar cells is a kind of 1D photonic crystal. Here we just introduce some important work using 2D and 3D photonic crystal that are much more difficult to fabricate than 1D one [57]. Those investigations provide one with important design guidelines for the fabrication of efficient nano solar cells and other optoelectronic devices such as photon detectors.

3.4.3 Light Trapping Using 2D Photonic Crystals

The gratings shown in Figs. 3.34 and 3.36 can be viewed as 1D photonic crystals. With some modification on the cell structures, the conversion efficiency can be further improved. Chutinan and co-workers introduced a 2D photonic crystals as the absorbing layer, as schematically shown in Fig. 3.37a, where the absorbing layer consists of 2D periodic square lattice of square dielectric rods [58]. The 2D photonic crystal has a photonic band diagram shown in Fig. 3.37b where the dielectric rods are in an air background. Compared to 1D case, 2D photonic crystals offer more room to tune the structural parameters to achieve high light trapping efficiency.

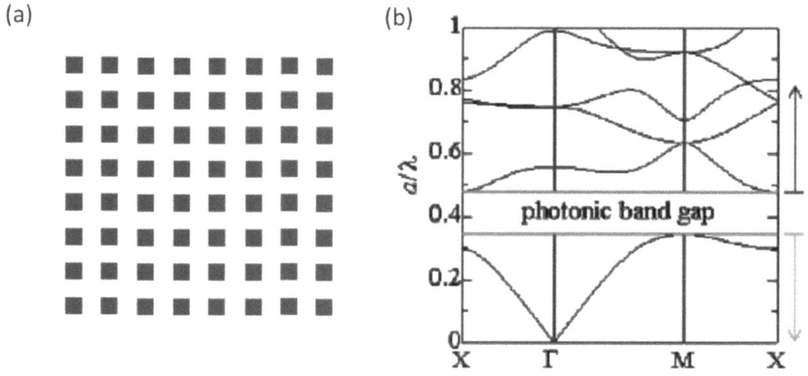

Figure 3.37 (a) A schematic 2D photonic crystal with a square lattice consisting of dielectric rods in an air background and (b) the corresponding photonic band structure where a photonic band gap is shown. Reproduced with permission from [58].

The photonic bandgap implies that light with wavelength within the bandgap is forbidden to travel in the material. For crystalline Si SCs, one could choose to place the photonic bandgap at 1.12 eV below which the c-Si is non-absorptive. One could expect, with the help of the high photon density of states of the photonic crystal at energies above the bandgap, the photons with energies larger than the Si bandgap (1.12 eV) of the active layer are able to be coupled into the SC.

Optical components, including photonic crystals (PCs), gratings, and distributed Bragg reflectors (DBRs), manipulate light propagation through three fundamental processes of reflection, refraction, and diffraction. Effective light trapping stems from proper combination of reflection, refraction, and diffraction. In comparison to conventional optical devices such as lens and mirrors, photonic crystals exhibit excellent performance. A well-designed photonic crystal could show reflectivity higher than that of aluminum reflectors that are superior among other metals for any angles, if the light frequencies and polarizations fall within the photonic bandgap [59]. DBRs, a one-dimensional PC consisting of two dielectrics with high index contrast, can have a broad reflection band of wavelength and broad range of incident angles [60].

3.4.4 Light Trapping Using 3D Photonic Crystals

Incorporating PCs into SCs, the structure and structural parameters are essential for the light trapping and coupling into active layers. For the fundamental SC structure as shown in Fig. 3.34, the major portions, anti-reflection coating, the active layer, as well as the back electrode, can adopt PCs for light trapping. As mentioned before, the anti-reflection part can be incorporated with 1D and 2D PCs for a wide spectral range anti-reflection.

Here the grating layer was used to scatter the light and the back reflector prevents the light penetrating out of the cell. From the uniform c-Si layer to the grating layer, the dielectric constant undergoes an abrupt change, which causes unwanted light reflection. The light scattering property thus improved by using a triangular-shaped grating that corresponds to a gradual change of dielectric constant from the uniform c-Si layer to the grating at the apex (see Fig. 3.36b). Note that in Fig. 3.36b the triangular gratings on the back reflector are used to avoid abrupt changes in the dielectric constant. The cell efficiency can be further improved by using 3D photonic crystals as shown in Fig. 3.37 [58]. The 3D photonic crystal makes it possible for the fine-tuning of the structure to enable optimized light trapping. For instance, a coupler could be introduced into the structure between the back reflector and the photonic crystal and/or between the anti-reflection coating and the photonic crystal, allowing for gradual change of dielectric constant, as shown in Fig. 3.38. The model calculation for a structure as shown in Fig. 3.38 demonstrates, for optimized structure parameters, a 20.92% conversion efficiency for the 2 μm (total thickness) cell and 24.21% for the of 10 μm cell. Thus, both the 2 μm and the 10 μm cells increase in conversion efficiencies, as compared to the corresponding cells with 1D photonic crystals.

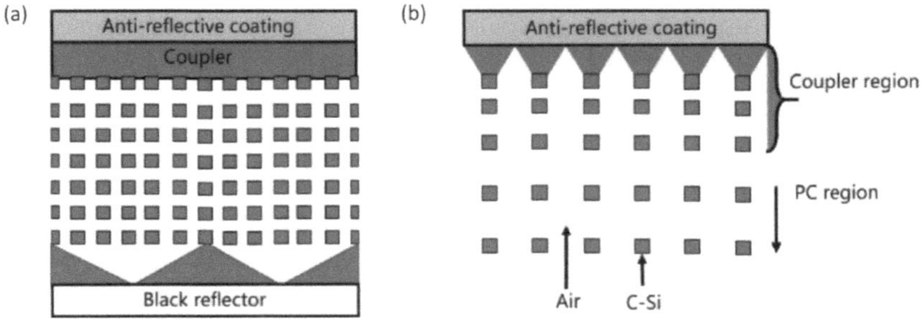

Figure 3.38 (a) Simple cell structures with an antireflective layer, a c-Si layer, a rectangular grating layer, and a back reflector with t = 70 (76) nm, t_g = 180 (720) nm, a = 620 (720) nm, and d_c = 0.5 (0.5). (b) Zoom-in of the front coupler region in (a).

3D nanostructured arrays exhibit high performance for suppression of light reflection and diffraction, so that they are receiving increasing attention for the use for solar energy harvesting. Hua and co-workers have investigated arrays of multi-diameter Ge nanopillars that showed wide band solar spectrum absorption using finite difference time domain simulations [61].

Light absorption capability of arrays of nanopillars, including multiple-diameter nanopillars and nanocones, depends on the wavelength of light, light absorption of the material, and the geometrical parameters that are material dependent. It was found that either material filling ratio or transverse resonance leaky modes dominates light absorption of nanopillars. Intuitively, a nanocone array should have the best light absorption capability due to its feature of gradually varied index of reflection. Actually, the same light absorption performance can be achieved on a well-designed multi-diameter nanopillar array in terms of broadband photon absorption capability.

3.4.5 Preparation of Photonic Crystal Used for Light Trapping

This kind of nanostructure is extremely effective for solar for cells with very thin active layers. For a 2D array of nanocones, the average material portion increases from the cone tips to the cone bases, leading to a gradual increase of dielectric constant and, as a result, zero reflection of can be achieved with such an efficient photon management design. The 2D nanocone array can be prepared in a scalable process at relatively low temperature [62, 63]. Both theoretical simulation and experimental optical measurement show that the thin-film solar cells incorporated with such nanocone array can achieve nearly perfect anti-reflection and optical absorption enhancement over abroad range of spectra and a wide range of angles of incidence. Accordingly, the device performance of solar cells with this design is significantly enhanced in comparison to solar cells with conventional design.

Zhu and co-workers presented a route to fabricate nanowire and nanocone arrays using silica nanoparticles as an etch-mask and the fabricating process is illustrated in Fig. 3.39 [64]. An a-Si:H film (see Fig. 3.39a) of 1 μm in thickness was grown on an indium tin oxide (ITO) layer coated on a quartz substrate and a close packed monolayer of the silica nanoparticles was assembled on top of the a-Si:H film by the Langmuir–Blodgett method. A reactive ion etching was used to etch the a-Si:H film. The silica nanoparticles have an etching rate that is much slower than that of the a-Si:H film. With controlled experimental condition, including etching time, nanowire array or nanocone array can be prepared [65].

Figures 3.40a,b present SEM images of a close-packed monolayer of silica nanoparticles on the a-Si:H thin film, with a uniform nanoparticles size of about 500 nm [63]. Figures 3.40c,d exhibit SEM images of a-Si:H nanocone arrays and Figs. 3.40e,f show those of a-Si:H nanowire arrays, prepared by reactive ion etching, with an average nanowire diameter of ~300 nm and nanowire length of ~600 nm. Panels e and f of the figure show SEM images of a-Si:H nanocone arrays, with an average nanocone length of ~600 nm, a tip diameter of ~20 nm and a base diameter of ~300 nm.

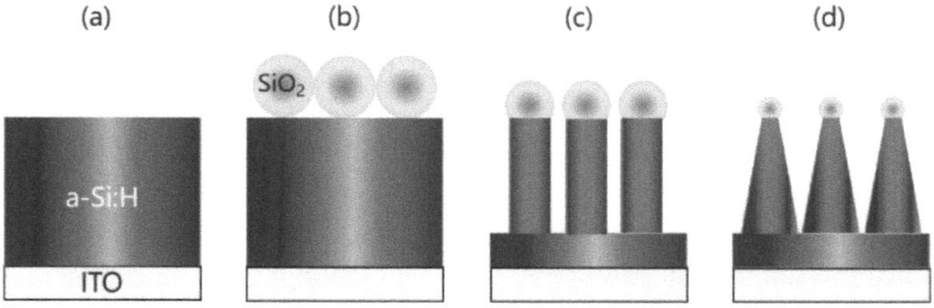

Figure 3.39 Schematics of (a) a-Si:H (1 μm in thickness) on ITO, (b) a monolayer of silica nanoparticles on top of a-Si:H, (c) nanowire array, and (d) nanocone arrays formed by reactive ion etching.

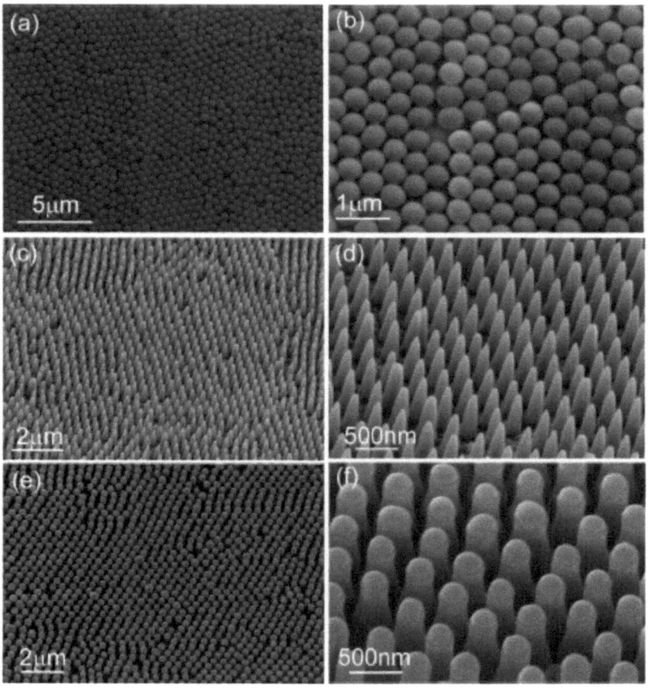

Figure 3.40 SEM images of silica nanoparticles (a), a-Si:H nanocone arrays (c), and a-Si:H nanowire arrays (e). Panels b, d, and f are the zoom-in images of panels a, c, and e, respectively. Reproduced with permission from [63].

With rapid advances in nanotechnology over the past decades, many approaches have been proposed and practiced to fabricate various nanomaterials and nanostructures for light trapping. Among them, quite a few of the methods have been demonstrated to be suitable for the fabrication of nanostructures on a large scale. Similar to previously mentioned nanomaterials and structures, the methods for light trapping nanomaterials and nanostructures belong to either the top-down or the bottom-up category. With a great controllability over feature sizes and shapes, photolithography and electron-beam lithography can fabricate nanostructures at a size below 100 nm and with large scale. The major problems for photolithography and electron-beam lithography are their high cost and low throughput, which limits their use for large-scale productions. Solution phase synthesis, including solvothermal and hydrothermal, is an alternative for the growth of QDs, nanorods, NW arrays, etc. The method has the advantages of low cost and large scale, but the control on feature sizes of nanostructures is less prominent. Vapor–liquid–solid (VLS) preparation has been proposed for more than 50 years and is now a major method for growing nanowire arrays. VLS has a good control over NW diameters, while the precise control over spacing between NWs remains to be a problem. People are still looking for ways to modify those approaches for the preparation of well-controlled nanomaterials and nanostructures in morphologies, including feature sizes and shapes. Some of the techniques appear to be promising [66, 67].

After the first demonstration of VLS-growth of Si nanowires by Wagner and Ellis, the VLS process has been used to grow many nanosized semiconductors. A successful preparation of ordered Si NW array using a photolithographically patterned VLS process was demonstrated by Kayes et al. [68]. The VLS process used Au and Cu particles as catalyst. With less than 5% areal fraction of Si nanowires, Kelzenberg et al. achieved 96% peak absorption and 85% above-bandgap sunlight absorption of day-integration [69]. Furthermore, the Si NW array showed greater overall AM 1.5D spectrum absorption than reflectance-free planer absorber with equivalent layer thickness, for photons with energy above the bandgap and at normal incidence. For all angle incidence, the 85% day-integrated absorption of the NW array is slightly higher than 82% of the planar case, demonstrating a clear effect that the Si NW array may lead to light trapping effect exceeding the theoretical absorption limit of an ideal Si planer surface without reflection. Figure 3.41 presents the calculated photon flux absorption spectra of three structures where the calculation assumes normal light incidence, and the absorption is spectrally weighted in terms of the AM 1.5D reference spectrum. A non-textured bare Si cell presents 34.7% absorption, i.e., only 34.7% of the solar radiation is absorbed and used for photovoltaic process. When a classical planar light trapping is used, the Si solar cell shows 80.8% light absorption. The use of Si nanowires leads to 84.4% absorption. The major advantage that the Si nanowire array exceeds the planar light trapping is in the infrared region with wavelength longer than 800 nm.

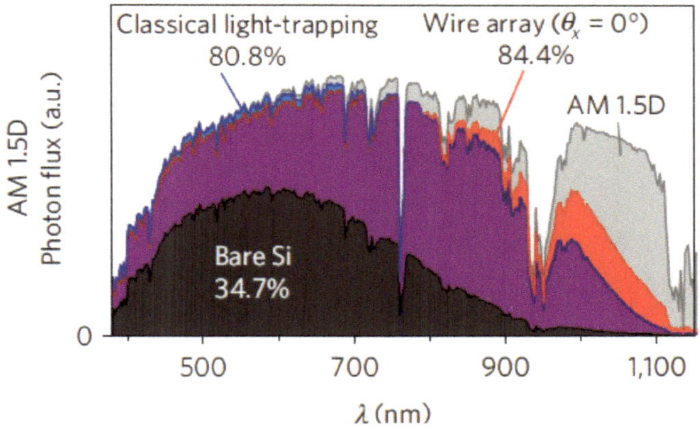

Figure 3.41 Absorbed photon flux spectra of the AM 1.5D solar radiation with spectrally weighting at normal incidence. The presented spectra correspond to those of bare Si, classical light trapping, and Si nanowire array with 4.2 filling ratio. Reproduced with permission from [71].

Fan et al. developed a catalytic VLS process to grow ordered, crystalline Ge NWs arrays on anodic alumina membrane (AAM) templates with controlled NW shapes and dimensions [70, 71]. The fact of AAM being optically transparent with a wide optical bandgap of 4.2 eV guarantees no light absorption in the AAM layer, which is ideal for the layer to be used as optical thin films for various purposes. The template-assistant VLS growth offers well-controlled tuning to the parameters of hexagonally ordered structures, so does to those of the Ge NWs. The diameter of Ge NWs, for instance, is determined by the pore size of the AAM template. The growth procedure of AAM is well established following simple multiple etching/anodization steps. What interesting is the adaptability of the VLS process for the growth of NWs with dual diameters. Figure 3.42a presents the SEM image of dual diameter Ge NWs, showing that each NW has a large-diameter base and a small-diameter tip on the top. Since the thinner part of the dual diameter NWs is on top, the dual-diameter NWs are closer to the gradual variation of effective refractive index from top to bottom than single diameter NWs. Better light trapping effect is expected for the dual diameter NWs. Indeed, the NWs show excellent photon trapping property—99% broad-band optical absorption, in contrast to 53% broad-band absorption of the planer Ge and to roughly 90% in the case of single diameter NWs, as shown in Fig. 3.42b.

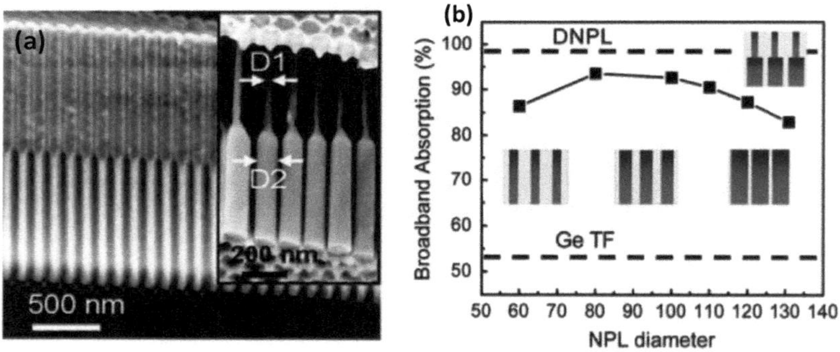

Figure 3.42 (a) SEM image of dual diameter NWs with the tip diameter (D1) ~50 nm and the base diameter (D2) ~130 and (b) average broadband absorption of the dual diameter NWs and a Ge thin film in comparison to broadband absorption of the single diameter NWs as a function of NW diameters. Reproduced with permission from [70].

3.5 Some Applications

3.5.1 Plasmonic Light Trapping for Thin-Film a-Si:H Solar Cells

Most work of plasmonic light trapping in solar cells is theoretical. The major difficulties encountered in experiments are the difficulties in incorporating nano-metal pattern in complicated structures of nano solar cells. Next, let's give two examples to illustrate the application. In the following, an example is given to describe the uses of plasmonic effect in light trapping for solar cells. The plasmonic light trapping in a-Si:H solar cells is on the back contact [72].

Due to very small overall thickness in a-Si:H solar cells, light trapping is challenging. As mentioned in Section 3.3.1, there are four different mechanisms for light-plasmon coupling effects. For applications of light trapping in nano solar cells, the designed structures ought to allow for strong coupling and a large fraction of the coupled modes should be distributed in the semiconductor. In the case of the thin-film solar cell, the waveguide modes and the standing wave modes are beneficial for enhanced absorption. Another important issue to note is that the enhanced absorption needs to cover a broad solar illumination spectrum and under wide light incident angle.

Ferry and co-workers fabricated a-Si:H thin-film solar cells on substrates that are nanopatterned using substrate conformal imprint lithography [43, 73]. The Ag nanoparticles in the plasmonic pattern have a hemispherical shape, with tested sizes of 200, 225, 250, and 275 nm. The SEM image of cross section of the a-Si:H solar cell is presented in Fig. 3.43 and the detailed fabrication of the structure is described in Ref. [43].

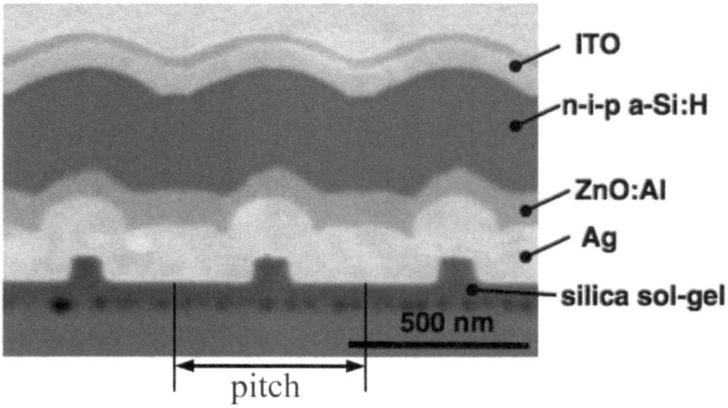

Figure 3.43 SEM image of cross section of an a-Si:H thin-film solar cell where the thickness of the a-Si:H layer is 340 nm. Reproduced with permission from [43].

Table 3.1 lists the measured parameters, namely, open-circuit voltage V_{oc}, short-circuit current density J_{sc}, filling factor *FF*, power conversion efficiency η at two different a-Si:H layer thickness of 160 and 340 nm and at two different pitch sizes of 500 and 700 nm. The results from a flat a-Si:H solar cell (without plasmonic pattern) are also listed in Table 3.2. These devices does not show much difference in V_{oc}, *FF*, and η. J_{sc} measured on the plasmonic a-Si:H cells is obviously higher than those obtained on other cells, an increasing by nearly 30%. J_{sc} also depends on the pitch size, the J_{sc} values of the cells with 500 nm pitch exceeding those of the cells with 700 nm pitch. As the result of J_{sc} enhancement, the relatively high conversion efficiency η is obtained on cells on the plasmonic a-Si:H cells. In Ferry and co-workers' work, the high J_{sc} value of 11.5 mA/cm^2 was measured on the plasmonic cell with 160 nm a-Si:H layer thickness and 500 pitch size, in comparison to 7.9 mA/cm^2 observed on the flat cell of the same geometry. The nearly same open-circuit voltages and fill factors of the cells imply the similar quality in the device fabrication. The apparently different J_{sc} is an indication that enhanced light absorption in the plasmonic cells due to the effective plasmonic light trapping. Note that the J_{sc} value of the cells with 160 nm a-Si:H layer thickness is smaller than those in cells with 340 nm layer, but their power conversion efficiencies are the same. Relatively high J_{sc} in the 340 nm cells are offset by relatively low V_{oc}, leading to their similar conversion efficiency.

Table 3.2 Major parameters of the a-Si:H solar cell with and without plasmonic light trapping layers

	Thickness	V_{oc} (V)	J_{sc} (mA/cm^2)	*FF*	η (%)
Flat	160	0.89	7.9	0.68	4.8
Asahi	160	0.87	10.8	0.64	6.0
500 nm pitch	160	0.89	11.5	0.66	6.6
700 nm pitch	160	0.88	10.4	0.65	5.6
flat	340	0.84	10.5	0.58	5.1
500 nm pitch	340	0.85	13.4	0.56	6.4
700 nm pitch	340	0.84	13	0.56	6.1

Recently, Wang and co-workers did a similar work in which an enhancement on conversion efficiency was demonstrated on Si thin-film solar cells [74]. In their device architecture, the Ag nanoparticle is separated from the Ag film and the trapping effect can be tuned by the separation gap from the Ag nanoparticle to the Ag film.

3.5.2 Light Trapping Layers Fabricated by Nano-Imprint Lithography

Nano-imprint is an efficient and cost-effective technology to prepare photonic crystal for light trapping. Han et al. developed a technique based on nano-imprint method to form nano-size dot pattern array on front surface of high efficiency GaInP/Ga(In)As/Ge tandem solar cells with a multi-junction structure [75]. The fabrication process is shown in Fig. 3.44a,b for the SEM images of Ni master mold and duplicated PVC mold, respectively. The imprinted pattern, made use of the PVC mold, shows very regular hexagonal distribution of the nanoparticles (Fig. 3.44c). Figures 3.44d,e present the atomic force microscopic (AFM) image of the imprinted pattern and its section analysis curve, respectively.

Since the dot size and period are smaller than the wavelength of incident visible light, photons "see" an average refractive index increasing continuously from that of air to that of the solar cell surface since the nano-sized dots have a cone shape. The nano-size moth-eye pattern suppresses the surface reflection in wide spectrum range, resulting in enhanced photoelectric conversion efficiency. The effectiveness of the light trapping is demonstrated by the measured reflectance spectra of the GaAs-based solar cell with and without nano-size moth-eye pattern shown in Fig. 3.45a and conversion efficiencies vs. light intensity (the number of suns) of the cell with and without the pattern as presented in Fig. 3.45b.

The multi-junction structured tandem solar cells exhibit a very high conversion efficiency up to 40% under 240 sun concentration [76]. The high conversion efficiency is achieved under the brighter irradiation condition by using Fresnel lens or other concentrators. Thus, the condensed light beam needs only solar cells with small area, which is helpful for cost reduction since the tandem solar cells are fabricated using delicate equipment such as MBE and MOCVD. For the high efficiency solar cells, proper measurement for reducing the surface reflection of light is extremely important.

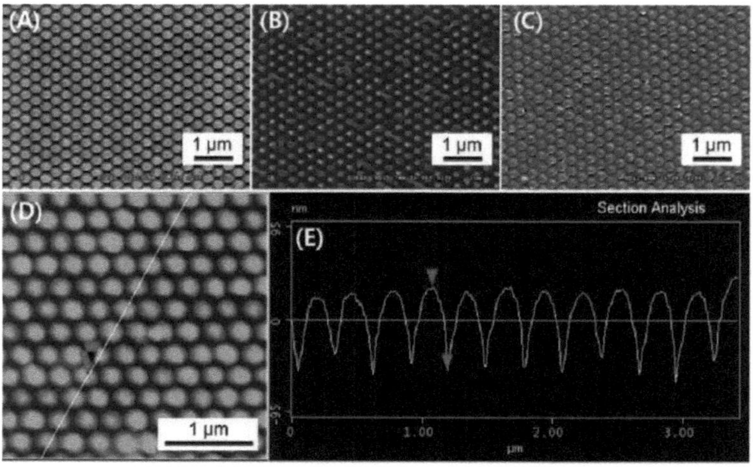

Figure 3.44 (a) SEM images of Ni master mold, (b) patterned PVC mold fabricated by hot-embossing process with the master mold, (c) imprinted pattern using the PVC mold, (d) AFM images of the imprinted pattern. (e) The section analysis on the imprinted pattern along the line in (d), showing that the diameter and the height of the imprinted nanostructures are 250 and 100 nm, respectively. Reproduced with permission from [75].

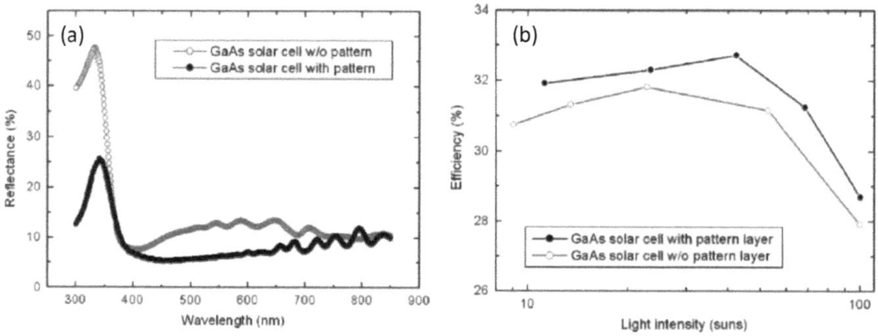

Figure 3.45 (a) Reflectance spectra of the GaAs-based solar cell with and without the moth-eye pattern. (b) Measured conversion efficiencies as a function of light intensity of the GaAs-based solar cell with and without the moth-eye pattern. Reproduced with permission from [75].

Surface texture, anti-reflecting layer, as well as the back reflecting layer can all be adopted using the nanomaterial or photonic crystal. In fact, the surface-texture/anti-reflecting-layer has limited performance. For instance, the ideal enhancement of light trapping is $4n^2$, while the currently adopted texture only leads to 50% of the ideal enhancement. The SiN_x layer has room to improve for both anti-reflection and passivation. On the other hand, the Al back reflecting layer, only reflect less than 80% of the incident light. With the architecture, there is almost no room to improve by adjusting the parameters such as the layer thickness and materials. Both the front and the back surface measurement can use the photonic crystals that would offer much better performance and big room for tuning the wavelength and travel direction of light, as well as photon density for the optimization of conversion efficiency. The adjustable parameters include the dielectric constant of the photonic crystal materials, shapes, and periodicities of the nanostructures in three dimensions, which offers much easier control over the light.

References

1. A. Givental, *Introduction to Quantum Mechanics*, published by Sumizdat (2019).
2. D. J. Lockwood, Rayleigh and Mie Scattering, in: R. Luo (Eds) *Encyclopedia of Color Science and Technology*. Springer, Berlin, Heidelberg (2015).
3. C. Battaglia, M. Boccard, F-J. Haug, and C. Ballif, Light trapping in solar cells: When does a Lambertian scatterer scatter Lambertianly? *J. Appl. Phys.* **112**, 094504 (2012).
4. C. I. Yeo, Y. M. Song, S. J. Jang, and Y. T. Lee, Optimal design of nano-scale surface light trapping structures for enhancing light absorption in thin film photovoltaics, *J. Appl. Phys.* **114**, 024305 (2013).
5. R. Yu, Q. Lin, S.-F. Leung, and Z. Fan, Nanomaterials and nano-structures for efficient light absorption and photovoltaics, *Nano Energy*, **1**, 57–72 (2012).
6. L. A. Dobrzański and A. Drygała, Surface texturing of multicrystalline silicon solar cells, *J. Achiev. Mater. Manuf. Eng.*, **31**(1), 77–82 (2008).

7. Y. Ein-Eli, N. Gordon, D. Starosvetsky, Reduced light reflection of textured multicrystalline silicon via NPD for solar cells applications, *Sol. Energy Mater. Sol. Cells*, **90**, 1764–1772 (2006).
8. C. Herzinger, B. Johs, W. McGahan, J. Woollam, and W. Paulson, Ellipsometric determination of optical constants for silicon and thermally grown silicon dioxide via a multi-sample, multi-wavelength, multi-angle investigation, *J. Appl. Phys.*, **83**, 3323–3336 (1998).
9. ASTMG173-03, *Standard Tables for Reference Solar Spectral Irradiances: Direct Normal and Hemisphericalon 37 Degree Tilted Surface* (ASTM International, West Conshohocken, Pennsylvania) (2005).
10. E. Yablonovitch and G. Cody, Intensity enhancement in textured optical sheets for solar cells, *IEEE Trans. Electron Devices*, **ED-29**, 300–305 (1982).
11. L. Cao, J. S. White, J.-S. Park, J. A. Schuller, B. M. Clemens, and M. L. Brongersma, Engineering light absorption in semiconductor nanowire devices, *Nat. Mater.*, **8**, 643–647 (2009).
12. A. Zhang, S. You, C. Soci, Y. Liu, D. Wang, and Y.-H. Lo, Silicon nanowire detectors showing phototransistive gain, *Appl. Phys. Lett.*, **93**, 121110-3 (2008).
13. J. Kupec and B. Witzigmann, Dispersion, wave propagation and efficiency analysis of nanowire solar cells, *Opt. Express*, **17**, 10399–10410 (2009).
14. M. Kevin, Y. H. Fou, A. S. W. Wong, and G. W. Ho, A novel maskless approach towards aligned, density modulated and multi-junction ZnO nanowires for enhanced surface area and light trapping solar cells, *Nanotechnology*, **21**, 315602 (2010).
15. A. G. A. Nisbet, G. Beutier, F. Fabrizi, B. Mosera, and S. P. Collins, Diffuse multiple scattering, *Acta Cryst.* A, **71**, 20–25 (2015).
16. R. A. Street, P. Qi, R. Lujan, and W. S. Wong, Reflectivity of disordered silicon nanowires, *Appl. Phys. Lett.*, **93**, 163109-3 (2008).
17. R. A. Street, W. S. Wong, and C. Paulson, Analytic model for diffuse reflectivity of silicon nanowire mats, *Nano Lett.*, **9**, 3494–3497 (2009).
18. O. L. Muskens, J. G. Rivas, R. E. Algra, E. P. A. M. Bakkers, and A. Lagendijk, Design of light scattering in nanowire materials for photovoltaic applications, *Nano Lett.*, **8**, 2638–2642 (2008).
19. O. L. Muskens, S. L. Diedenhofen, M. H. M. V. Weert, M. T. Borgström, E. P. A. M. Bakkers, and J. G. Rivas, Epitaxial growth of aligned

semiconductor nanowire metamaterials for photonic applications, *Adv. Funct. Mater.*, **18**, 1039–1046 (2008).

20. J. Wang, M. S. Gudiksen, X. Duan, Y. Cui, and C. M. Lieber, Highly polarized photoluminescence and photodetection from single indium phosphide nanowires, *Science,* **293**, 1455–1457 (2001).
21. O. L. Muskens, M. T. Borgstrom, E. P. A. M. Bakkers, and J. G. Rivas, Giant optical birefringence in ensembles of semiconductor nanowires, *Appl. Phys. Lett.*, **89**, 233117-3 (2006).
22. M. D. Kelzenberg, S. W. Boettcher, J. A. Petykiewicz, D. B. Turner-Evans, M. C. Putnam, E. L. Warren, J. M. Spurgeon, R. M. Briggs, N. S. Lewis, and H. A. Atwater, Enhanced absorption and carrier collection in Si wire arrays for photovoltaic applications, *Nat. Mater.*, **9**, 239–244 (2010).
23. S. Frank, J. P. D. Vanmaekelbergh, J. V. D. Lagemaat, and A. Lagendijk, Strongly photonic macroporous gallium phosphide networks, *Science,* **284**, 141–143 (1999).
24. S. E. Han and G. Chen, Toward the lambertian limit of light trapping in thin nanostructured silicon solar cells, *Nano Lett.*, **10**, 4692–4696 (2010).
25. V. K. Narasimhan and Y. Cui, Nanostructures for photon management in solar cells, *Nanophotonics,* **2**(3), 187–210 (2013).
26. S. Jeong and S. Wang, Nanoscale photon management in silicon solar cells, *J. Vac. Sci. Technol.* A, **30**(6), 060801 (2012).
27. L. Hu and G. Chen, Analysis of optical absorption in silicon nanowire arrays for photovoltaic applications, *Nano Lett.*, **7**, 3249 (2007).
28. Y. Tian, C. Hu, Y. Xiong, B. Wan, C. Xia, X. He, and H. Liu, ZnO pyramidal arrays: Novel functionality in antireflection, *J. Phys. Chem. C,* **114**, 10265–10269 (2010).
29. P. Hiralal, C. Chien, N. N. Lal, W. Abeygunasekara, A. Kumar, H. Butt, H. Zhou, E. U. H. E. Unalan, J. J. Baumberg, and G. A. J. Amaratunga, Nanowire-based multifunctional antireflection coatings for solar cells. *Nanoscale,* **6**(23), 14555-14562 (2014).
30. S. E. Han and G. Chen, Optical absorption enhancement in silicon nanohole arrays for solar photovoltaics, *Nano Lett.*, **10**, 1012–1015 (2010).
31. S. Jeong, E. C. Garnett, S. Wang, Z. Yu, S. Fan, M. L. Brongersma, M. D. McGehee, and Y. Cui, Hybrid Silicon Nanocone–Polymer Solar Cells, *Nano Lett.*, **12**, 2971–2976 (2012).

32. K. X. Wang, Z. Yu, V. Liu, Y. Cui, and S. Fan, Absorption enhancement in ultrathin crystalline silicon solar cells with antireflection and light-trapping nanocone gratings, *Nano Lett.*, **12**, 1616–1619 (2012).

33. H. W. M. Fung, S. So, K. Kartub, G. Loget, and R. M. Corn, Ultra-Antireflective Electrodeposited Plasmonic and PEDOT Nanocone Array Surfaces, *J. Phys. Chem C*, **121**, 22377–22383 (2017).

34. B. Hua, B. Wang, M. Yu, P. W. Leu, Z. Fan, Rational geometrical design of multi-diameter Nano pillars for efficient light harvesting, *Nano Energy*, **2**, 951–957 (2013).

35. Z. Fan, R. Kapadia, P. W. Leu, X. Zhang, Y.-L. Chueh, K. Takei, K. Yu, A. Jamshidi, A. A. Rathore, D. J. Ruebusch, M. Wu, and A. Javey, Ordered arrays of dual-diameter nanopillars for maximized optical absorption, *Nano Lett.*, **10**, 3823–3827 (2010).

36. H. Savin, P. Repo, G. von Gastrow, P. Ortega, E. Calle, M. Garín, and R. Alcubilla, Black silicon solar cells with interdigitated back-contacts achieve 22.1% efficiency, *Nat. Nanotech.*, **10**, 624 (2015).

37. K. R. Catchpole and A. Polman, Plasmonic solar cells, *Opt. Express*, **16**(26), 21793 (2008).

38. W. L. Barnes, A. Dereux, and T. W. Ebbesen, Surface plasmon subwavelength optics, *Nature*, **424**, 824 (2003).

39. H. A. Atwater and A. Polman, Plasmonics for improved photovoltaic devices, *Nat. Mat.*, **9**, 205 (2010).

40. S. Mokkapati, F. J. Beck, R. de Waele, A. Polman, and K. R. Catchpole, Resonant nano-antennas for light trapping in plasmonic solar cells, *J. Phys. D: Appl. Phys.*, **44**, 185101 (2011).

41. Y. Zhang, Z. Ouyang, N. Stokes, B. Jia, Z. Shi, and M. Gu, Low cost and high performance Al nanoparticles for broadband light trapping in Si wafer solar cells, *Appl. Phys. Lett.*, **100**, 151101 (2012).

42. K. R. Catchpole and A. Polman, Design principles for particle plasmon enhanced solar cells, *Appl. Phys. Lett.*, **93**, 191113 (2008).

43. V. E. Ferry, M. A. Verschuuren, H. B. T. Li, E. Verhagen, R. J. Walters, R. E. I. Schropp, H. A. Atwater, and A. Polman, Light trapping in ultrathin plasmonic solar cells, *Opt. Exp.*, **18**(102), A237 (2010).

44. X. Chen, B. Jia, J. K. Saha, B. Cai, N. Stokes, Q. Qiao, Y. Wang, Z. Shi, and M. Gu, Broadband enhancement in thin-film amorphous silicon solar cells enabled by nucleated silver nanoparticles, *Nano Lett.*, **12**(5), 2187–2192 (2012).

45. S. Pillai and M. A. Green, Plasmonics for photovoltaic applications, *Sol. Energy Mater. Sol. Cells*, **94**, 9, 1481–1486 (2010).

46. A. Wokaun, J. P. Gordon, and P. F. Liao, Radiation damping in surface-enhanced raman scattering, *Phys. Rev. Lett.*, **48**, 957–960 (1982).

47. H. R. Stuart and D. G. Hall, Absorption enhancement in silicon-on-insulator waveguides using metal island films, *Appl. Phys. Lett.*, **69**, 2327–2329 (1996).

48. H. R. Stuart and D. G. Hall, Island size effects in nanoparticle-enhanced photodetectors, *Appl. Phys. Lett.*, **73**, 3815–3817 (1998).

49. Y. Zhang, T. Wei, W. Dong, K. Zhang, Y. Sun, X. Chen, and N. Dai, Vapor-deposited amorphous metamaterials as visible near-perfect absorbers with random non-prefabricated metal nanoparticles, *Sci. Rep.*, **4**, 4850 (2014).

50. K. Nakayama, K. Tanabe, and H. A. Atwaterb, Plasmonic nanoparticle enhanced light absorption in GaAs solar cells, *Appl. Phys. Lett.*, **93**, 121904 (2008).

51. M. L. Brongersma, Y. Cui, and S. Fan, Light management for photovoltaics using high-index nanostructures, *Nature Mater.*, **13**, 451 (2014).

52. S. G. Johnson and J. D. Joannopoulos, *Photonic Crystals: The Road from Theory to Practice*. Boston: Kluwer (2002).

53. A. Birner, R. B. Wehrspohn, U. M. Gösele, and K. Busch, Silicon-based photonic crystals, *Adv. Mater.*, **13**(6), 377 (2001).

54. E. Pavarini, L. C. Andreani, C. Soci, M. Galli, and F. Marabelli, Band structure and optical properties of opal photonic crystals, *Phys. Rev. B*, **72**, 045102 (2005).

55. M. Beye, M. E. Faye, A. Ndiaye, F. Ndiaye, and A. Seidou Maiga, Optimization of SiN_x single and double layer ARC for silicon thin film solar cells on glass, *Res. J. Appl. Sci. Eng. Technol.*, 6(3), 412–416 (2013).

56. J. Q. Xi, M. F. Schubert, J. K. Kim, E. F. Schubert, M. Chen, S.-Y. Lin, W. Liu, and J. A. Smart, Optical thin film materials with low refractive index for broadband elimination of fresnel reflection, *Nature Photon.*, **1**(3), 176–179 (2007).

57. P. Bermel, C. Luo, L. Zeng, L. C. Kimerling, and J. D. Joannopoulos, Improving thin-film crystalline silicon solar cell efficiencies with photonic crystals, *Opt. Exp.*, **15**(25), 16986 (2007).

58. A. Chutinan, N. P. Kherani1, and S. Zukotynski, High-efficiency photonic crystal solar cell architecture, *Opt. Exp.*, **17**(11), 8871 (2009).

59. J. Joannopoulos, R. Meade, and J. Winn, *Photonic Crystals: Molding the Flow of Light*, (Princeton University Press, Princeton, New Jersey, 2008).

60. T. Ogawa and Y. Kanemitsu, eds., *Optical Properties of Low-Dimensional Materials* (World Scientific, Singapore, 1995).

61. B. Hua, B. Wang, M. Yu, P. W. Leu, Z. Fan, Rational geometrical design of multi-diameter nanopillars for efficient light harvesting, *Nano Energy*, **2**, 951–957 (2013).

62. Y.-F. Huang, S. Chattopadhyay, Y.-J. Jen, C.-Y. Peng, T.-A. Liu, Y.-K. Hsu, C.-L. Pan, H.-C. Lo, C.-H. Hsu, Y.-H. Chang, C.-S. Lee, K.-H. Chen, L.-C. Chen, Improved broadband and quasi-omnidirectional anti-reflection propertieswith biomimetic silicon nanostructures, *Nat. Nanotech.*, **2**, 770 (2007).

63. J. Zhu, Z. Yu, G. F. Burkhard, C.-M. Hsu, S. T. Connor, Y. Xu, Q. Wang, M. McGehee, S. Fan, and Y. Cui, Optical absorption enhancement in amorphous silicon nanowire and nanocone array, *Nano Lett.*, **9**(1), 279 (2009).

64. J. Zhu, Z. Yu, S. Fan, and Y. Cui, Nanostructured photon management for high performance solar cells, *Mater. Sci. Eng.*, **R70**, 330–340 (2010).

65. H. Ching-Mei, S. T. Connor, M. X. Tang, and Y. Cui, Wafer-scale silicon nanopillars and nanocones by Langmuir–Blodgett assembly and etching, *Appl. Phys. Lett.*, **93**, 133109 (2008).

66. C. L. Haynes and R. P. Van Duyne, Nanosphere lithography: A versatile nanofabrication tool for studies of size-dependent nanoparticle optics. *J. Phys. Chem. B*, **105**(24), 5599–5611 (2001).

67. A. Kosiorek, W. Kandulski, P. Chudzinski, K. Kempa, M. Giersig, Shadow nanosphere lithography: Simulation and experiment, *Nano Lett.*, **4**(7), 1359–1363 (2004).

68. B. M. Kayes, M. A. Filler, M. C. Putnam, M. D. Kelzenberg, N. S. Lewis, and H. A. Atwater, Growth of vertically aligned Si wire arrays over large areas (>1 cm^2) with Au and Cu catalysts, *Appl. Phys. Lett.*, **91**, 103110–103113 (2007).

69. M. D. Kelzenberg, S. W. Boettcher, J. A. Petykiewicz, D. B. Turner-Evans, M. C. Putnam, E. L. Warren, J. M. Spurgeon, R. M. Briggs, N. S. Lewis, and H. A. Atwater, Enhanced absorption and carrier collection in Si wire arrays for photovoltaic applications, *Nat. Mater.*, **9**, 239 (2010).

70. Z. Fan, R. Kapadia, P. W. Leu, X. Zhang, Y.-L. Chueh, K. Takei, K. Yu, A. Jamshidi, A. A. Rathore, D. J. Ruebusch, M. Wu, and A. Javey, Ordered

arrays of dual-diameter nanopillars for maximized optical absorption, *Nano Lett.*, **10**, 3823–3827 (2010).

71. R. Yu, Q. Lin, S.-F. Leung, and Z. Fan, Nanomaterials and nanostructures for efficient light absorption and photovoltaics, *Nano Energy*, **1**, 57–72 (2012).
72. V. E. Ferry, M. A. Verschuuren, H. B. T. Li, E. Verhagen, R. J. Walters, R. E. I. Schropp, H. A. Atwater, and A. Polman, Plasmonic light trapping for thin film a-Si:H solar cells, in *Program: 35th IEEE Photovoltaic Specialists Conference, PVSC*, pp. 760–765 (2010).
73. M. Verschuuren, H. van Sprang. 3D photonic structures by sol-gel imprint lithography, *Mater. Res. Soc. Sym. Proc.*, **1002**, N03–N05 (2007).
74. P. H. Wang, M. Theuring, M. Vehse, V. Steenhoff, C. Agert, and A. G. Brolo, Light trapping in a-Si:H thin film solar cells using silver nanostructures, *AIP Adv.*, **7**, 015019 (2017).
75. K.-S. Han, J.-H. Shin, W.-Y. Yoon, and H. Lee, Enhanced performance of solar cells with anti-reflection layer fabricated by nano-imprint lithography, *Sol. Energy Mater. Sol. Cells*, **95**, 288–291 (2011).
76. R. R. King, D. C. Law, K. M. Edmondson, C. M. Fetzer, G. S. Kinsey, H. Yoon, R. A. Sherif, and N. H. Karam, 40% efficient metamorphic GaInP/GaInAs/Ge multi-junction solar cells, *Appl. Phys. Lett.*, **90**, 183516 (2007).

Chapter 4

Transparent Conducting Electrodes and Dye-Sensitized Solar Cells

As the fundamental parts in nano solar cells (NSCs), the performance of electrodes plays an important role in determining photoelectronic conversion efficiency. Good electrodes should be highly conductive and highly transparent in the solar spectrum range. The best conductors are metals. Metals are also transparent when they are very thin (below 10 nm). However, the conductivities of thin metal layers are too low. Fortunately, our nature provides us with many choices of materials that are both conductive and transparent, most of them being metal oxides. Some materials are used in NSCs as photo-anodes and the fundamental requirements for this kind of materials are highly transparent and conductive, and adaptable to photon absorbers (QDs, dye, etc.) in energy band alignment.

Dye-sensitized solar cell (DSSC) was proposed by Gratzel and co-workers, which is now also called "Grätzel Cell" [1–3] DSSCs. This device imitates the natural processes that plants convert solar radiation into energy—photosynthesis. In their DSSCs, a nanocrystalline TiO_2 film was sensitized by Ru bipyridl complex and the charge separation is due to kinetic competition of charges. In DSSC architecture, each functional part has distinct boundaries with other parts and the picture how device works is clear. Transparent conductive film is an important part of nano solar cells including DSSCs, and the requirements of transparent

Introduction to Nano Solar Cells
Ning Dai

ISBN 978-981-4877-49-7 (Hardcover), 978-1-003-13198-4 (eBook)
www.jennystanford.com

conductive film for different solar cells are basically the same. In this chapter, DSSCs are used as a prototype photovoltaic cell to illustrate the basic characteristics, preparation and application of transparent conductive films.

Many routes have been developed to fabricate the transparent conductive films (TCFs) and the preparation of photoanodes usually uses chemical ways. The performance of various TCFs is judged according to the conductivity and transparency. In the meanwhile, the cost is an important factor one has to face in large-scale production. Popular TCF materials are metal oxides—transparent conductive oxides (TCOs), including ZnO, TiO_2, In_2O_3, SnO_2, CdO, etc. [4–7]. The alloy of In_2O_3 and SnO_2, indium-tin-oxide, is widely used in flat panel display due to its excellent performance as an alloy TCO [8]. Sometimes, multi-layer films are used to improve the performance [9, 10]. Other than those metal oxide thin films, carbon-based materials are also the candidates for TCO such as graphene and carbon nanotubes [11, 12], as well as metal mashes [13–15].

4.1 Dye-Sensitized Solar Cells

Working on the principle of plant photosynthesis, dye-sensitized solar cells (DSSCs) are new concept of SCs that bear specialty different from traditional SCs [16, 17]. Each part of dye-sensitized battery has clear function, well-separated by clear boundaries. The electrodes of DSSCs are not only used for conducting electricity, but also for splitting electrons and holes. Figure 4.1 shows schematically the work principle of the DSSC, where the device is comprised of a photoanode consisting of a semiconducting scaffold and sensitizers, a redox electrolyte containing iodide and triiodide ions, and a counter electrode, as the major parts. The energy levels of the functional parts are also shown. Upon absorbing a photon, an electron in dye is excited and injected into the conduction band of titanium dioxide TiO_2. Subsequently, an electron, donated by a redox couple in electrolyte, is regenerated back to the ground state. Thus, in a DSSC, the semiconducting scaffold is used as a photoelectron transporter only and the photoelectrons are provided by the sensitizers, while in a traditional solar cell the Si semiconductor acts as source of photoelectrons, the provider of the built-in electric field for the electron/hole separation, and the transporter of photoelectrons and photoholes.

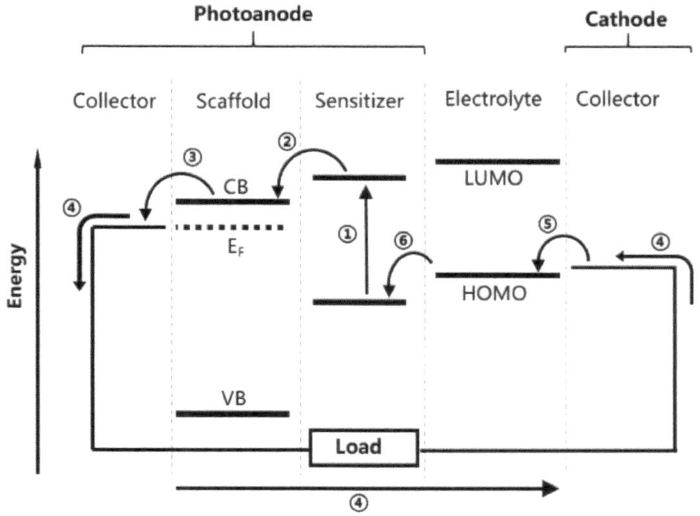

Figure 4.1 The working principle of a dye-sensitized solar cell, where electrons are excited by absorbing photons and follow the route from ① → ⑥.

4.1.1 Photoanodes of Dye-Sensitized Solar Cells

As depicted in Fig. 4.1, the dye-sensitized electrode absorbs photons and collects electrons as negative pole of the cell, so that it is termed as the photoanode. The photoanode is constructed with TCO glass and a thin layer of, for instance, TiO_2 nanomaterial attached with dye molecules or semiconductor quantum dots (termed as quantum-dot-sensitized solar cells). The counter electrode consists of conductive glass coated with noble metal platinum, acting to collect electrons from out-circuit and acting as reduction catalyst for the regeneration of I^-/I_3^- redox couples. One of the most attractive aspects of DSSCs is their potentially low cost if the expensive materials in DSSCs (e.g., platinum) are replaced with cheap nanomaterials. Since almost all the components in DSSCs are made from nanomaterials, the discussion on nanomaterials and structures presented in this chapter will be based on their roles in DSSCs.

The semiconducting scaffold is a transparent conducting film allowing solar light to penetrate through. Upon absorption of a photon with energy larger than the energy spacing between the energy spacing of the ground state and the excited state in a quantum dot or between the highest occupied molecular orbital (HOMO) and the lowest unoccupied molecular orbital (LUMO) energy levels of a dye molecule, electron will be excited from the ground (LOMO) state to the excited (HUMO) state. Thus, QDs or dye molecules act as photon antennas. The conduction band energy of TiO_2 is lower than the excited state level. As a result, the electron transfers from the QD to TiO_2 and then diffuses to the TCO film, so that the electron is separated from the leftover hole spatially, the quantum dot being oxidized. The TCO film is the electron-collecting electrode that is connected to the external circuit. The counter electrode, a glass substrate coated with a catalyst layer, is placed parallel to the front electrode (photoanode) in a face-to-face arrangement with a typical spacing of a few tens of micrometers. The spacing between the photoanode and the counter electrode is filled with electrolyte as conducting media for carriers. The oxidized QD is reduced by obtaining an electron from an iodide/triiodide (I^-/I_3^-) redox couple in electrolyte and the redox couple is regenerated by diffusing toward the counter electrode where it obtains an electron from the external circuit.

In quantum-dot-sensitized solar cells (QDSSCs), semiconductor quantum dots, such as CdSe, CdS, CdTe, PbS, PbTe, SnS, or Sb_2S_3, replace dye molecules in traditional DSSCs, as sensitizers. Semiconductor QDs have excellent photon harvesting ability and tunable bandgaps that are QD size dependent. The device architectures of QDSSCs are similar to their DSSCs counterparts, which consist of a nanomaterial TiO_2 deposited on TCO glass and loaded with QD sensitizers, a Pt counter electrode, as well as an electrolyte solution with redox couples. Due to serious photodegradation caused by I^-/I_3^- electrolyte [18], QDSSCs usually use polysulfide couple (S_n^{2-}/S^{2-}) electrolyte suitable for the stability of those semiconductor QDs [19]. The photoanodes of QDSSCs consist of transparent conductive electrode, conductive scaffold, as well as QD sensitizers. The sensitizers are mixed with the conducting scaffold. A carton picture of a QDSSC is shown in Fig. 4.2.

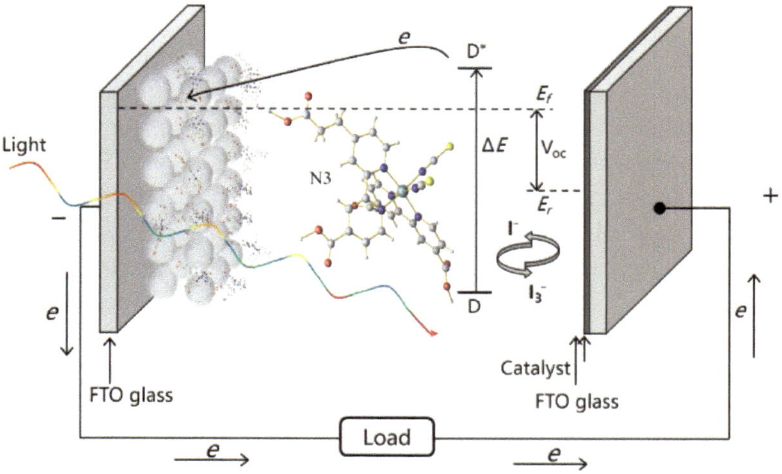

Figure 4.2 The schematic diagram of the DSSC which consists of an N3 dye, front and back FTO glasses, nano TiO_2, as well as the catalyst layer.

4.1.2 Transparent Conducting Electrodes

Transparent conducting electrode (TCE), or transparent conducting film (TCF), are widely used in various thin-film solar cells and other electronics such as flat panel displays and touchscreens. The materials used for TCFs are mostly metal oxides in thin-film forms, which are optically transparent and electrically conductive. In literatures, they are termed as transparent conductive oxide (TCO). The fundamental requirements for TCOs are highly conductive, highly transparent, and low cost. Other requirements, depending on applications, include chemically and mechanically stability, as well as flexibility in terms of application features. A number of oxide materials, including In_2O_3, SnO_2, ZnO, TiO_2, and CdO, show good transparent conducting properties. Among them, Cd is toxic and the TiO_2 layer requires rather high temperature thermal treatment. Very often, ternary or quaternary compounds of the metal oxides are formed to achieve optimized transparency and conductance, together with doping manipulation. Alternative materials and structures for electrodes include conductive polymers, metal grids, and carbon nanotube (CNT), graphene and various other nanomaterials.

TCOs used for solar cells are typically grown on transparent substrates such as glass and polyimide (PI) as supporting materials. The substrates are infrared-opaque, which has the benefit of preventing the infrared radiation to heat up the device. The preparation routes of TCOs include metal organic chemical vapor deposition (MOCVD), molecular beam epitaxy (MBE), spray pyrolysis, dc or rf magnetron sputtering, pulsed laser deposition (PLD), as well sol gel, among which magnetron sputtering has been popularly used. Magnetron sputtering needs targets that are usually inexpensive. As a result, the microstructures of those films are polycrystalline, nanomaterials, or amorphous.

As an important component in a NSC, TCO is used as a window allowing for light to pass through and reach to the active layer where photo-generated carriers are excited. Thus, it must be transparent at least to the photons with energies above the bandgap of the active material. TCO also acts as an electrode in a NSC to collect carriers and provide carriers for out-circuit, so that its Ohmic type of contact to the underneath cell is essential for carrier transport. It must be conductive then. For solar cell uses, transparent conducting electrodes should have a light transmittance greater than 85% for major solar spectrum and resistivity below 20 $\Omega/\square$ allowing for efficient carrier transport. Thus, TCO materials should have a bandgap larger than 3 eV allowing for most part of the solar spectrum to penetrate through. The transmittance of TCO electrodes is limited by scattering to incident light by dislocations, defects, and grain boundaries which rely strongly on the fabrication techniques. In order to have a good electric performance (low resistivity), TCOs should have low densities of dislocations, defects, and grain boundaries that also act as scattering centers for carriers for achieving high carrier mobility of above ~30 $cm^2/(V \cdot s)$. A high carrier concentration on the order of 10^{20} to 10^{21} cm^{-3} is needed for achieving low resistivity. Currently, the popular TCO are all *n*-type conductors in which the *n*-type carriers are due to the interstitial metal ions and oxygen vacancies both as donors, as well as doping ions. Typically, magnetron sputtering is adopted for preparing the

TCO electrodes. As a result, growth for TCO is often carried in a reducing environment for the formation of oxygen vacancies that act as donors.

Currently, the TCO of best performance is indium tin oxide (tin-doped indium-oxide, ITO), an alloy consisting of indium, tin and oxygen in varying compositions. A typical formulation is, by weight, 74% In, 18% O_2, and 8% Sn. ITO gives a sheet resistance of 5 Ω/□ (Ohms per square meter) at an average 90% transmittance in the visible solar spectrum range, or a low bulk resistivity of $\sim 10^{-4}$ Ω·cm [20]. Increasing the thickness of the TCO layer leads to the decrease of sheet resistance, but also decrease in transmittance. A trade-off between sheet resistance and light transmittance is usually made to meet a specific application. Thus, a compromise must be made for a given application. Further overall improvements on both resistance and transmittance are still possible so that the effort is still on their ways. ITO is, however, a bit mechanically brittle, which limits its use in flexible electronics. In addition, ITO is unstable chemically with respect to acids and at high temperature. ITO is easy to fabricate by several deposit techniques but expensive since indium is a rare-earth metal, which is no doubt a problem for large-scale uses. One is therefore looking for its substitution using other materials or multi-component oxides that use less indium and, very often, are composed of a combination of In_2O_3, SnO_2, and ZnO. Doped binary compounds, namely, aluminum-doped zinc oxide (AZO) and fluorine-doped tin oxide (FTO) are two widely used alternative materials.

Figure 4.3a presents evolutions of transmittance spectra measured on ZnO and Al-doped ZnO (AZO) grown by atomic layer deposition (ALD) at several DEZ-H_2O/TMA-H_2O cycle ratios of 50:1, 20:1, 10:1, 5:1 corresponding to the increase of Al content in ZnO [21]. Apparently, Al doping improves the film transmittance in the whole solar spectrum range. However, AZO films tend to be unstable at high Al concentration. In_2O_3 or In_2O_3-based films also show good TCO performance. Figure 4.3b depicts the evolutions of transmittance spectra with annealing conditions [22]. In_2O_3 film was deposited on quartz substrate by chemical spray pyrolysis (CSP) and then annealed at 550, 600, 650, and 700°C. The annealed samples are apparently superior to the as-deposited one also presented in Fig. 4.3b. The sample annealed at 600 displays the best performance. Unfortunately, the abundance of indium on earth is limited, so that SnO_2 film is much more expensive than ZnO films.

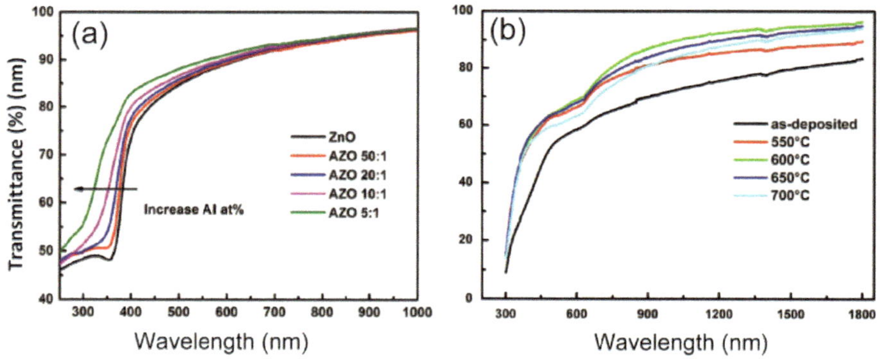

Figure 4.3 Evolutions of transmittance spectra of ZnO:Al (AZO) ALD-grown at 150°C with various doping levels (a) and In_2O_3 deposited by the chemical spray pyrolysis (CSP) method at 350°C for 2 h at several annealing temperatures (b). The films are all prepared on quartz substrates with the thickness of ~0.05 and ~1 μm for ZnO and In_2O_3, respectively. (a) Reproduced with permission from [21]; (b) reproduced with permission from [22].

Another TCO that gains widespread attention is ZnO:X (X is a dopant, usually, Al and Ga). ZnO:Al is composed of two inexpensive materials, Zn and Al. After doped with Al, the overall performance of a ZnO electrode is improved significantly. ZnO has a bulk bandgap of 3.37 eV, large enough for transparency in the major solar spectrum range from near infrared to near ultraviolet, and a low bulk resistivity below 10^{-3} Ω cm. Al-doped ZnO (ZnO:Al or AZO) and Ga-doped ZnO (ZnO:Ga or GZO) thin films can be prepared by magnetron sputtering. In preparing for AZO, for instance, two separate targets of ZnO and Al_2O_3 or of ZnO and Al are often used. Using those inexpensive targets, the magnetron sputtering for preparing AZO is very cost-effective and suitable for mass production.

AZO is not very stable though it is cheap and has satisfactory performance. Fluorine-doped tin oxide (SnO_2:F or FTO) is thermally and chemically more stable than ITO, which is compatible to solar cell technologies. FTO is cheaper, but it has an overall higher resistance and lower optical transmittance, 15 W/□ at an 85% transmittance in visible solar spectrum.

4.1.3 Nanomaterial Scaffold

The nanomaterial scaffold is part of the photoanode with which sensitizers are incorporated. Similar to TCE, the scaffold is usually made up of metal oxides, including zinc oxide (ZnO), titanium dioxide (TiO_2), indium oxide (In_2O_3), and the hybrids, that are conductive and transparent to visible light. The nano forms of the scaffold offer large surface area for dye attachment. The most widely used scaffold material is TiO_2 prepared to form nano (porous) structure for large surface area to adsorb as many dye molecules or QDs as possible. However, the conductivity of nano TiO_2 is usually unsatisfactory, therefore a composite film consisting of nano TiO_2 attached on TCO is often a good solution.

TiO_2 is a promising material that is highly expected to play important roles in accessing to clean energy and solving many environmental problems. Bearing the advantages of non-toxicity and environmental compatibility, TiO_2 is well known as one of the most promising photo catalysts for water splitting, beneficial from its transparency to sunlight and appropriate conduction band minimum located right at the level of the H_2/H_2O potential. Other applications include batteries, optical devices, photonic crystals, chemical sensors, environmental purification, as well as solar cells. For instance, TiO_2 electrodes are widely used for electrochromic devices and lithium batteries. As an oxide semiconductor, TiO_2 is stable in air and in contacting with many other materials. Metal Ti is relatively abundant and widely distributed on earth. However, the material is difficult to extract. For solar cell applications, TiO_2 is probably the most widely used photo-anode material in various nano forms with which they are both conductive and transparent in visible. The TiO_2 electrodes in anatase phase have an average visible light transmittance of 70% and a resistivity of 0.08 $\Omega \cdot$ cm.

Titanium dioxide (TiO_2) has a bulk bandgap of 3.0 to 3.2 and is transparent to visible. It is difficult to measure the bandgap of TiO_2 exactly since the material in bulk is not in single crystalline structure and the bandgap is modified by impurities. There are eleven TiO_2 crystalline phases, namely, rutile, anatase, brookite, TiO_2(B), TiO_2(H), TiO_2(R), TiO_2(II), baddeleyite, OI, OII, and cubic as identified by x-ray diffraction and pair-distribution function analysis. Among them, rutile, anatase, and brookite are the mineral forms, with rutile being the most common natural form (about 95%) and the transitions between the phases of TiO_2 are possible to happen under pressure or heat and the transitions are size and shape dependent. Figure 4.4 illustrates the unit cells of rutile, anatase, and brookite phases of TiO_2.

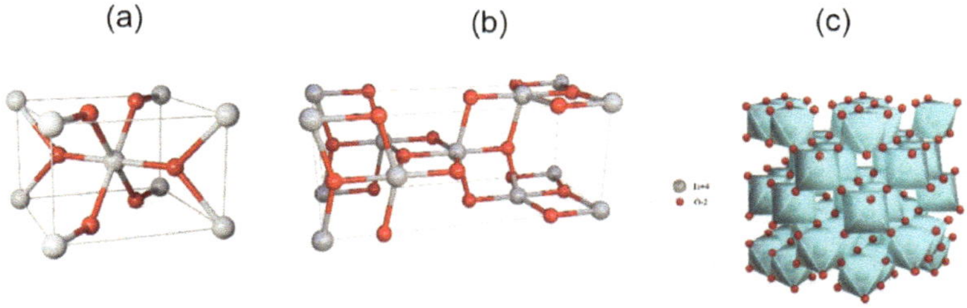

Figure 4.4 The unit cell of rutile (a), anatase (b), and brookite (c). Ti atoms are gray; O atoms are red.

TiO_2 is conductive due to auto-doping, though a wide bandgap semiconductor. TiO_2 is *n*-type due to unintentionally *n*-type doping and relatively low conductor band position with respect to vacuum energy level. Though the growth of pure single crystal is quite difficult, the crystal is easy to obtain in nano forms. In fact, many approaches have been developed to grow TiO_2 in the nano forms of nanopipes, nanodots, nanoshells, etc. The easiness in preparing the material with various nano forms, together with its transparent property in visible and good conductivity offer the material with good opportunity to be used as photoanodes for solar cells, especially dye-sensitized SCs.

4.1.4 Examples of Photoanodes

A photoanode is composed of TCE (TCO), conducting scaffold, and sensitizers, as shown in Fig. 4.1. The variety of the photoanodes correspond to the combination of the TCO, scaffold, and sensitizer, there being a large number of combinations. In this section, several photoanodes and corresponding devices are presented, namely, photoanodes with TiO_2-passivated 3D AZO, SnO_2, or TiO_2 backbone infiltrated with TiO_2 nanoparticles, with TiO_2/multi-walled carbon nanotubes, and with ZnO nanowires.

TiO_2-passivated 3D AZO

TiO_2 is a widely used material for photoanodes but its electron mobility is low. The electron mobilities in AZO and SnO_2 are about 6 and 5 orders of magnitude higher than that of anatase TiO_2 nanoparticles. Grätzel's group used high-mobility AZO and SnO_2 backbones passivated with TiO_2 for improving surface [23]. They reported the fabrication of a 3D photoanode, with a host-passivation-guest (H-P-G) concept according to the material functions. The photoanode consists of an FTO front electrode on glass, TiO_2 passivated 3D AZO backbone, as well as Z709 dye. The scaffold layer was fabricated by using disordered 2.2 μm polystyrene spheres prepared on the FTO as template followed by a layer-by-layer ALD deposition of Al, ZnO, or TiO_2. Such a sequential deposition guarantees the good quality of interfaces between Al and ZnO and between ZnO and TiO_2. Specifically, the template was made up of large monodispersed polystyrene spheres doctor-bladed on FTO top electrode. The large size of the spheres made it easy for the formation of large interconnecting pores. The polystyrene spheres were then coated with 90 nm AZO (~10% Al), SnO_2, or TiO_2 by ALD. A dry reactive ion etching was then performed to remove the top surface oxide layer, followed by an annealing at 360°C for 15 min to remove the polystyrene template, giving rise to a 3D host backbone on FTO glass as shown in Fig. 4.5a. The 3D backbone was then passivated by dense TiO_2 layer of 25 nm thickness (see Fig. 4.5b) on both sides of the shells. The TiO_2-passivated 3D backbone took only ~10% volume fraction. The passivation is important, since the interface between SnO_2 (or AZO) and electrolyte leads to very high carrier recombination rate, in comparison to the interface between TiO_2 and electrolyte, although electrons in ZnO and SnO_2 have much higher mobility and conductivity than in TiO_2. By calcinating backbone materials at 500°C for 15 min, X-ray measurements demonstrated the wurtzite, cassiterite, and anatase crystal structures of the AZO, SnO_2, and TiO_2, respectively.

The backbones were infiltrated with anatase TiO_2 nanocrystal paste (the guest) followed by calcination, which was made easy by their large pore sizes [24]. This step is necessary for the formation of large surface area for the effective loading of large number of sensitizers.

To improve electronic connection among the guest TiO_2 nanoparticles, the TiO_2 and SnO_2 H-P-G photoanodes were immersed in an aqueous solution of 40 mM $TiCl_4$ at 70°C for 30 min followed by water and ethanol washes. This procedure was not performed on the AZO H-P-G photoanode, since AZO was chemically unstable when treated $TiCl_4$ acid. $TiCl_4$ treatment was found significantly effective in increasing the photo-current and photo-voltage via improving electron hopping cross the boundaries between TiO_2 nanoparticles.

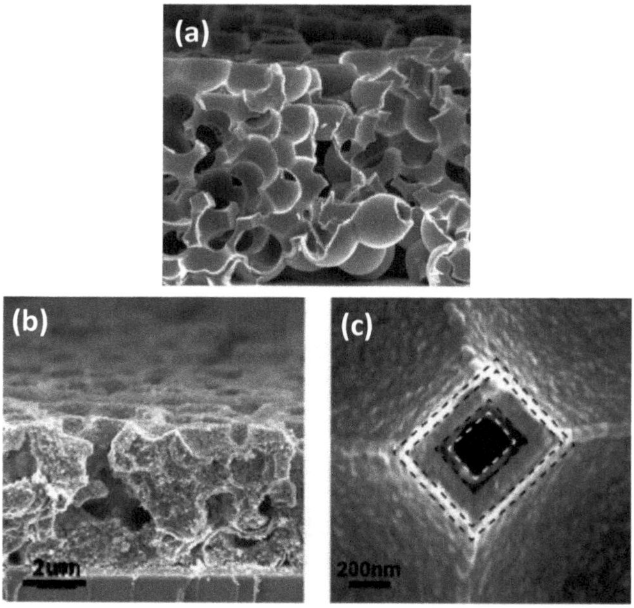

Figure 4.5 An SEM image of the TiO_2-passivated 3D AZO backbone (a) and the zoom-in image showing that the 3D AZO backbone covered with a 25 nm thin layer of dense TiO_2 on both sides of the wall, as shown by the dashed lines (b), and the completed photoanode consisting of the TiO_2-passivated 3D AZO backbone infiltrated with TiO_2 nanoparticles (c). Reproduced with permission from [23].

Figure 4.5c depicts the SEM image of the completed photoanode with AZO backbone, showing that about 50% of the backbone internal volume is filled with the TiO_2 nanoparticles. Three kinds of DSSCs devices were constructed, with different 3D photoanodes of TiO_2-passivated AZO, TiO_2-passivated SnO_2, and TiO_2-passivated TiO_2 backbones loaded with TiO_2 nanoparticles. Electrically, the photoanodes offer good control on the electron extraction and transport by reducing the recombination. Optically, the nanostructure photoanodes enable significant reduction on light reflection by increasing optical scattering. The photoanodes were calcined at 500°C for 15 min and then immersed in a Z907 solution (0.3 mM) for 10 min. The dye-sensitized electrodes were obtained after they were rinsed with acetonitrile followed by air drying. The electrolyte of 85:15 acetonitrile/valeronitrile mixture with I_3^-/I^- redox couples was used for the DSSCs, with the counter electrode being platinized FTO. Figure 4.6 presents the *J–V* curves for the DSSCs. Obviously, the DSSCs show very large open-circuit voltages of 842, 804, and 791 for the DSSCs with AZO, SnO_2, and TiO_2 backbone hosts, respectively, which are well above the best result of 730 mV obtained on DSSCs using Z907. Thus, one could expect reduced carrier recombination and reduced dark current that was experimentally proven by measured results presented in the low portion of Fig. 4.6.

From TiO_2, AZO, to SnO_2 backbone hosts, the presented photo-current increase is an indication of improvement in carrier transport. The electron mobility in SnO_2 host is about four orders of magnitude higher than that of TiO_2 host, which has a direct consequence for high current density as shown in Fig. 4.6. The AZO host has even a higher electron mobility than SnO_2 but it does not present a higher short-circuit current due to poor electronic connection between TiO_2 nanoparticles, since the AZO H-P-G photoanode was not treated by $TiCl_4$ acid. Yet the device with the AZO host still exhibits higher current than that with TiO_2 host due to the superior mobility of AZO in comparison to TiO_2. Thus, the use of high-mobility host backbone materials is beneficial for the enhancement of photo-current, while the use of TiO_2 as passivation layers is evidently effective for the reduction of carrier recombination at interfaces. A double layered TiO_2 photoanode consisting of hierarchical nanowire arrays and nanoparticles showed improved device performance [25].

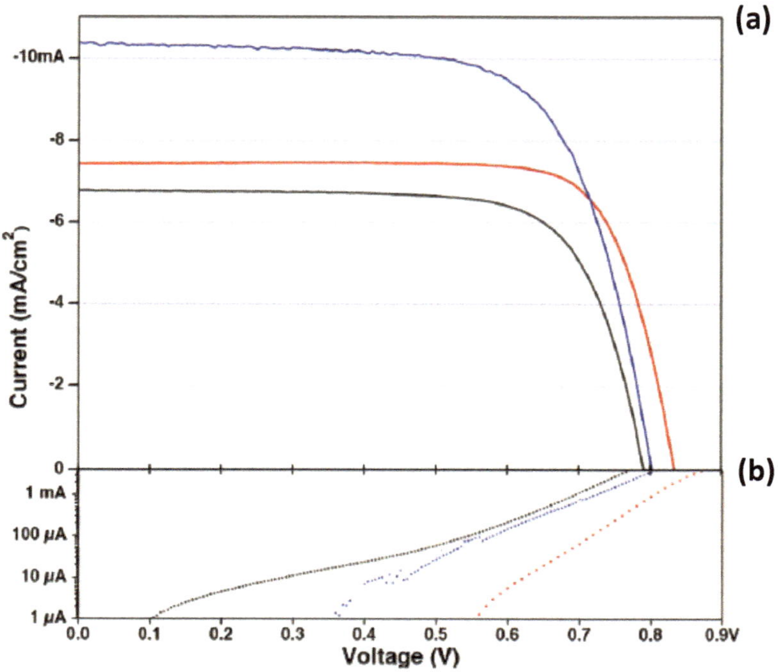

Figure 4.6 (a) J–V curves of the 3D host-passivation-guest DSSCs under AM 1.5 illumination. The measured parameters are J_{sc} = 7.5 mA/cm^2, V_{oc} = 842 mV, FF = 0.77, η = 4.9% for TiO_2-passivated Al/ZnO (red), J_{sc} = 10.4 mA/cm^2, V_{oc} = 803 mV, FF = 0.70, η = 5.8% for TiO_2-passivated SnO_2 (blue), and J_{sc} = 6.9 mA/cm^2, V_{oc} = 791 mV, FF = 0.73, η = 4.0% for TiO_2-passivated TiO_2 (black) backbone hosts; (b) the lower portion of the figure presents the corresponding dark currents (the dashed lines). Reproduced with permission from [23].

TiO_2/multi-walled carbon nanotubes

The conversion efficiency of NSCs is still significantly low compared to wafer-based Si SCs. The major obstacles in achieving NSCs with high conversion efficiency stems from the carrier recombination occurring majorly at grain boundaries among nanoparticles and the number of grain boundaries are huge. Thus, photon-excited electrons are likely to recombine with holes and be slowed down when travelling through the nanostructured network. TiO_2 is, on the other hand, not a very good conductor for electrons. To reduce the electron–hole recombination and accelerate the electron transport, Du et al. added multi-wall carbon nanotubes (MWCNTs) into nanostructured TiO_2 in order to improve the electronic properties of the composite by means of excellent electrical conductivity of MWCNTs [26].

Du and co-workers fabricated DSSCs with photoanodes based on TiO_2/MWCNTS composite nanofiber thin films with the TiO_2: MWCNTS weight ratios of 0, 0.1, 0.3, 0.5, and 1 wt.% [26]. The TiO_2/MWCNTS nanofiber thin films were prepared using electrospinning, a technology developed by Kim and co-workers on the FTO conductive glass substrate [27]. The TiO_2/MWCNTS nanofiber thin films were then sensitized using 0.5 mM N719. The DSSC was completed by assembling the sensitized composite Pt-coated counter electrode with 60 μm sealing spacer. The electrolyte was I^-/I_3^- acetonitrile (0.1 M lithium iodide, 0.03 M iodine, 0.5 M 4-tert-butylpyridine, 0.1 M guanidine thiocyanate, and 0.6 M 1,2-dimethyl-3-propylimidazolium iodide).

As shown in Fig. 4.7, the DSSC gives a photo-conversion efficiency of 4.46%, at zero MWCNTs content. Adding 0.1 wt.% MWCNTs into the electrode gives rise to an increase of conversion efficiency, indicated by a substantial increase of short-circuit current J_{sc} from 10.6 to 11.0 mA·cm^{-2} and a slight increase of open-circuit voltage V_{oc}. Increasing the MWCNTs content to 0.3 wt.% leads to significant enhancement of both J_{sc} and V_{oc}, corresponding to the conversion efficiency of 5.63%, a 26% increase with respect to the case of zero MWCNTs content with a 4.46% conversion efficiency. Further increasing the MWCNTs content, however, gave rise to deterioration of the conversion efficiency. Thus, adding a proper amount of MWCNTs in TiO_2 nanofibers enhances both short-circuit current and open-circuit voltage and, in turn, the power conversion efficiency.

MWCNTs have high electron affinity, which makes it easy for electrons in TiO_2 to transfer into MWCNTs. Thus, the electrons travel in MWCNTs move faster than they do in TiO_2 due to reduced scattering, i.e., the MWCNTs offer high conductive channels for photoelectrons. The electrochemical impedance curves were measured on the samples and the results presented in Fig. 4.8 clearly demonstrate the one-to-one correspondence between the impedance curves and the conversion efficiencies shown by the *J–V* curves in Fig. 4.7.

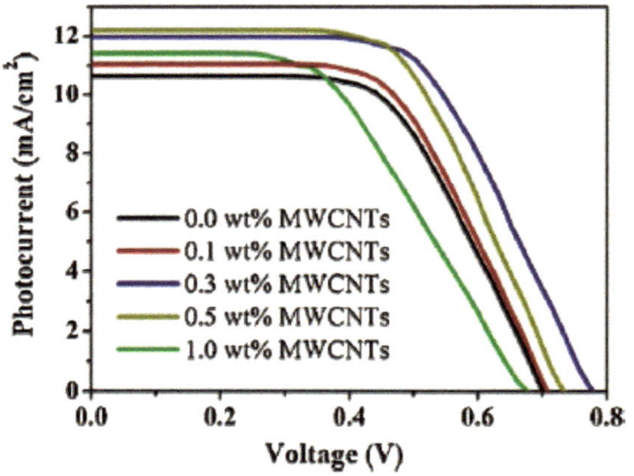

Figure 4.7 Measured *J–V* curves of the DSSCs with TiO_2/MWCNTs electrodes of different TiO_2: MWCNTs weight ratios. Reproduced with permission from [26].

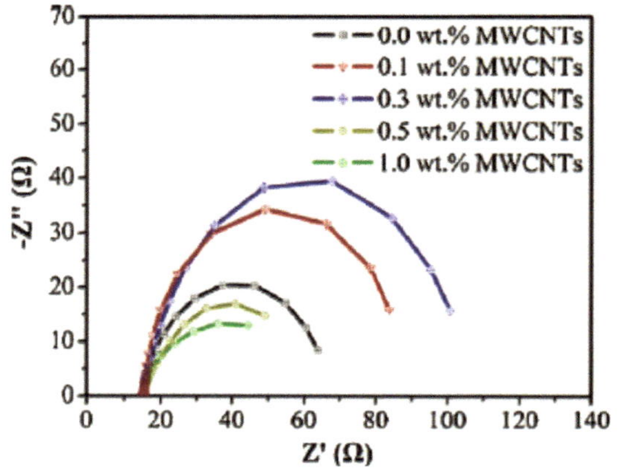

Figure 4.8 Measured electrochemical impedance curves of the DSSCs with TiO_2/MWCNTs electrodes of different MWCNTs contents. The measurement was carried on in the dark. Reproduced with permission from [26].

ZnO nanowires

Compared with TiO_2 nanostructures, ZnO nanostructures have much higher electron conductivity and a large electron diffusion coefficient, making it easy for photo-excited electrons to be collected by the electrode. Additionally, ZnO nanowires can be grown easily by many routes, as seen by a large number of publications. Among nanostructured materials, nanowires patterns offer the easiest infiltration for electrolyte due to their one-dimensional configuration, in comparison to mesoporous films. Good contacts between electrolyte and sensitizers can thus be made.

Yin and co-workers prepared photoanodes consisted of ZnO nanowires films, co-sensitized by CdS–CdSe, on FTO conductive glass substrates [28]. In the fabrication, a ZnO seed layer was initially grown on FTO glass slab based on a sol gel method. ZnO nanowires were then prepared by using a precursor containing 0.04 M $Zn(NO_3)_2$ and 0.8 M NaOH or by using a precursor containing 0.05 M $Zn(NO_3)_2$, 0.05 M hexamethylenetetramine, and varied amounts of polyetherimide (PEI). Experimental conditions were varied in order to obtain ZnO nanowires with different lengths. Figure 4.9 presents the cross-sectional SEM image of the ZnO NWs. The ZnO NWs were then coated with CdS and/or CdSe sensitizer layers, following the SILAR procedure [29]. The counter electrode used Au-sprayed FTO substrate and the electrolyte was a mixture of 1 M Na_2S, 2.5 M S, and 0.2 M KCl in 7:3 water/methanol of a volume ratio. A transparent hot-melt film was used to hot-pressing seal the device.

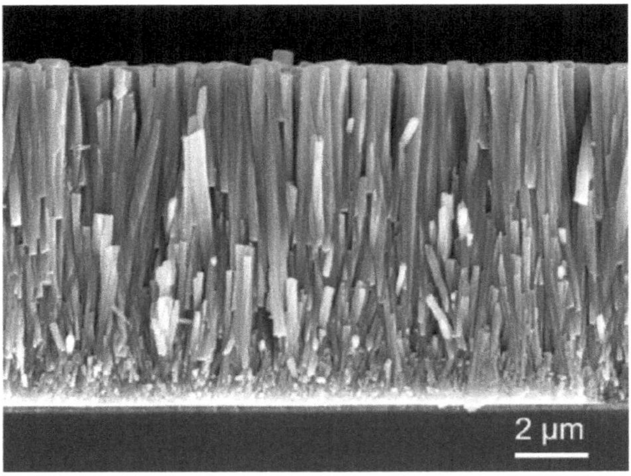

Figure 4.9 A cross-sectional SEM image of ZnO NWs grown using precursor solution. Reproduced with permission from [28].

SCs with the photoanodes consisting of ZnO NWs arrays co-sensitized by CdS/CdSe layers were fabricated for different ZnO NWs lengths and aspect ratios and different CdS, CdSe, and CdS/CdSe layer thickness. Those parameters were found to affect the device performance significantly. A best efficiency of 3.06% corresponding to a J_{sc} of 18.63 mA/cm^2 and V_{oc} of 480 mV was achieved on an optimized device, which was due mainly to the relatively strong light absorption and better electron injection efficiency.

4.2 Carbon-Based Transparent Conducting Electrodes

4.2.1 Carbon Nanotubes

Metal oxide films have great performance as electrodes for photovoltaic devices. However, they are mechanically fragile, which limits their uses in some cases. For instance, it has been shown that ITO tends to degrade with time when it is subjected to stress. In some applications where a continuous bending is required, TCO could break down due to fatigue. Carbon-based thin-film electrodes are mechanically strong and tolerate mechanical bending. They are, therefore, a potential alternative to TCO for the use in many special cases. Flexible solar cells, for instance, are bendable devices which open new possibilities for applications in flexible electronics as well as in integration with buildings especially on uneven surfaces such as tiles. In addition, flexible solar cells are very thin and light in weight, so that they are attractive for space uses.

Carbon nanotubes (CNTs) have been intensively studied aiming at a wide spectrum of applications. CNTs are excellent in many materials properties, specially the mechanical properties including a high elastic modulus of ~2 TPa, a high tensile strength of ~50 Gpa. Other than the mechanical properties, the electric characteristic is also excellent. Theoretically, it has been shown that metallic CNTs could tolerate an electric current density of 4×10^9 A/cm^2, about ~1000 times that of a good metal conductor like copper) [30]. Currently, CNT in thin-film forms is used as transparent electrodes in thin-film transistors (TFTs) and displayers.

There is a great deal of reports on the growth of CNTs, including various chemical vapor deposition (CVD), arc discharge, laser ablation, etc. CNT thin films are usually prepared using a precursor of CNTs in solution. It is important to separate the CNTs in solution for good transparency and conductance. A proven effective way is to mix the CNTs with surfactant and solution. A thorough sonication is very helpful for the separation of CNTs. The precursor is then spin-coated or dip-coated on a substrate, followed by thermal treatment to remove the solvent. Other methods include spray deposition using an ultrasonic nozzle. After the formation of thin films, the surfactant can be removed by water rinsing. In the prepared films, CNTs with different lengths, diameters, and chiralities are entangled together via van der Waals interaction. The van der Waals connecting between CNTs is unfavorable for achieving low resistance. Other measures are often taken to reduce the resistance.

The conductance and the transmittance of the CNT thin films are determined by diameters and chiralities of the CNTs. One could expect that pure and long CNTs give rise to high conductivity. The conductance and the transmittance of CNT thin films could also vary with tube diameters and depend on whether they are single wall or multi walls in geometrical configurations. If CNTs form ordered patterns, instead of randomly oriented bundles, one could expect increases for both conductance and transmittance, due to expected reduction in random scattering to electrons and better contact between carbon nanotubes. Patterned CNTs offer possibility to enhance transmittance by means of light interference. Performance of the prepared film also depends on the film preparation parameters, such as spinning speed, amount of precursor dropped, precursor viscosity, etc. At present, the overall performance of CNT thin-film electrodes for thin-film solar cells is not as good as that of traditional TCO, AZO and ITO materials. However, CNT thin films do have some advantages superior to other TCOs. For instance, CNT thin films show 90% transmittance in the infrared spectral range. This excellent performance shows that the

CNT electrode has high heat dissipation, which makes the CNT electrode very useful in some special applications.

4.2.2 Graphene

Graphene has a two-dimensional honeycomb lattice made of carbon atoms, i.e., an allotrope of carbon, as shown in Fig. 4.10a. Other allotropes include diamonds, graphite, carbon nanotubes, and fullerenes. Note that graphene is a single atomic sheet of graphite with the sheet surface normal along (0001) orientation. In graphene, each carbon atom has three nearest neighbors at a distance of 0.142 nm apart. As a group IV atom, each carbon atom has four bonds, three σ bonds, respectively, connecting to three nearest neighbor atoms and one π-bond pointing out the graphene sheet. The σ bond is formed by sp^2 orbital hybridization, i.e., a mixture of s, p_x, and p_y orbitals. The leftover p_z orbit composes the π-bond. The π-band and $\pi*$-bands then take their shapes through the π-bonds hybridization. The band structure of a single graphite sheet can be calculated by a tight-binding model [31]. The electronic and optical properties of graphene largely rely on the details of the σ-band and the π-band ($\pi*$-band). For instance, the free electrons in graphene are due to half-filled bands. The tightly packed carbon atoms and the covalent bonding feature determine that the graphene sheets are mechanically very robust.

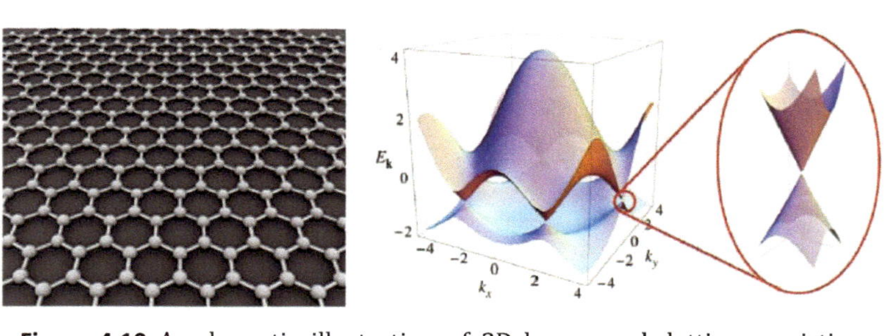

Figure 4.10 A schematic illustration of 2D honeycomb lattice consisting of carbon atoms (a) and the E–$\boldsymbol{k}$ dispersion in reciprocal space where the zoom-in at one of the Dirac points is also shown (b).

Graphene's hexagonal honeycomb lattice was revealed by transmission electron microscopy (TEM) in reciprocal space (momentum space) and scanning tunneling microscopy (STM) in real space, both having atomic resolutions. Graphene has a zero bandgap, with its conduction and valence bands connect at the Dirac points on the boundary of its Brillouin zone, as shown in Fig. 4.10b. Electrons present in pristine graphene behave as massless particles. There are six Dirac points due to the symmetry. Thus, as a gapless material, graphene does not show an optical absorption edge as other semiconductors do. In other words, graphene absorbs photons of any energy.

Currently, the reported room-temperature electron mobility is as high as 200000 $cm^2/V\cdot s$ for intrinsic graphene [32], while, depending on the use of substrates, most reported results fall around 10000 $cm^2/(V\cdot s)$. However, the very high mobility for intrinsic graphene does not result in low resistance due to low carrier concentration. The sheet resistance of a single layer of intrinsic graphene is around 6 kΩ/□ at an average solar spectrum absorption of 2.3%. Great deal of efforts has been made in reducing the resistance. Intuitively, doping has been tried intensively to reduce the resistance and it has been proven to be effective. Punckt et al. prepared functionalized graphene sheets by thermal exfoliation and graphite oxide reduction, and found that resistance of functionalized graphene sheets was tunable by verifying carbon to oxygen ratio [33]. The sheet resistance was reduced by nearly two orders of magnitude from 500 kΩ/□ for a carbon-to-oxygen ratio of ~9 to 7.7 kΩ/□ for a ratio of ~340.

There are many reported graphene structures, which depend on the ways of the preparations. The different preparations result in different forms of nanographene. On the other hand, doping studies have been carried out intensively for multiple purposes such as controlling the *n*- or *p*-type. Readily obtainable graphene and graphene-based material includes pristine graphene, graphene oxide, as well as reduced graphene oxide which is prepared by reducing graphene oxide via chemical or thermal methods. Reduced graphene oxide has structure and physical properties between pristine graphene and graphene oxide. Also classified by shapes, graphene also takes the forms of graphene nanoplates, nanolatelets, and even graphene quantum dots. From the 2D atomic configuration in Fig. 4.10a, it is easy to see that every carbon atoms in graphene are available for functionization from both sides. Many atoms and functionized groups can attach to graphene with high densities. Indeed, it was found that single-layered graphene sheets react more strongly with hydrogen atoms than multi-layer sheets. As a result, a great many of graphene-based materials can be design with broad spectrum of properties and fabricated with their electronic properties and biology properties, for instance, modifiable in a large extent.

Pristine graphene are currently prepared by mechanical exfoliation, chemical vapor deposition (CVD) [34–36], as well as epitaxial growth [37]. CVD preparation uses a gaseous carbonaceous precursor that flows over a copper and nickel substrate at high temperatures (~1000°C). The metals are used both as substrates and catalysts for carbon nucleation that leads to the growth of graphene sheets. The epitaxial growth for graphene is quite similar to the growth for other semiconductors. Although CVD and epitaxial growths lead to monolayer graphene sheets, their throughputs are very low. The current research efforts for the preparation of pristine graphene focus on reducing the number of defects and grains in CVD- and epitaxial grown graphene.

In liquid exfoliation, graphite is sonicated in a solvent with a surface energy close to that of graphite [38, 39]. To minimize the interfacial energy, the competing between the graphene/graphene interface and the graphene/solvent interface results the breaking of the van der Waals forces between graphene sheets. Liquid exfoliation gives rise to a mixture of graphene in the form of single, two, and multi-layer sheets and prolonged sonication increases the fraction of single layer graphene in the mixture. For practical uses, many efforts have been put to improve the concentration of single layer graphene.

Reduced graphene oxide, which is the mixed graphene oxide and pristine graphene mixture, can be prepared by chemical, electrochemical, and thermal method through reduction of graphene oxide in chemical environments. Graphene oxide is obtained through oxidizing graphite in using strong oxidants and intercalating materials such as HNO_3, H_2SO_4, $KClO_3$, etc. [40, 41]. As a result, there are many defects and functional groups on the reduced graphene oxide. One of the existing problems in preparing graphene oxide is unbalanced stoichiometry of oxygen and carbon atoms. There has been a reported C/O value of only 1.3 and it was shown by Punckt et al. that with this composition graphene oxide is stable [33].

4.2.3 Polymer Transparent Conducting Electrodes

In a large variety of polymers, there are a small group of them that are both conductive and transparent, suitable for the use as electrodes in photovoltaic devices. Although as polymers, their conductivities are low in comparison to those of TCO and AZO, their optical absorptions in visible solar range are comparable. Conductive polymers are mostly derivatives of polyacetylene, polyaniline, polypyrrole or polythiophenes. A disadvantage is that the transparent conductive polymers absorb quite a lot of the mid to near infrared solar radiation, leading to poor performance for opto-electronic conversion in the infrared regions. However, transparent conductive polymers are flexible, rendering them competitive with respect to inorganic TCO in some applications including but not limiting to flexible solar cells and flexible electronics.

4.3 Synthesis of TCF

Like other nanomaterials, the physical and chemical properties of nano TiO_2 depend not only on the nano sizes, but also on the shapes and even the routes of material preparation. In present, growth techniques of nano TiO_2 have advanced so that the nanomaterial can be grown with controlled sizes and shapes.

4.3.1 Wet-Chemical Preparation

TiO_2 nanomaterials can be synthesized with a wet chemical process during which hydrolysis and polymerization reactions of inorganic titanium salts or titanium organic compound precursors, such as titanium alkoxide, occurs via an acid-catalysis followed by annealing at moderate temperature and condensation. The annealing leads to the formation of solid gel after complete polymerization and solvent evaporation. A thin-film form of TiO_2 nanomaterial can be prepared by spin coating or dip coating the precursor on a substrate prior to annealing and condensation. Zinc oxide (ZnO) is another wide bandgap semiconductor that has good conductivity and is *n*-type, with the native dopants being oxygen vacancies or zinc interstitials. ZnO can be grown into various nano forms such as nanorods, nanowires, nanopipes, and nanopins. The growth techniques include solvothermal, sol gel, etc.

Figure 4.11 presents the SEM images of TiO_2 nanoparticles with various shapes tunable by controlling synthesis conditions. The TiO_2 nanoparticles can be used for photoanode. As a cost-effective method, the sol gel route has been used to synthesize many forms of nano TiO_2 from a precursor of a titanium alkoxide solution (or its sol—a colloidal suspension) [42, 43]. The precursor is then spin-coated or dip-coated on a substrate to form a wet gel thin film. A following-up drying and annealing procedure produces a thin film of nano TiO_2 of various sizes and shapes that depend on the details of the precursor and the other experimental parameters, such as annealing temperature, spinning rate, etc. TiO_2 is consisting of a Ti–O–Ti chain that is favored to form when, in the precursor, the hydrolysis rate and water content are low while titanium alkoxide content is rich. The development of the Ti–O–Ti chains gives rise to the growth of closely packed 3D polymetric nanoparticles. Increasing water content and hydrolysis rate favors the formation of $Ti(OH)_4$ that is beneficial for the growth of loosely packed nanoparticles at insufficient development of 3D polymeric skeletons.

Condensing titanium alkoxide in the presence of tetramethylammonium hydroxide, highly crystalline TiO_2 nanoparticles with an anatase crystal structure were prepared [44]. The growth route gives nano TiO_2 with controllable sizes and shapes. Chemseddine et al. prepared TiO_2 nanomaterials by adding titanium alkoxide into a three-neck container filled with alcoholic solvents [45]. The container was then warmed up

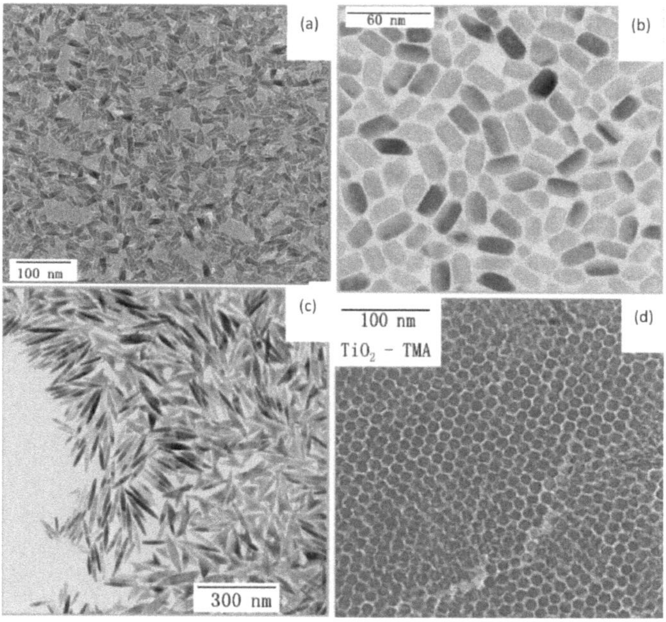

Figure 4.11 SEM images of TiO_2 nano particles in a triangular shape (a), a shape between triangular to rectangular (b), an enlongated shape (c), and a hexagonal shape organized in a 2D superlattice (d). Reproduced with permission from [45].

to 50°C for more than 10 days or up to 100°C for 6 h, followed by a secondary thermal treatment at 175 and 200°C in an autoclave. The elongated TiO_2 nanoslabs were characterized by TEM which shows images of good material crystallinity. Controlled growth of nano TiO_2 thin film can be realized by using templates. Anodic alumina templates have been used to grow TiO_2 nanorods and nantubes by the sol-gel method [46, 47]. During the synthesis, the porous anodic alumina templates were dipped into a boiled TiO_2 sol, TiO_2 nanorod thin film forming after a combined drying and heating process. In the synthesis of TiO_2 nanorods, the TiO_2 precursor was the mixture of titanium tetraisopropoxide with a solution of water, acetyl acetone, and ethanol. For the growth of TiO_2 nanotubes, the sol is the mixture of titanium tetraisopropoxide with 2-propanol and 2,4-pentanedione [48, 49]. Note that the TiO_2 nanoparticles are themselves single crystalline, as shown by the high resolution TEM image in Fig. 4.12 for particles in an enlongated shape.

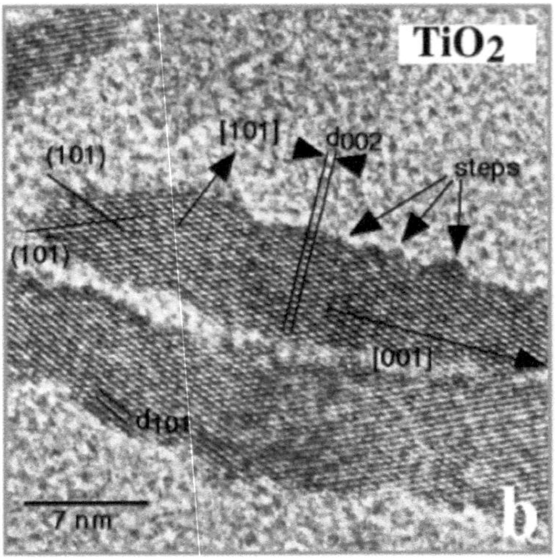

Figure 4.12 High-resolution TEM of enlongated TiO_2 nanoparticles in an shape. The nanoparticles is lying in the [0 1 0] plane and elongated in the [0 0 1] direction. The occurrence of corrugation on sides (denoted by the arrows) are due to alternate repetition of the [1 0 1] and [1 0 –1] planes. Reproduced with permission from [45].

4.3.2 Anodic Oxidation

When a piece of metal is immersed into a container filled with a suitable electrolyte, an oxide reaction takes place at a sufficient electric potential applied on the metal. The oxidation process is expressed by

$$M \rightarrow M^{+n} + ne, \tag{4.1}$$

where M is the metal, n the number of electrons lost during oxidation. After the initiation of the oxidation process, two reactions with possible competing take place. One reaction is that the metal ions chemically solvatize in electrolyte. Another reaction is that the metal atoms adjacent to the electrolyte form metal oxide that is chemically stable to the electrolyte. Depending on the electrochemical conditions such as the electric potential, the two reactions compete and, at certain conditions, result in the growth of porous TiO_2. Figure 4.13 shows schematically the anodic oxidation process. In fact, the reactions of Ti solvatization and TiO_2 formation at Ti/electrolyte interface lead to the synthesis of various TiO_2 nanomaterials, including nanotubes, nanorods, nanoparticles, etc. [50].

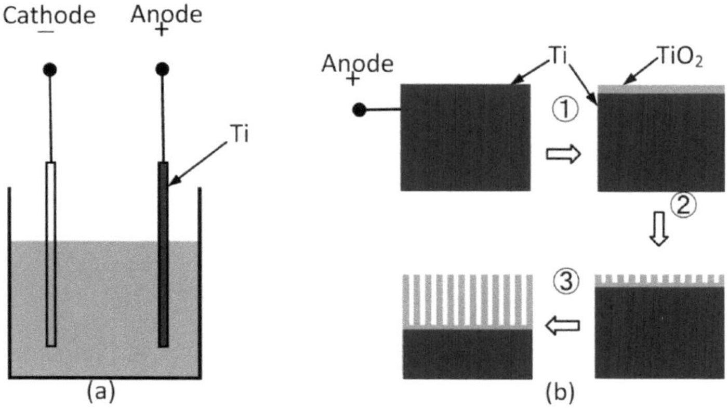

Figure 4.13 (a) Schematic apparatus setup for anodic oxidation. (b) Schematic process of anodic oxidation to fabricate TiO_2 nanotubes as shown by the open arrows: when a voltage is applied, the surface of Ti foil is first oxidized (step ①) followed by formation of TiO_2 nanotubes with continuous oxidation (step ②), and long nanotubes can be prepared by prolonged oxidation (step ③).

A great deal of attention has been given to ZnO and SnO_2 due to their higher electronic conductivity and electron mobility, as well as relatively low cost, in comparison to TiO_2. ZnO is a group II–VI wide bandgap semiconductor with many interesting properties. With a bandgap of 3.37 eV, ZnO is highly transparent to a wide range of the sun spectrum. The excitons in ZnO have an extremely large binding energy of 60 meV that offers the excitons with high stability, leading to strong room-temperature luminescence. However, the excitons in ZnO present very weak size confinement effect due to very small exciton Bohr radius. ZnO is natively *n*-type due to oxygen vacancies and zinc interstitials, yet has high electron mobility. However, its *p*-type doping is proven extremely difficult, although a great deal of efforts have been made in basic and applied research. The major applications of the material lie in transparent electrodes in solar cells, liquid crystal displays, heat-protecting windows, and in electronics including ultraviolet LED (light-emitting diode), LD (laser diode), piezoelectric devices, chemical sensors, and TFT (thin-film transistors). Allowing for low temperature growth, ZnO can be prepared with many different kinds of morphologies using thermal evaporation, wet chemical, laser ablation, hydrothermal, vapor transport, etc. ZnO nanomaterials, including nanoparticles, nanorods, nanowires, nanocages, nanobelts, nanoneedles, nanotubes, nanocombs, nanorings, nanohelixes, have been synthesized. Figure 4.14 shows the nanotubes prepared by anodic oxidation.

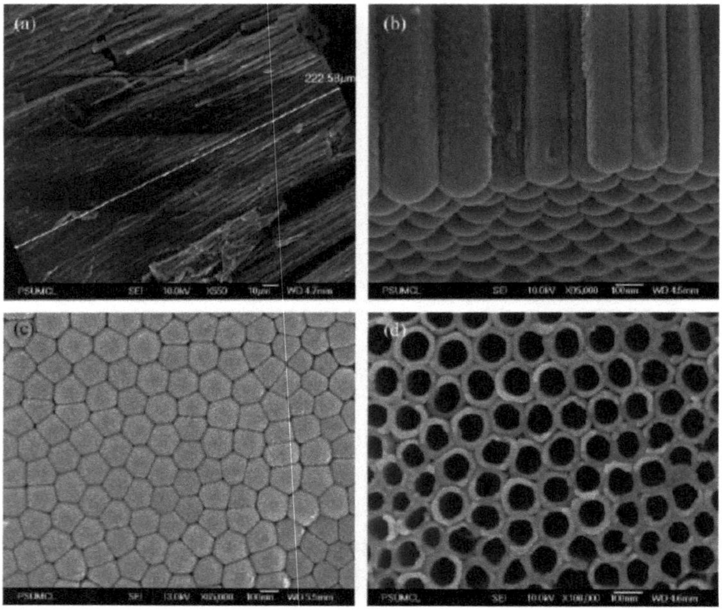

Figure 4.14 SEM images of TiO_2 nanopipes: side view (a), tilted view (b), bottom view (c) and top view (d), where the nanotubes have been stripped from the substrate. The inner diameter of the nanotubes is 50–100 nm with the wall thickness around 20 nm. Those parameters are tunable by experimental conditions. Reproduced with permission from [50].

4.3.3 Vapor Transport Synthesis

ZnO nanostructures can be prepared by many routes. Among them, vapor transport methods are the most widely used preparation routes. In the vapor transport process, Zn and oxygen vapor flows (or oxygen mixture vapor flow) react with each other during transport, leading to formation of ZnO nanostructures. The Zn vapor can be generated either by heating up Zn powder or by decomposing ZnO directly. In former case, a high temperature of ~1400°C is required, while in later case, a relatively low growth temperature of 500~700°C is needed. High quality material growth requires a fine control on the ratio of the Zn vapor pressure and the oxygen pressure.

Vapor transport synthesis includes the catalyst-free vapor–solid synthesis and the catalyst assisted vapor–liquid–solid (V-L-S) synthesis. During the catalyst-free vapor–solid synthesis, ZnO nanostructures are grown directly through condensing and nucleation of Zn and O_2 vapors. Sometimes, O_2 vapor is replaced by CO- and CO_2-mixed vapor. The growth method varies, depending on ways to generate Zn vapor and O_2 vapor. For instance, in metal-organic vapor phase growth, organometallic Zn compound (diethyl-zinc, for instance) react with O_2 or N_2O to form ZnO nanomaterial. The great advantage of the catalyst-free vapor–solid synthesis is that diverse ZnO nanostructures can be obtained, including nanowires, nanotubes, nanoneedles, nanorods, nanobelts, nanotetrapods, etc., depending on starting materials and growth parameters. However, the catalyst-free vapor–solid synthesis offers very weak control over location and alignment of the nanomaterial.

Kong and Wang reported the preparation of complex ZnO nanohelixes and nanobelts on an alumina substrate using vapor–solid (VS) synthesis by decomposing ZnO powder into Zn^{2+} and O^{2-} at ~1350°C under Ar flowing gas [51]. The result is presented in Fig.4.15a, where the nanostructures are dominated by nanobelts of a uniform size distribution. The image shows a large fraction of helical nanostructures and nanorings as denoted by arrows. The structure of the ZnO nanobelts is controllable. ZnO nanostructures have divergent surface energies largely associated with intrinsic polarization. By varying experimental conditions, various nanobelt structures can be grown with ±(0001) polarized facets. The polarized zinc- and oxygen-terminated surfaces lead to the rolling of single-crystal nanobelts and, as a result, the formation of a right-handed helical structures and nanorings driven by minimization of the total energy contributed by polarization and elasticity.

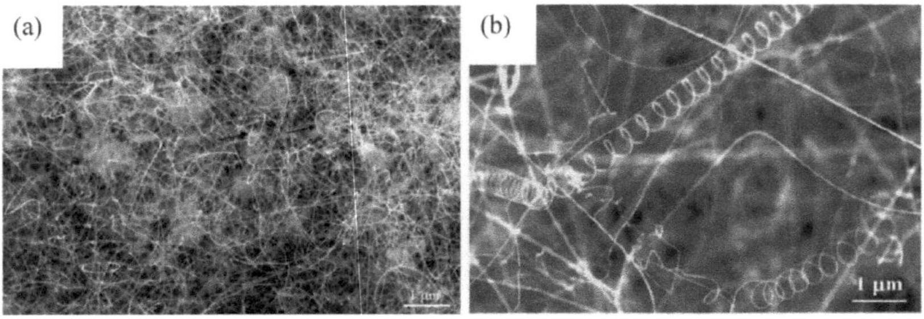

Figure 4.15 SEM image of the ZnO nanobelts with the sizes of 10–60 nm in widths, 5–20 nm in thickness and several hundreds of micrometers in thickness (a), and that shows the helical nanostructure (b). Reproduced with permission from [51].

In a similar synthesis, Wen and co-workers prepared hierarchical ZnO nanostructures in vapor transport and condensation route by heating up mixed powder of ZnO, In_2O_3, and graphite to ~850°C in a controllable manner. Highly regular nanostructures of nanoribbons, nanocombs, and nanorotors with four- and sixfold symmetries were synthesized, with SEM images as shown in Fig. 4.16a–c [52]. A simplified method to achieve nanowires, nanoribbons and nanorods was reported by Yao et al. who synthesized ZnO nanorods, nanoribbons, and nanowires using the mixture of ZnO powder and graphite [53]. They grew the ZnO nanomaterial by thermal evaporation in a quartz tube with one end closed. The source material, the mixture of ZnO and graphite powders with 1:1 molar ratio, was placed in a quartz tube at the closed end. Growth was carried out by inserting the quartz tube into a horizontal furnace with a temperature gradient, where the closed end of the quartz tube was heated to 1100°C and the open end to 500°C. The growth was stopped after 30 min evaporation. ZnO nanomaterials were found on the inner wall of the quartz tube, deposited at the temperature range of 800 to 500°C. SEM measurements on the as-grown ZnO nanomaterials show three morphologies of needle-like nanorods, nanoribbons, and nanowires with good crystalline quality, as shown in Fig. 4.17, corresponding to the deposition temperature ranges of 800–750, 750–650, and 650–500, respectively.

In addition to VS synthesis, a VLS process is often used to synthesize various nanomaterials. A number of catalysis assisted VLS methods have been proposed. The VLS process offers more control over the alignment and locations of the nanomaterial. In most cases, VLS preparations use metal catalysts such as Au, Co, Sn, Cu, etc. [54, 55]. and together with other variation of growth paramenters, various ZnO nanomaterials, including nanowires, nanotubes, nanoneedles, nanorods, etc., have been obtained. A schematic VLS procedure is depicted in Fig. 4.18.

Figure 4.16 An SEM image of ZnO nanorotors with 4- and 6-fold symmetries (a) and those at higher magnifications (b and c). Reproduced with permission from [52].

Figure 4.17 SEM images of nano ZnO: (a) needle-like nanorods, (b) nanoribbons, and (c) nanowires. Reproduced with permission from [53].

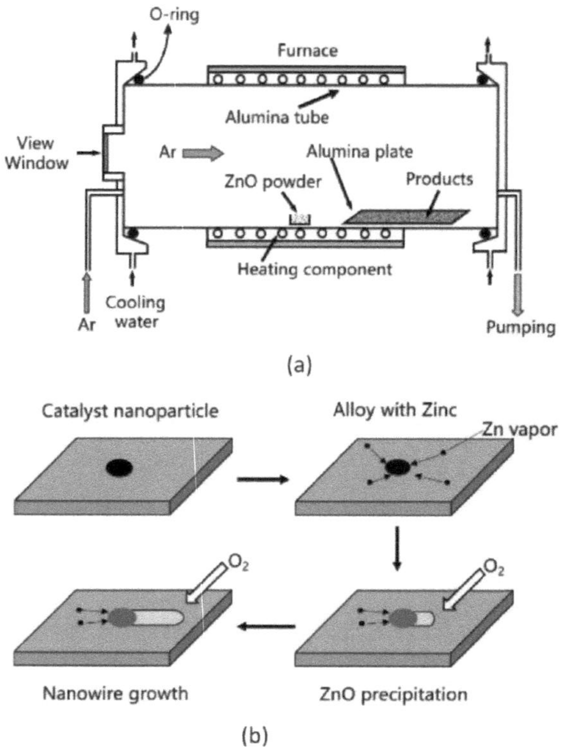

Figure 4.18 A schematic description of a VLS apparatus (a) and growth process (b).

As shown in Fig. 4.18, a typical VLS synthesis uses a substrate deposited with Au nanoparticles as catalytic metal and Zn power as the source material [56]. Both the substrate and the source material are loaded in a furnace. During the growth, Zn powder and the substrate are heated up to 700°C and an appropriate O_2 gas flows through the furnace. ZnO nanowires (or nanorods, etc.) grow on the sites where the Au nanoparticles are located. The growth of the ZnO nanomaterial is due to the supersaturation of Zn in the Au-Zn liquid alloy droplet which causes Zn precipitation. Zn and O_2 then form ZnO, as schematically shown in Fig. 4.18b. The growth occurs at the interface of Au droplet and ZnO, which pushes the droplet up continuously during the growth and the ZnO nanowires (or nanorods) are terminated with Au particles.

4.3.4 Hydrothermal Process

Very often, a VLS process needs a noble metal as catalyst and the metal nanoparticles on the tip of the ZnO nanomaterial need to be eliminated. A hydrothermal method does not require the use of catalytic metals. Other than this, the hydrothermal synthesis bears the advantages of low cost, low temperature, and high possibility of scale-up.

Wang and co-workers have developed an approach to grow patterned ZnO nanomaterial without the use of catalytic metal particles, in which, by means of electron beam lithography, patterned and aligned ZnO nanowires were grown on Si and GaN substrates at temperature below 100°C [57]. The growth could be carried out on a Si substrate on which a designed polycrystalline ZnO pattern were generated by electron beam lithography. The patterned ZnO was used as seeds and the growth was initiated at the seed sites. Interestingly, they could control one or multiple ZnO nanowire growth at the seed sites, as shown in Fig. 4.19a,b. When choosing a relatively low growth temperature at 70°C, multiple nanowires were grown at one single site, while at elevated temperature of 95°C, the multiple nanowires at each site merged to form thicker nanowires shown in Fig. 4.19c,d.

With some modifications, the hydrothermal process can fabricate ZnO nanomaterials on a relatively large scale. Hua et al. fabricated the mixture of 3:2 ZnO nanowires and ZnO nanobelts via a one-pot hydrothermal route in a stainless steel autoclave [58]. In the preparation procedure, 0.2 g $ZnCl_2$, 1.5 g sodium dodecyl sulfonate (SDSN), and 20 g Na_2CO_3 were added into a 50 mL autoclave that was then filled with distilled water up to 90% volume. The reaction solution was stir-mixed for 30 min before the autoclave was sealed. The growth was carried out at 140°C for 12 h. The Na_2CO_3 was used as both the alkaline source and the controllable reagent for preparation of ZnO nanomaterials. It was found that adding Na_2CO_3 affects the length/diameter aspect ratios of 1D ZnO nanowires and nanobelts significantly. The use of surfactant SDSN was essential for the controllable growth of ZnO nanobelts, while the use of Na_2CO_3 was found critical for the growth of ZnO nanowires. The use of high supersaturation solution gave rise to the growth of ZnO nanowires. At high Na_2CO_3 concentration, an supersaturate $Zn(OH)_4^{2-}$ precursor leaded to the formation of very small nuclei in a short instant during the initial stage of hydrothermal decomposition in a homogeneous nucleation process. The lateral dimension of the ZnO nanomaterials was determined by the size of the initial nuclei. In following up growth, $Zn(OH)_4^{2-}$ ions were incorporated into the nuclei continuously along the *c*-axis of ZnO crystal. Thus, prolonged growth gave rise to long ZnO nanowires.

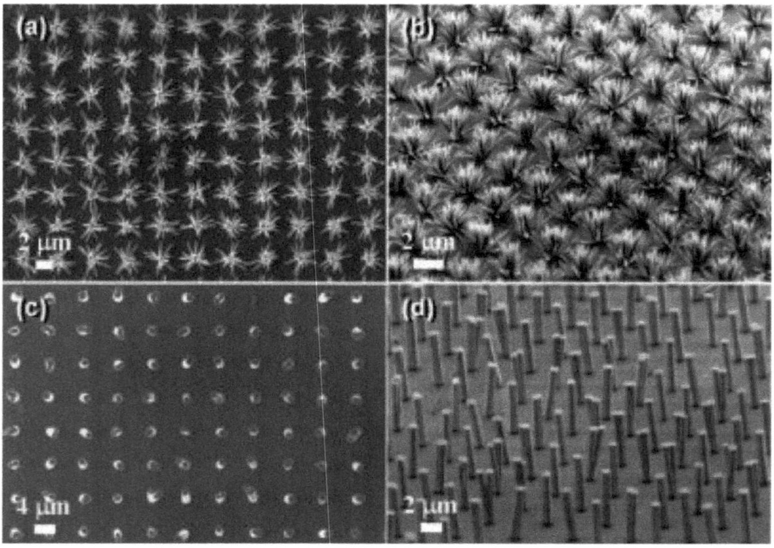

Figure 4.19 SEM images of patterned multiple ZnN nanowires at each seed site (a and b) and top view (c) and 60° tilt view (d) of patterned ZnO nanowires. The ZnO nanowires were grown on Si wafers. Reproduced with permission from [57].

The nanomaterial scaffolds are used to load various sensitizers. Thus, it is essential for the scaffolds to meet the two fundamental requirements. First, the scaffolds should have large surface areas in order to accommodate high density of sensitizers. Second, they should offer good electrical connection to the sensitizers and to the TCO layer in order for the photo-excited electrons in sensitizers to transport to the TCO electrode via scaffold easily. The second requirement implies that conduction band level of the scaffold be higher than that of TCO and lower than the LUMO in sensitizers. Very often, the interface between a sensitizer and a scaffold need to be modified for good electric connection.

4.4 Sensitizers

In dye-sensitized solar cells, sensitizers act to absorb photons and transfer photon energy to an electron that becomes active to contribute the photovoltaic effect. The properties of the sensitizers are critical to the overall performance of DSSCs and QDSSCs. Ideal sensitizers should absorb all the photons in solar radiation spectrum and enable to inject photon-excited electrons to the scaffold (semiconductor oxide), with a quantum yield of 100%. Thus, the excited state (HUMO level) of QD (or organic dye molecule) sensitizers should be sufficiently higher than the conduction band level of the scaffold. In addition, the redox potential should be high enough so that the sensitizers can be quickly regenerated via receiving electrons from the electrolyte. There are two kinds of sensitizers used in DSSCs, one being the organic molecule dyes and another being the inorganic sensitizers. Organic dyes can be found from nature (in plants, for instance) or synthesized in lab. Inorganic dyes include semiconductor nanostructures (nanoparticles, nanowires, etc) and metal complex (polypyridyl complexes of ruthenium and osmium, metal porphyrin, etc.).

4.4.1 Ruthenium(II) Sensitizers

The first successful DSSC used polypyridyl ruthenium(II) dyes as sensitizers [1]. Since then, the conversion efficiencies of the DSSCs based on metal-complex sensitizers increase steadily. Currently, high power conversion efficiencies of 11.2% and 10.4% have been achieved by using Ru bipyridyl complexes (N3 and N719) and the black ruthenium dyes, respectively [59, 60]. The molecule structures of N3, N719, and the black ruthenium dyes are depicted in Fig. 4.20. Other sensitizers, including, indoline, triphenylamine, bis (dimethylfluorenyl) aminophenyl, coumarin, carbazole, merocyanine, hemicyanine, phenothiazine, dialkylaniline, and tetrahydroquinoline, have been studied and their effectiveness has been demonstrated. Refs. [61–63] presented reviews on dye materials used for DSSCs.

Except the conversion efficiency, the stability and cost effectiveness should also be taken into consideration. The noble metal Ru, for instance, is not an earth-abundant element and is expensive. Ruthenium(II) sensitizers is also toxic. In current research, one is still looking for new sensitizers for DSSCs. Some metal-free organic dyes have high light extinction coefficients and are low cost.

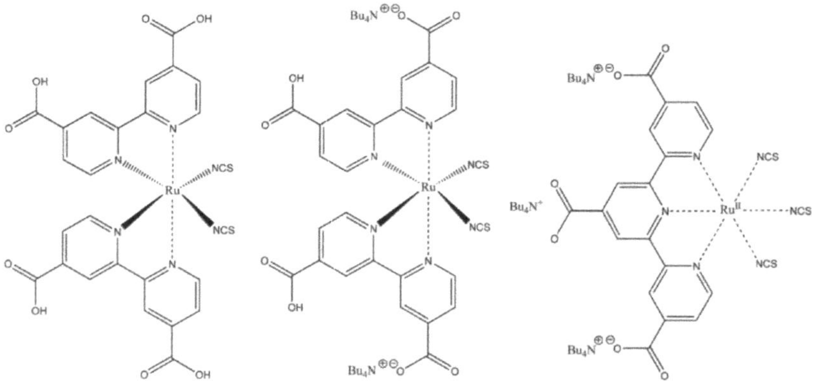

Figure 4.20 Molecular structures of N3 (left), N179 (middle) and black dye (right).

4.4.2 Metal-Free Tetrathienoacene Sensitizers

Metal-free organic photosensitizers have the advantages of facile preparation, good chemical versatility, low cost, as well as tunable energy state and structure for matching to the solar spectrum. The major disadvantages are relatively low PCEs, compared to their metal-complex counterparts, due mainly to their high recombination losses and low open-circuit voltages. In some works, however, PCEs using metal-free organic dyes such as YA442 has reached above 10%, with a Co(II/III) redox elements [64–69].

The major principles of design and selection for dye molecules are band alignment with respect to TiO_2 and optical absorption coefficient. Zhou and co-workers studied the design principle and fabricated four metal-free DSSC dyes (dyes A–D) with chemical structures shown in Fig. 4.21a [70]. Figure 4.21b presents the HUMO and LOMO energy positions of the dyes, together with those of TiO_2 and (I^-/I_3^-) redox potential, which shows clearly that the band alignments are suitable for DSSCs. Figure 4.22 presents the absorption spectra measured on the four dyes in 1,2-dichlorobenzene (o-DCB) solution. All the dyes show two strong absorption peaks due to the charge-transfer band in the spectrum range of ~480–515 nm and the $\pi \rightarrow \pi^*$ transition within 340–405 nm. Apparently, the absorption peak position can be tuned by inserting thiophenes in the dyes. More importantly, All these dyes, at their absorption maxima, exhibit molar extinction coefficients ~5 × 10^4 $M^{-1} \cdot cm^{-1}$ (see the inset in Fig. 4.22), which is roughly four times of that of the typical Ru(II) polypyridyl complexes N719.

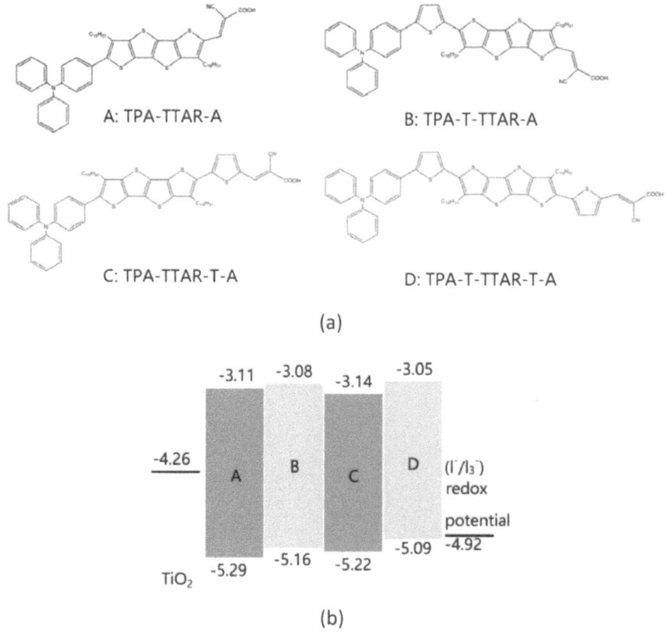

Figure 4.21 Schematic diagrams of the chemical structures of dyes 1–4 (1: TPA-TTAR-A, 2: TPA-T-TTAR-A, 3: TPA-TTAR-T-A, and 4: TPA-T-TTAR-T-A) (a) and the HUMO and LOMO energy positions of the dyes with respect to those of TiO_2 and (I^-/I_3^-) redox potential (b).

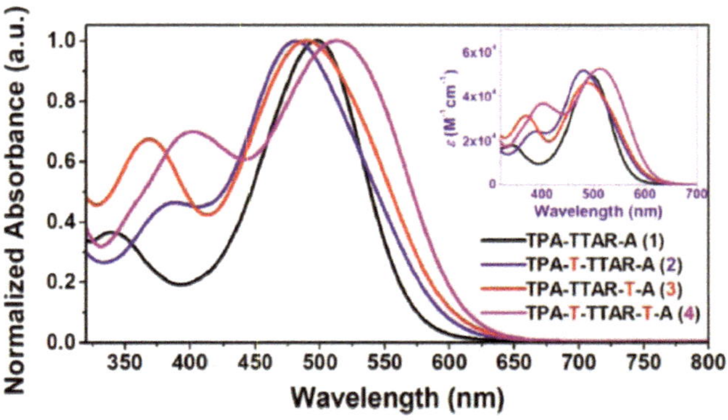

Figure 4.22 Absorption spectra of the four dyes in 1,2-dichlorobenzene (o-DCB) solution. Reproduced with permission from [70].

Zhou and co-workers fabricated DSSCs using the four dyes in tetrahydrofuran (THF) and EtOH mixed (1:1 v/v) solutions. The PCE of exceeding 10% was obtained on the dye-3 DSSC wining the champion mainly by its J_{sc} that is nearly 1.4 to 2.2 times of those of the other three cells. The charge transport dynamics of the DSSCs were then studied and it was found that under one sun condition, the dye-3 DSSC showed the lowest recombination rate and device resistance enabling large photocurrents in the DSSC device. The result is consistent with its measured highest J_{sc}, shown in Table 4.1.

Table 4.1 Performance of the DSSCs fabricated using the four dyes

Dye	V_{oc} (V)	J_{sc} (mA/cm^2)	FF (%)	PCE (%)
1	0.946	7.66	65.8	4.76
2	0.893	10.1	68.1	6.15
3	0.833	16.5	73.7	10.1
4	0.832	11.8	70.3	6.91

4.4.3 Porphyrin-Based D–π–A Sensitizers

Porphyrins play an important role in electron transfer and transfer during photosynthesis, implying that porphyrins could be used in DSSCs as sensitizers. In early trying, the PCEs of porphyrin-based DSSCs have been reported to be in the range of

5–7%. A breakthrough was achieved with a tailored porphyrin dye, YD-2, with its molecule structure as shown in Fig. 4.23a, and the reported 11% PCE (under AM 1.5 condition) was then the highest DSSC using a ruthenium-free sensitizer [71].

In the YD-2 porphyrin (see Fig. 4.23a), a diarylamino donor group connected to the porphyrin ring acts as a donor and the ethynylbenzoic acid moiety as an acceptor, while the porphyrin chromophore serves as the π bridge. The three parts form the D-π-A structure. However, the YD-2 itself has a green color, indicating that it has poor light harvesting ability in the green spectral band. Indeed, its measured incident photon to current conversion efficiency (IPCE) exhibits two main absorption bands denoted as the Soret and Q-bands. The pronounced valley between the Soret and Q-bands (at around 530 nm), however, leads to rather poor J_{SC} and, in turn, poor PCE value. Exhibiting a light absorption maximum at 532 nm, D-205 dyes show a complementary spectral responses in the visible spectral range with respect to YD-2. Bessho and co-workers fabricated DSSCs using cosensitized dyes of YD-2 and N205. The optimization on the cosensitized dye DSSC leads to an 11% PCE.

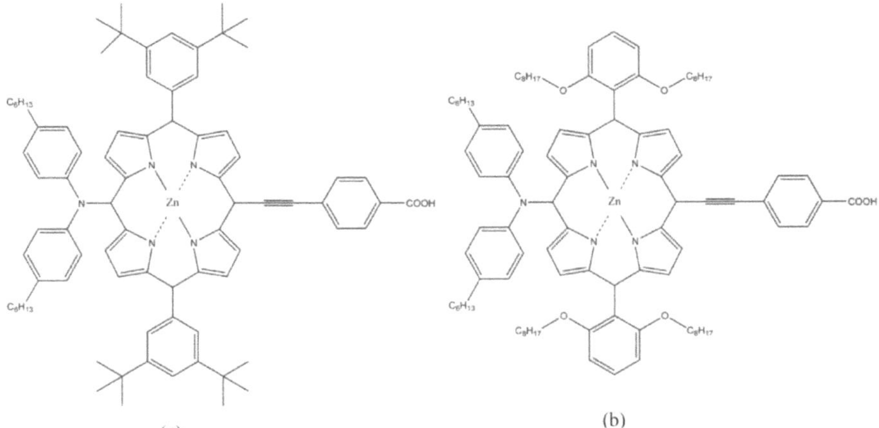

Figure 4.23 The molecular structures of YD2 (a) and YD2-*o*-C8 (b) porphyrin dyes.

The YD2 dye has the problem of insufficient light absorption between the Soret and Q-bands. From this perspective, a D–π–A zinc (Zn) porphyrin dye, denoted as YD2-*o*-C8, have been synthesized [72]. YD2-*o*-C8, with its molecular structure shown in Fig. 4.23b, has strong light absorption in the entire visible range. In addition, its alkoxy groups with long chains limit the interfacial back electron transfer reaction that leads to annihilation of photo-generated carriers. As shown in Fig. 4.23, YD2-*o*-C8 is a tailored variant of YD2, formed by attaching two octyloxy groups in the ortho positions of the meso-phenyl ring. DSSCs were fabricated with YD2-*o*-C8 in conjunction with Co(II/III)tris(bipyridyl)-based redox electrolyte and an obvious improvement in light absorption and charge separation efficiency was achieved. Figure 4.24 presents the *I–V* curves for DSSCs based on YD2 and YD2-*o*-C8 sensitizers. The use of YD2-*o*-C8 improves the short-circuit current and the open-circuit voltage values significantly and gives to a PCE value of 11%. Furthermore, the PCE value was increased to above 12% under full sun illumination through improving the light harvesting, using the cosensitization of YD2-o-C8 and Y123, an organic D–π–A dye denoted as Y123.

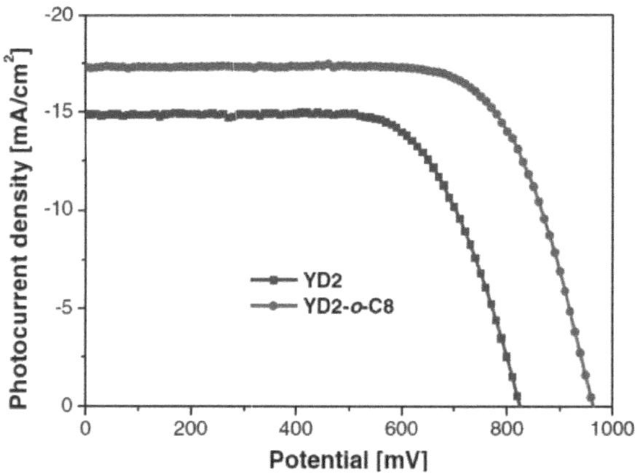

Figure 4.24 *I–V* curves of DSSCs using YD2 and YD2-*o*-C8 porphyrin dyes as sensitizers. Reproduced with permission from [72].

Mathew and co-workers modified the structure of D–π–A porphyrins and the compatibility between the cobalt–electrolyte and the sensitizers to improve the light-harvesting properties, leading to a PCE of 13% [73].

4.4.4 Natural Dyes

As bionic solar cells in working mechanism, one would expect that natural dyes could have a position in DSSCs as sensitizers. Because natural dyes are extracted from plants grow on the ground or in the sea, they are environmentally friendly and completely renewable. Other advantages include low fabrication cost, flexibility, Rich material resources, etc. In fact, all leaves, flowers petals, and fruits contain active ingredients that can be used as sensitizers. Many natural dyes, extracted from lily, perilla, petunia, rhododendron coffee, marigold, mangosteen pericarp, begonia, etc., have been tested for DSSCs that show varying PCEs [74, 75]. However, all the DSSCs based on natural dyes exhibit very low values of PCE, typically below 2%.

Figure 4.25 Photo picture of Basella alba plant (a) and *I*–*V* curve of the DSSC based on Basella alba dye (b).

Basella alba, as shown in Fig. 4.25, is a popular plant widely grows in as tropical and non-tropical area. Dixit and co-workers reported the fabrication of DSSCs using basella alba (red vine spinach) as sensitizers [76]. The photoanode is FTO (fluorine doped tin oxide) glass coated with TiO_2 film sensitized by Basella alba dyes. The photoanode (Basella alba dyes + TiO_2) shows strong light absorption in the entire visible range (see Fig. 4.25b). The counter electrode is FTO glass coated with graphite. The DSSC has a measured open-circuit voltage of 0.8 mV, a short-circuit current of 10 mA, a fill factor of 0.63, and conversion efficiency of 0.50%. Apparently, the low short-circuit current, currently a common shortcoming of bionic solar cells, is the main reason for the low PCE of the DSSCs. The lack of an effective current flow channel gives rise to large internal resistance of the bionic solar cells.

In other works, Zou and co-workers have used mangosteep pericap as a sensitizer in making DSSCs and obtained a PCE of 1.17% [77]. Calogero et al. fabricated DSSCs using betaline pigment containing red turnip as sensitizers and achieved an efficiency of 1.70% [78].

4.5 Counter Electrodes

In conventional SCs, electrodes are thin film of network wires for collecting the photo-generated current. Nanomaterials offer large surface area, which is beneficial to the carrier collection in NSCs. The counter electrode in NSCs, if it is used as rear electrode, has no requirement for good optical transparency. Thus, one can have more options to choose the materials and configurations to meet the needs for good conductance. In the case of dye-sensitized SCs, it is required that the counter electrode provide enough catalyst sites for the iodides to be regenerated by reduction of triiodides. A number of materials in their nano forms have been reported to be used as counter electrodes.

4.5.1 Metal Nanoparticles

Counter electrodes should have active catalytic properties and good corrosion resistance towards iodine and electrolyte. Rough/porous electrodes are beneficial for high density of catalytic sites. At the same time, the counter electrodes must be highly electronic conductive to allow for high current.

The metal nanoparticles on counter electrodes are usually prepared in top-down approach started from spin-coating, thermal evaporation, sputtering, vapor phase deposition, etc., followed by subsequent subdivision of the initial thin films by thermal treatment or photolithography. In bottom-up category, the preparation route starts from atomic and molecular precursors.

The growth of metal nanomaterials is completed through self-assembling and chemical reactions with their structures controlled by growth parameters and the precursor parameters. Platinum has been widely used as the counter electrodes for DSSC due mainly to its excellent catalytic properties for reducing triiodide. A typical sputtering growth system is shown in Fig. 4.26a. Sputtering gas such as Ar^+ ions is used to bombard the target to eject the target atoms. Sputtered ions are then deposited onto the substrate surface, forming a thin film of metal. Subsequent subdivision gives rise to formation of the metal nanomaterials, which serves as the catalyst layer on counter electrode shown in Fig. 4.26b.

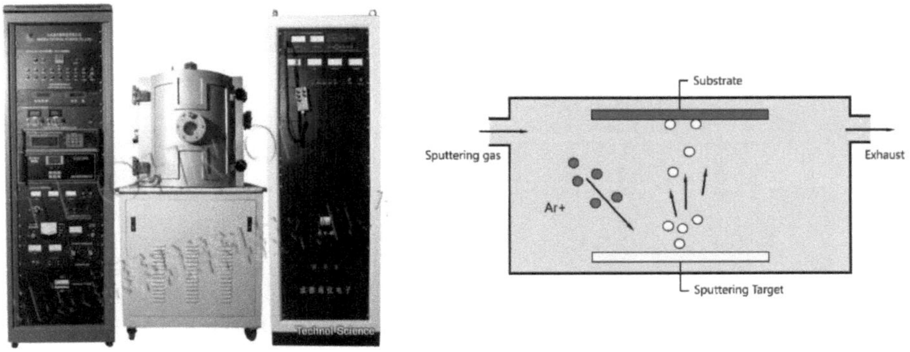

(a) (b)

Figure 4.26 A sputtering system (a) and schematic diagram (b) for metal deposition.

4.5.2 Carbon Nanomaterials

Carbonaceous materials are referred to carbon nanotubes (single wall and multi-wall), graphene, carbon black, graphite, fullerene, etc. Carbonaceous materials are earth-abundant and low cost. Their nano forms are highly active and have high catalytic performance. Being highly conductive and having high surface area, carbon nanomaterials are thus used in DSSCs as catalytic counter electrode materials [79]. In particular, CNTs are good catalyst for reducing tri-iodides to iodides. Thus, CNTs are good candidates to replace platinum in DSSCs. The good catalytic property of CNTs is guaranteed both by the highly active material and by large surface area. CNTs have very large aspect ratios of ~10^3. When conductive CNTs form thin film, CNTs could provide with efficient pathways for charge transfer, which gives rise to high carrier mobility. Other carbon nanomaterials are also used for counter electrodes. For instance, introducing carbon black powder in graphite enhances the catalyst property of the counter electrode [80].

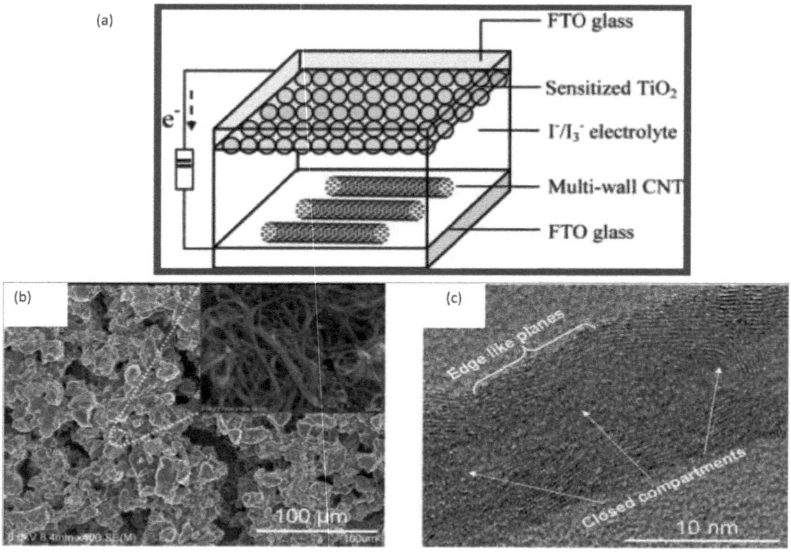

Figure 4.27 (a) Schematic device structure of a DSSC using multi-wall carbon nanotubes as the catalytic material, (b) the SEM image of the composite film of CNTs and (carboxymethyl)-cellulose sodium salt, and (c) the enlarged view of (b). Reproduced with permission from [79].

Lee et al. fabricated DSSCs using CNTs as the catalytic layer of counter electrode and achieved a power conversion efficiency of 7.7%. The schematic diagram of the DSSC is presented in Fig. 4.27a [79]. The CNT catalytic layer was prepared with a mixture consisting of 2.25 g multi-wall CNTs mixed in a solution of 0.16 g (carboxymethyl)-cellulose sodium salt dissolved in 19.84 ml water. The CNTs are ~20 nm in diameter and 5 μm in length. The mixture was then ball-milling for 1 h at the milling rate of 100 rpm and ended up with homogeneous CNT slurry which was casted onto FTO glass (fluorine-doped SnO_2-conducting glass substrate). The CNT-film/FTO/glass-substrate counter electrode was completed after it was dried at 50°C for 10 h in atmospheric environment. Note that the catalytic layer is actually the composite film of CNTs and (carboxymethyl)-cellulose sodium salt. Figure 4.27b,c shows the SEM images of the composite film. The DSSC gives a conversion efficiency of 7.7%.

4.5.3 TiN Nanotube Arrays

It has been demonstrated that, due to their similar electronic structures, a few transition metal nitrides show noble metal-like behavior, of the metal nitrides to that of the noble metals [81, 82]. It is therefore possible to use some metal nitrides for the replacement of platinum as counter electrode catalyst in DSSCs. Among the metal nitrides, titanium nitrides (TiN) have low resistivity $\sim 10^{-2}$ Ωm, good chemical stability and catalytic activity. TiN nanotube arrays have been prepared using TiO_2 arrays as precursor. The preparation was initiated by anodization of metallic Ti foil substrate followed by subsequent calcination at 450°C in air to obtain TiO_2 nanotube arrays. The TiO_2 nanotube arrays were transformed into TiN nanotube arrays by simple nitridation in an ammonia atmosphere [81]. Figures 4.28a and b show the as-prepared TiN nanotubes on the metallic Ti foil substrate. Quite surprisingly, the results from TiO_2 and TiN are very similar, both exhibiting highly ordered nanotube arrays. Using the TiN nanotube film as the catalyst layer of the counter electrode, the DSSC shows performance slightly better than that of DSSCs with traditional Pt counter electrode. This is clearly demonstrated by the *I–V* measurements where the results on DSSCs made from TiN nanotube array, Pt/FTO, and bare FTO are presented, as shown by the solar cell parameters listed in Table 4.2. The electrolyte in the DSSCs is the solution mixture of 0.1 M LiI, 0.05 M I_2, 0.6 M 1,2-dimethyl-3-propylimidazolium iodide (DMPII), and 0.5 M 4-tert-butyl pyridine in acetonitrile solvent. The three DSSCs are the same except their counter electrodes. The measured *I–V* curves of the DSSCs were presented in Fig. 4.28c.

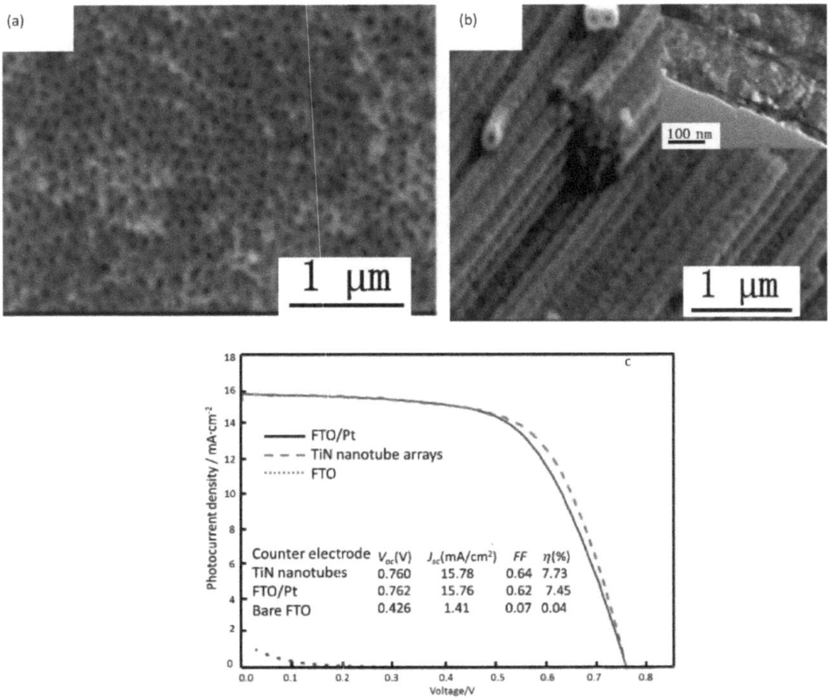

Figure 4.28 SEM top view (a) and cross-section (b) of SEM images of TiN nanotube arrays in which the inset presents the a TEM image of TiN nanotubes ultrasonically treated. (c) *I*–*V* of the DSSCs with different counter electrodes: TiN nanotube, FTO/Pt, and bare FTO. The measurements were carried out under AM 1.5 radiation. Reproduced with permission from [81].

Table 4.2 Cell parameters of DSSCs with TiN nanotube arrays, Pt/FTO, and bare FTO counter electrodes

Counter electrode	V_{OC} (V)	J_{SC} (mA cm^{-2})	*FF*	η
TiN/Ti-substrate	0.760	15.78	0.64	7.73
Pt/FTO-glass	0.762	15.76	0.62	7.45
FTO-glass	0.426	1.41	0.07	0.04

4.5.4 Porous Silicon

In searching for catalytic materials to replacing expensive Pt in counter electrodes, many materials have been tested, including the most popular semiconductor—Si. Si is earth-abundant and low cost. However, Si is a highly active material, with its surface properties easily affected by its ambient environment. Thus, if Si has good catalytic properties and one could passivate Si surface, Si could be good candidate for counter electrodes. Porous silicon, usually obtained by electrochemically etching the bulk silicon, has large surface area where a large number of sites are available for the I^-/I_3^- redox reaction. The key issue in passivating the silicon surface is to terminate dangling bonds in the silicon surface with oxygen atoms or carbon atoms, i.e., Si–O or Si–C.

Erwin and co-workers fabricated DSSCs with porous Si counter electrodes and compared the performance to a DSSC with a counter electrode based on Pt catalyst [83]. The carbon passivated porous silicon was prepared by etching heavily boron-doped silicon wafers for 180 s with a current density of 45 mA/cm^2 using a 3:8 solution volume ratio of HF (50% H_2O by volume) and ethanol [84, 85]. The resulting porous Si has 75% porosity with an average pore size near 25 nm. Carbon passivation was then initiated by placing the as-grown porous silicon wafer into a tube furnace. After the tube furnace was evacuated, 1000 sccm of Ar and 200 sccm of H_2 gases were inlet into the furnace followed by ramping the furnace to the initial temperature of 650°C for three different samples. Here H_2 acts to stabilize the porous Si surface, due to the reducing function of H_2 gas. Carbon passivation was then carried out by introducing the acetylene (C_2H_2), the carbon source for carbon-passivation, and heating up the furnace to 750°C for 10 min and then 800°C for another 10 min. For comparison, another two samples were prepared with an initial temperature of 550 and 750°C for Ar and He treatment, as well as 650 and 850°C followed by 700 and 900°C for carbon passivation, respectively. Note that H_2 treatment is critical for reducing porous Si from oxidation, which strongly affects the device performance.

The photoanode of the DSSCs consists of dye-absorbing TiO_2 active layer deposited on FTO glass, followed by the deposition of a TiO_2 scattering layer. The dye molecules loaded in nano TiO_2 active layer are Di-tetrabutylammonium *cis*-bis (isothiocyanato) bis(2,2′-bipyridyl-4,4′-dicarboxylato) ruthenium(II) (N719).

Figure 4.29 presents *I–V* curves of the DSSCs with carbon-passivated porous silicon counter electrodes fabricated with different temperature profiles. For comparison, the *I–V* curves of DSSCs with a Pt counter electrode and a pristine porous Si (without passivation) counter electrode are also shown. As clearly demonstrated, carbon passivation is essential for the enhancement of device performance, as shown by the DSSC with pristine Si counter electrode that has extremely poor device performance, expected from the measured *I–V* curve. Device performance critically depends on the passivation conditions. The DSSC manufactured with the temperature ramping condition of 650–800°C gives the best performance close to the DSSC with Pt counter electrode, except a bit lower open-circuit voltage. Erwin and co-workers also investigated and characterized the carbon passivation effects on porous Si. Raman measurements demonstrated that the pristine porous Si shows only a single Raman peak in the vicinity of 520 cm^{-1} corresponding to Si. Carbon passivation results in other two peaks at 1325 and 1602 cm^{-1}, indicating the formation of graphitic carbon coatings in porous Si surfaces.

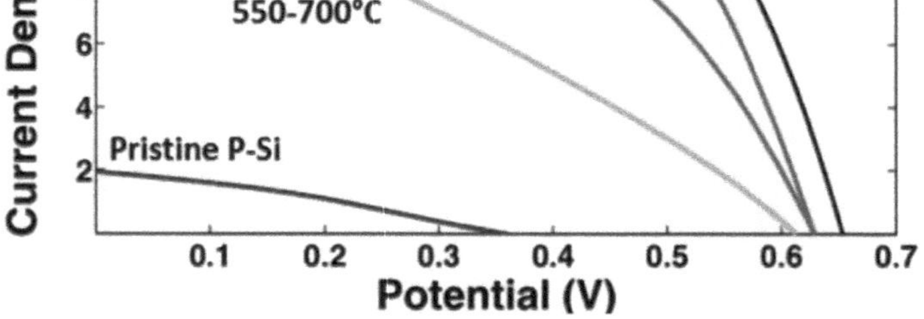

Figure 4.29 *I–V* curves of the DSSCs with counter electrodes of carbon-passivated porous silicon fabricated at different temperatures. The *I–V* curve of the DSSC with the Pt counter electrode is also presented for comparison. The measurements were carried out under AM 1.5 condition. Reproduced with permission from [83].

4.5.5 $Cu_2ZnSn(S_{1-x}Se_x)_4$

$Cu_2ZnSn(S_{1-x}Se_x)_4$ (CZTSSe) was also tested as a catalytic material for counter electrodes in QDSSCs and it was found superior to Pt counter electrode in several aspects [86]. First, CZTSSe is composed of earth-abundant elements much cheaper than Pt. Second, CZTSSe shows stronger catalytic properties in reducing polysulfide S_n^{2-} to sulfide S^{2-}. Third, the composition of CZTSSe offer more room to tune the catalytic properties through both material compositions and nanocrystal sizes to match with a variety of QDs for achieving high performance.

Cao and co-workers investigated quantum dot sensitized solar cells using CZTSSe as a counter electrode [86]. The QDSSC device configuration is shown in Fig. 4.30. The photoanode consisted of TiO_2 nanomaterial spin-casted from TiO_2 paste on FTO glass followed by 30 min thermal treatment at 450°C. The resulting TiO layer is roughly 10 μm in thickness. For CdS-QD attachment on TiO_2, the as-prepared TiO_2-on-FTO-glass was immersed in 0.5 M $Cd(NO_3)_2$ solution for 5 min, rinsed with ethanol, immersed in 0.5 M Na2S solution for another 5 min, and then rinsed with methanol. The CdS-QD attachment process was repeated for 3 times. For CdSe-QD loading on TiO_2, Na_2SeSO_3 was used as the Se source and process of CdSe-QD attachment was similar to that for CdS QDs except a longer time for 1 h and repeating times of 4. The final step for preparing the photoanode is the growth of a ZnS layer for passivation, which gives rise to a photoanode configuration of FTO/TiO_2/CdS/CdSe/ZnS.

OLA-Se and OLA-S solutions together with OLA-metal-precursor were used to prepare for nano CZTSSe. OLA-Se solution was prepared by dissolving Se powder in a container filled with OLA at high temperature. OLA-S solution was prepared by sonicating S powders in OLA. Copper acetate, zinc acetylacetonate, tin acetate, and OLA were mixed to obtain the OLA-metal-precursors solution. The injection of the OLA-S solution into the OLA-Se container was done at 240°C, followed by the injection of OLA metal-precursors 5 s later. CZTSSe nanocrystals grow at this temperature for additional 30 min. The CZTSSe nanocrystals were then obtained after purification and they were capped with OLA. The CZTSSe nanocrystal inks were prepared by a necessary ligand exchange procedure including two steps. First, the OLA-capped CZTSSe nanomaterials were dispersed in 1:10 (volume ratio) hexanethiol/toluene solution, resulting in hexanethiol-capped CZTSSe. Second, the hexanethiol-capped CZTSSe nanocrystals were mixed with toluene to obtain the ink. Catalytic CZTSSe nanocrystal films were prepared by casting the CZTSSe nanocrystal ink on FTO glass, followed by a thermal treatment at 350°C for 40 min in an Ar flow environment. Like other DSSC configurations, the photoanode and CZTSSe counter electrode were assembled in parallel form and the electrolyte is a polysulfide solution, the mixture of 0.5 M Na_2S, 0.125 M S, and 0.2 M KCl in 7:3 volume ratio of water/methanol.

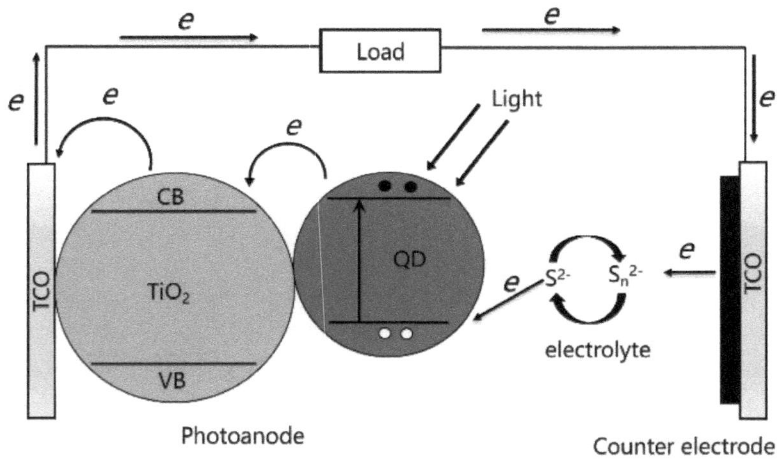

Figure 4.30 Schematic working principle of the DSSC (a) and I-V curves of the DSSCs with CZTSSe and Pt counter electrodes.

In Cao and co-workers' work, the counter electrodes were made by a number of $Cu_2ZnSn(S_{1-x}Se_x)_4$ nanocrystals with varied S/Se ratios corresponding to x = 0, 0.2, 0.5, 0.85, and 1. The performance of the counter electrodes was investigated. In Fig. 4.31, *I–V* curves of the QDSSCs with counter electrodes using CZTSSe of different S/Se ratios as catalytic layers. For comparison, a DSSC with Pt counter electrode is also presented. The *I–V* study indicates that the QDSSC fabricated using CZTSSe of S/Se-ratio = 1 ($Cu_2ZnSn(S_{0.5}Se_{0.5})_4$) gave a highest conversion efficiency of 3.01%, much higher than the value of 1.24% obtained on a QDSSC using Pt counter electrode. In fact, the counter electrodes made by any S/Se ratio have slightly higher performance than that of Pt counter electrode, at least comparable. The compositional variations from Cu_2ZnSnS_4 (x = 0) to $Cu_2ZnSnSe_4$ (x = 1) lead to the change in degree of electrocatalytic activities, which is clearly demonstrated by the varied counter electrode performance in Fig. 4.31.

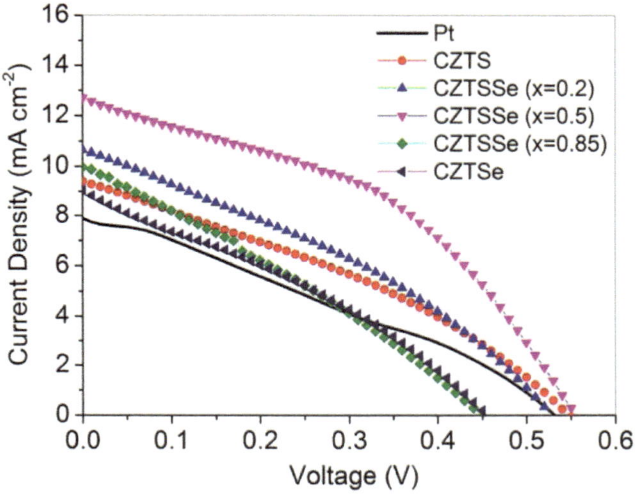

Figure 4.31 *I–V* curves of QDSSCs made by counter electrodes of Pt and $Cu_2ZnSn(S_{1-x}Se_x)_4$ with x = 0, 0.2, 0.5, 0.85, and 1. The measurements were carried out under AM 1.5G illumination (~100 mW/cm^2). Reproduced with permission from [86].

Except for the above nanomaterials, various other non-platinum-based materials have been tried for the catalytic materials of QDSSCs counter electrodes, including Cu_2S [87, 88]. CoS, NiS [89], PbS [90], reduced graphene oxide (RGO)–Cu_2S composite [91], and hollow carbon [92, 93]. Those are alternative candidates for the replacement of traditional Pt counter electrode in quantum-dot sensitized solar cells.

References

1. B. O'Regan and M. Grätzel, A low cost, high-efficiency solar cell based on dyesensitized colloidal TiO_2 films, *Nature*, **353**, 737–739 (1991).
2. M. K. Nazerruddin, A. Kay, I. Ridicio, R. Humphry-Baker, E. Mueller, P. Liska, N. Vlachopoulos, and M. Gratzel, Conversion of

light to electricity by *cis*-X_2Bis (2,2′-bipyridyl-4,4′-dicarboxylate) ruthenium(II) charge-transfer sensitizers ($X=Cl^-,Rr^-,I^-,CN^-$, and SCN^-) on Nanocrystalline TiO_2 electrodes, *J. Amer. Chem. Soc.*, **115**, 6382–6390 (1993).

3. M. Grätzel, Dye-sensitized solar cells, *J. Photoch. Photobio. C*, **4**, 145–153 (2003).
4. K. Ellmer, A. Klein, and B. Rech eds., *Transparent Conductive Zinc Oxide: Basics and Applications in Thin Film Solar Cells* (Springer Series in Materials Science) (2010).
5. K. Fleischer, E. Arca, and I. V. Shvets, Improving solar cell efficiency with optically optimized TCO layers, *Sol. Energy Mater. Sol. Cells*, **101**, 262–269 (2012).
6. K. Ellmer, Past achievements and future challenges in the development of optically transparent electrodes, *Nature Photon.*, **6**, 809 (2012).
7. S. Jin, Y. Yang, J. E. Medvedeva, J. R. Ireland, A. W. Metz, J. Ni, C. R. Kannewurf, A. J. Freeman, and T. J. Marks, Dopant ion size and electronic structure effects on transparent conducting oxides. Sc-doped CdO thin films grown by MOCVD, *J. Am. Chem. Soc.*, **126**, 13787–13793 (2004).
8. T. Minami, Present status of transparent conducting oxide thin-film development for indium-tin-oxide (ITO) substitutes, *Thin Solid Films*, **516**, 5822–5828 (2008).
9. C.-H. Lee, R. Pandey, B.-Y. Wang, W.-K. Choi, D.-K. Choi, and Y.-J. Oh, Nano-sized indium-free MTO/Ag/MTO transparent conducting electrode prepared by RF sputtering at room temperature for organic photovoltaic cells, *Sol. Energy Mater. Sol. Cells*, **132**, 80–85 (2015).
10. Y. (Michael) Yang, Q. Chen, Y.-T. Hsieh, T.-B. Song, N. D. Marco, H. Zhou, and Y. Yang, Multilayer transparent top electrode for solution processed perovskite/Cu(In,Ga)$(Se,S)_2$ four terminal tandem solar cells, *ACS Nano*, **97**, 7714–7721 (2015).
11. S. J. Wang, Y. Geng, Q. Zheng, and J.-K. Kim, Fabrication of highly conducting and transparent graphene films, *Carbon*, **48**, 1815–1823 (2010).
12. L. Peng, Y. Feng, P. Lv, D. Lei, Y. Shen, Y. Li, and W. Feng, Transparent, conductive, and flexible multiwalled carbon nanotube/graphene hybrid electrodes with two three-dimensional microstructures, *J. Phys. Chem. C*, **116**, 4970–4978 (2012).

13. R. Zhu, C.-H. Chung, K. C. Cha, W. Yang, Y. B. Zheng, H. Zhou, T.-B. Song, C.-C. Chen, P. S. Weiss, G. Li, and Y. Yang, Fused silver nanowires with metal oxide nanoparticles and organic polymers for highly transparent conductors, *ACS Nano,* **5**, 12, 9877–9882 (2011).

14. J.-Y. Lee, S. T. Connor, Y. Cui, and P. Peumans, Solution-processed metal nanowire mesh transparent electrodes, *Nano Lett.,* **8**, 2, 689–692 (2008).

15. A. R. Rathmell, S. M. Bergin, Y.-L. Hua, Z.-Y. Li, and B. J. Wiley, The growth mechanism of copper nanowires and their properties in flexible, transparent conducting films, *Adv. Mater,* **22**, 3558–3563 (2010).

16. K. Sharma, V. Sharma, and S. S. Sharma, Dye-sensitized solar cells: Fundamentals and current status, *Nanoscale Res. Lett.,* **13**, 381 (2018).

17. M. K. Nazeeruddin, E. Baranoff, and M. Grätzel, Dye-sensitized solar cells: A brief overview, *Sol. Energy,* **85**, 1172–1178 (2011).

18. H. J. Lee, P. Chen, S. J. Moon, F. Sauvage, K. Sivula, T. Bessho, D. R. Gamelin, P. Comte, S. M. Zakeeruddin, S. I. Seok, M. Grätzel, and M. K. Nazeeruddin, Regenerative PbS and CdS quantum do sensitized solar cells with a cobalt complex as hole mediator, *Langmuir,* **25**, 7602–7608 (2009).

19. V. Chakrapani, D. Baker, and P. V. Kamat, Understanding the role of the sulfide redox couple (S^{2-}/S_n^{2-}) in quantum dot-sensitized solar cells, *J. Am. Chem. Soc.,* **133**, 9607–9615 (2011).

20. K. L. Chopra, S. Major, and D. K. Pandya, Transparent conductors—a status review, *Thin Solid Films,* **102**(1), 1–46 (1983).

21. C. H. Zhai, R. J. Zhang, X. Chen, Y. X. Zheng, S. Y. Wang, J. Liu, N. Dai, and L. Y. Chen, Effects of Al doping on the properties of ZnO thin films deposited by atomic layer deposition, *Nanoscale Res. Lett.,* **11**, 407 (2016).

22. M. Kraini, N. Bouguila, I. Halidou, A. Timoumi, and S. Alaya, Properties of In_2O_3 films obtained by thermal oxidation of sprayed In_2S_3, *Mater. Sci. Semicond. Process,* **16**, 1388–1396 (2013).

23. N. Tetreault, E. Arsenault, L.-P. Heiniger, N. Soheilnia, J. Brillet, T. Moehl, S. Zakeeruddin, G. A. Ozin, and M. Grätzel, High-efficiency dye-sensitized solar cell with three-dimensional photoanode, *Nano Lett.,* **11**, 4579–4584 (2011).

24. S. Ito, T. Murakami, P. Comte, P. Liska, and C. Gratzel, Fabrication of thin film dye sensitized solar cells with solar to electric power

conversion efficiency over 10%, *Thin Solid Films,* **516**, 4613–4619 (2008).

25. W.-Q. Wu, J.-Y. Liao, H.-Y. Chen, X.-Y. Yu, C.-Y. Su, and D.-B. Kuang, Dye-sensitized solar cells based on a double layered TiO_2 photoanode consisting of hierarchical nanowire arrays and nanoparticles with greatly improved photovoltaic performance, *J. Mater. Chem.,* **22**(34), 18057–18062 (2012).
26. P. Du, L. Song, J. Xiong, N. Li, L. Wang, Z. Xi, N. Wang, L. Gao, and H. Zhu, Dye-sensitized solar cells based on anatase TiO_2/multi-walled carbon nanotubes composite nanofibers photoanode, *Electrochim. Acta,* **87**(1), 651–656 (2013).
27. M. Y. Song, D. K. Kim, K. J. Ihn, S. M. Jo, and D. Y. Kim, Electrospun TiO_2 electrodes for dye-sensitized solar cells, *Nanotechnology,* **15**(12), 1861 (2004).
28. X. Yin, W. Que, D. Fei, H. Xie, Z. He, and G. Wang, Strategies to prepare an efficient photoanode for ZnO nanowires-based CdS–CdSe co-sensitized solar cells, *Electrochim. Acta,* **89**, 561–570 (2013).
29. X. K. Qiu, W. X. Que, X. T. Yin, J. Zhang, and J. Chen, ZnO/CdS/CdSe core/double shell nanorod arrays derived by a successive ionic layer adsorption and reaction process for quantum dot-sensitized solar cells, *Semicond. Sci. Technol.,* **26**, 095028 (2011).
30. S. Hong and S. Myung, Nanotube electronics: A flexible approach to mobility. *Nature Nanotech.,* **2**, 4, 207–208 (2007).
31. C. R., Daniel, B. D'Anjou, N. Ghattamaneni, B. Harack, M. Hilke, A. Horth, N. Majlis, M. Massicotte, L. Vandsburger, E. Whiteway, and V. Yu, Experimental review of graphene (PDF). *ISRN Condensed Matter Physics* (International Scholarly Research Network) **2012**: 1–56 (3 November 2011).
32. J.-H. Chen, C. Jang, S. Xiao, M. Ishigami, and M. S. Fuhrer, Intrinsic and extrinsic performance limits of graphene devices on SiO_2, *Nature Nanotech.,* **3**, 206 (2008).
33. C. Punckt, F. Muckel, S. Wolff, I. A. Aksay, C. A. Chavarin, G. Bacher, and W. Mertin, The effect of degree of reduction on the electrical properties of functionalized graphene sheets, *Appl. Phys. Lett.,* **102**(2), 023114 (2013).
34. K. Novoselov, A. Geim, S. Morozov, D. Jiang, Y. Zhang, S. Dubonos, I. Grigorieva, and A. Firsov, Electric field effect in atomically thin carbon films, *Science,* **306**, 666 (2004).
35. Q. Yu, L. A. Jauregui, W. Wu, R. Colby, J. Tian, Z. Su, H. Cao, Z. Liu, D. Pandey, D. Wei, T. F. Chung, P. Peng, N. P. Guisinger, E. A. Stach, J. Bao,

S.-S. Pei, and Y. P. Chen, Control and characterization of individual grains and grain boundaries in graphene grown by chemical vapour deposition, *Nat. Mater.*, **10**, 443–449 (2011).

36. A. Reina, X. Jia, J. Ho, D. Nezich, H. Son, V. Bulovic, M. S. Dresselhaus, and J. Kong, Large area, few-layer graphene films on arbitrary substrates by chemical vapor deposition, *Nano Lett.*, **9**(1), 30–35 (2009).
37. Y. M. Lin, C. Dimitrakopoulos, K. A. Jenkins, D. B. Farmer, H. Y. Chiu, A. Grill, and P. Avouris, 100 GHz transistors from wafer-scale epitaxial graphene, *Science*, **327**, 662–662 (2010).
38. Y. Hernandez, V. Nicolosi, M. Lotya, F. M. Blighe, Z. Sun, S. De, I. T. McGovern, B. Holland, M. Byrne, Y. K. Gun'Ko, J. J. Boland, P. Niraj, G. Duesberg, S. Krishnamurthy, R. Goodhue, J. Hutchison, V. Scardaci, A. C. Ferrari, and J. N. Coleman, High-yield production of graphene by liquid-phase exfoliation of graphite, *Nat. Nanotechnol.*, **3**, 563–568 (2008).
39. J. N. Coleman, Liquid exfoliation of defect-free graphene, *Acc. Chem. Res.*, **46**, 14–22 (2012).
40. B. C. Brodie, On the atomic weight of graphite, *Philos. Trans. R. Soc. London*, **149**, 249–259 (1859).
41. D. C. Marcano, D. V. Kosynkin, J. M. Berlin, A. Sinitskii, Z. Sun, A. Slesarev, L. B. Alemany, W. Lu, and J. M. Tour, Improved synthesis of graphene oxide, *ACS Nano*, **4**(8), 4806–4814 (2010).
42. Y. Bessekhouad, D. Robert, and J. V. Weber, Synthesis of photocatalytic TiO_2 nanoparticles: Optimization of the preparation conditions, *J. Photochem. Photobiol. A Chem.*, 157(1), 47–53 (2003).
43. I. N. Kuznetsova, V. Blaskov, I. Stambolova, L. Znaidi, and A. Kanaev, TiO_2 pure phase brookite with preferred orientation, synthesized as a spin-coated film, *Mater. Lett.*, **59**, 3820–3823 (2005).
44. T. Moritz, J. Reiss, K. Diesner, D. Su, and A. Chemseddine, Nanostructured crystalline TiO_2 through growth control and stabilization of intermediate structural building units, *J. Phys. Chem. B*, **101**(41), 8052 (1997).
45. A. Chemseddine and T. Moritz, Nanostructuring titania: Control over nanocrystal structure, size, shape, and organization, *Eur. J. Inorg. Chem.*, **2**, 235–245 (1999).
46. L. Miao, S. Tanemura, S. Toh, K. Kaneko, and M. Tanemura, Fabrication, characterization and Raman study of anatase-TiO_2 nanorods by a heating-sol–gel template process, *J. Cryst. Growth*, **264**(1–3), 246–252 (2004).

47. L. Miao, S. Tanemura, S. Toh, K. Kaneko, and M. Tanemura, Heating-sol–gel template process for the growth of TiO_2 nanorods with rutile and anatase structure, *Appl. Surf. Sci.*, **238**(1–4), 175–179 (2004).

48. Y. Chen, J. C. Crittenden, S. Hackney, L. Sutter, and D. W. Hand, Preparation of a novel TiO_2-based *p–n* junction nanotube photocatalyst, *Environ. Sci. Technol.*, **39**(5), 1201–1208 (2005).

49. S. Lee, C. Jeon, and Y. Park, Fabrication of TiO_2 tubules by template synthesis and hydrolysis with water vapor, *Chem. Mater.*, **16**(22), 4292–4295 (2004).

50. X. Wen, J. C. Tao, Y. H. Sun, Y. Sun, and N. Dai, TiO_2 Nanotube Arrays for Quantum Dots Sensitized Solar Cells, *Proc. of SPIE* **7381**, 73810Z (2009).

51. X. Y. Kong and Z. L. Wang, Spontaneous polarization-induced nanohelixes, nanosprings, and nanorings of piezoelectric nanobelts, *Nano. Lett.*, **3**(12), 1625–1631 (2003).

52. J. G. Wen, J. Y. Lao, D. Z. Wang, T. M. Kyaw, Y. L. Foo, and Z. F. Ren, Self-assembly of semiconducting oxide nanowires, nanorods, and nanoribbons, *Chem. Phys. Lett.*, **372**, 717–722 (2003).

53. B. D. Yao, Y. F. Chan, and N. Wang, Formation of ZnO nanostructures by a simple way of thermal evaporation, *Appl. Phys. Lett.*, **81**(4), 757–759 (2002).

54. P. Yang, H. Yan, S. Mao, R. Russo, J. Johnson, R. Saykally, N. Morris, J. Pham, R. He, and H.-J. Choi, Controlled growth of ZnO nanowires and their optical properties, *Adv. Funct. Mater.*, **12**(5), 323–331 (2002).

55. H. T. Ng, J. Han, T. Yamada, P. Nguyen, Y. P. Chen, and M. Meyyappan, Single crystal nanowire vertical surround-gate field-effect transistor, *Nano. Lett.*, **4**(7), 1247–1252 (2004).

56. Z. Fan, D. Wang, P. Chang, W. Tseng, and J. G. Lu, ZnO nanowire field-effect transistor and oxygen sensing property, *Appl. Phys. Lett.*, **85**(24), 5923–5925 (2004).

57. S. Xu, Y. Wei, M. Kirkham, J. Liu, W. Mai, D. Davidovic, R. L. Snyder, and Z. L. Wang, Patterned growth of vertically aligned ZnO nanowire arrays on inorganic substrates at low temperature without catalyst, *J. Am. Chem. Soc.*, **130**, 14958–14959 (2008).

58. H. Hua, X. Huang, C. Deng, X. Chen, and Y. Qian, Hydrothermal synthesis of ZnO nanowires and nanobelts on a large scale, *Mater. Chem. Phys.*, **106**(1), 58–62 (2007).

59. M. K. Nazeeruddin, F. De Angelis, S. Fantacci, A. Selloni, G. Viscardi, P. Liska, S. Ito, B. Takeru, and M. Grätzel, Combined experimental and DFT-TDDFT computational study of photoelectrochemical cell ruthenium sensitizers, *J. Am. Chem. Soc.,* **127**(48), 16835–16847 (2005).

60. M. K. Nazeeruddin, P. Péchy, T. Renouard, S. M. Zakeeruddin, R. Humphry-Baker, P. Comte, P. Liska, L. Cevey, E. Costa, V. Shklover, L. Spiccia, G. B. Deacon, C. A. Bignozzi, and M. Grätzel, Engineering of efficient panchromatic sensitizers for nanocrystalline TiO_2-based solar cells, *J. Am. Chem. Soc.*, **123**(8), 1613–1624 (2001).

61. K. Cao and M. Wang, Recent developments in sensitizers for mesoporous sensitized solar cells, *Front. Optoelectron.,* **6**, 373–385 (2013).

62. F.-T. Kong, S.-Y. Dai, and K.-J. Wang, Review of recent progress in dye-sensitized solar cells, *Adv. Optoelectron.,* article ID 75384, 13 p. (2007).

63. R. M. El-Shishtawy, Functional dyes, and some Hi-tech applications, *Int. J. Photoenergy,* article ID 434897, 21 p. (2009).

64. K. Kakiage, Y. Aoyama, T. Yano, T. Otsuka, T. Kyomen, M. Unno, and M. Hanaya, An achievement of over 12 percent efficiency in an organic dye-sensitized solar cell, *Chem. Commun.*, **50**, 6379 (2014).

65. Y. Cao, N. Cai, Y. Wang, R. Li, Y. Yuan, and P. Wang, Modulating the assembly of organic dye molecules on titania nanocrystals via alkyl chain elongation for efficient mesoscopic cobalt solar cells, *Phys. Chem. Chem. Phys.,* **14**, 8282 (2012).

66. J.-H. Yum, T. W. Holcombe, Y. Kim, K. Rakstys, T. Moehl, J. Teuscher, J. H. Delcamp, M. K. Nazeeruddin, and M. Grätzel, Blue-coloured highly efficient dye-sensitized solar cells by implementing the diketopyrrolopyrrole chromophore, *Sci. Rep.*, **3**, 2446 (2013).

67. J.-H. Yum, E. Baranoff, F. Kessler, T. Moehl, S. Ahmad, T. Bessho, A. Marchioro, E. Ghadiri, J.-E. Moser, C. Yi, M. K. Nazeeruddin, and M. Grätzel, A cobalt complex redox shuttle for dye-sensitized solar cells with high open-circuit potentials, *Nat. Commun.*, **3**, 631 (2012).

68. M. Zhang, Y. Wang, M. Xu, W. Ma, R. Li, and P. Wang, Design of high-efficiency organic dyes for titania solar cells based on the chromophoric core of cyclopentadithiophene-benzothiadiazole, *Energy Environ. Sci.*, **6**, 2944 (2013).

69. J. Yang, P. Ganesan, J. Teuscher, T. Moehl, Y. J. Kim, C. Yi, P. Comte, K. Pei, T. W. Holcombe, M. K. Nazeeruddin, J. Hua, S. M. Zakeeruddin, H.

Tian, and M. Grätzel, Influence of the donor size in D–π–A organic dyes for dye-sensitized solar cells, *J. Am. Chem. Soc.*, **136**, 5722 (2014).

70. N. Zhou, K. Prabakaran, B. Lee, S. H. Chang, B. Harutyunyan, P. Guo, M. R. Butler, A. Timalsina, M. J. Bedzyk, M. A. Ratner, S. Vegiraju, S. Yau, C.-G. Wu, R. P. H. Chang, A. Facchetti, M.-C. Chen, and T. J. Marks, Metal-free tetrathienoacene sensitizers for high-performance dye-sensitized solar cells, *J. Am. Chem. Soc.*, **137**, 4414–4423 (2015).
71. T. Bessho, S. M. Zakeeruddin, C. Y. Yeh, E. W. G. Diau, and M. Grätzel, Highly efficient mesoscopic dye-sensitized solar cells based on donor-acceptor-substituted porphyrins, *Angew. Chem. Int. Ed.*, **49**, 6646 (2010).
72. A. Yella, H.-W. Lee, H. N. Tsao, C. Yi, A. K. Chandiran, M. K. Nazeeruddin, E. W.-G. Diau, C.-Y. Yeh, S. M. Zakeeruddin, and M. Grätzel, Porphyrin-sensitized solar cells with cobalt (II/III)–based redox electrolyte exceed 12 percent efficiency, *Science*, **334**, 629 (2011).
73. S. Mathew, A. Yella, P. Gao, R. Humphry-Baker, B. F. E. Curchod, N. Ashari-Astani, I. Tavernelli, U. Rothlisberger, M. K. Nazeeruddin, and M. Grätzel, Dye-sensitized solar cells with 13% efficiency achieved through the molecular engineering of porphyrin sensitizers, *Nat. Chem.*, **6**, 242 (2014).
74. O. Adedokun, K. Titilope, and A. O. Awodugba, Review on natural dye-sensitized solar cells (DSSCs), *Int. J. Eng. Tech.*, **2**(2), 34 (2016).
75. M. R. Narayan, Review: Dye sensitized solar cells based on natural photosensitizers, *Renew. Sustain. Energy Rev.*, **16**, 208–215 (2012).
76. M. S. Dixit, K. Sofi, P. Sonone, G. T. Tayade, and A. U. Ubale, Dye sensitized solar cell using basella alba (red vine spinach) sensitizer nanocrystalline TiO_2 photoanode, *Int. J. Sci. Res. Comput. Sci., Eng. Inform. Tech.*, **4**(4), 97–101 (2019).
77. H. Zhou, L. Wu, Y. Gao, and T. Ma, Dye-sensitized solar cells using 20 natural dyes as sensitizers, *J. Photochem. Photobiol. A: Chem.*, **219**, 188–194 (2011).
78. G. Calogero, G. D. Marco, S. Cazzanti, S. Caramori, R. Argazzi, A. D. Carlo, and C. A. Bignozzi, Efficient dye-sensitized solar cells using red turnip and purple wild sicilian prickly pear fruits, *Int. J. Mol. Sci.*, **11**, 254 (2010).
79. W. J. Lee, E. Ramasamy, D. Y. Lee, and J. S. Song, Efficient dye-sensitized solar cells with catalytic multiwall carbon nanotube counter electrodes, *ACS Appl. Mater. Interfaces*, **1**(6), 1145–1149 (2009).

80. A. Kay and M. Grätzel, Low cost photovoltaic modules based on dye sensitized nanocrystalline titanium dioxide and carbon powder, *Sol. Energy Mater. Sol. Cells,* **44**(1), 99–117 (1996).

81. Q. W. Jiang, G. R. Li, and X. P. Gao, Highly ordered TiN nanotube arrays as counter electrodes for dye-sensitized solar cells, *Chem. Commun.,* **44**, 6720–6722 (2009).

82. E. Furimsky, Metal carbides and nitrides as potential catalysts for hydroprocessing, *Appl. Catal. A General,* **240**(1), 1–28 (2003).

83. W. R. Erwin, L. Oakes, S. Chatterjee, H. F. Zarick, C. L. Pint, and R. Bardhan, Engineered porous silicon counter electrodes for high efficiency dye-sensitized solar cells, *ACS Appl. Mater. Interfaces,* **6**(12), 9904–9910 (2014).

84. L. Oakes, A. Westover, J. W. Mares, S. Chatterjee, W. R. Erwin, R. Bardhan, S. M. Weiss, and C. L. Pint, Surface engineered porous silicon for stable, high performance electrochemical supercapacitors, *Sci. Rep.,* **3**(43), 3020 (2013).

85. G. Gaur, D. S. Koktysh, and S. M. Weiss, Immobilization of quantum dots in nanostructured porous silicon films: Characterizations and signal amplification for dual-mode optical biosensing, *Adv. Funct. Mater.,* **23**(29), 3604–3614 (2013).

86. Y. Cao, Y. Xiao, J.-Y. Jung, H.-D. Um, S.-W. Jee, H. M. Choi, J. H. Bang, and J.-H. Lee, Highly electrocatalytic $Cu_2ZnSn\ (S_{1-x}Se_x)_4$ counter electrodes for quantum-dot-sensitized solar cells, *ACS Appl. Mater. Interfaces,* **5**(3), 479–484 (2013).

87. S. Giménez, I. Mora-Seró, L. Macor, N. Guijarro, T. Lana-Villarreal, R. Gómez, L. J. Diguna, Q. Shen, T. Toyoda, and J. Bisquert, Improving the performance of colloidal quantum-dot-sensitized solar cells, *Nanotechnology,* **20**(29), 295204 (2009).

88. J. F. Shi, Y. Fan, X. Q. Xu, G. Xu, and L. H. Chen, Influence of preparation conditions on the properties of Cu_2S photocathodes, *Acta Phys. Chim. Sin.,* **28**, 857–864 (2012).

89. Z. S. Yang, C. Y. Chen, C. W. Liu, and H. T. Chang, Electrocatalytic sulfur electrodes for CdS/CdSe quantum dot-sensitized solar cells, *Chem. Commun.,* **46**, 5485–5487 (2010).

90. Z. Tachan, M. Shalom, I. Hod, S. Rühle, S. Tirosh, and A. J. Zaban, PbS as a highly catalytic counter electrode for polysulfide-based quantum dot solar cells, *Phys. Chem. C,* **115**, 6162–6166 (2011).

91. J. G. Radich, R. Dwyer, and P. V. Kamat, Cu_2S reduced graphene oxide composite for high-efficiency quantum dot solar cells. Overcoming

the redox limitations of S^{2-}/S_n^{2-} at the counter electrode, *J. Phys. Chem. Lett.*, **2**, 2453–2460 (2011).

92. G. S. Paul, J. H. Kim, M. S. Kim, K. Do, J. Ko, and J. S. Yu, Different hierarchical nanostructured carbons as counter electrodes for CdS quantum dot solar cells, *ACS Appl. Mater. Interfaces,* **4**, 375–381 (2012).
93. J. H. Dong, S. Jia, J. Z. Chen, B. Li, J. F. Zheng, J. H. Zhao, Z. J. Wang, and Z. P. Zhu, Nitrogen-doped hollow carbon nanoparticles as efficient counter electrodes in quantum dot sensitized solar cells, *J. Mater. Chem.*, **22**, 9745–9750 (2012).

Chapter 5

Quantum Dot Solar Cells

Quantum dots (QDs) are zero-dimensional semiconductors in which carriers receive quantum confinement in all three space directions. A quantum dot solar cell (QDSC) has semiconductor QDs used as the active material. Due to the small sizes of QDs, QDSCs use only very small amount of the material.

Nano solar cells (NSCs) have multiple advantages. In constructing the devices, lattice matching to the substrate is not required, which offers great room for device architecture and material selections. The devices are usually fabricated at near-ambient conditions. Using no vacuum systems helps a lot in the reduction of device cost. Additionally, the device architecture puts almost no limitations on material combinations. Together with the quantum effect that can be tuned by QD sizes and surface ligand exchange, device structures can be designed with a great amount of choices for various materials (including ligands) and material sizes. QDSCs are thus a hopeful architecture for power conversion in spectral range and some advanced ideas like multiple exciton generation.

QDs are the most widely prepared nano form of semiconductors in which carriers receive quantum confinement in all three directions in space. QDs are usually referred to semiconductors, while for other materials they are called nanoparticles. The QDs can be fabricated by many technologies and in the case of solar cells (SCs) they are usually prepared by

Introduction to Nano Solar Cells
Ning Dai

ISBN 978-981-4877-49-7 (Hardcover), 978-1-003-13198-4 (eBook)
www.jennystanford.com

chemical routes. The solution-processed QDs are suspended in liquids—colloidal quantum dots (CQDs). The CQD-based SCs aims at low cost and high efficiency, implemented for the full solar spectrum harvesting. The QDs have sizes in the range of 2 to 10 nm in diameter. Thus, each QD contains about 100 to 100,000 atoms.

Due to the confinement on carriers in all three directions, the electron density of states exhibits δ-like distribution, as shown in Fig. 1.19 in Chapter 1, which is beneficial to the enhancement of opto-electronic device performance. QDs used for solar cells are mostly prepared by wet chemical routes, so that the QDs are dispersed in chemical solutions—so-called colloidal quantum dots (CQDs) and coated with various ligands depending on the preparation. The wet chemical preparation routes are low cost and easy to be scaled up. A large variety of semiconductor QDs of the materials including groups IV, III–V, and II–VI compounds, as well as semiconductor oxides have been prepared. With controlled fabrication conditions, QDs with various shapes can be prepared. Usually, semiconductors with cubic lattice are easily fabricated with the shapes of round particles, while those with hexagonal lattice, for instance, can be grown with the shapes of nanorods, nanoneedles, etc.

The fundamental requirements for QDs used for SCs are high crystal quality, low surface state density, highly monodispersive in solution, narrow size distribution, and controllable size. The high crystal quality guarantees the high efficiency photo-generation of excitons and long exciton life time. The low surface state density reduces the carrier trapping and non-radiative carrier recombination. A precursor of monodispersive QDs in solution gives rise to uniform QD thin film prepared by spin coating, drop casting, or dip coating. Narrow size distribution makes it easy to pin point the targeted wavelength of absorbed solar radiation, while the controllable QD size makes it possible for the absorption of selected wavelength, aiming at optimized device architecture. In addition, the QDs need to be attached with surface ligands that are highly conductive allowed for electron transport in between.

5.1 General Properties of Semiconductor Quantum Dots

5.1.1 Quantum Dot Composite Thin Films

By the name, quantum dots are scattered zero-dimensional semiconductor materials. In the use for solar cells, quantum dots are either embedded in other materials or inter-connected to form thin films. A number of routes have been developed to fabricate compositional quantum dot thin films for various practical uses. Well-studied QD composite thin films include layered QDs/graphene [1], QDs/Polymer [2], QDs/P3HT [3], etc. The materials that wrap QDs need to be conductive, so that carriers could transport through them to the electrodes. Figure 5.1 presents TEM image of CdSe QDs dispersed in host P3HT in 1:1 weight ratio [2]. A high-resolution TEM image is shown in the inset on top right of the figure. The average size of the CdSe QDs is ~5 nm. In order to obtain conductive CQD thin films, it is necessary to replace the long alkyl ligands with short ligands.

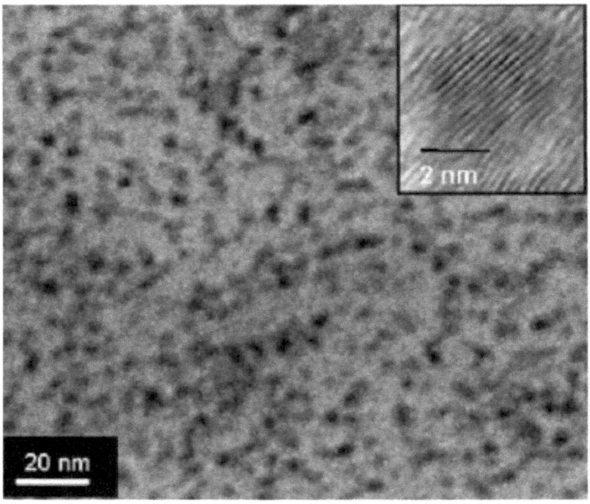

Figure 5.1 TEM image of CdSe QDs dispersed in host P3HT in a weight ratio of 1:1 and high-resolution TEM image (inset) of a CdSe quantum dot in host. Reproduced with permission from [3].

The major components of conventional crystalline Si solar cell are the absorber that consists of a *p*-phase and an *n*-phase material parts forming a *p–n* junction. Figure 5.2 shows schematically the active part of a Si solar cell. When photons are absorbed in *p–n* junction region, photon-excited electrons and holes drift toward opposite directions driven by the built-in electric field. If the photons are absorbed in the regions away from the junction region, photon-generated electrons and holes will have to move to their respective electrodes relying on diffusion.

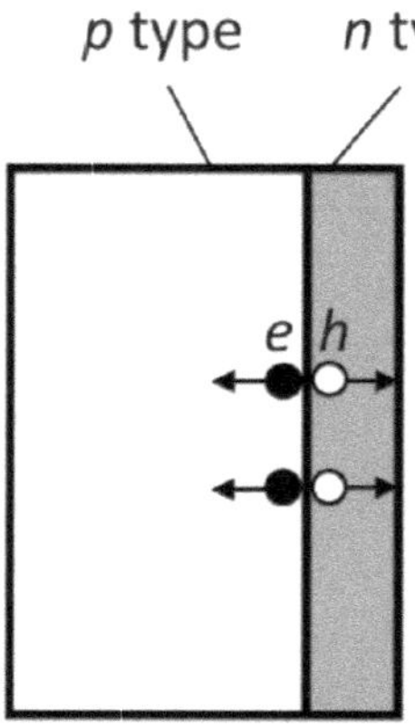

Figure 5.2 The schematic illustration of a *p–n* junction and the drifting electrons and holes driven by the built-in electric field.

In the case of nano solar cells, the active light absorption layer consists of a phase separated bi-layer heterojunction, as shown in Fig. 5.3a. Photons are absorbed by the absorber layer while the *n*-phase and the *p*-phase materials are used as transport layers for electrons and holes, respectively. Nano solar cells can be viewed as a continuous change of the *p–n* junction interface from the conventional cell, as shown in Fig. 5.3b,c for 3D heterojunctions. However, any NSC can be viewed as it is consisted of many small flat-band SCs, as shown by the zoom-in areas in Fig. 5.3b,c. All nano photovoltaic devices, including Schottky junction, depleted heterostructures, quantum dot dye-sensitized, hybrid heterojunction, extremely thin absorber solar cells, are the material and geometric variation of flat band solar cells.

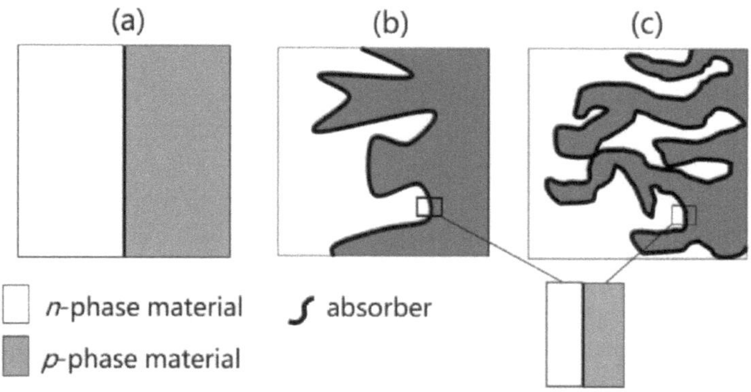

Figure 5.3 Different configurations of heterojunctions, (a) phase separated bi-layer heterojunction, and (b) and (c) 3D bulk heterojunctions. The bulk heterojunction offers more interfacial area between the two phases, helpful for charge transport.

5.1.2 Electronic Properties of Quantum Dots

In preparing a semiconductor heterojunction, an epitaxial technique is used to grow one semiconductor on top of another one. In a more general sense, a heterojunction is formed by bringing two different materials into physical contact with an atomic smooth interface. The properties of heterostructures rely not only on the parameters of the two materials but also on the band edge alignment (or energy discontinuity at the band edges). In fact, one knows quite well the energy positions of the valence and the conduction band edges of a lot of materials, with respect to vacuum energy level and normalized hydrogen electrodes (NHE), as shown in Fig. 5.4 for Si, Ge, CdTe, CdSe, ZnSe, ZnS, GaAs, InP, GaP, Ta_3N_5, WO_3, TiO_2, ZnO, PbO, etc. Unfortunately, the band discontinuities at the heterointerfaces cannot be predicted simply in terms of the band edge positions prior to the formation of the heterojunctions. It is rather difficult to determine the band edge discontinuities experimentally and accurately. Yet, the parameters are of critical importance for heterostructure devices.

The zero-dimensional quantum dots (QDs) receive confinement in all three directions in space. QDs have an enhanced density of states (DOS) at some eigen energy positions, which leads to their superior optical and transport properties. As a result, QD-based light-emitting diodes, laser diodes, and other opto-electronic devices show great performance. When an electron is excited into the excited state in the conduction band, a hole is left in the valence band and the electron and the hole are still attracted to each other by Coulomb interaction. The electron/hole pair is called an exciton. An exciton has a size characterized by the exciton Bohr radius that is close to the size of the electron wave packet. The hole wave packet is usually much smaller due to its large effective mass. In a QD with its radius comparable or smaller than that of the exciton Bohr radius, the excitons in the QD are squeezed by the confinement due to the potential barrier that is applied on the excitons.

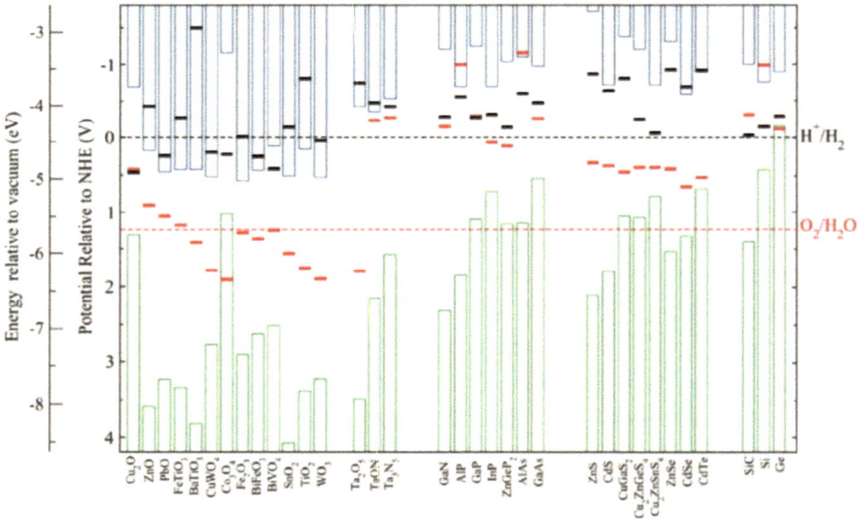

Figure 5.4 The conduction and the valence band edge positions of some semiconductors relative to normalized hydrogen electrodes (NHE) and vacuum energy. The energy levels of redox couples of H^+/H_2 and O_2/H_2O are denoted by the horizontal dashed lines.

In a QD, both the conduction and the valence bands become discrete energy levels, which can be simply modeled with "the particle in a box" model described in Chapter 2. The degree of the confinement depends on the QD size with respect to the Bohr radius of the exciton (or the electron). A strong confinement occurs in the case that the QD size is comparable to that of the exciton wave packet. When the size of quantum dot is much larger or much smaller than the Bohr radius of electron and hole, the electron and hole are in weak confinement. In semiconductors, a situation that corresponds to "an intermediate confinement" is very often encountered, where the QD radius is larger than the Bohr radius of the hole but smaller than the Bohr radius of the electron, since the hole mass is usually much larger than the electron mass. For QDs with their diameters comparable to or smaller than the Bohr radius of electrons, the bandgaps of the QDs become larger than those of their bulk due to the strong quantum confinement applied on the electrons. In a semiconductor, an exciton Bohr radius has the form of

$$a_{\text{ex}} = \varepsilon_r \left(\frac{m_e}{\mu} \right) a_B, \tag{5.1}$$

where $a_B = {}^{4\pi\varepsilon_0\hbar^2}\!/_{m_e e^2}$ (= 0.053 nm) is the Bohr radius that equals to the most probable distance between the proton and electron in a hydrogen atom at the ground state, ε_r the relative permittivity, m_e and μ are the bare electron mass and the reduced mass, respectively. Thus, an exciton in a semiconductor behaves like a big "hydrogen atom" with an orbit radius a_{ex} much larger than a_B. A quantum confinement acts to squeeze the orbit of the big "hydrogen atom", which causes the increase of energy levels. The exciton energy takes the form of

$$E_{\text{ex}} = -\frac{\mu}{\varepsilon_r^2 m_e} R_y = -R_y^*, \tag{5.2}$$

and the confinement energy is

$$E_{\text{conf}} = \frac{\hbar^2\pi^2}{2\mu a_{\text{ex}}^2} = \frac{\hbar^2\pi^2}{2a_{\text{ex}}^2}\left(\frac{1}{m_e} + \frac{1}{m_h} \right), \tag{5.3}$$

where $R_y \sim 13.6$ eV is Rydberg energy for hydrogen, m_e and m_h are the effective masses of the electron and the hole. The energy for opto-electronic transition will be

$$E = E_g + E_{ex} + E_{conf} = E_g - R_y^* + \frac{\hbar^2 \pi^2}{2a_{ex}^2}\left(\frac{1}{m_e} + \frac{1}{m_h}\right). \quad (5.4)$$

E_{conf} depends on the orbit radius a_{ex}, that is limited by the size of the quantum dots. a_{ex} should be replaced by the QD radius for a QD with its size smaller than the exciton Bohr radius. Thus, the exciton transition energy in a QD, corresponding to the energy of light emission or light absorption, increases with the decrease of the QD size, with respect to the transition energy $E_g - R_y^*$ in bulk.

5.1.3 Optical Properties of Quantum Dots

Semiconductor QDs have unique optical properties due to their limited sizes, which lead to not only the change of the energy bands, but also the charge distribution affected by band alignment. For instance, charge distribution, depending on band alignment, could result in distinct light emission behavior. As described in Chapter 2, photo-generated electrons and holes are both confined in QDs with type-I ban alignment. For type-II QDs, however, electrons are confined in the dots and the holes are in the surrounding barriers, or vice versa. Apparently, the electron–hole wavefunction overlap and, as a result, the electron–hole Coulomb interaction are different in type-I and type-II QDs, which leads to their different optical properties.

Having unique optical properties—very strong light absorption (when photon energies are equal to or larger than the bandgap energy), tunable absorption wavelength, high brightness, long fluorescence lifetime, etc., QDs are superior materials for solar cells. QDs can be excited by photons with energies equal and larger than the bandgap and QDs are referred as photon antenna due to their very strong light absorptivity. QDs have broad excitation profiles and very narrow emission spectra. As shown by Eq. (5.4), the optical transition energy is dependent on the orbit radius a_{ex} that is limited by QD size. When a QD is struck by a photon with energy equal to or larger than E in Eq. (5.4), an electron in the QD is excited to the conduction band, left with a positively charged hole in the valence band. As shown by Eq. (5.4), decreasing the size of the QD causes the increase of transition energy. Figure 5.5a presents the optical absorption spectra of the colloid CdSe quantum dots of different sizes (as labeled on the spectra). The QDs are hydrothermally prepared in an aqueous suspension. Figure 5.5b shows schematically that the optical transition energy varies with the size of a QD.

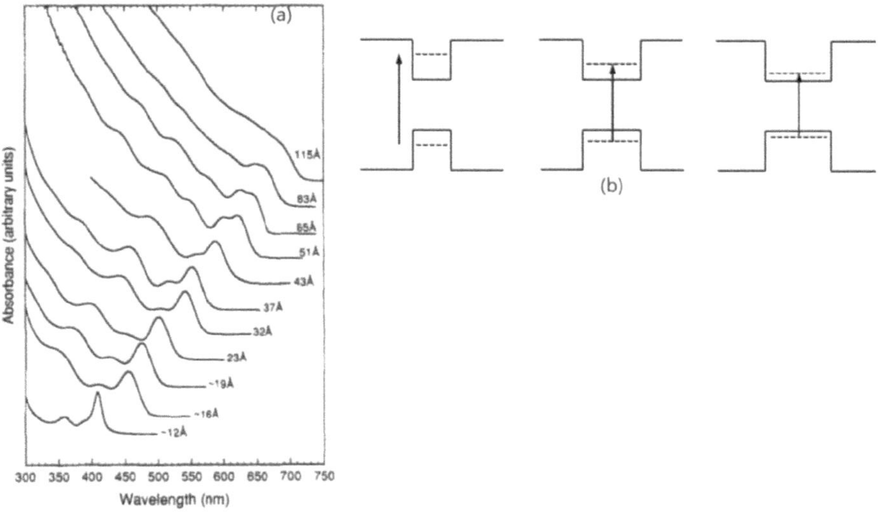

Figure 5.5 Optical absorption spectra of colloid CdSe quantum dots of various sizes (a) and schematic band profiles of the quantum dots of different sizes showing that the optical transition energies decrease with increasing sizes (b).

The quantum confinement pushes the electron eigen state upward and the hole eigen state downward in energy. The quantum confinement effect is used to tune both the optical absorption edge and the luminescence wavelength. Figure 5.6 presents the CdSe QDs of different radii in solutions which emit light with different colors from red to blue because of the quantum size effect. The photoluminescent radiations are excited by ultraviolet illumination.

Thus, the photon-absorption spectra of QDs can be engineered by tuning their geometrical sizes, shapes, as well as surface coating materials (by varying the strength of the confinement potential). Due to their larger sizes than atoms, it is relatively easier to form a close contact between, for instance, QDs and other nanomaterials and to make modifications on QD surfaces. In dealing with QDSCs, one could still adopt the parameters in semiconductors, such as mobility and carrier lifetime, by simply viewing the colloidal quantum dot films as an effective semiconductor medium. This offers easiness in analyzing and discussing the working process using the concept adopted in semiconductors.

Figure 5.6 Luminescence of CdSe QDs of different particle sizes. The red luminescence corresponds to the QD radius of 4 nm and the blue one to 1 nm.

5.2 Growth of Quantum Dot

Interests in semiconductor QDs were strongly stimulated by the discovery of various preparation routes including MBE, LPE, MOCVD, as well as solvothermal technology. Among them, the solvothermal preparation of colloidal quantum dots bears the advantages of simple, fast, cheap, and large area deposition, which is essential for the materials used for solar cell purpose. The resulting QDs prepared by the cost-effective solvothermal route are usually dissolved in solution. In the synthesis of colloidal QDs, precursors, organic surfactants, and solvents are usually used, as shown by the pioneer work done by Murray et al. [4]. Solvothermal preparation can be classified into three categories: hot-injection, non-injection heat-up, and flow reactor. In solar cell devices, the QDs should be prepared to form thin films. The typical methods for thin-film fabrication include spin coating, dip coating, and drop casting. In all the fabrication routes, the nucleation and the growth are separated, which is beneficial for obtaining good uniformities in QD sizes and shapes.

5.2.1 Hot-Injection Synthesis

Figure 5.7 presents schematically the three synthesis routes of colloidal QDs. In hot-injection synthesis, nucleation occurs upon the injection of a precursor (usually at room temperature) into surfactant at high temperature. The growth is companied by reaction cooling [5, 6]. In non-injection heat-up method, the precursors are pre-prepared prior to reaction. Growth process is initiated by heating the precursors together and the temperature then triggers the decomposition, nucleation and growth phases sequentially [7]. In flow reactor synthesis, growth process is carried out through separating synthetic phase by multi-stage and continuous flow synthesis reactor [8].

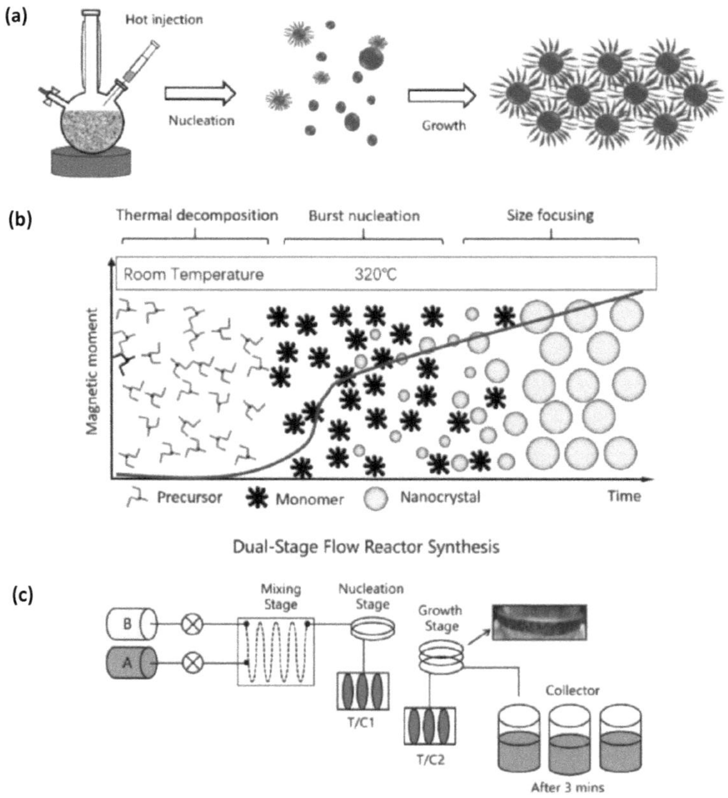

Figure 5.7 Synthesis routes of colloidal QDs: (a) Hot-injection synthesis, (b) non-injection heat-up method, (c) Flow reactor synthesis.

In solvothermal preparations, usually two effects compete for the growth, one being diffusion-controlled and another being reaction-controlled. In diffusion-controlled synthesis, growth rate is reversely proportional to particle size. Thus, particles with larger sizes grow slower than those with smaller sizes. This leads to narrow size distributions. In reaction-controlled synthesis, however, the solubility of the QDs depends on QD sizes. QDs with smaller sizes dissolve due to their higher chemical potential, while QDs with larger sizes keep grown, a phenomenon known as Ostwald ripening. In solvothermal synthesis, the diffusion-controlled growth occurs first, which gives rise to a narrow QD size distribution. The reaction-controlled growth dominates when the QD sizes exceed a critical value. The reaction-controlled mechanism thus leads to broad QD size distribution and decreased number of QDs due to particle dissolution.

In hot-injection synthesis, nucleation occurs immediately after the injection of a room temperature precursor into a high-temperature surfactant. The mix of the low-temperature precursor and the high-temperature surfactant results in supersaturate solution, creating the condition for nucleation. The nucleation rate is determined mainly by solvent temperature, degree of supersaturation and stoichiometric components in solution, as well as interfacial tension. Nucleation is terminated either automatically due to concentration reduction of elements, consumed by nucleation, to a level below the critical value, or artificially by fast injection of a room-temperature precursor that quickly bring down the reaction temperature. The mixed solvent are then at a right temperature allowed for QD growth. The growth of QDs is characterized by molecular addition of the monomers remaining in the solution. A typical example of hot-injection preparation is the growth of CdSe colloidal QDs, which has been widely reported.

5.2.2 Non-Injection Heat-Up Synthesis

In solvothermal preparation of QDs, nucleation and growth occur at different temperatures so that the nucleation and growth stages are separated [9]. The concurrence of nucleation and growth leads to broad size distributions. In a hot-injection route, nucleation occurs right upon the injection of a room temperature precursor. Nucleation keeps in progress due to the drop of the temperature and the precursor concentration below threshold values. The nucleation time is controlled by rapid precursor injection and strong stirring, which gives rise to very short nucleation time. A short nucleation time is critical for the separation of the nucleation stage from the growth stage. Growth takes place right after the termination of nucleation. Clearly, the injection and stirring feature of the hot injection route is not suitable since the rapid precursor injection is impossible to achieve for large-scale preparation of the QDs.

In non-injection heat-up synthesis, all precursors are prepared prior to the initiation of the growth. Nucleation and growth are initiated by controlling and changing the temperature. Kwon and co-workers have studied the growth of iron oxide nanocrystals based on the solution phase thermal decomposition of iron-oleate complex via the "heating-up" process [9]. In the non-injection heat-up synthesis, a pre-prepared solution of iron-oleate complex in long-chain hydrocarbon solvents (1-octadecene or 1-eicosene) was heated up to a temperature higher than 300°C. Thermal decomposition reaction occurred, which gave rise to the formation of thermal free radicals. Figure 5.8 presents schematically a two-temperature-step process, where the nucleation is initiated by heating the precursor to a given temperature. Growth of QDs starts after the temperature is further raised to a higher value. Controlling the temperatures at both stages is critical for the growth of high-quality quantum dots.

When the solution was heated from room temperature to about 230°C, the magnetic moment of the solution changed very little, as shown in Fig. 5.8a. The magnetic moment then started to increase quickly with increasing temperature. Figure 5.8b presents the same result but as a function of time, in terms of experimental temperature rising curve (experimentally recorded solution temperature as a function of time). As shown in the inset of Fig. 5.8b, the measured magnetic moment of the solution (as a function of time) has two rises separated by the inflection point of the magnetic moment curve. The observed abrupt increase of the magnetic moment indicates that the structural change is associated with the thermal decomposition of iron-oleate complex. The similarity of the magnetic moment curve and the reaction extent curve (see the blue curve in Fig. 5.8b) suggests that the first increase of the magnetic moment is due to the increase in the concentration of the thermal decomposition product, since the recoded data showed that the magnetic moment did not increase linearly with the concentration of the intermediate species. In order to examine how the increase in the magnetic moment in the first rise is correlated with the thermal decomposition reaction, Kwon and co-workers also studied the reaction extent of the thermal decomposition process that was obtained by integrating the CO_2 evolution rate as an approximation. Intermediate species must form between the thermal decomposition of iron-oleate complex and the nucleation/growth of the iron oxide nanocrystals. Figure 5.9 shows the TEM images of iron oxide QDs drawn from the solution at 320°C after aged for 20 and 60 min. The QDs has a square-circle shape with a quite uniform size distribution.

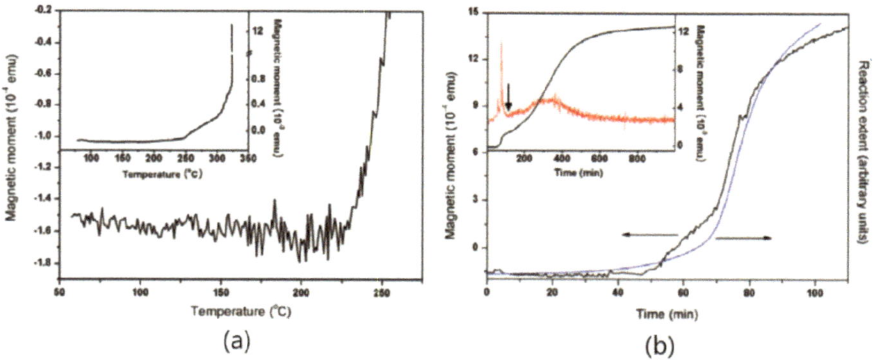

Figure 5.8 (a) Measured magnetic moment of the iron-oleate solution as a function of temperature (the inset: the same curve for a larger scale) and (b) the same measurement as a function of time (black curve) and the curve for the integral of ion current intensity at the mass charge ratio of m/z = 44 (blue curve) (the inset: the magnetic moment curve (black) shown in a larger time range and its corresponding derivative curve (red). The inflection point of the magnetic moment curve in the inset in (b) is marked by the arrow. Reproduced with permission from [9].

Cao and Wang proposed a one-pot colloidal synthesis route with which the CdS nanocrystals have the quality comparable to that of nanocrystals grown by hot-injection synthesis [10]. In the one-pot reaction system, the nucleation is separated from the growth by adding two reagents, namely, tetraethylthiuram disulfides and 2,2′-dithiobisbenzothiazole that act as nucleation initiators. Figure 5.10 presents results of absorption spectroscopic measurements on the colloidal QDs during the growth. As clearly indicated, no nucleation occurs at temperature of 200°C, as shown by the featureless 200°C curves. The curves with absorption lines were measured on colloidal QDs grown at the reaction temperature of 240°C and the curves show temporal evolution of the synthesis. The absorption lines become more and more distinct with time approaching 8 min (as the particles grew). The distinct absorption lines that even show absorptions associated with excited states is an indication of high quality and narrow size distribution of the QDs. The 8 min curve and the 12 h one show essentially no difference, indicating that the quality and the narrow size distribution can be maintained for long time as the nanocrystals grow. In the growth stage, the QD concentration in the solution remained unchanged and no nucleation events were found. Thus, in such a one-pot nanocrystal synthesis the nucleation and the growth stages are separated, which is a fundamental requirement for the synthesis of QDs with narrow size distributions.

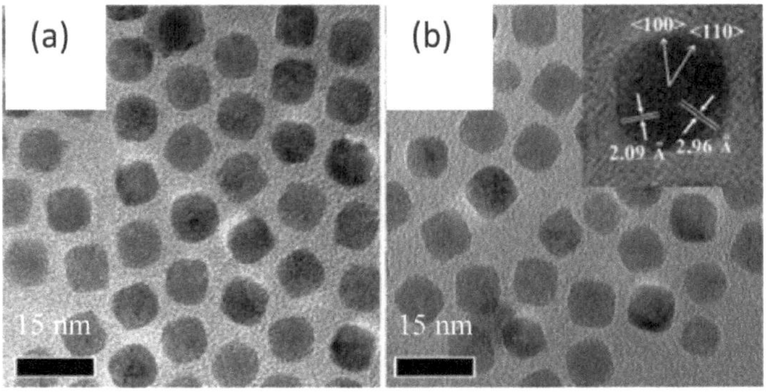

Figure 5.9 TEM images of iron oxide QDs drawn from the solution at 320°C, aged for 20 (a) and 60 min (b). The inset in (b) presents the high resolution TEM image of a QD with the *d*-spacing values of 2.09 and 2.96 Å as labeled, corresponding to (400) and (220) planes with the zone axis along [001]. The iron oxide QD has a spinel crystal structure. Reproduced with permission from [9].

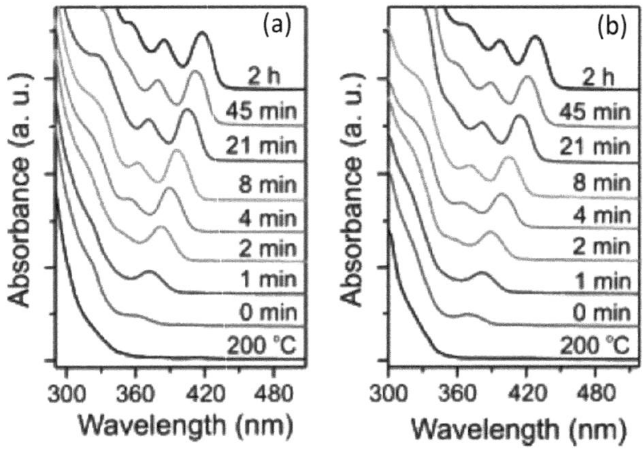

Figure 5.10 Absorption spectra that indicate the temporal evolution of synthesis for colloidal QDs grown in precursors with sulfur: 2,2′-dithiobisbenzothiazole ratios of 1:1/16 (a) and 1:1/8 (b). Reproduced with permission from [10].

The observed QD absorption lines in Fig. 5.10 correspond to transitions between discrete states of the electron and the hole. The discrete states are due to the quantum confinement. The discrete states can be described by an electron being confined in a QD of a nano-size box, similar to an electron being confined to an atom. A QD is therefore called an artificial atom.

5.2.3 Flow Reactor Method

Both hot-injection and non-injection heat-up methods are essentially limited to lab-scale synthesis of colloidal QDs, due mainly to the difficulties, encountered in the case of a very large reactor, in increasing chemical inhomogeneity and terminating the reaction in a large reactor over a short time interval. An interesting route based on the idea of a flow setup in which QDs were grown following the consecutive steps of solution mixing, nucleation, and growth [11]. The flow reactor synthesis uses a multi-stage and continual flow synthesis reactor, where used materials are in separating phases. The continuous flow setup renders the synthesis route an automated, in-line, and scalable feature, suitable for sensor integration in in-situ growth monitoring of large-scale production. Other advantages of the continuous flow method include improved control in growth temperature, cooling and heating rates, heat and mass transfer, efficient reagent/solvent mixing, as well as low reagent consumption. As a result, it also offers improved reproducibility due to easiness in controlling the experimental parameters such as temperature profile in the whole line. The major experimental parameters for the continuous flow method are the temperature of the nucleation reactor, the temperature of the growth reactor, and the residence times in each of the reactors. The interplay between the experimental parameters determines quality of the particles, including crystalline quality, particle sizes, and particle size distribution.

A challenging issue is the efficient mixing of regents, which is troublesome in a continuous flow path where the QD size distribution, for instance, could be broadened by the velocity distribution and the residence time distribution in flow driven by pressure. In a single-phase case, the velocity of the fluid takes a parabolic form, as shown in Fig. 5.11a, where the fluid near wall moves more slowly than that near the center. As a result, the particles close to the wall spend longer time than those in the center, leading to a broad particle size distribution. Compartmentalization of reagents in nanoliter volumes, which can be realized by introducing gas bubbles in flows, has been proven to be effective in enhancing reaction and controlling the growth [12, 13]. As shown in Fig. 5.11b, the single phase is broken into segments by introducing gas bubbles. In such a two-phase system, liquid recirculation within each segment brings material exchange between the wall and the center. The enhancement of liquid mixing gives rise to narrow residence time distributions of the particles and, as a result, a narrow particle size distributions.

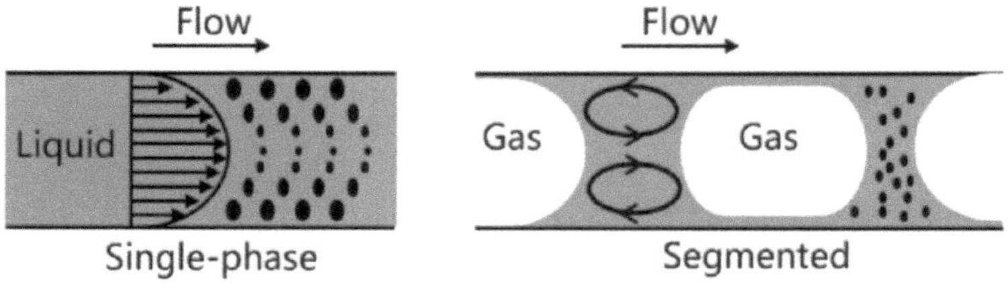

Figure 5.11 Schematic diagram of parabolic velocity-profile of a single-phase fluid flow (a) and a two-phase gas-liquid flow (b).

Pan and co-workers proposed a dual-stage continuous flow setup, shown in Fig. 5.7c, and demonstrated the growth of PbS colloidal QDs with quality comparable to those prepared by the traditional batch synthesis (hot-injection and non-injection heat-up) in the monodispersity and surface quality [11]. The major experimental parameters are the nucleation temperature, growth temperature, and total residence time. In the setup, the precursor A comprises lead oxide, oleic acid, and octadecene (ODE) and the precursor B consists of bis(trimethylsilyl) sulfide (TMS) mixed with ODE. The whole preparation procedure is initiated by mixing the two precursors that are pumped into the mixing stage under N_2 overpressure. In this stage, the temperature is low enough to ensure no chemical reactions among reactants. The mixed precursors A and B continue to proceed to the nucleation stage where the temperature is set (T/C 1, as shown in Fig. 5.7c) so that nucleation takes place. The nucleation takes very short time to complete and it is controlled in the way similar to conventional batch synthesis. The solvent that contains PbS nucleation seeds further flows to the growth stage and starts to grow at a given growth temperature (T/C 2 as marked in the figure). The dual-stage continuous flow setup was developed from a simple standard hot-injection setup shown in Fig. 5.12.

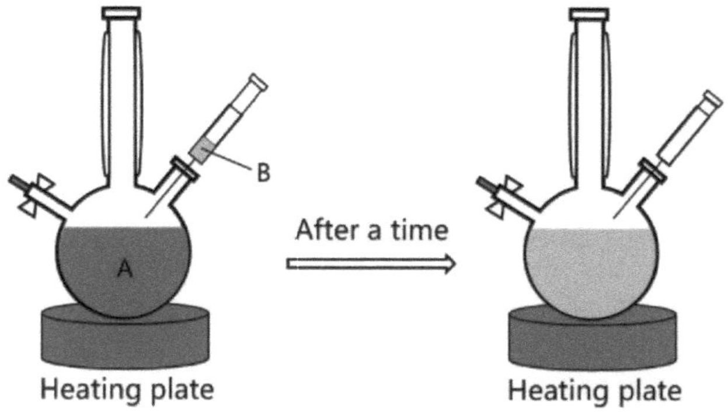

Figure 5.12 Schematic diagram for a standard hot-injection setup. A solution containing precursor A is heated to a chosen temperature and room-temperature precursor B is injected into the solution. Sudden nucleation occurs to form colloidal quantum-dot seeds and the seeds grow into quantum dots with mono-dispersed sizes if the experimental parameters are controlled precisely.

Controlling the residence time in the nucleation stage is critical for obtaining narrow size distribution. For very short residence time, the fraction of subcritical particles is very high because there is not enough time for the particles to become critical or supercritical. Thus, it is helpful to choose a residence time longer than the time required for the particles to reach to their critical or supercritical sizes. However, prolonged residence time might cause broad size distribution due to the growth of supercritical particles.

The hydrothermal growth is mostly performed on II–VI semiconductors. The bonding in III–V QDs is more covalent than that in II–VI ones. As a result, separating the nucleation and growth via high temperature dual injection is more difficult for the preparation of III–V QDs. The "molecular seeding" approach offer an alternative route that does not require the high temperature injection procedure. This approach utilizes identical molecules in the form of molecular clusters as the nucleation sites for nanoparticle growth. At a moderate temperature, precursors are added alternatively in a growth container until QDs are at the right size. Another advantage of the "molecular seeding" process is that it can be used to grow high-quality quantum dots on a large scale.

5.3 Thin-Film Preparation

In the usage for solar energy conversion, QDs should be in the form of thin films themselves or incorporating with other nanomaterials. It is thus critical to fabricate QD thin films as demanded by the device architecture of solar cells. Several methods have been developed to prepare QD thin films and some of them are suitable for solar cells, bearing the feature of low cost, large area, and low temperature.

5.3.1 Spin Coating

In the spin-coating process, a droplet of precursor solution is deposited on a substrate mounted on a rotating stage. The substrate is spun at the speed of 1000 to 6000 rpm right after the deposition of the solution droplet. The droplet then spreads and covers the substrate, due to the adhesive force between the substrate surface and the solution, to form a uniform thin film. This process can be repeated to obtain desired film thickness and even a multi-layer film if two or more precursor solutions are used. The thickness of the film in one spin-coating step is mainly determined by the viscosity of the precursor solution and the spinning speed. High solution viscosity and low spinning speed lead to a thick film. A spin-coating experimental setup is schematically shown in Fig. 5.13.

This method is used to prepare high-quality QD thin films with high uniformity and smooth film surfaces. In addition, the film properties can be modified by selectively adding other elements and doping. One could expect reproducible film parameters if the experimental conditions are controlled precisely, such as ambient conditions (temperature and humidity, for instance). Since most QDs are unstable at high temperature, solvents with relatively low boiling point and quick evaporation are usually preferentially considered. Therefore, solvents such as toluene and octane are very often used to disperse QDs.

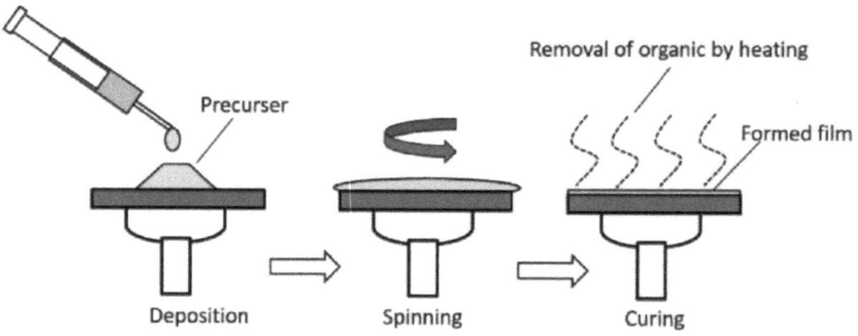

Figure 5.13 Schematic diagram of a spin-coating setup.

Spin coating is a widely used technology for thin-film preparation, especially for nano solar cells. It has also been adopted to prepare periodic multi-layer of 1D photonic crystal with good optical properties [14]. The major disadvantage of the method is the insufficient use of QDs, since usually most of the precursor solution is wasted during spin coating. As a film method in lab, spin coating is powerful for small-scale thin-film preparation, but it could face some difficulties for scaling up the production.

A colloidal suspension of a given quantum dots is usually spin-casted onto a suitable substrate that is mostly a thin glass slide coated with a transparent and conductive oxide (TCO) thin film. Unlike epitaxial growth of thin films, the spin coating preparation of QD thin films does not require an atomic smooth substrates. A good electric contact of the nanomaterials with TCO is essential for charge collection and, as result, high conversion efficiency.

5.3.2 Dip Coating

Thin films can also be prepared by dipping a substrate into a container filled with precursor solution where QDs are dispersed, followed by withdrawing the substrate at a constant speed. The dip coating setup and process is schematically shown in Fig. 5.14. Due to surface tension and viscosity of the solution, a liquid thin film is deposited on the substrate surface. After solvent evaporation, a solid thin film consisting of QDs coated with ligands forms on the substrate. Here a constant withdrawing speed is essential for the uniformity of the thin film. Obviously, this process can be easily carried out automatically by machines, which offers better control on repeatability and precision. The dip coating is more probable than spin coating for large-scale production.

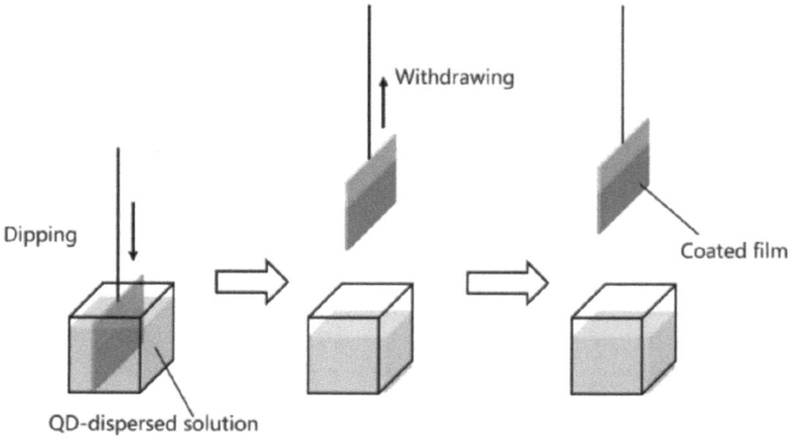

Figure 5.14 Schematic diagram of a dip-coating setup.

Compared to spin coating, dip coating wastes less precursor solution. However, dip coating usually leads to less uniform films and control on the film thickness.

5.3.3 Drop Casting

In the drop-casting process shown in Fig. 5.15, a drop of solution is dropped on a substrate and the droplet spread out to form a layer that covers part or whole surface of the substrate [15]. The layer is then allowed to dry to form a thin film. The solution is prepared such that QDs are dispersed uniformly in a liquid that are usually organic solvent such as dimethyl sulfoxide. Thus, the uniformity of the solution is essential to the quality of the thin film. The organic solvent is only used as such solvents are generally only used to carry the QDs for drop casting after which the organic solvent is evaporated out. The thin film is thus composed of disperse quantum dots capped with ligands. The whole process is carried out on a hot plate at a temperature set for solvent evaporation. The temperature could also affect the surface morphology as well as chemical, optical, and electric properties characteristics of the thin film [16].

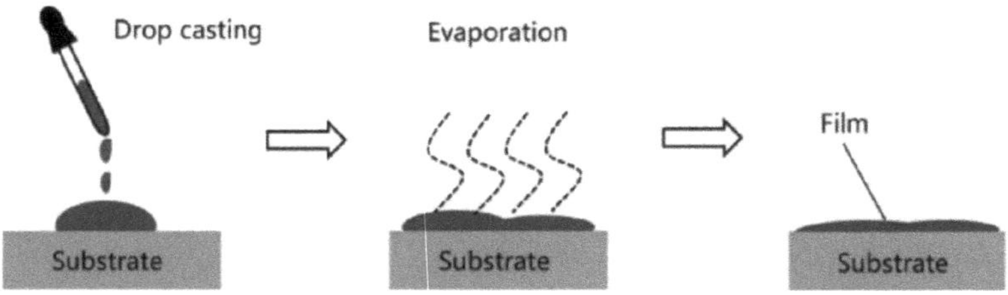

Figure 5.15 Schematic diagram of a drop-casting setup.

5.3.4 Spray Coating and Inkjet Printing

The batch processes for QD thin-film preparation discussed above are only suitable for small-scale fabrication. Fortunately, scaled-up deposition techniques can be possible through some modifications on those batch processes. The fundamental requirements for the scaled-up deposition are automatic, low solution waste, and large areas. Kramer et al. proposed a spray-coating method for the fabrication of colloidal QD thin film and used the thin film for solar cells [17]. Guo and co-workers suggested an inkjet printing method [18]. Both spray coating and inkjet printing are designed as automated apparatuses to achieve high reproducibility, as shown in Fig. 5.16 for spray-coating installation. Small amount of solution with colloidal QDs is almost all used to produce large-area thin films, so that the setups waste very less amount of solution material. The designs also use a continuous rolling substrate beneficial for high throughput.

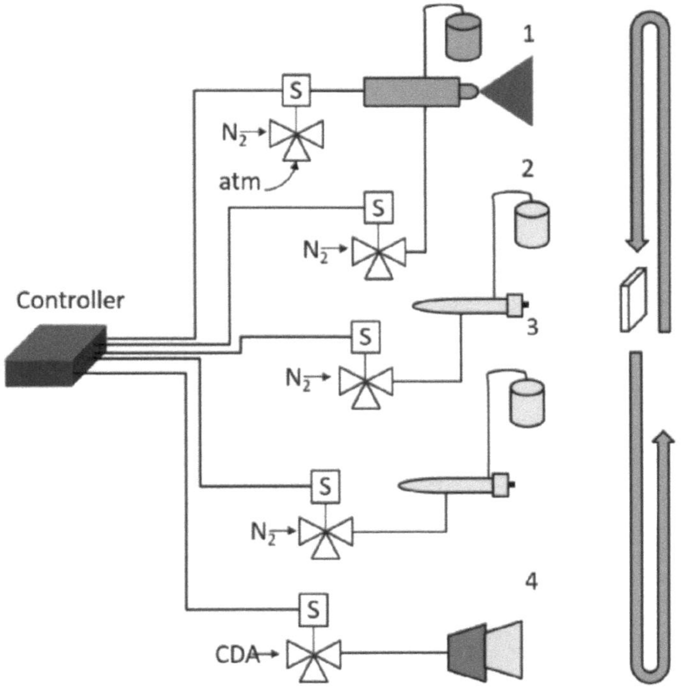

Figure 5.16 Schematic setup of a spray-coating system for the fabrication of colloidal QD films.

5.4 Surface Ligand Exchanges and Shell Layer Growth

Surfaces of colloidal QDs are covered with organic ligands of large molecules that are usually long alkyl chains terminated by acids, phosphines, amines, and phosphonic acids, etc. The use of the organic capping ligands in synthesizing QDs is beneficial for narrow size distribution and good shape control. Those ligands are, however, bad for the performance of QD solar cells due to the formation of a potential barriers between QDs. They are highly resistive and often incomplete in passivating the QD surfaces. Surface ligand exchange is thus a necessary step for the use of QDs in solar cells and a critical process for achieving high device performance. The original ligands need to be exchanged to obtain QD thin film with better electric performance. The new ligands are usually smaller molecules for the enhancement of the charge transport.

Surface ligands exchange (or a shell layer growing on the surface of QDs) could varies the functionality and reactivity of the QD surfaces. For instance, after exchange of surface ligands, QDs can become hydrophobic from original hydrophibic, or vice versa. Suitable ligand exchange could also improve the dispersability and stability against environmental chemical and physical infringement. More importantly, well-prepared surface ligands or shell layers could efficiently protect the QDs from oxidation and reduce the density of surface states that trap the photo-carriers, leading to enhanced optical and electrical properties of QDs. Typically, protic solvents are used to aid to desorb the original ligands such as the alkyl long ligands on solution-prepared colloidal QDs, making unbounded sites to bind with the desire ligands.

QDs act as photon antenna in QD solar cells. The photovoltaic effect, however, requires both the photoexcitation and migration of electrons and holes toward opposite directions. Thus, QDs should allow for the photoexcited electrons or/and holes to escape from them. In solvent preparation of QDs, surface ligands play the key role in controlling the nucleation and growth of QDs. The surface ligands are also required for the chemical stability of QDs solutions when individual QDs are assembled into a QD thin film. The QD thin film carries the migration of the carriers and, as a result, the surface ligands determine the electronic properties of the QD thin film. The surface ligands on QDs can be changed or modified so that interparticle spacing and interaction, which have extremely strong impact on the electronic properties of QD thin film, can be tuned. The tunable properties include carrier transport, interfacial potential profile, relaxation rate, etc. The tunability stems largely from the high surface-to-volume ratio inherent to the QDs.

5.4.1 CdSe QDs Capped with Fullerene Derivatives

Josep Albero and co-workers used C_{70} to replace the original ligands on CdSe QDs fabricated by a wet chemical synthetic method and the ligand exchange procedure is shown in Fig. 5.17 [19]. The QD thin film was made by C_{70}-capped CdSe blended with the polymer poly-3-hexyl thiophene (P3HT). In making the QD solar cells, an aqueous PEDOT:PSS dispersion was deposited on indium tin oxide (ITO) substrates by spin coating, followed by annealing at 120°C for 30 min in nitrogen atmosphere. The photoactive layer was made from QD:polymer solution that was spin-coated onto the PEDOT:PSS layer. The SC cathodes was 80 nm-thick aluminum (Al) thermally deposited onto the QD:polymer blend. The whole structure was thermally annealed at 150°C for 15 min in nitrogen atmosphere. The SCs have a bulk heterojunction configuration.

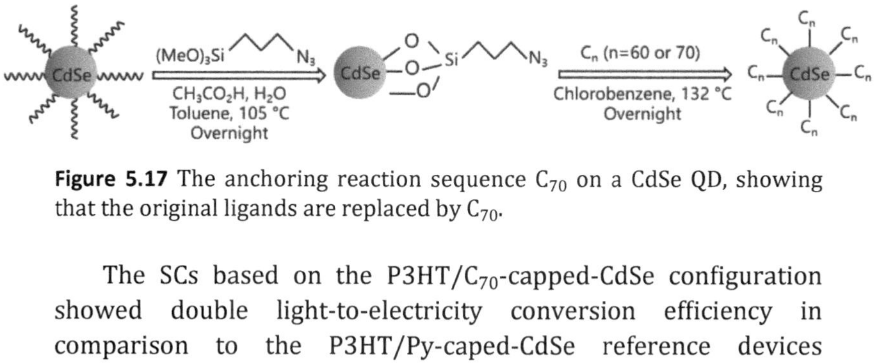

Figure 5.17 The anchoring reaction sequence C_{70} on a CdSe QD, showing that the original ligands are replaced by C_{70}.

The SCs based on the P3HT/C_{70}-capped-CdSe configuration showed double light-to-electricity conversion efficiency in comparison to the P3HT/Py-caped-CdSe reference devices prepared using pyridine. The significant enhancement in conversion efficiency is due to an increase in photocurrent and fill factor, benefited from the improved charge transport in the P3HT/C_{70}-capped-CdSe solar cells, though introducing C_{70} capping layer on CdSe QDs gave rise to the increase in carrier recombination that was made clear by transient absorption spectroscopy and transient photovoltage measurements.

5.4.2 Air Stabilization of Colloidal Quantum Dots in Solid Matrixes

Colloidal quantum dots, in both solutions and solid matrixes, are easy to get oxidized without proper protection. Particularly for *n*-type QDs, oxidation is the major problem that causes the QDs to be unstable since electrons in *n*-type QDs are rather active. Intuitively, people would look forward to finding a material to completely cover the quantum dots and the material is completely anti-oxidant at direct exposure to oxygen. Ning and co-workers used halides to cover the PbS QDs [20]. Previous studies showed that the halides act to passivate the surface of colloidal QDs quite effectively [21]. For *n*-type QDS, halides have the advantages of their *n*-type doping properties and complete coverage due to small steric hindrance. Among halides (iodide, bromide, and chloride), iodide has the largest atomic radius, which appear to give the best protection to PbS CQDs. They used transient photovoltage to study the density of trap states within energy gap of QD solids. A notably higher defect density was found in chloride- and bromide-terminated samples than in the iodide-treated PbS QDs.

PbS QD SCs were fabricated, with the QDs being treated with iodide, bromide, and chloride for comparison. The architecture of the device consists of a strongly *p*-type PbS-CQD layer by hydroxide exposure, followed by growth of a thicker *n*-type PbS-CQD capping layer. The top and the bottom electrodes are MoO_3/Au/Ag and TiO_2/FTO. Figure 5.18a presents the conversion efficiencies of the PbS CQD solar cells measured in air versus time (up to four days). Obviously, the bromide- and the chloride-treated devices show low initial conversion efficiencies and time-unstable characteristics, while the solar cell with PbS CQD terminated with iodides has a relative higher efficiency and the device is very stable. Figure 5.18b shows that the oxygen contents in the bromide- and chloride-treated PbS CQD solids are much higher than that in the iodide-treated material, which clearly shows that the solid film with iodide-treated PbS QDs is well-passivated against oxygen.

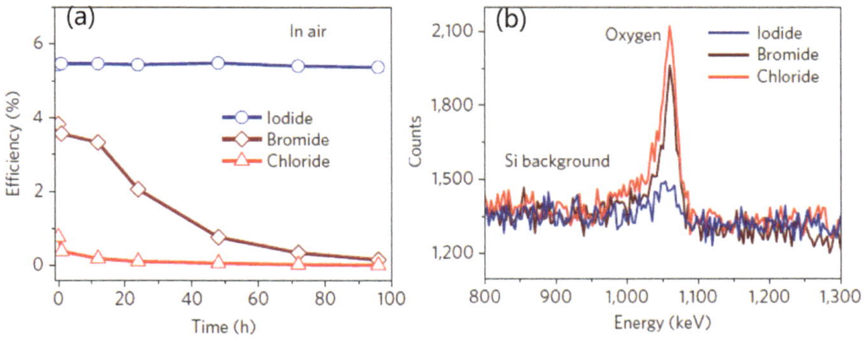

Figure 5.18 (a) Conversion efficiency as a function of time of three PbS CQD solar cells, with PbS CQD treated with iodides, bromides, and chlorides. The measurement was performed in air and a time scale of four days. The solar cell with iodide-treated PbS QDs shows the best stability, while both bromide- and chloride-treating present significant instability with prolong air exposure. (b) Rutherford backscattering spectrometry shows large amounts of oxygen contents in the bromide- and chloride-treated PbS CQD solids and weak trace of oxygen in the iodide-treated material. Reproduced with permission from [20].

5.4.3 CdS Quantum dots with Tunable Surface Composition

Not only ligands, but also surface stoichiometry has a strong impact on the photoelectric characteristics of the quantum dots. The stability of ligand connection is even related to the surface stoichiometry—the top-most layer of QDs. Therefore, the synthesis of high-quality QDs not only needs to ensure the crystal quality, but also to control the stoichiometry of the surface layer. Wei and co-workers synthesized colloidal CdS QDs, being able to tune the surface stoichiometry [22]. High-quality CdS QDs were synthesized using DPP-S and Cd-stearate in tetradecane with a narrow size distribution and controllable sizes from 2.8 to 5.2 nm in diameter. A S-terminated surface was realized by injecting diphenylphosphine sulfide into the solution with a complete consumption of the injected material, while a Cd-terminated QD surface are formed by injecting Cd-stearate. Here the S- and the Cd-terminated surfaces mean a single monolayer of S and Cd, respectively. The growth of the S- and Cd- terminated surface can be repeated to obtain CdS QDs with a desire size. CdS QD surface composition could be tuned from all Cd to all S termination.

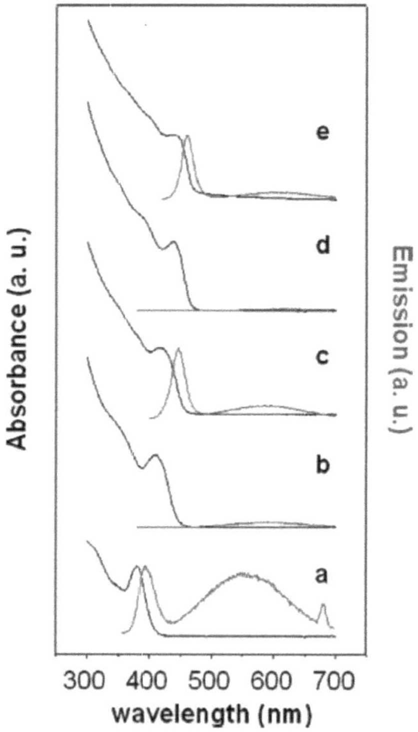

Figure 5.19 Photoluminescence spectra (red line) and absorption (black line) of (a) as-prepared CdS QDs, (b–e) QDs with subsequent and alternative growth of the S-terminated monolayer and Cd-terminated monolayer (c). The small peak at 680 nm in (a) is from the overtone of the excitation laser. The strong peak nearby the absorption edge is band-to-band photoluminescence emission and the broad band around 550–600 nm is due to defect band emission. Reproduced with permission from [22].

Figure 5.19 shows the photoluminescence spectra (red curves) where the peak in 400 to 475 nm corresponds band-to-band emission. The measured photoluminescence strongly depends on the surface stoichiometry of the QDs. Noticeably, the PL spectra of the as-prepared, S-terminated and Cd-terminated CdS QDs show significantly different behaviors. The absorption spectra (black curves in Fig. 5.19) measured on those CdS QDs with different surface stoichiometry do not exhibit too much difference, which is understandable, since the light absorption spectrum is not sensitive to the surface on the QDs with those sizes. The as-prepared QDs present a strong band-to-band emission at 400 nm and a quite strong defect band at around 550 nm. The PL spectrum changes significantly with the coverage of the first S-terminated monolayer on the as-prepared CdS QDs, with the band-to-band emission almost completely quenched and the defect band weakened significantly (see curve b). The PL emission is recovered by a subsequent growth of a Cd monolayer (curve c). The PL emission is quenched and recovered again by covering the QDs with S and Cd monolayers (curves d and e), respectively.

Both S- and Cd-termination on the QD surfaces leads to significant suppression of the defect band, implying that S and Cd tend to passivate the surface to a certain extent, which is possible since the defect band and the PL quench are due to the different kinds of defects. The observed red-shift of the PL emission peak and the absorption edge is due to the increase of the QD size—a prominent quantum confinement effect.

5.5 Device Architectures of QD Solar Cells

The key part of a photovoltaic device is the built-in potential that separates the electron and the hole in an exciton created via absorbing a photon. The first colloidal QD solar cell, proposed by McDonald and co-workers, utilizes the potential difference between the transparent ITO electrode and Mg metal contact, where the QD thin film is sandwiched in between, to construct a built-in field [23]. Excitons are photo-generated in the QD thin film and charges transport cross the QD thin film toward electrodes.

5.5.1 Schottky Junction

A Schottky barrier is a potential energy barrier formed for electrons or holes at a metal–semiconductor junction [24]. The colloidal quantum dots can be fabricated into "quantum dot semiconductor" where "atoms" are quantum dots. Various techniques have been developed for the preparation of this kind of QD thin films. The fundamental device structure of the Schottky QD SCs is constructed by a thin QD film sandwiched between a transparent-conductive oxide (TCO) electrode for electron extraction and a metal film electrode for hole collection, as shown in Fig. 5.20a for an early version of Schottky junction SCs [25]. In this device architecture, the CQD thin film acts as both the photon absorbers and the charge transport medium. The transparent conductive oxide (TCO), such as indium tin oxide (ITO) and ZnO, has a relatively large work function. On the interface between the TCO and a *p*-type CQD film, Ohmic contact is formed. The back-side electrode of the device uses a metal with a shallow work function, such as magnesium or aluminum. The built-in electric field, as shown in Fig. 5.20b, is established via a self-consistent redistribution of electrons in the QD-film/metal interface due to the contact of QD film and the metal electrode. The built-in potential is roughly equal to the difference between the Fermi levels of the QD film and the metal. The built-in electric field is located in the depletion region, as shown in Fig. 5.20b, established during the self-consistent process where electrons are extracted and holes are repelled from the interface.

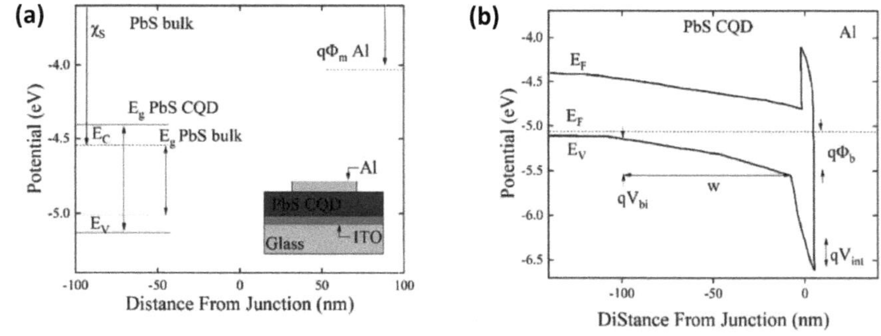

Figure 5.20 (a) Schematic architecture of a PbS-CQD solar cell with an PbS CQD film sandwiched between a TCO electrode for electron extraction and a metal film electrode for hole collection, together with the energy level description of the CQD and PbS bulk; (b) Energy band profile of the a PbS-CQD solar cell, showing Schottky-like heterojunction. Reproduced with permission from [25].

A charge-separating junction is thus established. Obviously, the device open-circuit voltage (V_{oc}) is determined mainly by the Fermi level difference between the TCO and the metal. The Schottky cell has the simplest device structure among all colloidal QDSCs and was the first structure that achieved a conversion efficiency exceeding 1%.

Photons enter the cell from TCO side and those arriving at the built-in field region might generate excitons if they are absorbed by electrons in QDs. The electrons and holes are separated by the built-in potential and transport to Al contact and TCO electrode, respectively, to complete the photovoltaic process. The charge separation is promoted by the built-in field in the depletion layer. It is essential that the rate of charge separation and transport to their respective electrodes outpace the rate of charge recombination. In the depletion layer, photo-generated electrons and holes carriers are quickly separated by the built-in electric field. In the charge-neutral region outside of the depletion layer, however, carrier transport relies on diffusion. In a QD film, carrier diffusion lengths are typically in the range of 10–100 nm. For a QD film with a diffusion length of 100 nm, the thicknesses of the active-layer should not exceed 200 nm. Otherwise, a dominant amount of photo-carriers are generated in the charge-neutral region that is too far away from the junction, so that carrier extraction cannot be completed efficiently, leading to low internal quantum efficiency. For the Schottky cell configuration with optimized structural design, an internal quantum efficiency of above 80% is achievable.

The SCs with Schottky junction configuration yield power conversion efficiencies (PCEs) typically less than 4%. The low conversion efficiency is due to several causes. First, the built-in electric field is on the Al contact side, i.e., incident photons have to travel through the QD film to be absorbed by QDs located in the built-in field. However, light absorption begins at QD thin film near the TCO electrode, where in the quasi-neutral region far away from the built-in field, electrons and holes are separated with low efficiencies, due to the poor electron mobility. The photo-generated electrons (minority carriers in the QD film) need to travel the entire QD film to arrive at the Al metal contact, during which the electrons have high probability to recombine with holes since the QD film is filled with defects and other imperfection for electron transport. Second, there is high density of defects states at the QD-film/metal interface. As a result, Fermi level is often pinned at the defect states [26]. This leads to a low V_{oc}, which is unfavorable for device efficiency. Third, the rather thick QD film in the Schottky junction SC acts also as the charge transport medium that has low conductivity, where charge transports are slow due to various scattering centers and the charges (electrons and holes) have high chance to recombine. Other disadvantages of the Schottky junction SCs include the shallow barrier for holes at QD-film/Al interface that is often ineffective for hole blocking.

The alternative configurations of Schottky junction SCs use semi-transparent metal contact and the light enters the device from the metal electrode, in order to improve the light absorption efficiency in the built-in potential region. The higher open-circuit voltage can be achieved by using metal with lower work functions, as shown in Fig. 5.21. It has been demonstrated that the use of Ca, Ag, or Mg indeed improves the open-circuit voltage V_{oc} of the device [27].

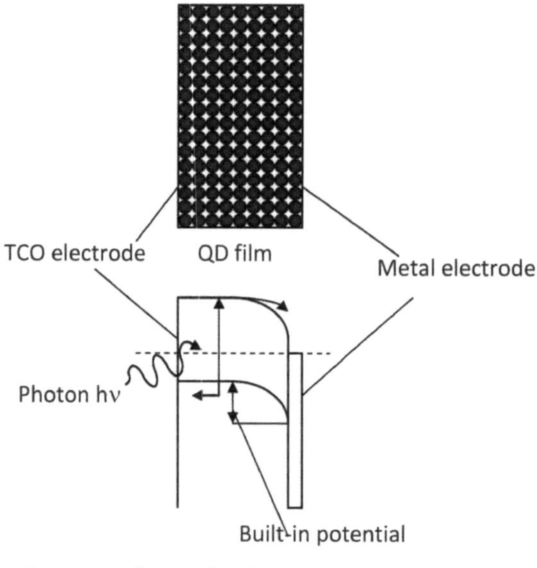

Figure 5.21 Scheme of a Schottky junction QD solar cell and the corresponding potential profile.

Sargent's group elucidate the physical picture of the Schottky cell, with a 0.4 eV Schottky barrier height formed at the junction between the metallic Al contact and the PbS colloidal QD thin film. Their devices display a built-in potential of 0.3 V, a depletion width from 90 to 150 nm corresponding to an acceptor density from 2×10^{16} to 7×10^{16} cm^{-3} [28]. Later on, Ca, Mg, and Ag were used as alternative metallic contacts, which showed improvement in open-circuit voltage and, as a result, power conversion efficiencies.

Progress has continuously been made on Schottky junction QDSCs through improving material quality of CQD films and using advanced device architectures. A 2% conversion efficiency was achieved in the year of 2008. The active layer of the cell is PbSe QD film deposited by layer-by-layer dip coating. The device displays AM1.5 conversion efficiency of 2.1%, with an EQE of 55–65% in the visible range and 25% in the infrared range [29]. In the year of 2010, Sargent's group reported a QDSC with the architecture of ITO/PbS-QD film/LiF/Al/Ag. The device was prepared in air. With a proper use of ligands, they obtained a conversion efficiency of 3.6% at AM1.5 condition [30]. Ma et al., studied Schottky cells with an ITO/PEDOT/PbSe-QDs/Al device configuration. It was found that the large QD bandgap, corresponding to a small PbSe QD size of ~2.3 nm (a ground exciton energy of 1.6 eV), gave rise to high open-circuit voltage of ~0.6 V. This device architecture displays a typical device efficiency of 3.5%, and a champion efficiency of 4.57% [31]. The recent record for Schottky solar cells, 5.2%, was set by Piliego and co-workers [32]. The record was obtained on a very simple device structure of an active PbS-QD film sandwiched between an ITO and a LiF/Al electrode, i.e., ITO/PbS-QDs/LiF/Al. The high conversion efficiency was achieved due to the careful ligand exchange and optimized processing parameters used for post-synthetic treatments.

5.5.2 Depleted Heterojunction

The depleted-junction solar cells are a bit more complicated than the Schottky-junction ones. A typical device architecture based on depleted heterojunction is shown in Fig. 5.22, where a large-bandgap and shallow-work-function electron acceptor layer, most commonly the transparent nano TiO_2 or ZnO layer, is placed between the QD film and the TCO electrode for electron extraction. The electron acceptor layer is selected so that its electron affinity favors electron extraction with high efficiency without having too much negative impact on the open-circuit voltage. Since TCOs, transparent and conductive oxides, are usually heavily *n*-type doped, the depleted heterojunction SCs use *n*-type metal oxide and *p*-type colloidal QD film to form a *p*–*n* heterojunction. The colloidal QD films are typically 50–500 nm in thickness. Thus, in a colloidal QDSC with a depleted heterojunction charge separation occurs at the front side of the cell where a junction is formed between the colloidal QD film (the active layer) and the nano TiO_2 layer that is deposited on top of a substrate. The substrate consists of a glass slab or a flexible transparent film coated with a thin TCO layer such as ITO or FTO (fluorine-doped tin oxide). The quality of the contact on the back side is also crucial for the whole device performance. For the active layer being large-bandgap QD film, an electrode with large work-function materials, such as gold or a highly doped oxide (MoO_3) followed by silver or aluminum, is usually required.

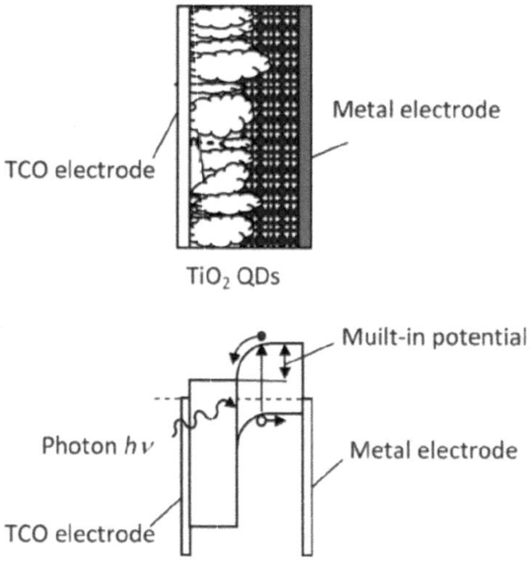

Figure 5.22 Scheme of a depleted-junction QD solar cell and the corresponding potential profiless.

A built-in electric field is induced due to hole depletion in the QD film close to the TiO_2 side. With light entering the solar cell from the TCO side, the photo-generated electrons and holes are separated by the built-in electric field (the band bending shown in the figure). The improved power conversion efficiency (PCE) for the depleted heterojunction SC is benefited from the efficient minority carrier separation. For instance, photo-generated electrons in the depleted junction are driven to nearby TiO_2 by the built-in potential efficiently, since the depleted junction is located on the illuminated side. With a bandgap of more than 3 eV, TiO_2/PbS interface introduces a very large discontinuity in the valence band, which effectively blocks hole diffusion to TCO where holes could recombine with electrons.

The simplest depleted heterojunction architecture uses planar QD films, as shown in Fig. 5.23a [33]. The typical 50–500 nm thickness of colloidal QD films is due to the absorption–extraction compromise. The thickness is limited by the short carrier transport length in colloidal QD films. However, the thickness optimized for visible light is not sufficient for infrared photons that have absorption lengths longer than the sum of depletion layer width plus diffusion region length. An alternative heterojunction, namely, bulk heterojunction was proposed to overcome the limitation [33]. In bulk heterojunction colloidal QDSCs, the *n*-type wide bandgap semiconductor (the metal oxide) and the colloidal QD film form an interpenetrating configuration, as shown in Fig. 5.23b. The interpenetrating configuration guarantees that the most carriers are photo-generated near the charge-separating region within their diffusion length. This is especially beneficial for minority carriers whose lifetime has strong impact on the device performance. The nano forms of the *n*-type wide bandgap metal oxides can be prepared in several ways. Bottom-up fabrication methods are cost-effective and can be used to obtain different forms of nanomaterials, including nanorods, nanopillars, nanoparticles, etc. Top-down methods, on the other hand, are used to lithographically prepare regular nano patterns, which offers better control on sizes and periodicities of nanomaterials, though lithographical etching is usually expensive.

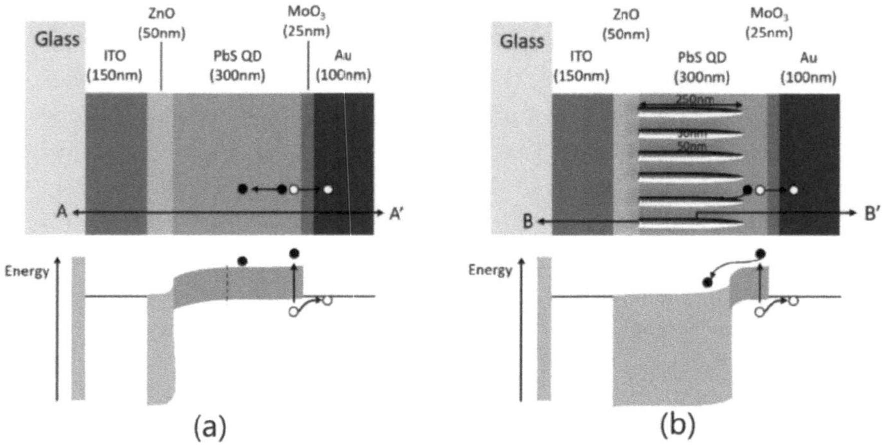

Figure 5.23 Schematic device architectures of a planar depleted heterjunction solar cell (a) and a bulk depleted heterojunction one (b), together with their band alignments.

Fortunately, with a nano form of TiO_2 and ZnO, colloidal QDs can easily infiltrate into the nanostructured materials to form a structured interface, as shown in Fig. 5.23b. Obviously, the structured interface cuts short the length through which the photo-generated carriers need to pass, which is beneficial efficient carrier collection. This was made very clear by the work down by Jean and co-workers, where QD solar cells with a planer heterostructure and a bulk heterostructure are compared, which showed a significant increase in photocurrent due to enhanced carrier collection as depicted in Fig. 5.24 for the *J*–*V* and EQE results [33]. As a result, the conversion efficiency increase from 3.2% of the planar case to 4.3% of the bulk case. Other advantage for the structured interface is the increasing interface area between TiO_2 (or ZnO) and the QDs. As a result, the increasing incorporation of the QDs on the nanomaterials leads to improvement in photon absorption.

The nano form of TiO_2 enables the formation of large TiO_2/QD interface area for photoelectronic conversion, termed as depleted bulk heterojunctions. The depleted bulk heterojunction strategy utilizes the bulk configuration to make almost the entire QD film depleted, so that a built-in potential is established in entire QD film. As a result, the trade-off between light absorbance and carrier extraction is thus reduced. Relative to device with depleted planar heterojunction, a device based on depleted bulk heterojunction could exhibit a nearly 30% enhancement on short-circuit current J_{sc} and a power conversion efficiency of 5.5% [34]. On the other hand, the large interface area of the heterojunction is troublesome due to the bimolecular recombination. In fact, both the device architectures based on Schottky junction and depleted heterojunction are susceptible, therefore, proper interface modifications are required for improving the device stability. A single-junction conversion efficiency of 6% has been achieved recently by modifying the device architecture and constituent materials [35].

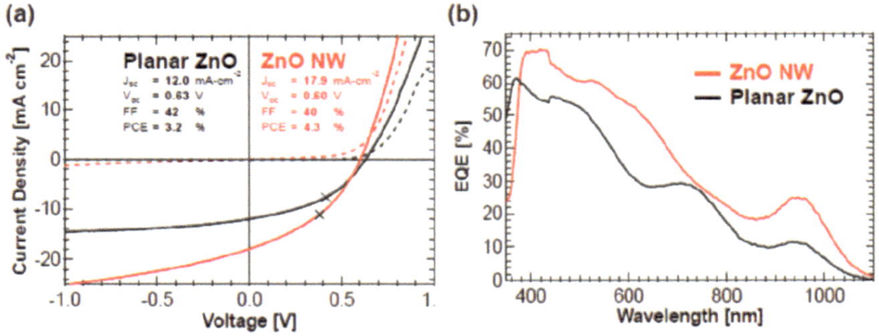

Figure 5.24 (a) J–V curves measured under AM1.5 condition for a representative planar (black) and a bulk (red) heterojunction colloidal QD solar cell. The dashed curves are the corresponding measurements in dark; (b) measured external quantum efficiencies for solar cells with the planar (black curve) and bulk (red curve) junctions. Reproduced with permission from [33].

5.5.3 Nanoheterojunction Colloidal QD Solar Cells

For QDSCs based on the architectures of the Schottky junction and the heterojunction, one of the main limitations on conversion efficiency stems from the low built-in potential. As a result, the open-circuit voltages of those devices are low, on which one does not have much room to work, since the low built-in potential is due to the intrinsic material properties. Additional difficulties include limited choices of the metal oxide materials that all have nearly the same work functions, and the high density of interface states between the metal oxide layer and the colloidal QD film.

Intuitively, high performance *p*–*n* junction of can be constructed theoretically using *p*-type and *n*-type QD layers, based on a homo-junction-like architecture. In another words, CQD solar cells can be benefited by QDs not only as absorbers but also as a whole junction material for high built-in potential. Theoretically, a high built-in potential is achievable by using large bandgap materials or QDs with small sizes, together with suitable densities of *p*-type and *n*-type doping. Additional advantage of the nanoheterojunction CQD solar cells is that the bandgap of the active material can be easily tuned by varying the QD sizes to match to the solar spectrum. Apparently, a nanoheterojunction colloidal QDSC demands both efficient *p*- and *n*-type QDs. Most colloidal QDs are *p*-type, originated from the oxygen atoms. Thus, the major challenge in constructing the SCs lies in good *n*-type doped QDs. There have been several methods that can be adopted for *n*-type doping and are compatible to the technology of colloidal QDSCs, including atomic diffusion and ligand exchanging. Hydrazine treatment, for instance, has been used and has converted *p*-type PbSe colloidal QDs to *n*-type [36].

The surfaces of colloidal QDs are covered with organic ligands that are usually large molecules. As a result, the large molecules are unable to cover every atomic site on the surfaces of the QDs, so that the ligands are imperfect in protecting the surfaces from oxygen. On the other hand, the small size ligands such as halogen ions (Br^- and Cl^-, for instance) could form a perfect coverage on all the atomic sites to protect the QD surfaces from oxygen attack. Each halogen ion provides an extra electron with respect to a chalcogen ion, in favor for *n*-type doping.

Tang and co-workers have reported the construction of all-colloidal-QD-junction solar cells based on colloidal QDSCs [37]. The single junction SC achieved, at the optimal material bandgap of 1.6 eV, a power conversion efficiency of 5.4% at AM1.5 condition. In their work, the *p*–*n* junction has the structure formed by the homo-junction between *p*-type and *n*-type PbS QD films, prepared by treating an *n*-type QD film with a halide carried in an inert environment. In their work, large-size organic ligands, the oleate ligands, on PbS QD surfaces were replaced with ultrasmall inorganic halogen ions. Iodide treatments were performed on the PbS QDs to remove the oleate ligands, which leads to nearly full iodine-ion coverage on the QD surfaces. In doing this, tetrabutylammonium iodide (TBAI) in methanol was used to prepare the CQD films. The formation of Pb-I bonding was demonstrated through elemental analysis by X-ray photoelectron spectroscopy (XPS), as shown by high iodide concentration in the iodide-treated PbS QD film [38].

Figure 5.25 Cross-sectional SEM image of a quantum junction solar cell based on the *n*-type-PbS-QDs/*n*-type-PbS-QDs structure. The white dashed line highlights the boundary between *n*-type and *p*-type regions. Reproduced with permission from [39].

For solar cell device purposes, a stable *p*–*n* junction with high built-in potential and good Ohmic electrode contacts to the *n*-type and the *p*-type colloidal QD films are essential. Tang et al. designed a SC device with the structure of glass/ITO/p-type-PbS/*n*-type-PbS/Al/Ag, as shown in Fig. 5.25 [39]. A heavily *p*-type PbS QD thin film was deposited on a glass slab covered with ITO. The contact between ITO electrode and the PbS QD is Ohmic. An iodide treatment was then performed on the *p*-type QD film, which resulted in an *n*-type layer on top of the original *p*-type film. A *p*–*n* junction was thus formed, which was characterized by a depleted region on both sides of the *p*-type and the *n*-type QD layers in the vicinity of the junction. For efficiency photocarrier extraction, high electron and hole mobilities in both the *n*-type and the *p*-type regions, respectively, are highly demanded in order to achieve high short-circuit current and maximized fill factor. Al/Ag contact extracts electrons from the *n*-type PbS QD layer and the ITO electrode collects holes from the *p*-type QD layer. Metal Al/Ag was chosen due to their shallow work function. The electron mobility is a few times larger than that of holes. The thickness of the *n*-type PbS QD layer is chosen to be 300 nm and that of the *p*-type layer is only 50 nm, in order for the optimal device performance. The best device displays a short-circuit current of 22.2 mA/cm^2, an open-circuit voltage of 0.52, and a fill factor of 47%, which yields a power conversion efficiency of 5.4%.

5.5.4 Quantum Dot–Sensitized Cells

A quantum dot–sensitized solar cell consists of a photoanode, that is formed by transparent and conductive scaffold loaded with photon-capturers—QDs on conductive glass, electrolyte, and a counter electrode. The transparent and conductive scaffold is usually a mesoporous wide-bandgap semiconductor, such as TiO_2, WO_2, and ZnO. The redox electrolyte acts as a hole conductor. Quantum dots are highly efficient photon capturers. In QD-sensitized solar cells, excitons are created in QDs after absorbing photons. Electrons are separated from holes by diffusing to a photo-anode electrode with the conduction band level below that of QDs. As shown in Fig. 5.26, the existing band discontinuity at the QDs/TiO_2 interfaces blocks the electrons from returning to QDs. The electrons flow to the electrode (transparent conducting oxide) where they are collected. Photo-generated holes in the valence band of QDs extracted by electrolyte and they diffuse to the metal electrode where they combine with electrons traveling to the metal electrode through external circuit. The QDs are usually loaded on 3D photo-anode consisting of nano TiO_2 on a TCO thin film. For high efficiency light absorption, the QDs need to be well-passivated, densely-packed, and highly-monodispersed on surfaces of nano TiO_2. The surfaces of the solution-based QDs are attached with ligands that are usually highly resistive. Thus, an essential issue is the establishment of good electric connections between QDs and nano TiO_2 in order for efficient electrons transport.

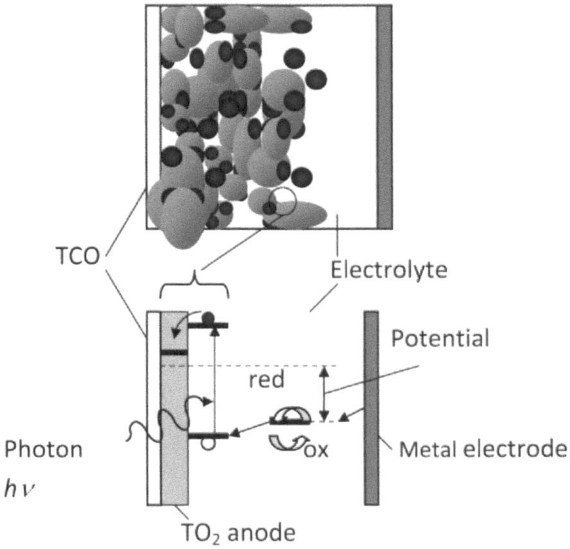

Figure 5.26 Scheme of a QD sensitized solar cell with depleted junction and the corresponding potential profile.

Assuming direct contact, TiO_2 and CdSe quantum dots display a potential arrangement as shown in Fig. 5.27 [40]. With such a staged type-II band alignment for both the conduction and the valence band, the CdSe/TiO_2 junction is similar to a *p*–*n* junction. Photo-generated electrons and holes could thus be separated and swept towards their respective electrodes. The wide-bandgap TiO_2 guarantees that such a type II band alignment remain unchanged while the bandgap of CdSe is tuned by varying the size of CdSe QDs. Many semiconductor QDs form such a type-II band alignment with TiO_2 and thus they can be used to construct QD-sensitized solar cells. Those QDs include CdSe, CdS, InP, PbSe, PbS, PbTe, etc.

As for nanostructure photo-anodes, not only TiO_2, some other metal oxides can also be used as photo-anode materials, such as ZnO. The 3D nanostructures might have many forms, including nanotubes, nanorods, nanowires, nanoparticles, as well as mesoporous films. The nanostructures offer large surface areas that can accommodate a large number of QDs. The counter electrode materials are usually the platinized TCO layers.

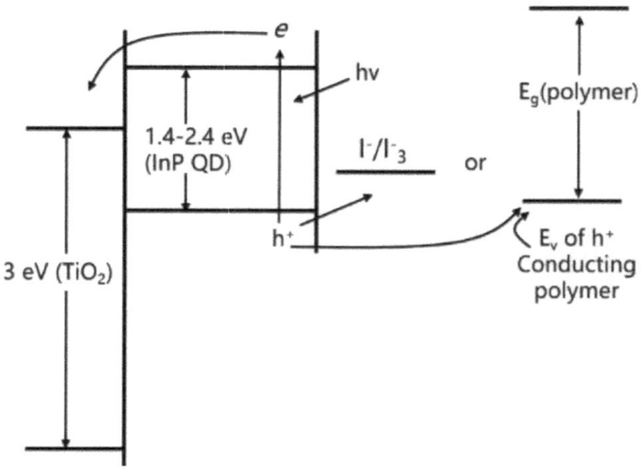

Figure 5.27 Schematic band alignment of the QD-sensitized solar cell.

5.5.5 Other Quantum Dot Solar Cell Configurations

Similar to atoms in crystals, QDs might form ordered 3D arrays. If the spacing between neighboring QDs is sufficiently small, electron wavefunctions of neighboring QDs might overlap to form minibands. The miniband offer channels for electrons, making it easy for electrons to transport in the 3D QD arrays, as the situation in semiconductors where electrons in the conduction band travel freely due to translational invariant of the crystals.

The ordered 3D QD arrays can be used to construct QD solar cells with expected high power conversion efficiencies. Very often, a *n*+*i*+*p* configuration is constructed, with "*i*" being an intrinsic 3D QD array, and "*n*" and "*p*" represent *n*- and *p*-type contacts, as shown in Fig. 5.28 [40]. Apparently, in this case a built-in electric field is established in *i*-region that is sandwiched between the *n*- and the *p*-region, with the magnitude of the electric field depending on doping levels and bandgaps of *n*- and *p*- contact materials, as well as the thickness of the 3D QD array.

Electrons in the minibands of the 3D QD arrays are delocalized. In fact, photo-generated hot electrons show slow cooling rate, allowing them to pass through the QD array and to be collected by contact electrodes. As a result, this kind of SCs might have a high open-circuit voltage. On the other hand, the hot electrons are more energetic than cold electrons in interacting with other electrons. They might release their energy through impact ionization. In this case, enhancement of short-circuit current could also expected.

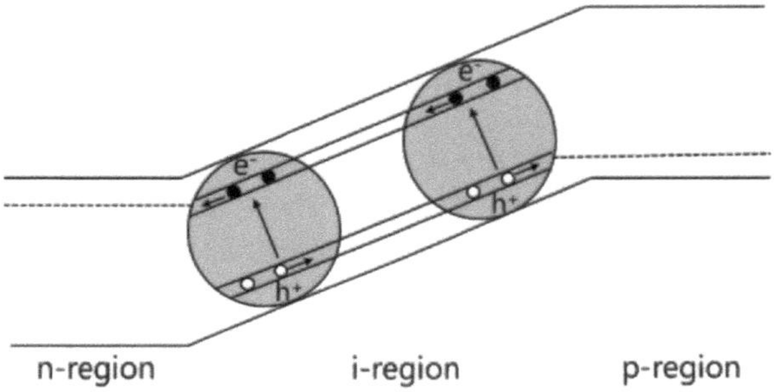

Figure 5.28 Scheme of a 3D QD array layer that is subjected to a built-in electric field.

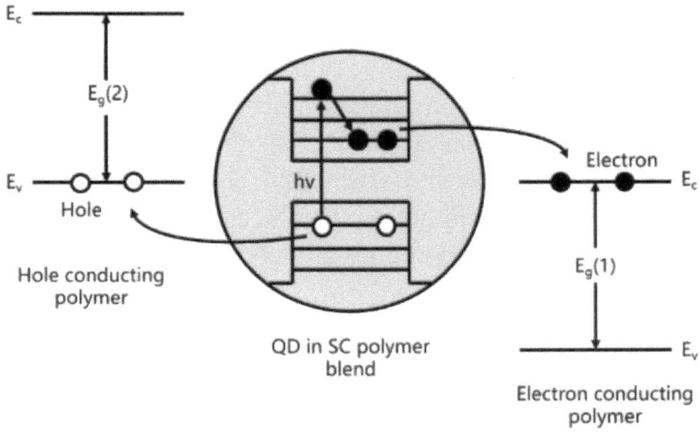

Figure 5.29 Scheme of energy band alignment of a QD array sandwiched between a hole-conducting and an electron-conducting polymers.

QDs might also be dispersed in organic semiconductor. For instance, CdSe QDs have been placed in a hole-conducting polymer, MEH-PPV (poly(2-methoxy, 5-(2-ethyl)-hexyloxy-pphenylenevinylene), in the form of a disordered array. The device configuration exhibits a band alignment as shown in Fig. 5.29 [40]. Thus, the device is constructed so that a built-in electric field is formed cross the QD array, the QDs being sandwiched between the hole-conducting and the electron-conducting polymers. Upon absorbing photons by the QDs, photo-generated electrons and hole are injected into the electron-conducting polymer and the hole-conducting polymer, respectively, and subsequently collected by their respective contact electrodes.

5.5.6 Extremely Thin Absorber Cells

Quantum dots are excellent photon capturers. Thus, a QD-based solar cell needs only a very thin layer to capture the incident photons. The so-called extremely thin absorber solar cell has a device configuration as shown in Fig. 5.30, which is characterized by a very thin absorption layer sandwiched between an electron-conducting layer and a hole-conducting layer. Generally, a compromise on device configuration has to be made for light absorption and carrier collection in a solar cell. Effective light absorption requires a rather thick absorption layer while a thin layer is beneficial for charge collection. The advantage of having a very thin absorber is the very thin overall device thickness [41]. In fact, both the electron and the hole-conducting layers can be made very thin and, as a result, the photo-generated electrons and holes can be collected efficiently. The thin absorption layer can be made three-dimensional, which increases the interface area and enhances light trapping, as shown in Fig. 5.30.

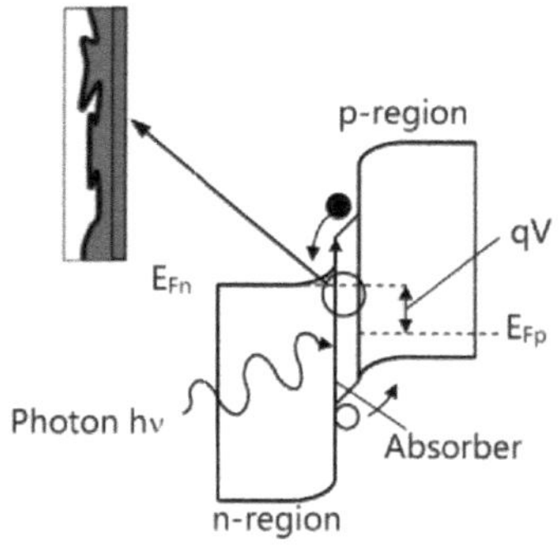

Figure 5.30 Schematic energy band alignment of an extremely thin absorber SC. The junction can have a 3D form, as shown by the zoom-in drawing of the junction region, for increasing the junction area.

5.5.7 Tandem Nano Solar Cells

The bandgap of semiconductor QDs, with their bulk bandgap at short-wavelength infrared, can be tuned to cover the whole visible spectrum by varying their sizes on the order of the exciton Bohr radius [42]. Tandem solar cells aim at photon absorption of the whole solar spectrum, which have the device architecture of a sequence of subcells with increasing bandgaps stacked on top of each other. The tandem solar cells are based on both III–V and II–VI semiconductors. Due to the efficient utilization of the whole solar spectrum, the device efficiency reaches to nearly 45%, more than two times of that of crystalline Si solar cells. The tandem architecture can also be constructed using nanomaterials, especially quantum dots. The device has the configuration of series of QD layers stacking together, with the layers with largest bandgap on the top, as shown in Fig. 5.31 [42]. The multiple QD-layer structure might have an absorption spectrum closely matched to the solar radiation spectrum.

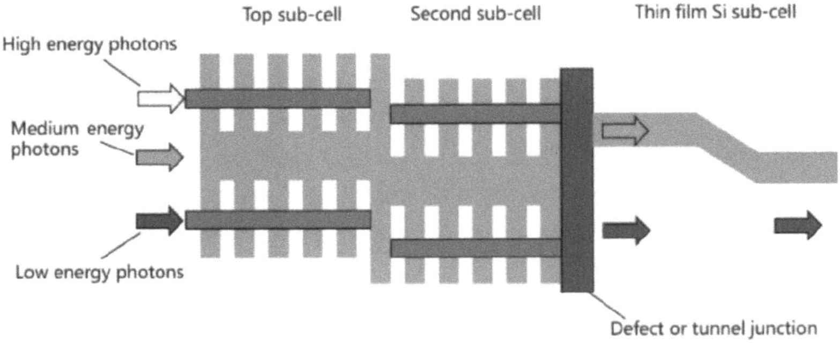

Figure 5.31 Schematic band alignment of a tandem QD solar cell, where the QD layers are sequentially stacked with the layer with largest bandgap on the top. High energy (H.E.) photons are absorbed by the top sub-cell and the medium energy photons absorbed by the second sub-cell. The thin film Si sub-cell uses the low energy photons for photo-voltaic conversion. Note that in the top sub-cell 2 nm QDs, having a bandgap of 1.87 eV, are used and in the second sub-cell the QDs have larger sizes for a narrower bandgap. The bandgap of thin film Si is 1.12 eV, lowest among all three sub-cells.

When photons are incident onto the solar cell, photons with energies equal to and larger than the bandgap of the top subcell are absorbed and excitons are generated in the top subcell. The photons, with energies smaller than the bandgap of the top cell but equal to and larger than the bandgap of the second subcell, penetrate through the top subcell and are absorbed and generate electron–hole pairs in the second subcell. Photons with energies equal to and larger than the bandgap of the bottom cell are used by the bottom subcell for photovoltaic conversion. Thus, all photons with energies larger than the bandgap of the bottom cell are absorbed by the tandem solar cell with their energies being used sufficiently for photovoltaic conversion due to the close matching of the solar spectrum and the bandgap distribution of the subcells.

The major advantage of the tandem nano solar cells (NSCs) lies in the prospect that the cells use quantum dots with inherent tunable bandgaps, corresponding to tunable sizes of single material QDs using only a simple manufacturing process. The construction of the tandem NSCs thus requires only a very simple technological process, which would lead to a great reduction of the cost in large-scale production.

The most critical challenge for high power conversion efficiency of a tandem SC is in-series device configuration that requires good matching of the photocurrent in those subcells. The overall current of the tandem cell is determined by the subcell with lowest photocurrent. Another challenge is the high density of defect centers like in other NSCs that act as recombination centers.

5.6 Recent Work in Quantum Dot Solar Cells

At their initial stage, solar cells based on semiconductor quantum dots show very low power conversion efficiencies typically below 3%. In recent years, the efficiency is significantly improved due to the use of many technologies such as surface modification, improved counter electrode, as well as the use of new materials such as perovskite quantum dots. Here some works done on QDSCs with conversion efficiency exceeding 10% are presented.

5.6.1 QD Solar Cells with Conversion Efficiency Over 12% Due to Improved Photoanodes and Counter Electrode

For NSCs, the property of photoanodes, functioning to separate photo-generated electrons and holes, is essential to the device performance. Counter electrodes are used for collecting holes, which also play a critical role in NSCs.

Wang and co-workers fabricated photoanodes consisting of ZnCuInSe and CdSe QDs and mesoporous TiO_2. 3-mercaptopropionic acid bifunctional linkers were used in the fabrication to control the connections between the QDs and the TiO_2 thin films [43]. ZnCuInSe QDs have a wide spectral range of light absorption and CdSe QDs are excellent photon antennas. The synergistic effect of the two kinds of QDs leads to improvement of the photon-harvest and the photon-to-electron conversion efficiency, where the performance of the photoanode was optimized by tuning CdSe QDs sizes. The counter electrode of the NSCs is mesoporous carbon supported by Ti meshes and modified polysulfide solution was used as the electrolyte. In optimized device structure parameters and preparation technology, the device achieved a short-circuit current J_{sc} of 27.39 mA cm^{-2}, an open-circuit voltage V_{oc} = 0.752 V, and a filling factor FF = 0.619, corresponding to a power conversion efficiency of 12.75%, which is the highest for the kind of NSCs at that time.

The role of counter electrodes is to collect electrons from external circuit and transfer the collected electrons back to electrolyte through catalysis of the reduction reaction of oxidized redox couples (Sn^{2-}/S^{2-}, for instance) in electrolyte. Counter electrodes usually consist of catalyst and conducting substrate (catalyst on a substrate). In QD-sensitized SCs, the basic requirements for counter electrodes are low resistance for charge transport and high electrocatalytic activity for catalysis of the reduction reaction.

Various technologies have been developed to modify the materials for counter electrodes that are very often nanomaterials themselves. In Jiao and co-workers' work, N atoms were incorporated into the lattices of mesoporous carbons and the N-doped material was used as counter electrodes [44].

The counter-electrode materials with several nitrogen contents were fabricated by a silica-template method [45]. The N-doped mesoporous carbon materials showed excellent catalytic activity, as demonstrated in ZnCuInSe QDSCs. It was found that N-doped mesoporous carbons with a N content of 8.58 wt % presents the strongest activity and, as a result, the corresponding QDSC gave the best device performance with a conversion efficiency of 12.23% that was the best value at the time.

Figures 5.32a,b present the *J–V* curves measured on the ZnCuInSe QD solar cells with counter electrodes of different N-doping contents and the certified solar cell based on 8.58% nitrogen content shows conversion efficiency of 12.07%, an open-circuit voltage V_{oc} of 0.765 V, a short-circuit current density J_{sc} of 25.21 mA·cm^{-2}, and a filling factor *FF* of 0.626. Du and co-workers adopted mesoporous carbons supported by Ti mesh substrate as the counter electrode material for $CdSe_{0.65}Te_{0.35}$ QDSCs and achieved a certified efficiency exceeding 11.16% [46]. The Ti mesh substrate provides electrons with efficient electron channels, leading to improved conductivity of the counter electrodes.

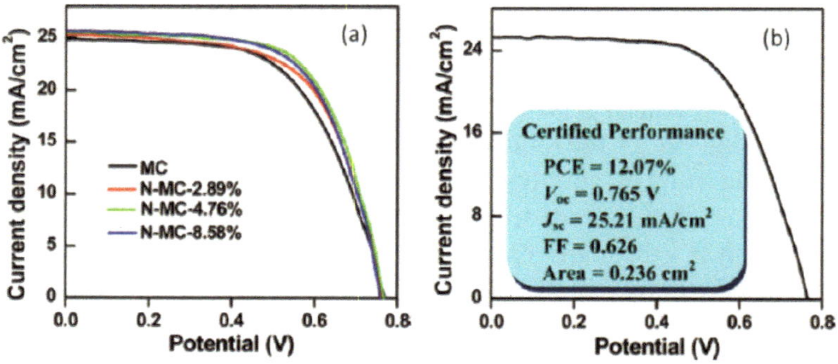

Figure 5.32 (a) *J–V* curves of the ZnCuInSe QD solar cells base on the mesoporous carbon material with different N-doped contents and (b) the certified device with listed parameters. Reproduced with permission from [44].

5.6.2 High Conversion Power Achieved Using Perovskite QDs

$CsPbI_3$, an inorganic analog to the hybrid organic cation halide perovskites, has an ideal bandgap for photovoltaic devices and other excellent opto-electronic properties. However, α-$CsPbI_3$ (the cubic phase of bulk $CsPbI_3$) is not stable at room temperature. Luther's group fabricated α-$CsPbI_3$ QD films that are phase-stable for months in ambient air [47]. Nano solar cells were fabricated based on the perovskite QD thin films, where the α-$CsPbI_3$ QD film was used as the active layer. The device has an open-circuit voltage of 1.23 V, short-circuit current density of 13.47 Am · cm^{-2}, a filling factor of 65%, and a conversion efficiency of 10.77%. Figure 5.33a presents the schematic structure of the α-$CsPbI_3$ QDSC, while Fig. 5.33b shows the cross-sectional TEM image of the device. The *J*–*V* curves of the α-$CsPbI_3$ QDSC measured within 15 days after fabrication are presented in Fig. 5.33c. It is interesting to see that the conversion efficiency improves with time in this course of 15 days. The same research group, with some improved technology for the perovskite QD film, the device conversion efficiency was raised to 13.47% [48].

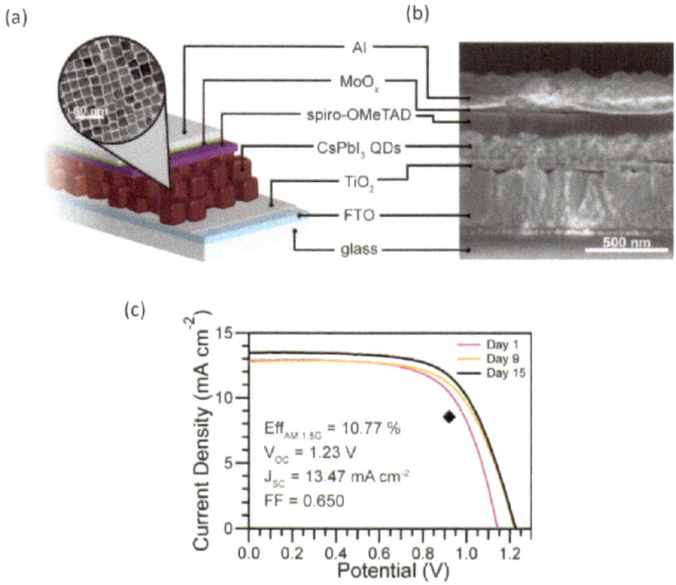

Figure 5.33 Schematic device structure (a) and cross-sectional TEM (b) of the α-$CsPbI_3$ QD SC, together with *JV* curves (c) of a device measured in ambient conditions over in 1, 9, and 15 days after fabrication. Reproduced with permission from [48].

Ma's group adopted a different colloidal perovskite QD film as an active layer and fabricated the solar cells with the structure of FTO/TiO_2/$CsPbI_3$ QDs/PTB7/MoO_3/Ag [49]. They used Yb to dope $CsPbI_3$ during synthesis, finding that Yb doping gave rise to reduction of defect and trapping state densities, as shown by the enhancement of photoluminescence quantum yield. In addition, crystallinity, thermal stability and carrier transport the QD films were also improved. The best conversion efficiency of 13.12% was achieved on the solar cells using 20% Yb-doped QDs, corresponding to V_{oc} = 1.25 V, J_{sc} = 14.18 Am·cm^{-2}, and FF = 74%. Other than those, the device showed excellent storage stability under ambient conditions.

Used as an active layer in NSCs, α-$CsPbI_3$ QDs has a bandgap energy of 1.75 eV that is too large for a single-junction photovoltaic devices. Formamidinium (FA) incorporation into α-$CsPbI_3$ forms formamidinium lead halide perovskite ($FAPbI_3$) with a reduced bandgap of 1.55 eV. Ma's group improved device efficiency by introducing the bilayer structure consisting of stacked α-$CsPbI_3$ and $FAPbI_3$ [50]. Figures 5.34a,b present the schematic structure and cross-sectional SEM image of the solar cell based on the α-$CsPbI_3$/$FAPbI_3$ perovskite QD bilayer. The whole solar cell was fabricated via a layer-by-layer deposition technology. The use of the bilayer broadens the light absorption and, as a result, enhances photo-generation of carriers. On the other hand, thermal annealing of the α-$CsPbI_3$/$FAPbI_3$ bilayer gives rise to graded energy band profile in the heterojunction region that is greatly beneficial for charge extraction by the electrode and the counter-electrode. The gains on both the light absorption and the carrier extraction result in obvious improvements on device performance. For the best solar cell based on the α-$CsPbI_3$/$FAPbI_3$ perovskite QD bilayer as the active layer, Ma's group achieved an open-circuit voltage V_{oc} = 1.22 V, a short-circuit current J_{sc} = 17.26 Am·cm^{-2}, a filling factor FF = 72%, and a PCE = 15.6%, with the major improvement lying in J_{sc}; 15.6% is one of the highest conversion efficiency at that time.

Comparatively, $FAPbI_3$ QD films are more stable than α-$CsPbI_3$ QD films. The device shown in Fig. 5.34 was constructed in such a way that the $FAPbI_3$ QD layer was on top of the α-$CsPbI_3$ QD layer. Thus, the α-CsPbI3 layer was protected by the more stable $FAPbI_3$ layer against ambient environmental impacts. In addition, the layer-by-layer deposition technology could easily be extended to triple- and multiple-layer structures, to construct multi-junction QDSCs for achieving higher conversion efficiencies.

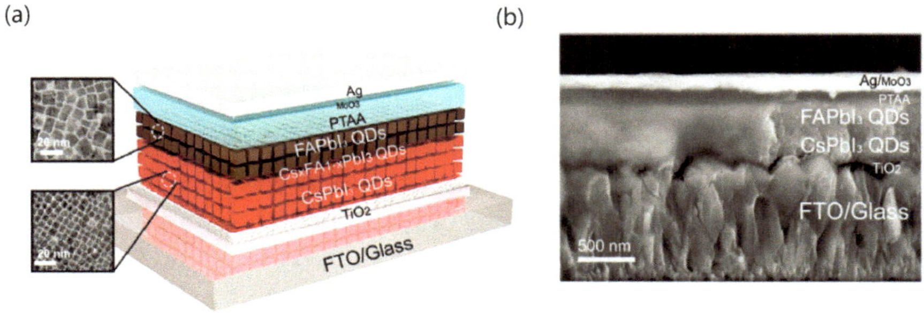

Figure 5.34 Schematic device structure (a) and cross-sectional SEM image (b) of the solar cell with an α-$CsPbI_3$/$FAPbI_3$ perovskite QD bilayer. Reproduced with permission from [50].

5.6.3 QD Solar Cells with Improved Quality of QD Films

Reported colloidal QD SCs have low power conversion efficiencies. Among many reasons, poor monodispersity, heterogeneous aggregation, large surface areas, and size polydispersity are the major problems. Those lead to the formation of band tail states in both the conduction and the valence bands and the tail states are the killers to open-circuit voltage and short-circuit current density.

Sargent's group was able to design a ligand exchange route and ligand-stabilization technology, where the ligand exchange was performed by using lead halide precursors and assisted by using ammonium acetate [51]. The colloidal QD is stabilized by both $[PbI]^+$ and $[NH_4]^+$. After ligand exchange and stabilization, the precipitation of the colloidal QDs was completed by adding toluene (an anti-solvent). $[PbI_3]^-$- and $[PbI]^+$-capped QDs were then obtained by centrifugation. The process of consequent ligand exchange, QD stabilization, precipitation, and separation is schematically presented in Fig. 5.35.

This ligand exchange route enables the complete replacement of the original long chain oleic acid and the obtained QD solid contains no organic residues. The high halide content offers excellent passivation on the QDs. Thick QD film can then be prepared for the QD solar cells, so that light absorption is enhanced and more solar irradiation can be utilized for photoelectric conversion. Electrically, the QD film prepared based on the $[PbI_3]^-$- and $[PbI]^+$-capped QDs show significantly reduced band tail states, which is beneficial for the enhancement of open-circuit voltage and improvement on charge transport and carrier collection. For the solar cell based on the $[PbI_3]^-$- and $[PbI]^+$-capped QD film (active layer) with thickness of 350 nm and a bandgap of 1.32 eV, the highest conversion efficiency of 11.28% was achieved under AM1.5 condition, corresponding to V_{oc} = 0.61 V, J_{sc} = 2.72 Am·cm^{-3}, and *FF* = 67.8%. Figure 5.36 plots the *J–V* curve of this best device with obtained device parameters. In addition, these solar cells have high stability. Only 10% of their best conversion efficiency value was lost after 1000 h usage even they were stored unencapsulatedly in air.

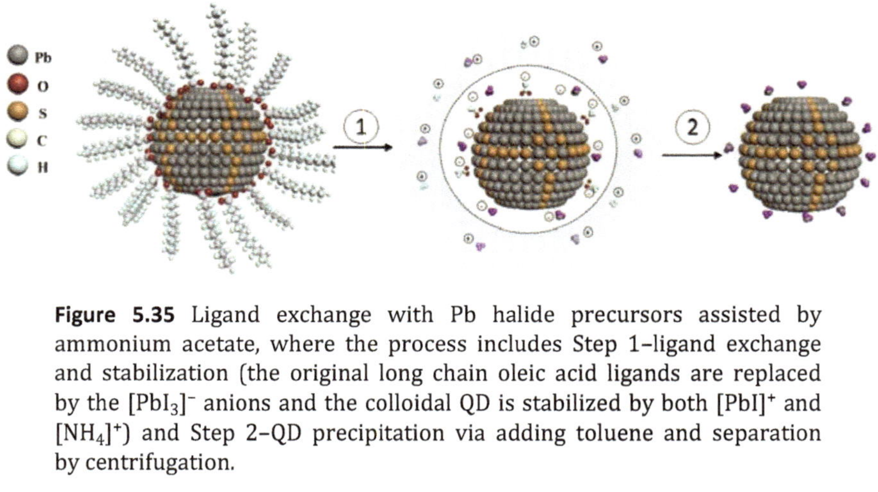

Figure 5.35 Ligand exchange with Pb halide precursors assisted by ammonium acetate, where the process includes Step 1–ligand exchange and stabilization (the original long chain oleic acid ligands are replaced by the $[PbI_3]^-$ anions and the colloidal QD is stabilized by both $[PbI]^+$ and $[NH_4]^+$) and Step 2–QD precipitation via adding toluene and separation by centrifugation.

Current solar cells are based on the generation of one exciton per photon. If an electron absorbs a photon with the energy exceeding the bandgap energy, the extra kinetic energy of the electron will be lost to lattice as heat. QDs offer great room for various SC device architectures and the use of QDs may be able to enhance the device efficiency through generation of more than one exciton from a high-energy photon. The so-called multiple exciton generation (MEG) has been demonstrated on some semiconductor QDs.

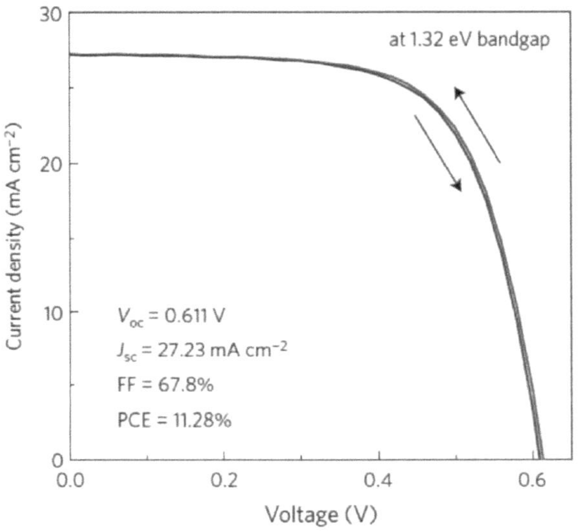

Figure 5.36 The *J–V* curve of the best device with 350 nm $[PbI_3]^-$- and $[PbI]^+$-capped QD active layer. The QD film has a bandgap of 1.32 eV. Reproduced with permission from [51].

5.7 Existing Problems for Quantum Dot Solar Cells

Quantum dot solar cells, with various device architectures, are in their developing stage, although a great deal of work has been done in material manufacturing and device structure design. Currently, the power conversion efficiency of QDSCs is still rather low. The value is, however, rising quickly, which offers hope for future applications.

The device efficiency of QDSCs is mainly limited by the following factors:

(1) The random dispersion of QDs in QDSCs gives rise to low carrier transport efficiency. In crystals, conduction band electrons act like free electrons due to the translational invariant of crystal lattices since, in this case, the scattering to electrons from crystalline potential are largely cancelled.

(2) The negative side of the extremely large interface area in QDSCs sometimes surpluses the positive side. The large number of defects in the interfacial region provides high density of centers for carrier recombination, leading to very short carrier lifetime. The interface could also slow down the carrier transport.

(3) The surface of the colloidal QDs is covered by organic ligands that are usually non-conductive so that carriers are unable to be extracted by electrodes. On the other hand, the surface ligands can also behave as carrier recombination centers.

(4) It is still rather difficulty to assemble QDs into mesoporous TiO_2 matrix. In another words, it is really a challenge to incorporate high density of QDs on nano metal oxide surface.

(5) Long-term stability is an important fundamental requirement for the use of solar cells. QDSCs are still unstable due to large interfacial area of the nanomaterials. The stable interface chemistry is challenged, especially, under solar radiation, which affects the device fill factor and hence the device efficiency. It is still a challenge to find an electrolyte in which QDs are stable.

Other limitations are common as to other solar cells. Like other materials, QDs cannot absorb photons with energy lower than their bandgaps and cannot use the excess part of photon energy larger than the bandgaps. Those problems are all material- and/or structure-related and can be eliminated or at least partially eliminated through quality control on material fabrication and device design. For instance, the formation of periodic QD pattern could be beneficial for the suppression of various scattering to carriers due to surface potential and interfacial defects.

References

1. C. X. Guo, H. B. Yang, Z. M. Sheng, Z. S. Lu, Q. L. Song, and C. M. Li, Layered graphene/quantum dots for photovoltaic devices, *Angew. Chem. Int. Ed.*, **49**, 3014–3017 (2010).

2. X. Cao, C. M. Li, H. Bao, Q. Bao, and H. Dong, Fabrication of strongly fluorescent quantum dot-polymer composite in aqueous solution, *Chem. Mater.*, **19**, 3773–3779 (2007).

3. K. Kumari, S. Chand, V. D. Vankar, and V. Kumar, Enhancement in hole current density on polarization in poly(3-hexylthiophene): Cadmium selenide quantum dot nanocomposite thin films, *Appl. Phys. Lett.*, **94**, 213503 (2009).

4. C. B. Murray, C. R. Kagan, and M. G. Bawendi, Synthesis and characterization of monodisperse nanocrystals and close-packed nanocrystal assemblies, *Annu. Rev. Mater. Res.*, **30**(1), 545–610 (2000).

5. C. B. Murray, D. J. Norris, and M. G. Bawendi, Synthesis and characterization of nearly monodisperse CdE (E = sulfur, selenium, tellurium) semiconductor nanocrystallites, *J. Am. Chem. Soc.*, **115**(19), 8706–8715 (1993).

6. S. G. Kwon and T. Hyeon, Formation mechanisms of uniform nanocrystals via hot-injection and heat-up methods, *Small*, **7**(19), 2685–2702 (2011).

7. W. Zhang, H. Zhang, Y. Feng, and X. Zhong, Scalable single-step noninjection synthesis of high-quality core/shell quantum dots with emission tunable from violet to near infrared, *ACS Nano*, **6**(12), 11066–11073 (2012).

8. K.-J. Kim, R. P. Oleksak, E. B. Hostetler, D. A. Peterson, P. Chandran, D. M. Schut, B. K. Paul, G. S. Herman, and C.-H. Chang, Continuous microwave-assisted gas–liquid segmented flow reactor for controlled nucleation and growth of nanocrystals, *Cryst. Growth Des.*, **14**, 5349–5355 (2014).

9. S. G. Kwon, Y. Piao, J. Park, S. Angappane, Y. Jo, N. M. Hwang, J. G. Park, and T. Hyeon, Kinetics of monodisperse iron oxide nanocrystal formation by "heating-up" process, *J. Am. Chem. Soc.*, **129**, 12571–12584 (2007).

10. Y. C. Cao and J. Wang, One-pot synthesis of high-quality zinc-blende CdS nanocrystals, *J. Am. Chem. Soc.*, **126**, 14336–14337 (2004).

11. J. Pan, A. O. El-Ballouli, L. Rollny, O. Voznyy, V. M. Burlakov, A. Goriely, E. H. Sargent, and O. M. Bakr, Automated synthesis of photovoltaic-quality colloidal quantum dots using separate nucleation and growth stages, *ACS Nano*, **7**(11), 10158–10166 (2013).

12. H. Nakamura, Y. Yamaguchi, M. Miyazaki, H. Maeda, M. Uehara, and P. Mulvaney, Preparation of CdSe nanocrystals in a micro-flow-reactor, *Chem. Commun.*, **23**, 2844–2845 (2002).

13. E. M. Chan, A. P. Alivisatos, and R. A. Mathies, High-temperature microfluidic synthesis of CdSe nanocrystals in nanoliter droplets, *J. Am. Chem. Soc.*, **127**, 13854–13861 (2005).
14. C.-H. M. Chuang, P. R. Brown, V. Bulovic, and M. G. Bawendi, Improved performance and stability in quantum dot solar cells through band alignment engineering. *Nat. Mater.*, **13**, 796–801 (2014).
15. A. Fischer, L. Rollny, J. Pan, G. H. Carey, S. M. Thon, S. Hoogland, O. Voznyy, D. Zhitomirsky, J. Y. Kim, O. M. Bakr, and E. H. Sargent, Directly deposited quantum dot solids using a colloidally stable nanoparticle ink, *Adv. Mater.*, **25**, 5742–5749 (2013).
16. N. T. K. Thanh, N. Maclean, and S. Mahiddine, Mechanisms of nucleation and growth of nanoparticles in solution, *Chem. Rev.*, **114**, 7610–7630 (2014).
17. I. J. Kramer, J. C. Minor, G. Moreno-Bautista, L. Rollny, P. Kanjanaboos, D. Kopilovic, S. M. Thon, G. H. Carey, K. W. Chou, D. Zhitomirsky, A. Amassian, and E. H. Sargent, Efficient spray-coated colloidal quantum dot solar cells, *Adv. Mater.*, **27**, 116–121 (2015).
18. Q. Guo, S. J. Kim, M. Kar, W. N. Shafarman, R. W. Birkmire, E. A. Stach, R. Agrawal, and H. W. Hillhouse, Development of $CuInSe_2$ nanocrystal and nanoring inks for low-cost solar cells, *Nano Lett.*, **8**, 2982–2987 (2008).
19. P. Riente, J. N. Clifford, M. A. Pericàs, and E. Palomares, Improving CdSe quantum dot/polymer solar cell efficiency through the covalent functionalization of quantum dots: Implications in the device recombination kinetics, *J. Phys. Chem. C*, **117**, 13374–13381 (2013).
20. Z. Ning, O. Voznyy, J. Pan, S. Hoogland, V. Adinolfi, J. Xu, M. Li, A. R. Kirmani, J.-P. Sun, J. Minor, K. W. Kemp, H. Dong, L. Rollny, A. Labelle, G. Carey, B. Sutherland, I. Hill, A. Amassian, H. Liu, J. Tang, O. M. Bakr, and E. H. Sargent, Air-stable *n*-type colloidal quantum dot solids, *Nat. Mater.*, **13**, 822–828 (2014).
21. A. H. Ip, S. M. Thon, S. Hoogland, O. Voznyy, D. Zhitomirsky, R. Debnath, L. Levina, L. R. Rollny, G. H. Carey, A. Fischer, K. W. Kemp, I. J. Kramer, Z. Ning, A. J. Labelle, K. W. Chou, A. Amassian, and E. H. Sargent, Hybrid passivated colloidal quantum dot solids, *Nat. Nanotech.*, **7**, 577–582 (2012).
22. H. H.-Y. Wei, C. M. Evans, B. D. Swartz, A. J. Neukirch, J. Young, O. V. Prezhdo, and T. D. Krauss, Colloidal semiconductor quantum dots with tunable surface composition, *Nano Lett.*, **12**, 4465–4471 (2012).

23. S. A. McDonald, G. Konstantatos, S. Zhang, P. W. Cyr, E. J. D. Klem, L. Levina, and E. H. Sargent, Solution-processed PbS quantum dot infrared photodetectors and photovoltaics, *Nat. Mater.,* **4**, 138–142 (2005).

24. B. L. Sharma, ed., *Metal-Semiconductor Schottky Barrier Junctions and Their Applications*; 1984 Plenum Press. New York.

25. J. P. Clifford, K. W. Johnston, L. Levina, and E. H. Sargent, Schottky barriers to colloidal quantum dot films, *Appl. Phys. Lett.,* **91**, 253117 (2007).

26. W. Monch, in *Semiconductor Surfaces and Interfaces*; Springer, p. 572 (2001).

27. J. M. Luther, M. Law, M. C. Beard, Q. Song, M. O. Reese, R. J. Ellingson, and A. J. Nozik, Schottky solar cells based on colloidal nanocrystal films. *Nano Lett.,* **8**, 3488–3492 (2008).

28. J. P. Clifford, K. W. Johnston, L. Levina, and E. H. Sargent, *Appl. Phys. Lett.*, **91**, 253117 (2007).

29. J. M. Luther, M. Law, M. C. Beard, Q. Song, M. O. Reese, R. J. Ellingson, and A. J. Nozik, Schottky solar cells based on colloidal nanocrystal films, *Nano Lett.*, **8**, 3488 (2008).

30. R. Debnath, J. Tang, D. A. Barkhouse, X. Wang, A. G. Pattantyus-Abraham, L. Brzozowski, L. Levina, and E. H. Sargent, Ambient-processed colloidal quantum dot solar cells via individual pre-encapsulation of nanoparticles, *J. Am. Chem. Soc.*, **132**, 5952–5953 (2010).

31. W. Ma, S. L. Swisher, T. Ewers, J. Engel, V. E. Ferry, H. A. Atwater, and A. P. Alivisatos, Photovoltaic performance of ultrasmall PbSe quantum dots, *ACS Nano,* **5**, 8140–8147 (2011).

32. C. Piliego, L. Protesescu, S. Z. Bisri, M. V. Kovalenko, and M. A. Loi, 5.2% efficient PbS, nanocrystal Schottky, solar cells. Energy environ, *Science,* **6**, 3054–3059 (2013).

33. J. Jean, S. Chang, P. R. Brown, J. J. Cheng, P. H. Rekemeyer, M. G. Bawendi, S. Gradečak, and V. Bulović, ZnO nanowire arrays for enhanced photocurrent in PbS quantum dot solar cells. *Adv. Mater.,* **25**, 2790–2796 (2013).

34. D. A. R. Barkhouse, R. Debnath, I. J. Kramer, D. Zhitomirsky, A. G. Pattantyus-Abraham, L. Levina, L. Etgar, M. Grätzel, and E. H. Sargent, Depleted bulk heterojunction colloidal quantum dot photovoltaics, *Adv. Mater.,* **23**, 3134–3138 (2011).

35. J. Zhang, J. Gao, C. P. Church, E. M. Miller, J. M. Luther, V. I. Klimov, and M. C. Beard, PbSe quantum dot solar cells with more than 6% efficiency fabricated in ambient atmosphere, *Nano Lett.,* **14**, 6010–6015 (2014).

36. D. V. Talapin and C. B. Murray, PbSe nanocrystal solids for *n*- and *p*-channel thin film field-effect transistors, *Science,* **310**, 86–89 (2005).

37. J. Tang, H. Liu, D. Zhitomirsky, S. Hoogland, X. Wang, M. Furukawa, L. Levina, and E. H. Sargent, Quantum junction solar cells, *Nano Lett.,* **12**, 4889–4894 (2012).

38. D. Zhitomirsky, M. Furukawa, J. Tang, P. Stadler, S. Hoogland, O. Voznyy, H. Liu, and E. H. Sargent, *n*-Type colloidal-quantum-dot solids for photovoltaics, *Adv. Mater.,* **24**, 6181–6185 (2012).

39. J. Tang, H. Liu, D. Zhitomirsky, S. Hoogland, X. Wang, M. Furukawa, L. Levina, and E. H. Sargent, Quantum junction solar cells, *Nano Lett.,* **12**, 4889–4894 (2012).

40. A. J. Nozik, Quantum dot solar cells, *Physica E,* **14**, 115–120 (2002).

41. D.-C. Nguyen, S. Tanaka, H. Nishino, K. Manabe, and S. Ito, 3-D solar cells by electrochemical-deposited Se layer as extremely-thin absorber and hole conducting layer on nanocrystalline TiO_2 electrode, *Nanoscale Res. Lett.,* **8,** 8 (2013) doi:10.1186/1556-276X-8-8.

42. G. Conibeer, I. Perez-Wurfl, X. Hao, D. Di, and D. Lin, Si solid-state quantum dot-based materials for tandem solar cells, *Nanoscale Res. Lett.,* **7**, 193 (2012).

43. W. Wang, W. Feng, J. Du, W. Xue, L. Zhang, L. Zhao, Y. Li, and X. Zhong, Cosensitized quantum dot solar cells with conversion efficiency over 12%, *Adv. Mater.,* **30**, 17057462018 (2018).

44. S. Jiao, J. Du, Z. Du, D. Long, W. Jiang, Z. Pan, Y. Li, and X. Zhong, Nitrogen-doped mesoporous carbons as counter electrodes in quantum dot sensitized solar cells with a conversion efficiency exceeding 12%, *J. Phys. Chem. Lett.,* **8**(3), 559–564 (2017).

45. H. Chen, F. Sun, J. Wang, W. Li, W. Qiao, L. Ling, and D. Long, Nitrogen doping effects on the physical and chemical properties of mesoporous carbons, *J. Phys. Chem. C,* **117**, 8318–8328 (2013).

46. Z. Du, Z. Pan, F. Fabregat-Santiago, K. Zhao, D. Long, H. Zhang, Y. Zhao, X. Zhong, J.-S. Yu, and J. Bisquert, Carbon counter-electrode-based quantum-dot-sensitized solar cells with certified efficiency exceeding 11%, *J. Phys. Chem. Lett.,* **7**, 3103–3111 (2016).

47. A. Swarnkar, A. R. Marshall, E. M. Sanehira, B. D. Chernomordik, D. T. Moore, J. A. Christians, T. Chakrabarti, and J. M. Luther, Quantum dot-induced phase stabilization of α-$CsPbI_3$ perovskite for high-efficiency photovoltaics, *Science*, **354** (6308), 92–95 (2016).

48. E. M. Sanehira, A. R. Marshall, J. A. Christians, S. P. Harvey, P. N. Ciesielski, L. M. Wheeler, P. Schulz, L. Y. Lin, M. C. Beard, and J. M. Luther, Enhanced mobility $CsPbI_3$ quantum dot arrays for record-efficiency, high-voltage photovoltaic cells, *Sci. Adv.*, **3**, eaao420 (2017).

49. J. Shi, F. Li, J. Yuan, X. Ling, S. Zhou, Y. Qian, and W. Ma, Efficient and stable $CsPbI_3$ perovskite quantum dots enabled by in situ ytterbium doping for photovoltaic applications, *J. Mater. Chem.* ***A***, **7**(36), 20936–20944 (2019).

50. F. Li, S. Zhou, J. Yuan, C. Qin, Y. Yang, J. Shi, X. Ling, Y. Li, and W. Ma, Perovskite quantum dot solar cells with 15.6% efficiency and improved stability enabled by an α-$CsPbI_3$-$FAPbI_3$ bilayer structure, *ACS Energy Lett.*, **4**(11), 2571–2578 (2019).

51. M. Liu, O. Voznyy, R. Sabatini, F. P. García de Arquer, R. Munir, A. H. Balawi, X. Lan, F. Fan, G. Walters, A. R. Kirmani, S. Hoogland, F. Laquai, A. Amassian, and E. H. Sargent, Hybrid organic–inorganic inks flatten the energy landscape in colloidal quantum dot solids, *Nat. Mater.*, **16**, 258–263 (2016).

Chapter 6

Solar Cells Based on One-Dimensional Nanomaterials

Semiconductor nanowires (NWs) exhibit novel optical, electronic, and thermal properties as a result of their unique geometric shapes and quantum confinement effects in two directions perpendicular to wires. Semiconductor NWs are the building blocks for a broad range of macroscopic devices, including solar cells (SCs) and key components of many nanodevices, based on their widely tunable band structures and broad selection of materials and compositions. Thus, the one-dimensional (1D) NWs represent a unique system for exploring novel physical phenomena at the nanoscale.

Semiconductor nanowires (NWs) are quasi-one-dimensional nanostructured materials, with diameters typically less than 500 nm and aspect ratios (length-to-diameter) typically greater than 10. The geometries of nanomaterial can be various shapes. The 1D or quasi 1D nanomaterials include nanowires, nanopillars, nanocones, nanodomes, and nanotubes depending on the aspect ratio and the geometric shapes of the nanowires. Nanotubes can be viewed as "anti-nanowires" based on their geometries. The major processes that form a photovoltaic process consist of photon absorption, electron/hole splitting, and photocarrier collection. Those major processes should be optimized before high conversion efficiency can be achieved. In comparison to conventional planar solar cells, the nanorod array solar cells, which are more like 3D solar cells, could enhance both the photon

Introduction to Nano Solar Cells
Ning Dai

ISBN 978-981-4877-49-7 (Hardcover), 978-1-003-13198-4 (eBook)
www.jennystanford.com

absorption due to enhanced light trapping with minimized angular dependence and the waveguiding effect, and photon-generated carrier collection. Nanowires are very promising class of materials for the exploration of new concepts and high efficiency devices for solar energy conversion.

As 1D materials, NWs offer 2D electronic quantum confinement on carriers. As a result, after electrons are separated from holes in the space charge region, the carriers move along the wires before they are collected by electrodes. Such a 2D confinement is beneficial for carrier diffusion since it reduces the probability that the carriers are captured by surface states in NWs, though, on the other hand, the surface states tend to minimize dark currents. Nanowire solar cells (NWSCs) are usually in the form of a NW array aligning in vertical geometry. As for the electric contact, both axial and radial junctions remain to be a challenge.

6.1 Fundamental Material Properties of Nanowires

Semiconductor NWs are the building blocks for NW-based solar cells in which NWs are used for photon trapping, electron–hole separating, and carrier transporting. Thus, the electronic and optical properties of NWs determine the performance of nanowire solar cells (NWSCs) to a large extent.

6.1.1 Basic Theory

Diameters of NWs used for solar cells are usually in sub-micron range, acting like optical cavities to support resonant modes in the high-quality faceted NWs. Indicated by theoretical calculations, resonance modes suggest that the optical field could be trapped in certain region in NWs and, as a result, interaction between light field and the electronic system is intensified, in favor of enhancement of power conversion efficiency (PCE). Apart from enhanced light absorption in comparison to bulk semiconductors, the small dimensions are also beneficial for efficient charge separation. The lengths of the wires could be a few microns to tens of microns, which does not usually support cavity modes in the axial direction of NWs. Cao and co-workers performed calculations on optical absorption efficiency Q_{abc} of Si NWs with several diameters together with the corresponding experimental results, and the results are presented in Fig. 6.1a, which are characterized by distinct line structures at some wavelengths. 2D plot of Q_{abc} versus wavelength and diameter, with several TE and TM modes, are depicted in the bottom panel of Fig. 6.1b. Correspondingly, photocurrent enhancement curve (short-circuit current density J_{sc}) shows resonant peaks as marked by the vertical arrows (see the top panel in Fig. 6.1b) [1]. The results clearly indicate that the geometry of the NWs strongly affect the optical and the optoelectric response.

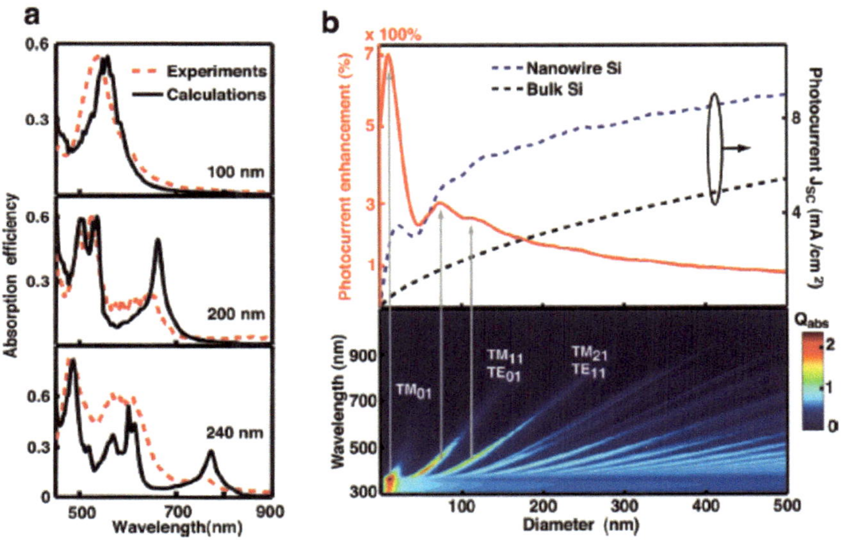

Figure 6.1 (a) Absorption efficiency Q_{abs} of Si nanowires with the diameters of 100, 200, and 240 nm, as a function of wavelength. (b) Calculated short-circuit current density J_{sc} for Si NWs (blue dash) and for Si bulk (dark dash). The photocurrent enhancement of the NWs per unit volume of Si material (red solid), calculated by $\left(\frac{J_{\text{sc,NW}}}{V_{\text{NW}}}-\frac{J_{\text{sc},B}}{V_B}\right)\Big/\frac{J_{\text{sc},B}}{V_B}$, and 2D plot of the calculated absorption efficiencies for the NW as functions of the wavelength and diameter with several TE and TM modes marked. The various leaky mode resonances are shown by the distinct streaks of high Q_{abs} and the correspondence between the leaky mode resonances and the photocurrent enhancement lines is indicated by the vertical arrows. Reproduced with permission from [1].

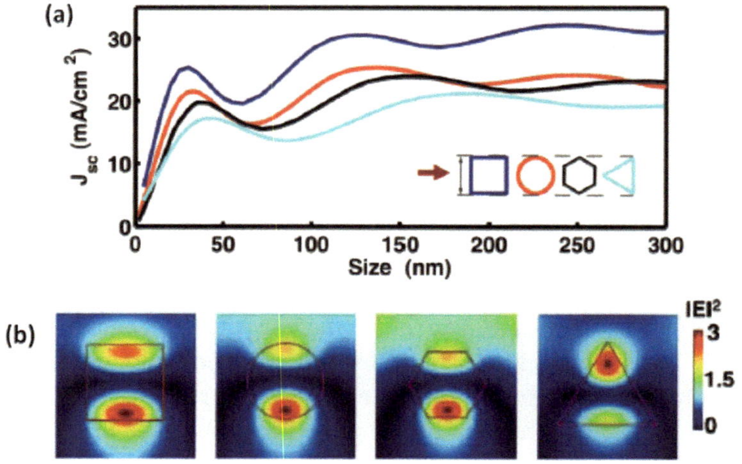

Figure 6.2 (a) Calculated short-circuit photocurrent density J_{sc} for Si NWs with circular, rectangular, hexagonal, and triangular cross section as a function of size (the inset shows the way that the sizes are measured and the illumination geometry that is indicated by the red arrow in the calculation; (b) calculated electric field distribution of these NWs upon the resonance at the absorption peak of 700 nm with the calculation, based on FDFD method, being normalized to the incidence light. Reproduced with permission from [1].

In fact, a slight tuning on the geometric parameters of the NWs often leads to dramatic alteration of their optical behaviors. Using the sensitive "quantum size" and "quantum shape" ("quantum geometry") effects, one now has the ability to significantly tailor light absorption and light trapping in NWs, vital for achieving high photovoltaic conversion efficiency. Conventional planar solar cells do not offer such flexibility. The high refractive index (with respect to the air) and subwavelength size allow for the NWs to collect light in their volumes where distinct optical resonant modes are formed. Light absorption in NWs depends on (i) optical antenna effect, (ii) absorption coefficient of the material, (iii) geometric sizes and shapes that determine cavity mode profiles, and (iv) wavelength matching between the cavity modes and spectrum of the solar radiation. All those can be controlled and tuned by geometric sizes and shapes which can be controlled by synthetic parameters of the NWs during preparation. Cao and co-workers also performed the simulations on J_{sc} as a function of cross-sectional sizes of the square, circle, hexagonal, and triangle NWs, together with the electric-field intensity distribution of the NWs upon the resonant at the absorption peak of 700 nm, as shown in Figs. 6.2a,b, respectively. The simulation was based on the finite difference frequency domain (FDFD) method [1].

6.1.2 Electrical Properties of Semiconductor Nanowires

In semiconductors, charged carriers experience various scatterings by other carriers, defects, impurities, surface adatoms, surface and interface states, surface and interface roughness, as well as acoustic and optical phonons. According to Sasaki, the carrier scattering effect could be significantly suppressed in NWs (1D semiconductors), compared to 2D and 3D cases [2], due to reduced *k*-space carrier accessible volume, as for electronic density of states in 1D. However, the surface situation, crystallinity, and even the nanowire shapes are unpredictable, so that, in a real situation, it is difficult to give a general idea about estimated values of material parameters such carrier mobility. Particularly when NWs have rather complicated structures like multiple heterojunctions or multi-component coaxial heterojunctions, one has to rely on experiments to extract trustable data, though this kind of experiments is often extremely difficult. It is encouraging that there are some reported works showing that carrier mobilities in nanomaterials exceed those in their bulk counterparts.

In fact, electronic properties of NWs can be traced from their corresponding bulk counterparts whose energy band structures and carrier mobilities are well known, on condition that the extrinsic effects, such as surface roughness and surface adatoms, are ignorable. Then, the electronic properties of a NW is intrinsically related to the material properties in its bulk form but the dimensionality effects, unique to the nanomaterial, can be used to tune the transport properties of nanowires. For solar cell fabrication, the problem lies in the interface between a NW and another material such as a metal. In most cases, transport through the interfaces appears to be more difficult.

The properties of Si NWs are mostly described by effective mass approach using $\boldsymbol{k}\cdot\boldsymbol{p}$ scheme by which many useful material parameters can be extracted. Other approaches include pseudopotential and tight-binding. Earlier, Zhao and co-workers performed a density-function simulation on electronic and optical properties [3]. Importantly, they showed that along the (110) direction, the bandgap of Si NWs is direct at the center of the Brilliouin zone at Γ point for the NWs with a diameter below 4.2 nm. This result is extremely important, because it means great promise for the application for Si NWs with a direct bandgap to be used for optoelectronic devices. For solar cell applications, the result means enhanced optical absorption for the coaxial NW heterojunctions.

In addition, the geometry of the NWs offers large room for the construction of many nanostructures and nanodevices. The growth of axial multiple heterojunction, for instance, enables the fabrication of a quantum dot or multiple quantum dots along the wire with controllable distance between quantum dots. Those architectures are greatly useful for quantum devices.

Assuming that the free direction of a nanowire is along z, the eigen energy E of an electron as a function of wavevector k_z takes the form of

$$E^e_{m,n}(k_z) = E_c + \frac{\hbar^2\pi^2}{2m_c^*}\left(\frac{m^2}{L_x^2} + \frac{n^2}{L_y^2}\right) + \frac{\hbar^2 k_z^2}{2m_c^*} = E_c + E_c^i + \frac{\hbar^2 k_z^2}{2m_c^*}, \tag{6.1}$$

where m_c^* is the effective mass of the electron, E_c the energy at the conduction band edge in 3D bulk, E_c^i the energy of the i-th electron subband, and $L_x(L_y)$ the size in the $x(y)$ direction. Electrons are confined in the x and y directions, due to limited sizes of L_x and L_y, so that m and n, are used to index quantum confinement along both directions. Correspondingly, the eigen energy E of a hole as a function of wavevector k_z is

$$E^h_{m,n}(k_z) = -\frac{\hbar^2\pi^2}{2m_v^*}\left(\frac{m^2}{L_x^2} + \frac{n^2}{L_y^2}\right) - \frac{\hbar^2 k_z^2}{2m_v^*} = E_v - E_v^i - \frac{\hbar^2 k_z^2}{2m_v^*}, \tag{6.2}$$

where m_v^* is the effective mass of the hole, E_v the energy at the valence band edge, E_v^i the energy of the i-th hole subband. Note that E_v and E_v^i are different for heavy and light holes due to different m_v^*.

It is crucial to control the electrical properties of NWs for achieving the full potential of the materials. The most essential parameters for characterizing electrical properties of semiconductor NWs include the carrier densities, carrier mobilities, lifetimes, and carrier recombination mechanisms, etc. One needs to know how to tailor the NWs in order to optimize the performance in engineering the NW-based devices. It is very difficult to perform accurate measurements on the electrical parameters of NWs in experiments.

6.1.3 Optical Properties of Semiconductor Nanowires

Conventional wafer-based solar cells or thin-film solar cells achieve good light trapping using anti-reflection techniques such as surface texturing and/or a quarter-wavelength dielectric thin-film coating. Those, of course, requires the use of additional materials and additional fabrication procedures which would likely lead to increasing cost. 3D arrays of nanostructures offers excellent properties of light trapping and carrier collection, critical for the enhancement of photoelectric conversion efficiency.

A polished planar Si structure shows significantly high reflection (typically ~ 35%) that is incident-direction (with respect to the surface normal) and light-polarization dependent (10% differences between *s*- and *p*-polarized lights). In contrast, both periodic and aperiodic Si nanowire array show optical reflectance insensitive to light incident and light polarization directions, beneficial for PV conversion efficiency. The nearly perfect suppression of optical reflectance is achieved on black silicon—the silicon nanopyramid array, when the array periodicity is much shorter than the wavelength of light, would have a continuum variation of dielectric constant from the surface at the pyramid tips to the bottom.

Nanowire arrays absorb light in a way that is significantly distinct from bulk material. They could have much better light absorption ability than bulk semiconductor material, with much less use of the material. The fundamental structural parameters for a semiconductor NW array include array periodicity a, wire diameter d, as well as wire length L, on which the light-trapping behavior depends. A NW array has excellent light-trapping properties in wide spectrum range if the parameters

a, d, and L are optimized. In the case of hexagonally arranged Si nanowire arrays, there appears strongly enhanced light absorption with d = 500 nm, a = 600 nm and d/a = 0.83 [4]. Li et al. also found that the light trapping does not enhance monotonically with the length L of the nanowires, since the surface recombination mechanism plays an important role.

Figure 6.3 presents three typical cases of incident light wavelength $\lambda >> a$, $\lambda \sim a$, and $\lambda << a$. At $\lambda >> a$, the light "feels" a thin film with an average dielectric constant of $n_0(1 - P) + n_sP$, where P is the ratio of the nanomaterial occupies, and n_0 and n_0 are the refractive indices of air and the semiconductor, respetively. When λ equals the nanostructure of the array periodicity, light experiences multiple scattering that is helpful for light trapping. In the case of $\lambda << a$, the light is reflected by uneven NW array and reflection is not strongly suppressed. It is extremely important that NWs are synthetically encoded with ideal electrical and optical properties for high efficiency solar power conversion.

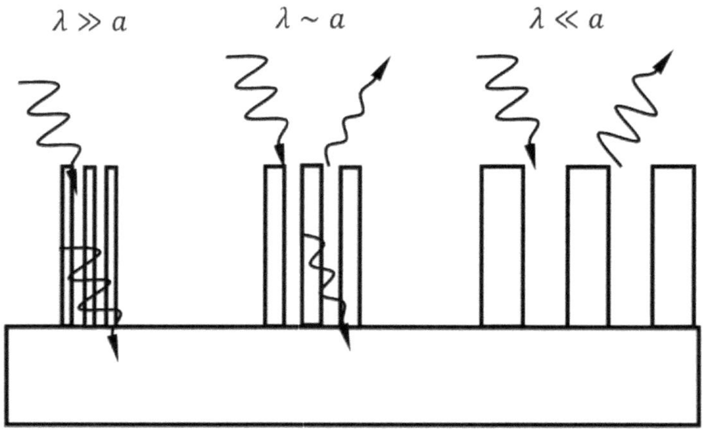

Figure 6.3 Schematic depiction of light incidence on semiconductor nanowire arrays in three typical cases: The light wavelength λ is much longer than the array periodicity a (left), where the incident light penetrate through the NW array and nearly no reflection on the NW/substrate interface; the light wavelength is comparable to the array periodicity (middle), where some light scattering occurs; λ is much shorter than a, where the incident light is strongly reflected since the "wires" behave like bulk materials (right).

In the case of NWSCs where NWs stand vertically on substrates, photons might be absorbed in NW surface with incident directions nearly perpendicular to surface normal and in any locations from the tip to the bottom of a NW. Intuitively, both the density and the geometry of NWs affect photon absorption. Garnett and Yang reported the controllable fabrication of Si NWs on Si substrates by deep reactive ion etching using a dip coated close-packed monolayer of silica beads as template [5]. Figure 6.4 presents their results, showing that the formation of NWs on Si substrate significantly suppresses light transmission in the spectrum range with strong solar radiation. Comparatively, the 5 μm NWs on Si substrate offer better light trapping than the 2 μm NWs.

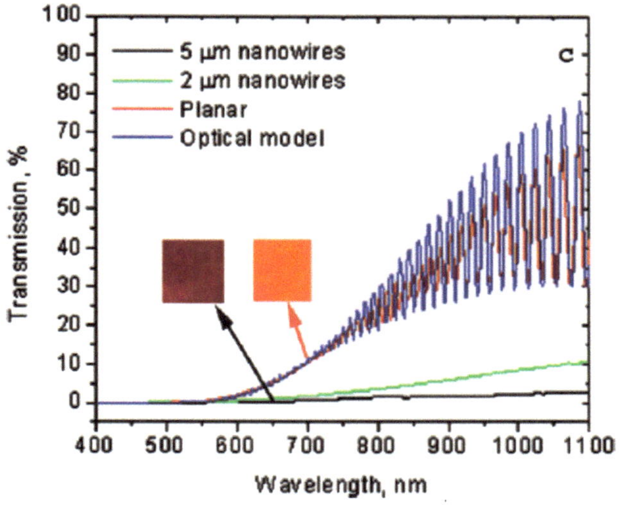

Figure 6.4 Optical transmission spectra of theoretical modeling (blue) and experimental measurement (orange) of a 7.5 μm (thickness) silicon substrate and experimental results on 2 μm (green) and 5 μm (black) nanowires etching formed on the 7.5 μm substrate. The insets show the backlit color images of the substrates prior to and after etching formation of 5 μm NWs. Reproduced with permission from [5].

6.2 Growth of 1D Nanomaterials

A large number of NW growth methods have been developed for various semiconductor materials. Technologies for controllable growth of NWs with difference doping, lengths, and morphologies have been reported. For instance, the growth of NWs with square or rectangular cross sections can be realized simply by using suitable catalyst patterns [6]. In addition, NW arrays can be prepared using position-dependent catalytic growth that adatom diffusion on NW at the catalytic tip, sidewall and substrate are made linear, exponential, and parabolic, respectively [7].

Si and GaAs nanowires have been prepared by many ways for solar cells [8–10]. As a dominant material for solar cell applications due to its low cost and earth abundance, Si can be easily doped and grown with many different forms of morphologies. Tian and co-workers fabricated single Si NW solar cells with the coaxial structure of *p*-type core/intrinsic shell/ *n*-type shell/ SiO_2 shell. The *p*-type core was prepared by vapor–liquid–solid (VLS) route. VLS is a cost effective approach suitable for the growth of many semiconductors, in which growth is driven by supersaturation due to the agents as sinks for atomic precipitation through interactions between metal catalyst and semiconductors. The intrinsic shell and the *n*-type shell were fabricated using chemical vapor deposition (CVD), and the subsequent SiO_2 shell was deposited by plasma-enhanced chemical vapor deposition (PECVD) [8]. The full Si NWSC yields a conversion efficiency of 3.4%. Although the conversion efficiency is not high, this Si NWSC resulted a high short-circuit current density J_{sc} due to improved light absorption of the nanocrystalline feature in the thin *n*-type shell and in the intrinsic layer. Single Si NW solar cells have also been prepared via VLS using rectifying contact consisting of metal Al on *n*-type Si [11].

The techniques widely used for III–V NW synthesis are the top-down and bottom-up methods including molecular beam epitaxy (MBE), metal–organic chemical vapor deposition (MOCVD), chemical beam epitaxy (CBE), CVD, VLS route, as well as photolithography and ion etching. MOVCD is sometimes called metal organic vapor phase epitaxy (MOVPE).

6.2.1 Vapor–Liquid–Solid Synthesis of III–V Nanowires

IV, III–V, and II–VI nanowires can be VLS-grown catalytically using a number of metals including Ag, Au, Al, Pt, Ni, and their alloys such as Au/Pd, the metal agents acting as sinks for atomic precipitation. The metal agents have been widely used in epitaxial growth such as MBE, MOCVD (or MOVPE), CBE, CVD, and laser ablation, especially for positioning growth. The Au-assisted MOCVD growth, for instance, is closely related to the VLS growth in growth mechanism. Figure 6.5 outlines the growth procedure that is similar for the growth of many groups IV and III–V, and other semiconductor compounds. As shown in Fig. 6.5a, the procedure starts with Au nanoparticle deposition on a semiconductor substrate, where lattice match is not required, followed by a thermal annealing to form Au-nanoparticle/substrate interface (Fig. 6.5b). Nucleation takes place at the Au-nanoparticle/substrate interface when vapor phase group-III and group-V precursors are introduced into the growth chamber (assuming the growth of III–V semiconductor nanowires) and the precursors decompose to provide with group-III and group-V elements (see Fig. 6.5c). As soon as the formation of III–V nucleation, the later-on growth occurs at the Au/III–V-compound interface that guides the preferential growth along the formed nanowire with prolong growth, as shown by Fig. 6.5d.

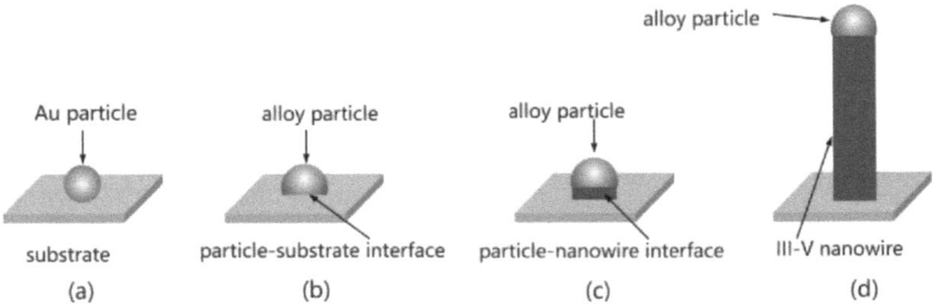

Figure 6.5 Schematic steps for the growth of semiconductor NWs: (a) an Au thin film was deposited on a substrate; (b) an Au nanoparticle with an Au/substrate alloy interface is formed after annealing the Au thin film; (c) growth of NW is initiated at the interface; (d) NW growth continuous at the interface of the Au-particle/NW.

For growth of III–V compounds, group-III element Ga and In, and group-V element Sb are used as vapor phase organometallic precursors, i.e., trimethylgallium ($Ga(CH_3)_3$), trimethylindium ($In(CH_3)_3$), and trimethylantimony ($Sb(CH_3)_3$), respectively, while group-V elements As and P are used as vapor phase hydrides, i.e., arsine (AsH_3) and phosphine (PH_3), respectively. The precursors are carried into the growth chamber by ultra-high purity hydrogen gas, where they release group III and group V elements through pyrolyzing.

The VLS synthesis offers preferential growth along a certain crystal axis with the use of metal catalysts (Au, for instance), which provides with option of choosing NWs with a proper crystal orientation. The technique also tolerates lattice mismatch with substrates. On the other hand, the use of metals as catalysts also brings with the problems such that the metals could introduce deep impurity levels in the bandgap, i.e., the metal catalysis could act as unwanted dopants in NWs. A catalyst-free VLS approach is, in this case, highly desired for the growth of NWs. So far, several methods without using conventional metal catalysts have been proposed, such as self-catalyzed VLS using group-III elements [12], group IV Si [13], oxide materials [14], as well as selective-area growth by metalorganic vapor phase epitaxy [15]. In VLS growth, NW orientation, morphology, and crystal quality are determined by growth parameters, including flux rate, element flux ratio, substrate temperature, substrate orientation, environment gas pressure, etc.

Upon photogeneration of an exciton, an electric field must be there to split the electron–hole pair as an essential step in photovoltaic process. The electric field is established by either doping to create a *p–n* junction or constructing a heterostructure to form a potential discontinuity. Controllable doping in semiconductor NWs is not trivial, the main challenges being to control the doping concentration and stability since, very often, the dopants tend to segregate on the surface of NWs. LaPierre's group demonstrated the preparation of GaAs core/shell NWs on GaAs substrates by switching dopants during growth [16]. Later, Morral and co-workers worked on doping in GaAs NWs during MBE growth of *p-i-n* GaAs core/shell NWs with Si-doped *p*-type core and *n*-type shell on (111) GaAs substrate [17].

Earlier, reported Si NW core–shell junctions were formed by preparing an oppositely doped Si layer on preformed Si NWs. For instance, solid-state phosphorus diffusion has been used to convert the shell layer of boron-doped *p*-type Si NWs to *n*-type, to form core–shell *p–n* junctions on vertically arrayed Si nanowire. Later, many doping techniques on nanomaterials are developed for optoelectronic devices.

A number of material systems have been studied for in situ doping, such as InP and GaAs [18, 19], although doping nanoscale materials is still challenging. At the same time, great progress has been made on constructing compound heterostructure, including GaAs/GaP and InP/InAs [15, 20].

6.2.2 Vapor–Liquid–Solid Synthesis of Si Nanowires

The most popular bottom-up growth method for Si NWs is the VLS process, but it involves the use of toxic gases and the growth requires high temperature and vacuum equipment. The growth technique was proposed long time ago by Wagner et al. [21, 22]. which utilizes silane (SiH_4), $SiCl_4$, or a Si-metal catalyst target for the silicon source. The VLS growth procedure is schematically shown in Fig. 6.6 where Au metal catalyst is used [23]. A metal layer is first coated on Si wafer by thermal evaporation and the following up thermal annealing results in the formation of individual metal catalyst droplets patterned on the Si wafer. The growth occurs at the metal catalyst sites where the gas phase precursor is absorbed. The absorption of Si atoms in the Au-Si alloy nanodroplets leads to supersaturated Au-Si alloy droplets upon continued exposure to the SiH_4 gas. The supersaturation in Au-Si alloy droplets is the driving force for the growth of Si nanowire, with the alloy droplets remaining at the tips of the nanowires.

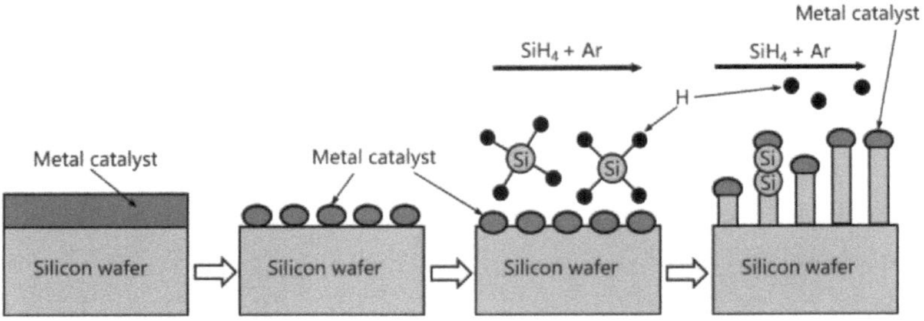

Figure 6.6 Schematic process of the vapor–liquid–solid NW growth.

"VLS" gets the name from vapor-SiH_4, liquid-Au-Si alloy droplets, and solid-nanowire-Si. An alternative approach is the solid–liquid–solid (SLS) growth, where no gas phase precursor is involved [23, 24]. In the SLS process shown in Fig. 6.7, the Si source is the Si substrate itself. Si nanowires are grown directly on metal catalyst droplet pattern that are formed by thermal deposition and annealing [23]. At selected growth temperature, Si atoms continuously migrate from the substrate into the Si-Au nanodroplets. The Si saturation in Au-Si alloy droplet results in precipitation of silicon atoms at the surface of the droplet. Si nanowires are then obtained by continuing the migration-precipitation process until reaching to the required nanowire length.

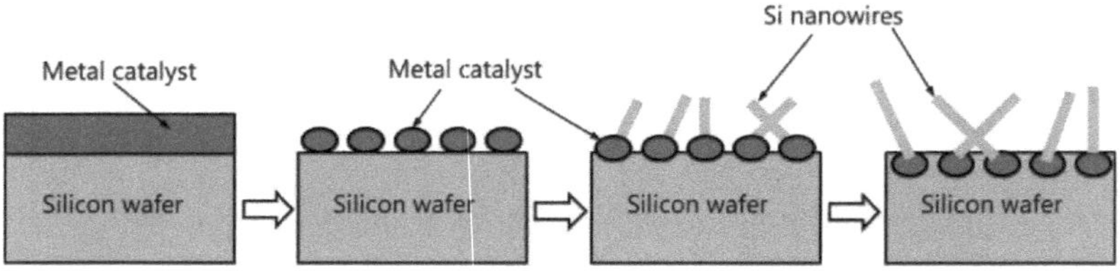

Figure 6.7 Schematic process of the solid-liquid-solid route for NW growth.

There are many reported methods for the preparation of NWs and NW arrays. Most of the methods are cost-effective and a large number of low-cost materials can be adopted to prepare, with top-down and/or bottom-up techniques, which is essential for the fabrication of solar cell devices [25]. NWs of many structures are now accessible via controlled synthesis and unique morphologies are obtainable through tailored synthesis. NWs with hexagonal, octagonal, and rectangular cross sections have been synthesized through variations of growth parameters including growth temperature and stoichiometry of starting materials.

Catalytic NW growth has also been carried out using more dedicate method such as MBE and MOCVD, with their growth mechanisms quite different from that of VLS. InAs and GaSb/InAs NWs, for instance, have been grown by MBE and MOCVD, respectively [26, 27]. The growth rate and the growth location of the NWs, which is often the main concern for the dedicated and expensive growth routes, was found to depend significantly on adatom diffusion on the tip of the catalytic metals. Although tremendous progresses have been made in NW material synthesis and device fabrication, many issues remain unsolved. The challenges include large-scale, low cost, high quality, uniformity, and reproducibility for material preparation.

6.2.3 Growth of Core/Shell NWs

The core/shell NW structure makes it easy to effectively separate electron holes and improve the efficiency of carrier collection. In addition, the coaxial geometry, together with the small junction area relaxes the lattice-matching requirement that is usually a severe condition for atomic smooth epitaxial growth. As a result, core/shell heterojunctions composed of materials of large lattice mismatch can be grown with high crystalline quality [28]. In terms of NW growth, the following points should be noted:

(1) Controllable doping and composition modulation in NWs is critical, yet challenging for PV applications.
(2) For GaAs and its alloys, the good part is that they have a large optical absorption coefficient, due to direct bandgap, and a nearly ideal bandgap for solar radiation harvesting. The bad part is their very high surface recombination velocities, indicating that good passivation is required.
(3) Controllable growth of NWs with radial and axial heterostructures. Specially, a growth technique needs to control the NW growth separately in axial or radial directions [29].
(4) Most NW heterostructures are based on III–V semiconductors. The optoelectronic properties of the III–V materials are highly sensitive to defects, dislocations, and impurities that could give rise to reduced carrier lifetime and mobility that are critical to device performance.

Due to the simplicity and potential large-scale production, VLS approaches have been intensively investigated for the growth mechanisms. It was found that surface diffusion rates of adatoms play a key effect in the determination of NW growth rate at catalytic tips and NW morphologies. One of the most important advantages is the controllable growth at specific locations to form arrays through preparing various patterns formed by catalyst [30]. In addition, NWs with various shapes of cross section areas can be grown easily by using controlled shapes of catalyst patterns [6].

6.2.4 Growth of NW with Axial Junction

NWs can be grown with stacked heterojunctions or graded composition along axial direction, with the structures including vertically stacked *p–n* junction, heterojunction, or varied bandgap nanowires. Those are extremely useful for solar cell construction based on NWs. Generally, preparation of high-quality junction with abrupt compositional variation relies on dedicate crystal growth techniques such as MOCVD and MBE. Many NW material systems with heterojunctions along the axial direction have been grown. Plissard and co-workers prepared GaAs/GaAsSb heterostructure NWs on Si substrate using MBE [31]. Figure 6.8 presents the TEM image of the NW with GaAs/GaAsSb heterojunctions. Other systems include GaAs/InAs and GaSb/GaAs [32, 33]. There are still challenges in improving

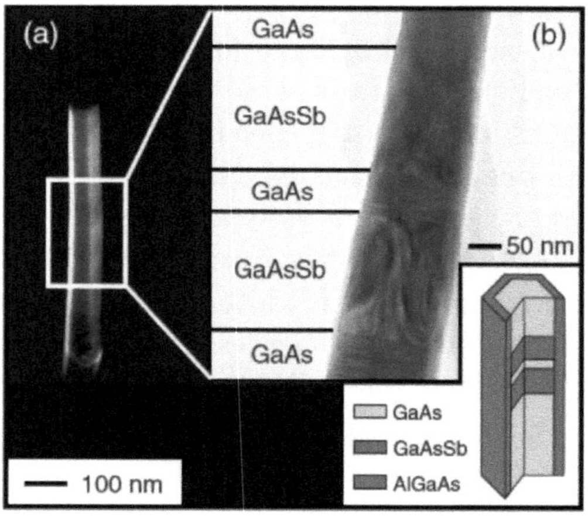

Figure 6.8 (a) SEM image of a catalyst-free GaAs/GaAsSb nanowire grown by gas source molecular beam epitaxy and (b) TEM image showing GaAs/GaAsSb heterojunctions. The nanowire is coated by an AlGaAs shell layer. Reproduced with permission from [31].

the crystalline qualities and heterojunction abruptness, and in reducing densities of various defects, dislocations, and micro-twins. For VLS growth, however, it is rather difficult to obtain high-quality junctions due to the difficulty in quickly switching the growth from one material to another.

For solar cell manufacturing, the most basic requirements are large area, low material cost, and cost-effective device-manufacturing technology. Specifically, the growth of large area III–V NWs on Si substrates is appealing. In the past decade, MOCVD as a technology for both material growth and device fabrication shows promising future to meet the basic requirements for III–V nano solar cells [13, 34, 35].

6.2.5 Catalyst-Free Growth

While catalyst-assisted growth offers semiconductor nanowires with high quality and controllable sizes and lengths, it uses noble or rare metals, making large-scale manufacturing unacceptable. In addition, the leftover metal elements could be a problem as defect centers that are not favored by solar cell devices. Au, for example, is believed to act as deep impurities in Si long time ago [36].

There are many reported instances of the growth of III–V nanowires using no metal particle catalysts. InAs nanowires have been grown by MOVPE, where a thin SiO_x layer was deposited on substrates prior to the nanowire growth [37]. By self-catalyzing of the In droplets, InAsP nanowires have been grown on silicon substrates. The nanowires exhibit a wurtzite crystalline structure with its *c* axis perpendicular to the nanowire axis [38]. Furthermore, the growth of GaAs/GaAsSb heterostructure nanowires has been reported. The synthesis of the nanowires is initiated with a high vertical yield of GaAs nanowires, followed by the growth of $GaAs_xSb_{1-x}$ segments along the nanowire axes as shells surrounding the GaAs core. The process can be repeated with varied growth conditions for desire structures. The synthesis was then finalized by passivating the nanowires using $Al_xGa_{1-x}As$ shells grown with the same growth technique.

6.2.6 Top-Down Etching

The most straightforward way of preparing 1D nanomaterial arrays is the top-down etching, as shown schematically in Fig. 6.9 for the process. Photolithography (or electron-beam lithography), after photoexposure (or electron-beam-exposure), gives rise to designed patterns of 1D nanomaterial arrays. When preparing a Si nanowire array, a well-polished Si wafer is first etched using HF to remove the natural surface oxide layer (SiO_2). A thin photoresist layer is then spin-coated on the Si wafer. The photoresist is under UV light exposure through a mask and a specific patter on the photoresist layer is defined by the mask that is patterned on the photoresist by lithography. A developing solution washes off the UV exposed part (for positive photoresist) or unexposed part (for negative photoresist).

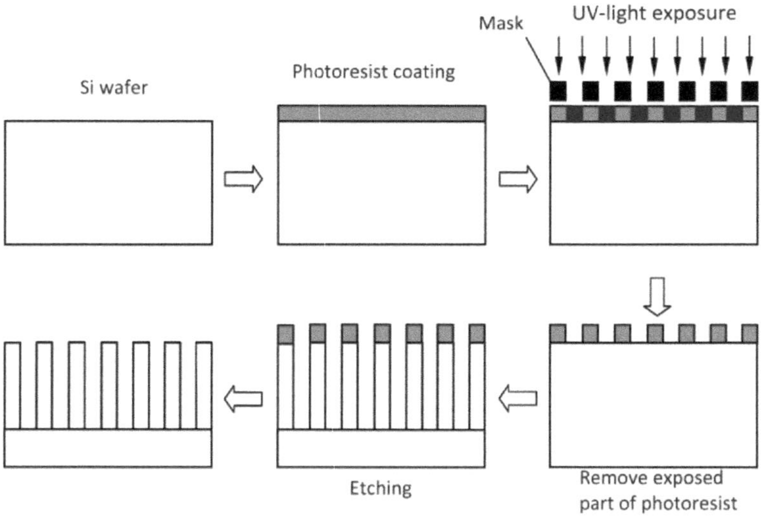

Figure 6.9 A top-down process for the fabrication of a NW array.

The unprotected (by photoresist) part of the substrate is etched away by either wet etching in which a specific liquid solution or etchant is used to remove materials in a wafer or dry etching in which plasmas or etchant gasses are used to remove the substrate material. After wet or dry etching, the remaining part on substrates form desired nanomaterial array on a substrate. The resulting nanomaterial array usually has a well-defined geometric pattern with distinct boundary between nanomaterials and environment. In Fig. 6.10 where cross-sectional SEM images of Si NWs with diameters of 1.8 μm and 480 nm are presented, the vertically aligned Si NWs are fabricated by wet chemical etching with various etching conditions [39].

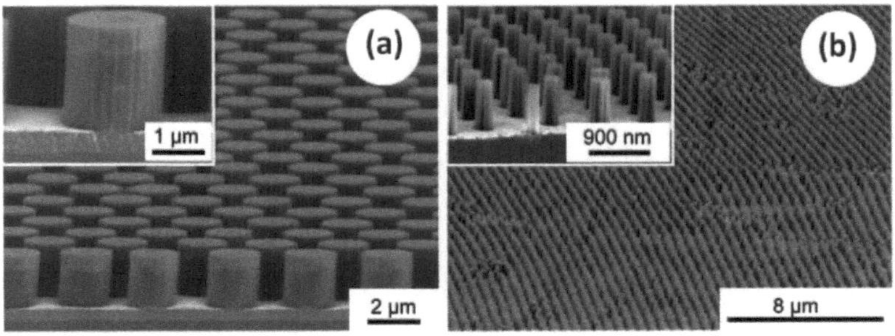

Figure 6.10 Cross-section SEM images of Si NW arrays fabricated by metal-assisted wet etching, with the NW diameters of 1.8 μm (a) and 480 nm (b). The NW diameters can be tuned by growth parameters. Reproduced with permission from [39].

Wet etching uses wet chemistries which, in the case of Si, are hydrofluoric-acid-based. Dry etching uses plasmas or etchant gasses to remove the substrate material through interaction between the plasmas or etchant gasses and the substrate atoms. When a material is etched by a liquid or vapor etchant, the etching could be isotropic (uniform in all directions) or anisotropic (non-uniform in all directions). Since the patter on photoresist is two-dimensional and the etching is three-dimensional (be either isotropic or anisotropic), a critical issue in top-down preparation of nanowires is the anisotropic etching rate. Wet etching can be highly anisotropic compared to dry etching and the etching rate depends on crystalline plane exposed to the etchants and on the solution used for etching. Potassium hydroxide (KOH), ethylenediamine pyrocatechol (EDP), and tetramethylammonium hydroxide (TMAH) are highly anisotropic etching solutions for Si. As a result, an array of NWs, particularly very long and thin NWs, is the quite difficult to fabricate using top-down method, since NWs have very high aspect ratio. Wire collapsing is thus very likely to happen if the etching conditions are not well-selected and controlled.

Metal-assisted etching in combination with some patterning was demonstrated as a promising way for fabrication of very long Si NWs and reported aspect ratio of Si nanowires is as high as 1700, as shown in Fig. 6.11 [40]. Dry etching, on the other hand, can be highly anisotropic, suitable for fabrication of Si NWs with high aspect ratio. Both methods have their own merits and demerits. Wet etching is faster and more sensitively dependent on temperature and the concentration of active species than dry etching. In addition, wet etching is more cost-effective for the perspective of mass production.

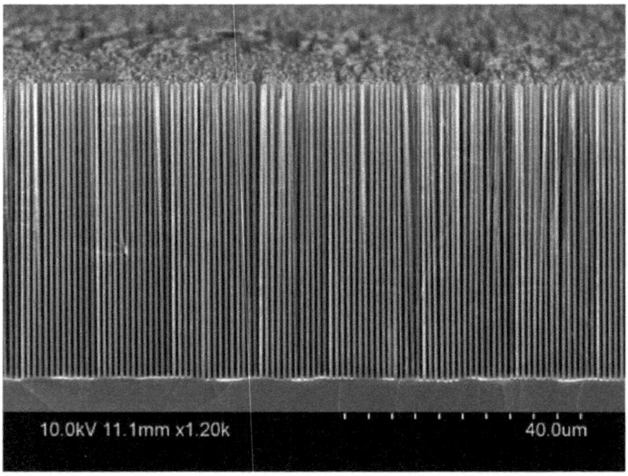

Figure 6.11 SEM image of Si nanowire array with ultra-high aspect ratio prepared by metal-assisted top-down process. Reproduced with permission from [40].

Garnett and Yang proposed a method to fabricate ordered Si NW arrays in which self-assembly of silica beads was used as a perfect template on a Si wafer and then deep reactive ion etching was performed to obtain ordered Si NW arrays. The major preparation steps are as shown in Fig. 6.12. The simple preparation technique requires only a silicon wafer, an aqueous suspension of given-sized silica beads, deep reactive ion etching (DRIE) facility. In the preparation for ordered Si NW arrays, the silicon wafer is dipped into the suspension to coat a close-packed monolayer of silica beads. The bead monolayer then serves as an etch mask for a follow-up dry-etching to form the NW arrays. The beads on top of the NWs can then be removed using HF. Boron diffusion was performed to form the radial *p*–*n* junction. Standard photolithography and metal sputtering were used for the top finger grid while metal evaporation provided the back contact (not shown). This is a preparation route that can produce NW array with a very large area [5].

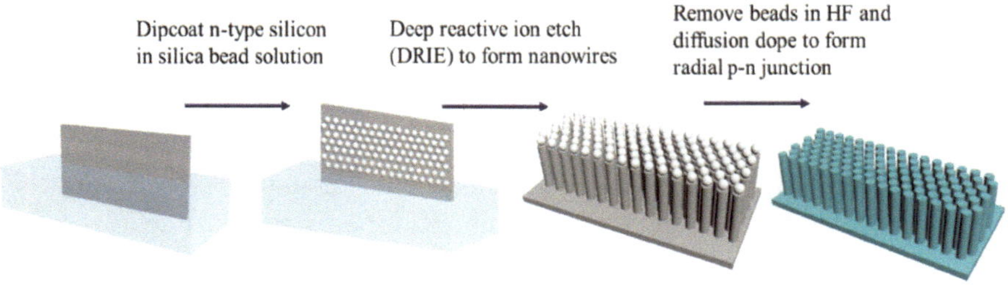

Figure 6.12 The fabrication process of Si NW array: dip coating an *n*-type Si wafer into silica bead suspension, followed by reactive ion etching, bead removal and boron diffusion and, finally, photolithography and metal sputtering to obtain NWSCs. Reproduced with permission from [5].

6.3 Device Architectures of Nanowire Solar Cells

A photovoltaic process includes photon absorption, carrier generation, electron–hole separation, charge collection. The device architectures of NW solar cells aim to complete those steps efficiently utilizing the fundamental properties of NWs.

6.3.1 Charge Generation and Separation in NWs

NWs ensure the high efficient light absorption by optimizing the geometrical sizes (diameters and lengths) of wires. Similar to other solar cells, electrons and holes are generated in NWs via photon absorption, and the electrons and holes need to be spatially separated and collected by respective electrodes to complete a photovoltaic process. The separation of electrons and holes relies on a built-in potential that can be formed by *p*–*n* junction in NWs based on *p*-type and *n*-type doping. The built-in potential is either axial along the nanowires with the *p–n* junction perpendicular to the wires or radial with the *p–n* junction parallel to the wires. However, controllable doping in NWs is extremely difficult mainly due to two challenges. One is to achieve desired doping concentration and the other is to control the sites of impurities that often tend to occupy the surface sites.

A potential discontinuity can be constructed using hetero-junction with which two different semiconductors form a staged type-II band alignment in the conduction band (CB) and the valence band (VB), as shown in Fig. 5.13a. A typical example is the type-II core–shell NWs constructed using III–V or II–VI semiconductors. The band alignments offer potentials featuring efficient charge separation which has been widely used in many applications such as long wavelength infrared detectors, dye-sensitized solar cells, photoelectrochemical water splitting for hydrogen generation, etc. Ideal heterojunction interface is discontinuous but a self-consistent process based on electrons redistribution tends to smear out the discontinuity (see the dashed lines in Fig. 6.13a). Resembling a *p–n* junction as shown in Fig. 6.13b, the smeared band discontinuity corresponds to a space charge region where the photogenerated electrons and holes are separated and driven to corresponding electrodes. However, the space potential is located in a very narrow region in the NWs, so that most photogenerated electrons and holes are photogenerated out of the potential region. This is unfavorable for high power conversion efficiency (PCE). The obstacle was overcome by Pan and co-workers, who demonstrated that NWs with a spatial composition-graded alloy could result in improved conversion efficiency [41].

The heterojunction potential plays the key role for the enhancement of charge separation and, as a result, the PCE. Except for using suitable materials or doping, the potential can also been established by strain in NWs. For Si NWs, for instance, a non-uniform distribution of strain along the NWs could lead to a potential-like space charge potential due to *p–n* junction. A simulation shows that, in Si NWs with 2.2 nm diameter, a 2% compressive (tensile) strain shifts both the conduction band minimum and the valence band maximum upward (downward) by 0.2 eV [42]. Thus, if a compressive strain is applied on one half and a tensile strain on another half of a Si NW, an electronic potential of 0.4 eV could be established, which could act to separate photogenerated electrons and holes for photovoltaic effect. In this case, doping and heterojunction are not required.

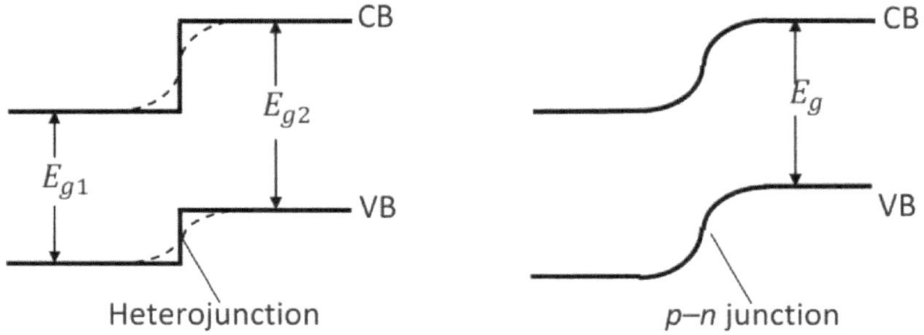

Figure 6.13 (a) Schematic heterojunction with Type-II band alignment for solar energy conversion, where the dashed lines show smeared interface due to self-consistent electron redistribution; (b) a *p–n* junction (homojunction) for comparison. E_{g1}, E_{g2}, and E_g are the bandgaps of the semiconductors.

Other ways of constructing a potential barrier include the use of narrow tapered NWs, surface morphology distribution, as well as different material coating along the NWs. All those create a different quantum confinement along the NWs to cause a proper variation of the conduction band minimum and the valence band maximum along the NWs.

Upon absorb a photon with energy equal to or larger than the bandgap, an electron will be excited into the conduction band, left with a positively charged hole in the valence band. As in a conventional solar cell, the way that a NWSC works requires a following-up separation of the photon-generated electrons and holes, prior to their recombination. It is thus essential for the device to keep a short charge separation feature time to ensure low recombination probability. Although NWSCs have good light absorption or light-trapping properties, complicated interface configuration, high density of various defects and dislocations, as well as the use of multiple materials complicate the charge generation. An important point that needs to be emphasized is the location of charge generation in nanowire solar cells—the photon-generated charges should be in the *p–n* (or *p-i-n*) junction region or heterojunction region for efficient electron–hole separation. Charges generated in other locations have low efficiency to be collected by electrodes. In order for efficient photovoltaic processes, quite a few of proposed active layer materials and junction configurations were proposed. Axial *p–n* junction (or heterojunction), radial *p–n* junction characterized by core/shell nanowires, and hybrid (inorganic and organic) structure are those among the proposed configurations. After excitation, photon-generated electrons and holes have a strong tendency to recombine prior to separation and transportation to contacts. It is of utmost importance to prevent the electron/hole recombination.

The charge separation process is closely related to the configuration of nanowire array SCs. As described in Chapter 1 and 2, photo-excited electron/hole pairs are separated either by an electrostatic force due to a built-in electric field across interface between the *p*-doped and the *n*-doped layer or a chemical potential discontinuity due to a type-II potential alignment. The effective separation of the photon-generated electron and hole requires good quality of the *p–n* or *p-i-n* junction that favors large Fermi level offset and abrupt *p–n* junction, which depends on the doping and the bandgap of the nanomaterials. The advantage of the nanowire array SCs is the much larger overall junction area than conventional solar cells, which is greatly helpful for the photon collection and electron/hole pair generation. The negative part is, however, the high density of surface and interface defects that might trap the photon-generated electrons and holes. This could be a very serious problem since in NWs photon-generated electrons and holes are far from their respective contacts. The electrons and holes are likely to recombine or are trapped during transportation.

The driving force for the charge separation can either be electrostatic due to the formation of *p–n* junction or non-electrostatic force due to heterostructures. Furthermore, the *p–n* junction and the heterojunction can be either axial or radial, depending on device configurations. During transporting to contacts to complete the photovoltaic process, photon-generated electrons and holes must pass through the junctions and through the regions with a large number of defects, dislocations, charged impurities, etc., where they might be captured. In NWSCs, photon-generated electrons and/or holes are provided with transport channels—the nanowires in which electron/hole transport is confined and could be made much easier if the nanowires are of good crystalline quality. As schematically shown in Fig. 6.14, photoexcited electrons and holes are driven and separated by built-in potential and they are directed by the nano-sized wires to transport in opposite directions along the wires. For Si NWs of high crystalline quality, electrons and holes could receive very few scattering events.

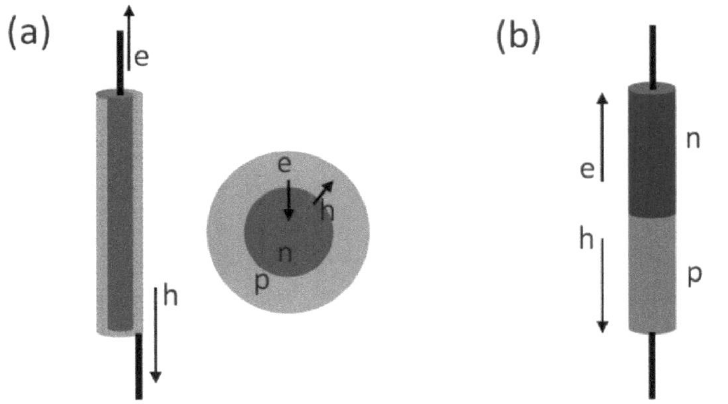

Figure 6.14 Schematic of a NW with radial *p–n* junction (a) and an axial *p–n* junction (b). The photon-generated electrons and holes are collected by the contact electrodes after going cross the junctions. The nanowires provide the channels for electrons and holes transportation.

As pointed out in Chapter 2, a heterojunction formed by two semiconductors has potential discontinuities in both the conduction and the valence bands. The potential discontinuities correspond to effective driving force to drive the transportation of the electrons and holes. In fact, not only the band discontinuity but also other discontinuities of material parameters could act as the quasielectric fields. A typical example is the effective mass discontinuities of electrons and holes at the hetero-interface. For core/shell NWs, the heterojunction with staggered potential discontinuity (type-II energy band alignment) is favored for photovoltaic effect since electrons and holes are separately confined, as shown in Fig. 6.15. Under well-controlled growth, the type-II junction of the core/shell NWs can be engineered with high quality and uniformity. This kind of junctions avoids the doping that is proven to be challenging for nanomaterials.

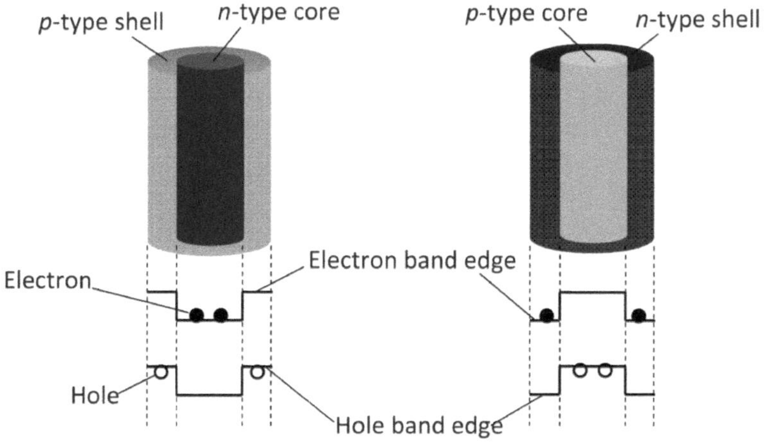

Figure 6.15 Core–shell nanowires with type-II band alignment in the conduction and the valence bands. Left: Electrons are confined in the core and holes in the shell. Right: Holes are in the core and electrons in the shell.

A quasielectric field can also be constructed using compositional semiconductor materials with graded bandgap [43, 44]. An advantage of the graded gap solar cells is the broadband absorption for photoelectric conversion. With the wide-bandgap side at the device surface, the device configuration is greatly effective for the suppression of surface recombination of photon-excited carriers. Wu and co-workers have proved by calculation that, based on the topological structure of the near-bandgap wave function, the axial carrier separation can be enhanced by the strain in the material [42]. Along NWs, quantum confinement strength and the surface morphology can be tuned through radial size. As a result, the eigen energy levels vary along the axial direction of the wires. Ideally, this could construct a type-II junction as the driving potential for electron–hole separation.

Another device configuration that deserves to be specially noted is the junction formed between a semiconductor and an electrolyte, similar to a core/shell junction. The liquid–solid junction bears the advantages of being conformal and controllable. As expected, one sees an enhanced photocurrent density in Si NW/electrolyte solar cells, in comparison to conventional planar devices. This device configuration is also known as a photoelectrochemical cell (PEC) that shows a potential for high-efficiency energy conversion.

6.3.2 Charge Collection

As a final step of the photovoltaic effect, photon-generated electrons and holes need to be collected by contact electrodes that connect to external circuit. Contacts are made of metals or highly conductive metal oxides. Usually, the interfaces between contacts and the semiconductors (or electrolyte) are not in good quality, so that the surface and interface recombination is very detrimental for obtaining high conversion efficiency. Charge collection in NWSCs is expected to be efficient due to the special geometrical configuration of the cells. In fact, the charge collection in NWSCs can be tuned by the device geometry. The NW configuration is, however, not superior in making high-quality electronic contacts, especially the top contact at the nanowire tips. For instance, the conventional comb-like electrodes for conventional Si wafer-based SCs are not suitable for NWSCs due to the very high surface/interface carrier recombination rates in later where the carrier collection distance is comparable to or even smaller than the diffusion length of the minority carriers.

Transparent conductive oxide (TCO) may be used as contact electrodes to cover the whole area of light incidence. However, the advantage of using uniform contact is counter balanced by the limited conductivity and the poor contact between the TCO and the electron/hole transporting layer, making it insufficient to solve the problem. Conductive polymers have been used to fill all space in the NW (nanopillar) layer. Though having a very large electrode contact area, the NWSCs still have low conversion efficiency due to the low conductivity and high carrier recombination rate at the NW/polymer interface [45].

During charge collection (or carrier transportation to the electrodes), photogenerated carriers need to pass across many surfaces and interfaces or pass by defects and impurities, where they could be plagued by recombination. NWs have a large surface-to-volume ratio. As a result, the surface recombination often remains a major challenge. For both the radial and axial junction NWSCs, photogenerated electrons and holes are separated by the junction potential into *n* and *p* regions, respectively. Before reaching at the electrodes, the electrons and holes have to travel long distance during which they have good chance to be trapped and recombined. In fact, carrier recombination remains to be one of the major mechanisms for low conversion efficiency of NWSCs.

Charge collection ability remains a major obstacle that limits PCE of NWSCs, with the use of traditional electrode configuration such as TCO thin films and interdigitated metal grid. Z. L. Wang's group has developed a device architecture in which Si NWs are completely covered by ITO top transparent electrode, as shown in Fig. 6.16, where Si NWs were prepared by templating-assistant chemical etching on *n*-type Si wafers [46]. In such a device configuration, Si NWs are in fact imbedded in ITO except for their bottom part that stands on Si substrate. Electric contact is made by pasting silver paste on ITO directly. Enhanced photoresponse has been clearly demonstrated due to high efficiency photon trapping and good electric contact between NWs and ITO.

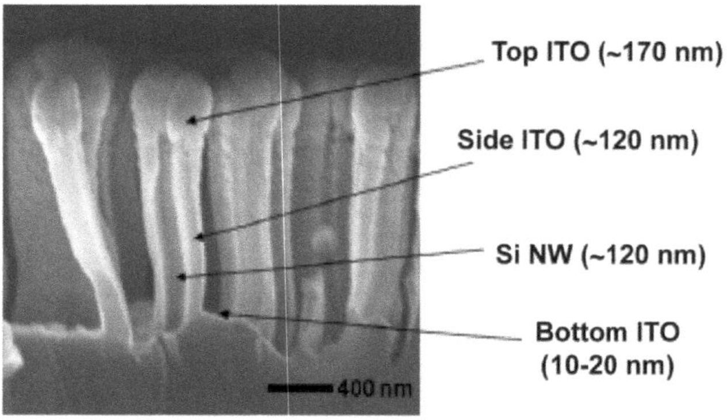

Figure 6.16 Si NWs that are buried in ITO. Si NWSCs can be fabricated by *i*-th type-II band alignment in the conduction and the valence bands. Reproduced with permission from [46].

6.3.3 Device Architecture of Nanowire Solar Cells

Nanowires offer specialties associated with photon trapping, exciton generation, and charge separation and transportation. NWs provide electrons (holes) with paths due to their 1D geometry and limit the electron–hole recombination due to the confinement effect in the wires. In addition, the very small contact areas between NWs and substrates tolerate large lattice mismatch. In the device architecture of NWSCs, the active regions are in the wires, allowing for free selection of low-cost substrates. The device structures of NWSCs differ in the geometries of *p–n* junctions, the ways that the photogenerated carriers are collected, and the material systems.

Nanowire *p–n* junctions

A *p–n* junction can be axial or radial. Figure 6.17 presents the schematic configurations of NWSCs based on axial (Fig. 6.17a) and radial (Fig. 6.17b) *p–n* junctions. Multiple junctions are possible with fine-tuning preparation technologies. In constructing NWSCs, a *p–n* junction can be prepared by opposite doping during nanowire growth, ion implantation, or using rectifying metal contacts [11, 47, 48]. In an axial *p–n* junction, the electrostatic force applied on carriers is along the nanowires, so does the travel of photogenerated carriers. Since the junction is perpendicular to the direction along the wires, the junction areas are thus quite small, which is one of the disadvantages for the axial *p–n* junction. The carriers transport along the wires so that the length of the nanowires cannot be too long, in order to reduce the carrier recombination during transportation. Thus, there is a trade-off between photon trapping, which might need long NWs, and carrier collection, which favors short NWs. In addition, most photons are absorbed in regions far away from the junction area. As a result, photocarriers generated in those regions are hard to be collected.

A radial *p–n* junction can be prepared with core/shell structure of nanowires, where the light absorption pathway and the carrier collection pathways are perpendicular to each other. As shown in Fig. 6.17b, the radial junction interface extends all the way along the nanowires, the geometry offering large junction areas and the carrier separation is highly efficient. Another advantage of the radial junction SCs is that, with light pathway almost along the nanowires, light is absorbed very near the surfaces of the nanowires. A shallow radial *p–n* junction leads to dominant light absorption nearby the *p–n* junction. The carrier collection distance could be smaller or at least comparative to diffusion length of minority carriers, which is greatly helpful for the solar cells to achieve high efficiency. Theoretical study shows that, although NWSCs with axial junctions are less tolerant to the surface states, they often have higher V_{oc} and large room in designing junction configurations [49].

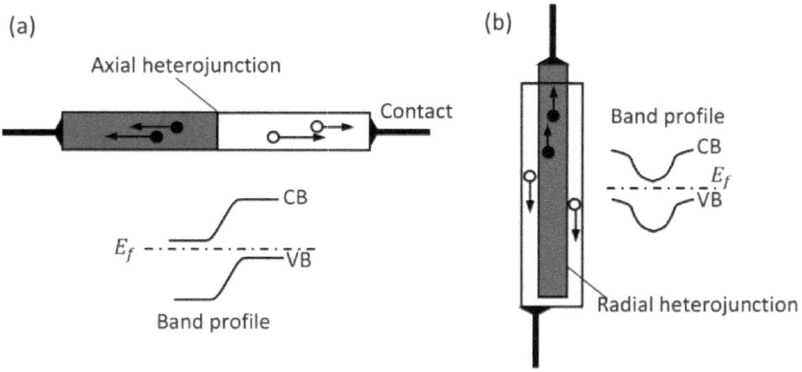

Figure 6.17 Schematic *p–n* junctions for axial (a) and radial (b) nanowires. The arrows show the directions of motion of electrons and holes. The driving force of the electron/hole separation and transport is due to the built-in potential and diffusion.

Nanowire heterojunctions

Doping in nanomaterials is still a big challenge. Without doping to form *p–n* junction, a heterojunction may be constructed by growing two different semiconductors with atomic smooth interfaces [50]. The staggered band alignment in the hetero-interface displays a non-electrostatic force for carrier separation, as shown in Fig. 6.18. The energy band profile of a staggered interface displays a non-electrostatic force that could separate the electron and the hole that make up the exciton. The effectiveness of the force is due both to the band discontinuities and diffusion.

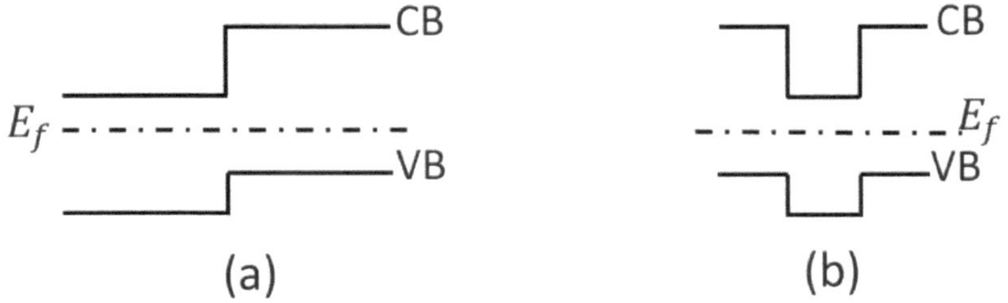

Figure 6.18 Type-II band alignments for axial (a) and radial (b) heterojunctions. The driving force of the electron/hole separation and transport is due to the band discontinuity and diffusion.

Device architecture

Typical NWSCs have device architectures as shown in Fig. 6.19 where NWs in vertical array geometry. Vertical-standing NWs offer high conversion efficiency through enhancement in light absorption, charge separation, and carrier collection. The enhanced light trapping is due to several effects, mainly including reduced light reflectance and waveguiding effect. Sharp needle-like NWs offer even better light-trapping effect than NWs with uniform diameters. Very importantly, the enhanced light trapping by covering NWs on solar cell surface has very weak angular dependence, making it practical for applications. Thus, NWSCs of the vertical array and varied diameter along the NWs should exhibit very high PCE. Due to the geometry and interfaces, carrier diffusion length is usually shorter in NWs than in bulk. Thus, the construction of strong built-in electric field in NWSCs is essential to the effective collection of carriers.

As discussed above, a junction can be constructed by performing *p*- and *n*-type doping or preparing a heterojunction. Radial junction interface extends along the length of a NW, featuring large junction area. Axial junction interface located at the cross section of a NW, with an area limited by the diameter of the NW. Clearly, a distinct advantage of radial junctions is their large junction area. Additional advantage is the orthogonalized pathways for light absorption and carrier collection, where a photon could pass through a distance in NWs long enough for effective absorption while a distance for carrier collection is comparable to the minority carrier diffusion length. Photogenerated carriers could reach to the *p–n* regions or electrodes with relatively high efficiency featured by low recombination rate due to short traveling distance. In conventional wafer-based solar cells, however, long traveling distance is the main factor that limits the conversion efficiency of the devices with planar junction.

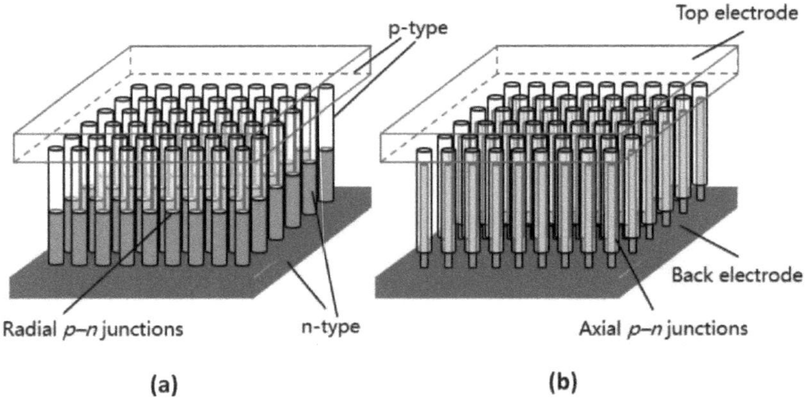

Figure 6.19 Schematic of NWSCs with axial *p–n* junctions (a) and radial *p–n* junctions (b).

The many varieties of 1D nanomaterials offer great deal of choices for the architectures of NSCs. In core/shell NWSCs with radial *p–n* junctions (or heterojunctions), the carrier collection distance can be made comparable to or smaller than the diffusion length of minority carriers, allowing photogenerated carriers to reach the *p–n* junction regions quickly and efficiently. According to theoretical modeling, core/shell NW structures offer superior light absorption and carrier collection abilities in comparison to planar PV devices, including those based on axial junction NWs, due to improved efficiencies for both carrier generation and separation [5, 51].

As depicted in Fig. 6.20, the performance of the planar junction NWSCs is determined by the geometry of the nanowires, carrier densities, energy band profiles, Fermi energy position, built-in potential height, diffusion lengths L_n (L_p) for photogenerated electrons (holes), width x_n (x_p) of depletion layer in *n* (*p*) region, etc. All those parameters need to be optimized for high device performance. The junction can be formed by introducing a heterojunction to form a potential to split the photogenerated an electron/hole pair axially [52, 53], or radially (core/shell) [50, 54]. The junction (for both axial and radial) can also be created by controlled doping along axial and radial directions [17, 47, 55]. The axial junction can also be formed by preparing metal contacts on unintentionally doped NWs with a rectifying property [11].

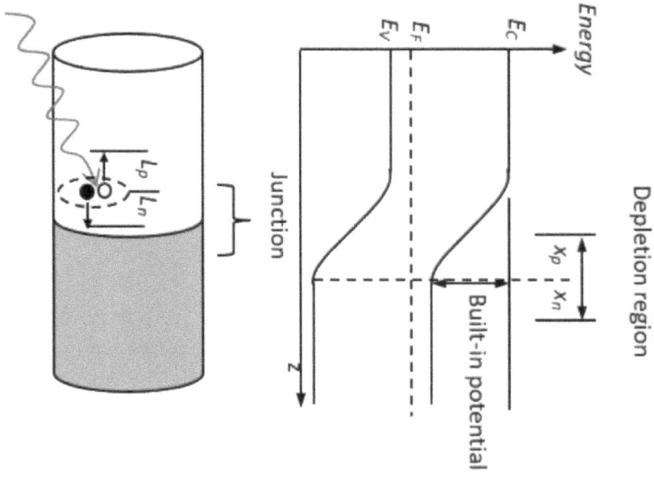

Figure 6.20 A nanowire with axial p–n junction (left) and potential profile of the junction (right), where L_p and L_n are diffusion length of photogenerated holes and electrons and x_p and x_n the widths of depletion layers in p-type and n-type regions, respectively.

The geometry of the vertical NWs enhances light trapping, makes it easy to optimize charge separation and collection. Many problems remain, including the large surface area, which introduces surface states that might trap carriers. Huang and Povinelli introduced passivation layers on GaAs NWSCs with axial junction and achieved conversion efficiency as high as 21% [56].

A radial junction is usually fabricated by growing core/shell structure and many core/shell structures of different semiconductor combinations have been grown. Figure 6.21 presents a core/shell NW with radial p–n junction where electrons and holes are photogenerated very nearby the junction. L_p and L_n (the diffusion lengths for holes and electrons, respectively) are in the same order of magnitude of the diameter of the core and the thickness of the shell. The built-in potential in the *p*–*n* junction region could separate the electrons and holes efficiently.

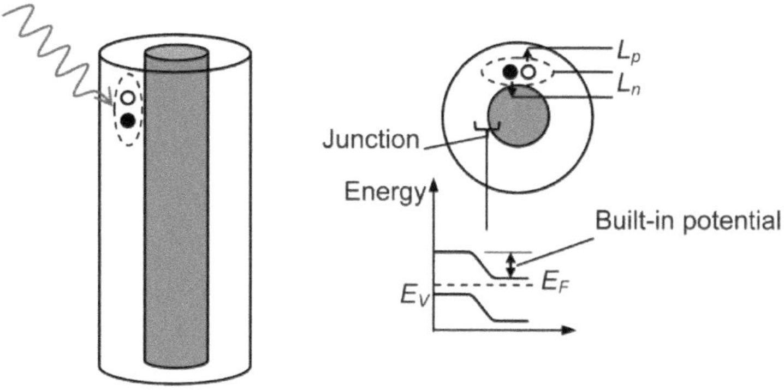

Figure 6.21 A nanowire with radial *p*–*n* junction (left), as well as potential profile of the junction (right-bottom) and cross section of the wire (right-top), where L_p and L_n are diffusion length of photogenerated holes and electrons, respectively.

Currently, NWSCs with radial junction still suffer from low conversion efficiency, though one expects efficiency carrier generation and separation processes in the devices. Improved design and optimization is required for better device performance of the radial junction NWSCs. A few theoretical studies indicated clearly that NWSCs with radial junction enhance both light absorption and carrier collection as compared to planar PV devices [5, 51, 57]. In Hu and Chen's work, for instance, simulation shows that, with moderate filling ratio, NW arrays could present much lower optical reflectance in comparison to thin films [57].

Another architecture of junction is the substrate junction located at the bottom of NWs, i.e., a junction between the nanowire and the substrate. The junction can be constructed simply by growing *n*-type NWs on *p*-type substrate as shown in Fig. 6.22, or vice versa. A well-defined nanoscale interface is vital for charge separation in NW solar cells. In NWSCs with axial junctions, the carriers travel along the direction of NWs before they are collected by the electrodes and the transport distance is dependent on the length of NWs. Thus, while relatively long NWs are beneficial for light trapping, they are unfavorable for carrier collection.

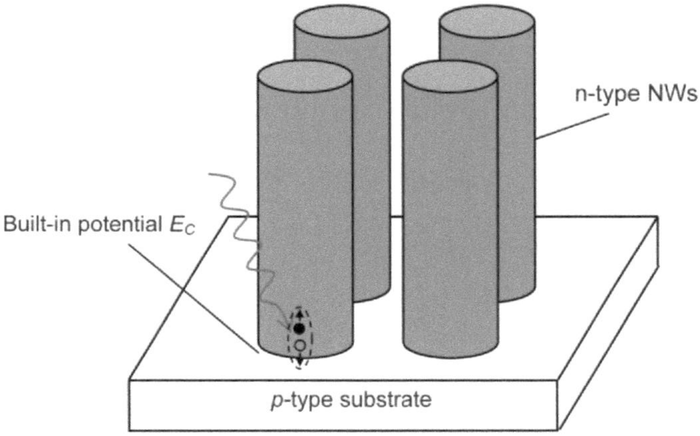

Figure 6.22 Schematic of *n*-type NWs grown on a *p*-type substrate. A *p–n* junction is formed at the bottom of the NW with the substrate. A heterojunction can also be formed where the NW and the substrate are different materials.

Kayes and co-workers studied and discussed NWSCs based on doped Si and GaAs using theoretical simulation [58]. They emphasized how device performance relies on material, structural, and device geometric parameters, including photocarrier concentration, carrier diffusion length, depletion width, doping level, the minority carrier lifetime, and layer thickness. Optimizing those parameters could lead to high device power conversion efficiency characterized by high short-circuit current, open-circuit voltage, and fill factor. Their work served as a guidance for later device fabrication [8, 16].

The concept of full solar spectrum conversion is essential for achieving high device efficiency, which is implemented by tandem stacking of sub solar cells of semiconductors of different bandgaps, realized by the growth of multiple materials of different bandgaps. The tandem configuration can also be adopted for NWSCs with axial and radial junctions, in which multiple *p–n* junctions or hetero-junctions give rise to maximized solar light absorption, beneficial for both the short-circuit current and open-circuit voltage. Based on the same consideration, varied material composition along the NW axis could offer light absorption in a wide solar spectrum range. In addition, multiple quantum wells or multiple quantum dots distributed along NWs could also enhance light absorption in wide spectrum range.

6.4 Some Nanowire Solar Cells

6.4.1 Group IV Nanowire Solar Cells

Si nanowire/nanopillar arrays

A number of studies have been reported that focus on Si NW solar cells. The SCs have high efficiency in optical management, but this is counter-balanced by their low electric efficiency, leading to overall low conversion efficiency. Currently, the highest achieved efficiency is about 10% so far and the NW solar cell consists of *n*-type Si NW arrays on a *p*-type Si substrates. The efficiency is still lower than that of the best thin-film Si solar cells and that of wafer-based solar cells [9]. Sun and co-workers listed some Si NW solar cells of various structures and compared the device parameters [59]. Several nano forms of the Si material have been adopted to fabricate nano solar cells, with the device configuration being designed to match to the optoelectronic properties unique to the nano forms of the material.

NWSCs with radial junction offer higher efficiency for carrier collection, as reported by Lieber's and Yang's groups, for instance [8, 60]. Figure 6.23 presents the array of Si NWs with the configuration of *n*-type core and *p*-type shell. The core/shell NWs are vertically aligned with an average length of 18 μm and high packing density. Strong light-trapping effect has been observed on arrays of 5 and 2 μm NWs with radial *p*–*n* junction, in comparison to a planar Si window of 7.5 μm in thickness. The transmission over the spectral range of 600 to 1100 nm is reduced by factors of 2.9–7.8 and 12–29 for the 2 and 5 μm NW arrays, respectively.

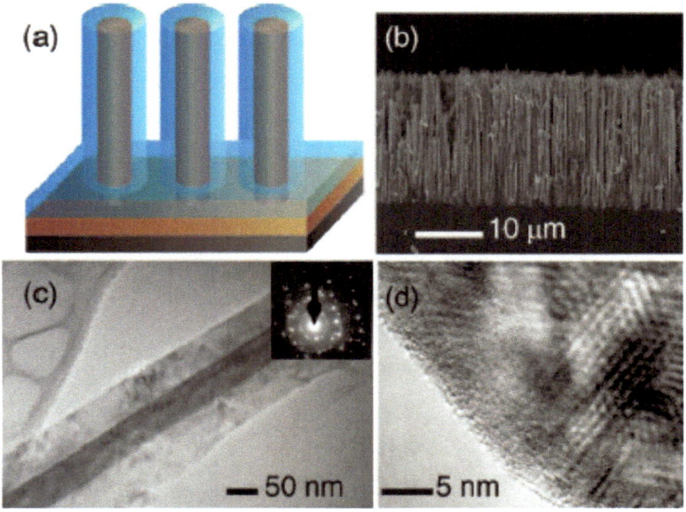

Figure 6.23 (a) Schematic structure of a Si NW solar cell with the single-crystalline *n*-type Si NW core (in brown), the polycrystalline *p*-type Si shell (in blue), and the back contact electrode (in black); (b) a cross-sectional SEM image of a Si NW solar cell showing dense and vertical alignment Si NWs; (c) a TEM image of the single-crystalline *n*-type Si NW core and polycrystalline *p*-type Si shell, and the selected area electron diffraction pattern is presented in the inset; (d) a TEM image showing nanocrystalline domains nearby the core–shell region of the nanowire. Reproduced with permission from [60].

In another work, Yang's group fabricated core/shell Si NWSCs and the 5 μm NWs were synthesized by using silica bead mask (formed by dip coating) and deep reactive ion etching on 8 or 20 μm thin *n*-type Si absorbing layers [5]. The radial *p*–*n* junction was formed by boron diffusion. The schematic of the device is shown in Fig. 6.24. The top contact electrode was the aluminum/palladium alloy filling completely between the nanowires, which was fabricated by sputtering (not shown in the figure). Measured at AM 1.5G condition, the solar cells give PCEs of 4.83% and 5.30% for the 8 μm absorber device and the 20 μm absorber one, respectively.

Compared to the Si ribbon solar cells, the 8 μm absorber Si NWSC has an efficiency of 4.85%, which is 20% higher than that of ribbon Si solar cells with a ribbon thickness of 8 μm. However the 20 μm absorber Si NW cells generate an efficiency of 5.30%, which is 35% lower than the ribbon solar cells with the same thickness [5]. Given the fact that the overall short-circuit current J_{sc} of the 8 μm NW cell is 4% higher than the 8 μm ribbon cells and J_{sc} of the 20 μm NW cell is 14% lower versus the 20 μm ribbon ones as shown in Fig. 6.25. The work clearly demonstrates that NWs offer strong light-trapping effect as their positive side and large surface area also enhance strong surface carrier recombination as the negative side.

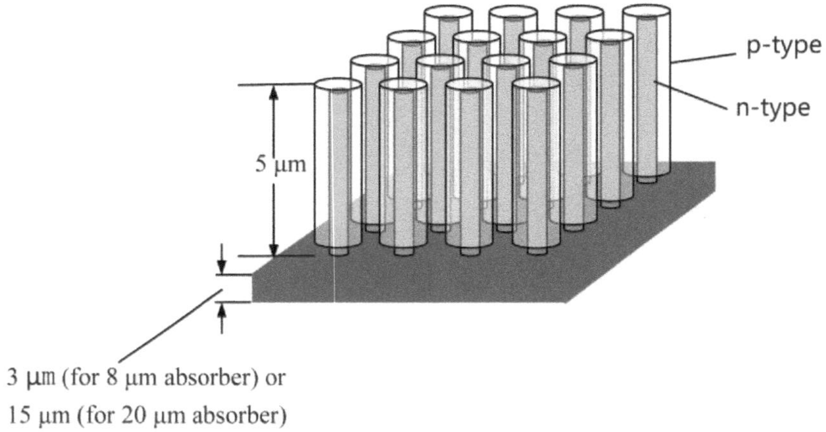

Figure 6.24 A device schematic of Si nanowire solar cells with radial *p–n* junction by Yang's group, fabricated by etching an *n*-type Si layer with the thickness of 8 and 20 μm using dip-coating silica bead mask followed by deep reactive ion etching and boron diffusion.

As far as the device structure is concerned, the coaxial NWSCs, which have a radial *p–n* junction, are very different from those with an axial *p–n* junction. They have the same open-circuit voltage, which is not surprising since open-circuit voltage V_{oc} is mainly determined by materials and doping. The major advantage of the radial junction nanowire devices is their much larger junction area that could be beneficial for the enhancement of short-circuit current density J_{sc}. Note that compared with conventional wafer-based Si solar cells that have a planar *p–n* junction, the axial junction devices have a very small total junction area and the radial junction ones have very large total junction area. Indeed as reported, the short-circuit current of a radial *p-i-n* nanowire solar cell is the order of 25 mA/cm^2.

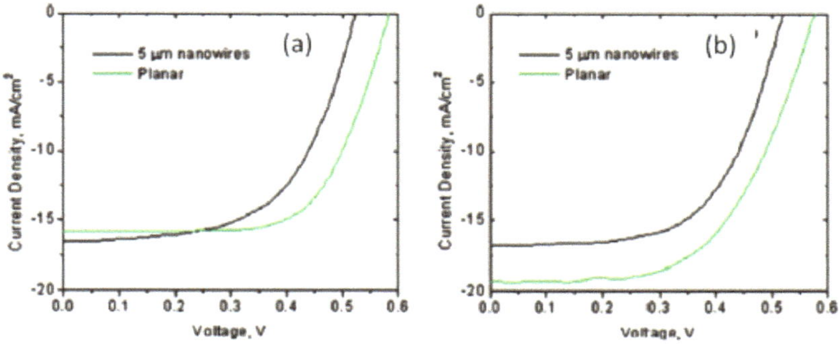

Figure 6.25 Measured *I–V* curve of an 8 µm absorber Si NW cell and a 8 µm ribbon Si solar cell (a) and the same comparison as the right panel but a 20 µm absorber NW cell and a 20 µm ribbon cell. Reproduced with permission from [5].

Currently, the increasing surface, junction, contact recombination, as well as complicated carrier travel paths in NWSCs are responsible for the overall low conversion efficiency in comparison to straightforward planar SCs. The existing problems of low conversion efficiency for the Si NWSCs, like other nano solar cells, originate from high density of various kinds of defects and interface/surface recombination. For the improvement of the device parameters, including the rather low V_{oc}, a careful device structure design and material and interface quality control are necessary. Further, the minority carrier lifetime needs to be improved.

Among various approaches proposed to enhance the photoelectric conversion efficiency of Si NWSCs, dispersing quantum dots (QDs) into the NW array is one of the effective ways for enhancing light scattering in the nanowires. This has been proven effective not only to Si NW devices, but also to NSCs based on other materials. The added-in quantum dots are usually transparent to the light so that they are only used to change the direction of photon traveling, not to absorb the light. It have been shown that the cost effective SiO_2 and Al_2O_3 are excellent light scatters and helpful for the enhancement of device current and, in turn, the conversion efficiency [61, 62]. One could also add metal nanoparticles in the Si NW array and the metal particles generate the plasmonic effect in addition to the light scattering effect. The nanoparticles mixed with NWs enhance the light absorption by dispersing and redirecting the light towards the Si NWs, making the light absorption insensitive to light incoming direction. Pt and Au nanoparticles have been introduced to intensify light scattering based on plasmonic effects [63–65].

Ge nanowire/nanopillar arrays

Staying in the shadow of Si, group-IV Ge is much less investigated for NWSCs, though it is another important model semiconductor. The growth of nano-sized Ge, such as nanowires, nanopillars, and quantum dots, is relatively mature. With bandgap of 0.66 eV at room temperature, Ge solar cells can be used for solar radiation conversion up to the wavelength of nearly 2 μm, covering quite a strong part of solar radiation in infrared. On the other hand, due to its bandgap that is significantly smaller than other popular materials such as Si (1.12 eV) and III–V GaAs (1.41 eV), Ge could be used with Si or GaAs in tandem SCs for the spectrum range uncovered by Si and GaAs.

Pan and co-workers reported radial NWSCs with the junction architecture of *n*-Si/*p*-SiGe. The large-area nanowire array SC on a 5 inch substrate was prepared by combining lithography and catalytic etching [66]. The preparation process was initiated by the growth vertically aligned single-crystal *n*-Si array, followed by epitaxially coating a single-crystal *p*-SiGe layers on the Si NWs by ultrahigh vacuum chemical vapor deposition. The *n*-Si/*p*-SiGe radial *p*–*n* junction solar cells were completed by depositing a Ti/Au layer as the back electrode. The NW array was buried by a PMMA layer, and a layer of ITO was deposited as the top electrode for the solar cell. The device gives an open-circuit voltage V_{oc} of 650 meV, a short-circuit current density J_{sc} of 8.38 mA/cm^2, a fill factor f of 0.60, as well as a conversion efficiency η of 3.26%. Apparently, V_{oc} of 650 meV is close to that of a conventional crystalline Si solar cell, but J_{sc} of 8.38 mA/cm^2 is much less. This architecture is potentially useful for application with the possibility of large-scale production and low cost. Yet, further enhancement of the short-circuit current, which would lead to improvement in the fill factor, is a critical issue. The relatively high conversion efficiency of this device architecture is associated with its excellent anti-reflection properties of the nanowire array and with relative high quality of the radial *p–n* junction.

6.4.2 III–V Nanowire Solar Cells

III–V semiconductor compounds are traditionally an expensive photovoltaic technology while they also make the most efficient solar cells. They are expensive owing to the relatively complex material synthesis and device fabrication and use of relatively rare elements, such as Ga and In. With a direct energy bandgap, group III–V semiconductors interact strongly with light, characterized by their strong photon absorption and excellent luminescence properties. They display excellent and efficient optoelectronic characteristics useful for devices and are therefore excellent candidates for solar cells for expected high conversion efficiencies. III–V semiconductors are a big material family with a number of compounds being nearly lattice-matched and direct bandgaps. This makes it possible to flexibly combine them for ternaries and even quaternaries with, importantly, tunable energy bandgaps to form tandem structures using energy band engineering. Most III–V compounds have larger bandgaps than that of Si, so that III–V solar cells may offer higher open-circuit voltage and lower excess reversed saturation current that is critical for device performance. In addition, the material growth technique for III–V compounds is quite mature and potentially low defect-density materials are widely available. Their bandgaps have wide spectrum coverage.

III–V semiconductors form a large number of heterostructures allowing for the preparation of multi-junctional tandem planar solar cells of very high PCE based on the GaInP/GaInAs/Ge system, for instance. For nano solar cells, it is also possible to construct tandem stacked devices using multiple *p–n* junctions in axial or radial configuration. It is expected that the tandem NWSCs could even show higher PCE. It has been demonstrated theoretically and experimentally that III–V materials make the most efficient photovoltaic cells. To date, the highest-efficiency solar cells are constructed by III–V compounds with the device architecture of tandem multi-junction. The device has an efficiency of 40.8% (Spectrolab, USA) [67] at concentrated sunlight of 326 suns and an efficiency of 41.1% at that of 454 suns (Institute of Solar Energy System, Germany) [68]. Later in May 2013, Soitec, a world leader in the manufacture of semiconductor materials, announced a new device record of 43.6% conversion efficiency based on a four-junction architecture at 500 suns, which was confirmed by the Fraunhofer ISE Calibration Laboratory. Lately, the conversion efficiencies of this type of solar cells have been raised above 46%.

A number of III–V semiconductor compounds have been used for nanowire solar cells, including GaN, InAs, GaAs, and InP. Although GaN has a wide bandgap transparent in visible range, GaN-based ternary might cover the solar radiation range. $In_{1-x}Ga_xN$, for instance, its direct gap extends continuously from the near infrared 0.7 eV (InN) to the mid-ultraviolet 3.4 eV (GaN). It is thus potentially possible to use this single ternary alloy for the fabrication of multi-junction solar cells.

GaN-based nanowire solar cells

Tang and co-workers designed a nano solar cell structure consisting of vertically aligned Mg-doped *p*-type GaN nanorod arrays on *n*-type Si, the *p*–*n* junction being located on the interfaces between the *p*-type nanorods and the *n*-type Si substrate [69]. The Mg-doped GaN nanorods were grown by CVD, with the tilt top-view SEM image as shown in Fig. 6.26a. The inset in Fig. 6.26a presents the cross-sectional view. The SEM images indicate that a gold nanoparticle is on each of the nanorods, which is a clear indication that the growth of the nanorods is governed by the vapor–liquid–solid process. The single-crystalline feature of the nanorods is made clear by the high resolution transmission electron microscope (HRTEM) image and its selected area electron diffraction (SAED) pattern shown in Fig. 6.26b and its inset, respectively. The atomic force microscope (AFM) image in Fig. 6.26c shows the tips of GaN nanorods.

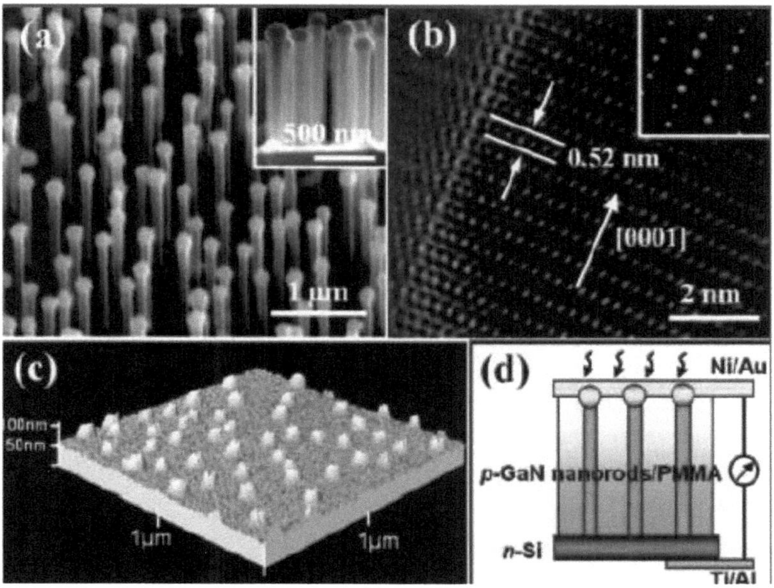

Figure 6.26 (a) Tilt top view and cross-sectional (inset) SEM images of GaN nanorod array doped with Mg; (b) HRTEM image and its SAED pattern (inset); (c) AFM image showing the tips of GaN nanorods; (d) a schematic of the solar cell. Reproduced with permission from [69].

The Ti/Al (30/50 nm) bottom contact and the Ni/Au (30/50 nm) top contact are prepared on the back side of the *n*-type Si substrate and on the *p*-type GaN nanorods, respectively, by electron beam evaporation. Figure 6.26d shows the schematic diagram of the device structure. With a wide bandgap being transparent to solar radiation, the GaN nanorods are here used as the window layer in the solar cells, the light-trapping layer, and the *p*-type layer for the formation of the *p*–*n* junction. The solar radiation generates electrons and holes in the GaN-nanorod/Si-substrate interfacial region where the photon-generated electrons and holes experience a strong electric field that drives the holes into the nanorods and the electrons into the Si substrate. The photovoltaic process is completed as the holes and the electrons transport to the top contact electrode and the bottom electrode, respectively. The well-defined rectification property of the nanorod solar cell, with a rectification ratio above 10^4 in dark and very low reverse currents, indicates the high quality of the *p–n* junction. The GaN nanorod SC exhibits a short-circuit current density of 7.6 mA/cm^2, a very high open-circuit voltage of 0.95 V under AM1.5 illumination at 100 mW/cm^2. The large open-circuit voltage is due to the large bandgap of the GaN nanorods. This device gives energy conversion efficiency of 2.73%, which is majorly limited by the relatively low short-circuit current due to the loss on defects and interfaces, though GaN is believed to have a low minority carrier recombination rate, helpful for the enhancement of the device conversion efficiency.

One might notice that the area of the *p–n* junction for the above-mentioned nanowire (nanorod) solar cells could be much smaller than those of conventional planar solar cells with the same macroscopic area. This would limit the short-circuit current of the NWSCs. Indeed, a short current density J_{sc} is observed to be typically one order of magnitude smaller than those of planar solar cells. To overcome the limitation, NWSCs with radial *p–n* junction are proposed. For the same device area, the overall *p–n* junction area of a radial *p–n* junction NWSC can be much larger than that of a conventional Si planar solar cell. However, the electrons and holes in this case are transported anti-parallel to each other in very thin nanowires, leading to the strong overlap of electron and hole wave functions. As a result, electrons and holes have high probability to recombine before they are extracted by their respective electrodes.

The architecture of the NWSC, consisting of an *n*-GaN core, *i*-$In_xGa_{1-x}N$ inner shell layer and a *p*-GaN outer shell layer, was proposed and fabricated using MOCVD by Dong et al. [28]. In the SC, the *i*-$In_xGa_{1-x}N$ is the active layer for photon absorption. The structure offers the flexibility that the bandgap of the active layer can be tuned by varying In concentration, theoretically from 2.25 to 3.34 eV, so as to control the wavelength of light absorption. At simulated one-sun AM1.5 illumination, experiments show that a single nanowire solar cell of *n*-GaN/*i*-$In_xGa_{1-x}N$/*p*-GaN core/shell architecture with x = 0.27 gives a conversion efficiency of 0.19% only. One of the causes for the low efficiency of the device is due to the large bandgap so that a fairly large part of the solar spectrum is not used.

GaAs-based NW solar cells

A *p–n* junction can also be prepared in a cross-sectional area of a NW, which has been done on GaAs NWs by MOVPE [70]. It was found that with controlled growth, the *p–n* junction was formed in a narrow area at the midpoint of the NWs. The NWs show efficient current-induced light emission even up to room temperature. The nanowire array with a *p–n* junction in the midpoint of the NWs is suitable for use in NWSCs. The device architecture differs from the above GaN nanorod SCs by the active region. Here the light absorption occurs in the midpoint of the nanowire where the *p–n* junction is located. The photon-generated electrons and holes are separated by the built-in static electric field and transport through half of the length of the NWs to their respective contact electrodes. Assuming that the substrate is *n*-type doped so that the *n*-type side of the NWs is grown on the substrate, electrons are transported to the substrate and the hole to the *p*-side electrode.

With nearly ideal bandgap and large photon absorption coefficient, GaAs is probably the best candidate for photovoltaic devices. In the work done by Colombo et al., a nanowire solar cell with GaAs *p-i-n* radial junctions was fabricated by MBE, as shown in Fig. 6.27 with a schematic device structure (Fig. 6.27a) and an SEM image of the GaAs NWSC [71]. Simulated at the illumination condition of AM1.5, the device gives the best conversion efficiency of 4.5%, which was among the best values in various structured nano solar cells at the time. Besides the optimal bandgap, the improvement in material and interface qualities, as demonstrated by photocurrent and electro-luminescence characterization, is also important for achieving relatively high conversion efficiency. The *i*-GaAs spacer layer in the *p-i-n* junction plays an important role for the enhanced conversion efficiency. Some other III–V compounds also have bandgaps matched to solar radiation and their nanowires can be growth with high quality. InP has bandgap of 1.35 eV that is better matched to solar spectrum than Si, and has a much higher optical absorption coefficient [72]. Goto et al. reported investigation on nano solar cells based on core/shell InP nanowire arrays. Their devices were grown by a selective-area MOCVD and they obtain a satisfactory device efficiency of 3.37% at that time [73].

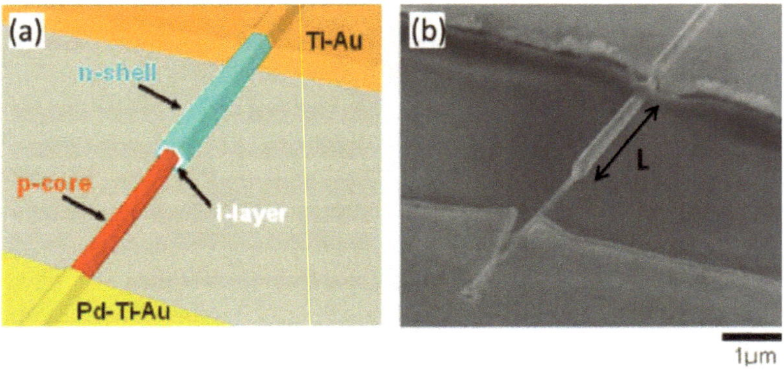

Figure 6.27 A schematic device structure (a) and a SEM image of the GaAs single NW solar cell with a *p–i–n* radial junction. Reproduced with permission from [71].

6.4.3 II–VI Nanowire Solar Cells

II–VI compounds are, compared to group-IV Si and Ge, and III–V ones, relatively less studied and they are not as robust as the Si, GaAs, and GaN. Yet the II–VI compounds have a wider coverage on bandgap energy than their IV and III–V counterparts. II–VI compounds include but not limited to the binaries such as CdTe, ZnTe, HgTe, CdSe, ZnSe, HgSe, CdS, ZnS, and HgS. Among them, HgTe has a "negative" bandgap of –0.3 eV, while ZnSe has a bandgap of 2.7 eV in blue spectrum and the well-known II–VI oxide compound ZnO has a bandgap of 3.37 eV in ultraviolet. Like some other metal oxides, ZnO is conductive to electrons and transparent to visible and near ultraviolet light, making it a great material for contact electrodes in thin-film and nano solar cells. Same as III–V compounds, II–VI ones are a big family with a large number of ternaries and quaternaries that can be derived from the above-listed binaries. Furthermore, II–VI materials carry the intrinsic benefits of III–V compound semiconductors, i.e., high light absorption coefficient (direct bandgaps), high carrier mobility, and good material stability.

As mentioned in Chapter 5, one of the unique advantages of the II–VI compounds is the flexibility in solution-based synthesis of abundant nanomaterials and nanostructures, much more solvent precursors being available for II–VI than for III–V nano synthesis. Some low-cost solvent synthesis routes have been demonstrated on II–VI nanowires, including CdSe, CdS, CdTe, ZnSe, ZnS, and ZnTe. The solvent-based growth for nanowires was also demonstrated on the group-II oxide ZnO, CdO [54, 74], and IV–VI PbSe, PbS and PbTe [75–77]. ZnO has simple preparation and excellent optical properties, but it is extremely difficult to dope with *p*-type. Various solar cell architectures based on II–VI nanowires, including axial and radial *p*–*n* junctions, have been reported.

CdTe/CdS nanopillar cells

Among thin-film solar cells, CdTe/CdS configuration has the lowest cost and takes the highest market share. The CdTe/CdS junction was also fabricated in the nanowire form. Highly regular and single-crystalline *n*-type CdS nanopillar arrays are grown embedded in polycrystalline thin films of *p*-type CdTe on aluminum substrates [78]. Both CdS nanopillars and CdTe thin films are grown by VLS. The light absorption is completed in *p*-CdTe adjacent to *n*-CdS where the wide bandgap CdS acts as windows for light and the nanopillar form of the *n*-CdS enables the enhancement of high absorption. As shown in Fig. 6.28, the photon-generated electrons and holes are separated on the interfaces of *p*-CdTe and *n*-CdS nanopillars and collected by the bottom and the top electrodes, respectively.

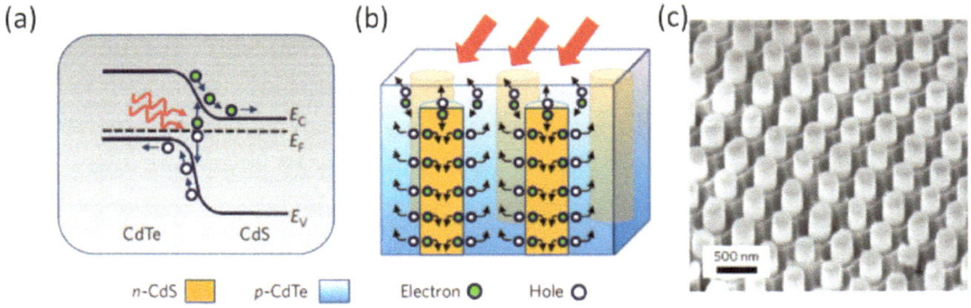

Figure 6.28 Schematics of the heterojunction energy band profile (a) and the device architecture (b) of the solar cell with the structure of *n*-type CdS nanopillar arrays grown embedded in polycrystalline thin films of *p*-type CdTe on aluminum substrates; (c) the SEM image of the highly regular CdS nanopillars. Reproduced with permission from [78].

The current major challenges for high efficiency II–VI nanowire solar cells are further lowering down defect density and better control on interface properties. Additionally, better doping techniques still appear to be technically challenging, particularly for those prepared by low-temperature synthesis routes. However, the II–VI nano solar cells are highly regarded for the expectation of realizing efficiencies >20% due to low recombination velocities of the materials at the interface [79].

ZnO nanowire solar cells

ZnO nanomaterials are probably the most well-studied metal oxide semiconductor in terms of growth and device fabrication. The reason why they have gained so much attention is their ease of growth with a variety of nano forms, including quantum dots, nanoneedles, nanorods, nanowires, and nanotubes., making them very promising for various optoelectronic applications. Having a wide bandgap of 3.37 eV, ZnO is transparent, intrinsically *n*-type and conductive due to its high free electron concentration. Despite all these interesting features, ZnO does not absorb visible light. ZnO has to "collaborate" with other semiconductors with bandgaps absorbing solar radiation for solar cells.

As a II–VI oxide semiconductor, ZnO closely matches a number of other II–VI compounds. The ZnO/CdSe heterojunction forms a type-II band alignment, rendering the structure a good candidate for solar cells. The type-II potential alignment is also helpful for slowing down the electron–hole recombination since the electrons and holes are confined in separate region. Tak et al. reported the fabrication of ZnO/CdS radial junction NWSCs using solution-based preparation route [54]. The coaxial ZnO/CdS nanowire is schematically shown in Fig. 6.29, together with its energy band profile. Incident photons are absorbed mostly by the CdS layer and electron/hole pairs are generated. The interfacial non-static forces separate the electrons and the holes. The photon-generated electrons are transported in ZnO and holes in CdS toward their respective electrodes, as shown in the figure. A short-circuit current density of 7.23 mA/cm^2, an open-circuit voltage of 1.5 V, and a conversion efficiency of 3.53% are achieved on this device. The large open-circuit voltage is due to the relatively large bandgaps of both ZnO and CdS (2.4 eV). The large bandgap limits short-circuit current since, obviously, quite big portion of the solar radiation with photon energy below 2.4 eV cannot be used by the device for photoelectronic conversion.

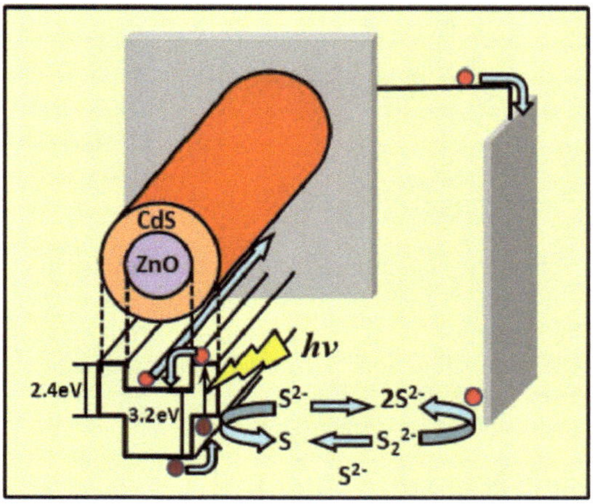

Figure 6.29 Schematic diagram of a ZnO/CdS core/shell NW and its type-II band profile. Reproduced with permission from [54].

More work on device structure requires to be done for increasing light absorption at photon energy below 2.4 eV. In fact, with optimization of structure design, it might be possible to utilize the step-like type-II band alignment to broaden the absorbed solar spectrum, as described in Fig. 6.30. If photons are absorbed within the two semiconductors composing the heterostructure, the photon energies need to be equal to or larger than the bandgaps of the semiconductors. Electron transition from the valence band of one semiconductor to the conduction band of another might occur in the vicinity of the hetero-interface, as shown by the tilted arrow in Fig. 6.30, although the real-space indirect transition probability is expected to be low.

In the work done by Tak et al., successive ion layer absorption and reaction was adopted to varying the CdS layer thickness to achieve optimized use of incident photons. Without CdS layers on ZnO nanowires, a short-circuit current J_{sc} of only 0.37 mA/cm^2 was measured. Using CdS shell layers with optimized thickness, J_{sc} boosted from 0.37 mA/cm^2 of a bare ZnO nanowire device to 7.23 mA/cm^2 with a CdS shell.

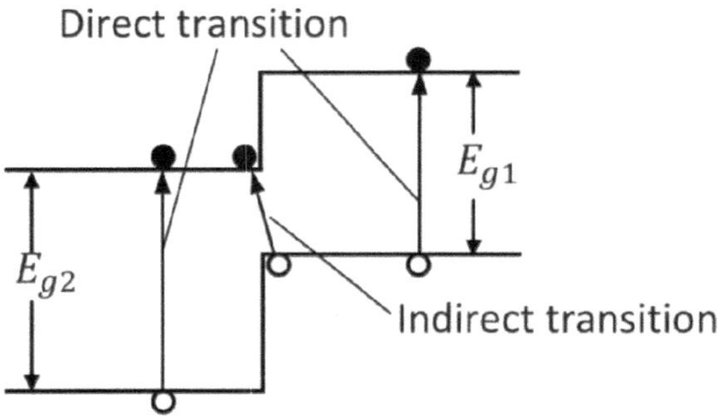

Figure 6.30 Schematic diagram of a type-II heterojunction, showing that a spatially indirect transition is possible to have transition energy smaller than those of direct transitions in the two semiconductors composing the heterojunction. E_{g1} and E_{g2} are the bandgaps of the two semiconductors.

6.4.4 Single Nanowire Solar Cells

Single NW solar cells are the building blocks of nanowire solar cells. Thus, single NWSCs represent an important platform in solar cells research. On the other hand, a single NWSC itself has special applications in a wide range as power suppliers for nano medical robots, integrating optoelectronic device, etc.

Si single NW solar cells

SCs based on vertical Si NW were fabricated by an Au-film-assisted electrochemical etching [80]. The core–shell *p*–*n* junction was formed via solid-state phosphorus diffusion that converts the boron-doped *p*-type shell to *n*-type. Diffusion leads to a pristine *p*–*n* junction comparing to junctions formed by doping during growth. Due to the excellent light harvesting properties over a wide range of spectrum, the SC exhibits a relatively high conversion efficiency of 1.47% at the time, which was a good improvement in comparison to the previously reported Si NWSCs with core–shell *p*–*n* junctions prepared by respective *p*-type and *n*-type doping.

Kempa and co-workers designed and prepared NWSCs with improved material and structural qualities configured by polymorphic core/multi-shell NWs of coaxial *p*/*i*/*n* junction and hexagonal cross section with diameters at 200–300 nm [81].

The silicon core/multi-shell NWs were prepared based on Au-nanocluster catalytic VLS growth for the *p*-type Si NW core followed by vapor–solid (VS) growth of the shell Si using SiH_4 as Si reactant. *P*-type and *n*-type doping was achieved during catalyzed growth process with the use of B_2H_6 and PH_3 dopants, respectively. The thicknesses of the *p*- and *n*-type doped layers, as well as that of the intrinsic layer were controlled by growth times. Four coaxial single NW core/shell structures, *p*-core/*n*, *p*-core/*i*/*n*, *p*-core/*p*/*n*, and *p*-core/*p*/*i*/*n* labeled according to their doping sequences, were prepared and studied, where *p* and *n* refer to *p*-type and *n*-type, respectively, and *i* refers to intrinsic. All those structures have a *p*-type core.

Interesting results were obtained on current density measurements performed on the four coaxial single NW core/shell structures with electrical contacts being selectively connected to the *p*-type core and the *n*-type shell (see Fig. 6.31a). The *J*-*V* curves were measured under AM 1.5 G and one sun illumination conditions and the current densities *J* were calculated using the projected area of the NWs. Figure 6.31b presents measured *J*–*V* curves of the four NW device geometries, where distinct changes in V_{oc} are clearly displayed. The *p*-core/*i*/*n* geometry gives the highest V_{oc} value of 0.48 *V* among the four devices with NW diameters as small as 200 nm. This device also exhibits a high fill factor (*FF*) of 73%. Together with the steep drop at the current density approaching to V_{oc}, the high *FF* is a clear indication that the series resistance in the devices is very small.

A properly inserted layer in the devices is very helpful for device performance, as shown by inserting an intrinsic layer (~30 nm in thickness) into *p*-core/*n* (forming *p*-core/*i*/*n*), which gives rise to 140 mV enhancement in V_{oc}, and into *p*-core/*p*/*n* (forming *p*-core/*p*/*i*/*n*) which leads to 40 mV improvement in V_{oc}. Another important parameter is the leakage current of the junction, which is a characterization of the junction performance. Figure 6.31c plots V_{oc} versus short-circuit current I_{sc} normalized by dark saturation current I_o (at the logarithm scale) for totally 16 devices with all four junction geometries. Clearly, V_{oc} versus the logarithm ratio of I_{sc}/I_o is linear and V_{oc} increases monotonically with decreasing I_o. The best results were obtained on the *p*-core/*i*/*n* device, which has a V_{oc} and J_{sc}

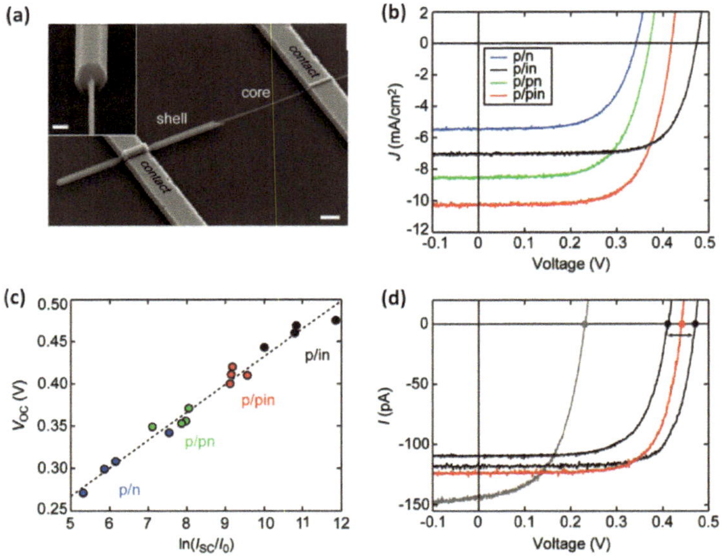

Figure 6.31 (a) SEM image of a core/shell NW structure with Ti/Pd contacts (prepared by photolithography) that connect the shell and core regions of the NW (scale bar: 1 μm. Inset shows the SEM image of a single core/shell NW on a Si_3N_4 substrate, which shows that the shell part of half of the NW is chemically etched away (scale bar: 200 nm). (b) Measured current-density–voltage (J–V) curves of single-NW solar cells of the four device structures. (c) Measured open-circuit voltage (V_{oc}) versus the ratio of short-circuit current (I_{sc}) to dark saturation current (I_o) (in logarithm scale) of the four device geometries. Dashed line is the fit to the ideal diode equation, on which an ideality factor of n = 1.28 is extracted. (d) Measured current–voltage (I–V) curves: a p-core/i/n NW device where the Au catalyst was wet-chemically etched away by aqueous KI/I_2 prior to growth of the shells (red), a p-core/p/i/n NW device grown with Al catalyst (gray), and a single Au-catalyzed p-core/i/n single NW devices (prepared with Au catalyst) measured prior to and after one-week oxidation in air. Reproduced with permission from [81].

as high as 0.48V and 0.34 fA/μm^2, respectively, and a dark saturation current I_o as low as 1.1 fA, which is a significant improvement with respect to their previous NW cells with nanocrystalline shells [8]. It is expected that the dramatic reduction in leakage current could account for a twofold enhancement in V_{oc}. Figure 6.31d presents measured I–V curves of four devices, the p-core/i/n NWSC with Au catalyst wet-chemically etched away by aqueous KI/I_2 prior to growth of the shells (red), the *p*-core/*p*/*i*/*n* NWSC grown with Al catalyst (gray), and the single Au-catalyzed *p*-core/*i*/*n* single NWSC, prepared with Au catalyst, that was measured prior to and after one-week oxidation in air.

Under one sun illumination condition, their devices show leakage current as low as 1 fA, V_{oc} up to 0.5 V and *FF* as high as 73%. Very importantly, measurements showed that the current densities of the single NW solar cells, calculated using the projected NW area, are twice those of silicon film solar cells with comparable thickness and the external quantum efficiencies were more than 100%. A photon antenna effect was suggested to explain the phenomena, which is supported by the results of finite-difference-time-domain (FDTD) simulations performed on the NW devices.

Experimental external quantum efficiency (EQE) of a *p*-*i*-*n* core/multi-shell NWSC was obtained and the results were compared with FDTD simulation as shown in Figs. 6.32a,b. Some important points can be made, based on a close look at the experimental result and simulation. The experimental and simulated spectra exhibit a number of resonant peaks, where there is a good agreement between experiment and theory in peak positions and amplitude. Analysis indicates that resonance modes 1, 2, 4, and 5 are due to Fabry–Perot effect, while modes 3 and 6 are assigned as whispering-gallery resonance due to the hexagonal cross section of the NW. Note that the FDTD simulations were performed without using adjustable parameters. In Fig. 6.32a,b, the diameters of the NWs are 240 and 305 nm, respectively. The internal quantum efficiency is assumed to be 100%. In calculating the short-circuit current densities, NW projected area was used.

The fact that EQE is close to or exceeds 100%, as shown in the spectrum range of 400–450 nm in Fig. 6.32a, is a clear indication of an optical antenna effect [82, 83], i.e., the absorption cross-sectional area of the NW exceeds its own physical cross-sectional area due to the subwavelength diameter of the NW, which occurs in the short wavelength range as shown in Fig. 6.32a by both the experimental results and the FDTD simulation curve. In fact, the antenna effect is essentially gone for NW with a diameter of 305 nm (see Fig. 6.32b). For comparison, a simulated EQE spectrum of a planar bulk Si solar cell with a top layer of the same thickness of 240 nm (dashed green curve in

Fig. 6.32a). This EQE spectrum shows no resonance structure and the amplitude is below those of the experimental and the FDTD simulated curves. The resonant peaks are due to strong interaction between light and the NW structure. The enhanced interaction leads to enhanced absorption, which is clearly illustrated in Fig. 6.32c by the spatial profiles of the resonant modes 1–3 for *p/in* (top panel) and 4–6 for *p/pin* (bottom panel) device configurations.

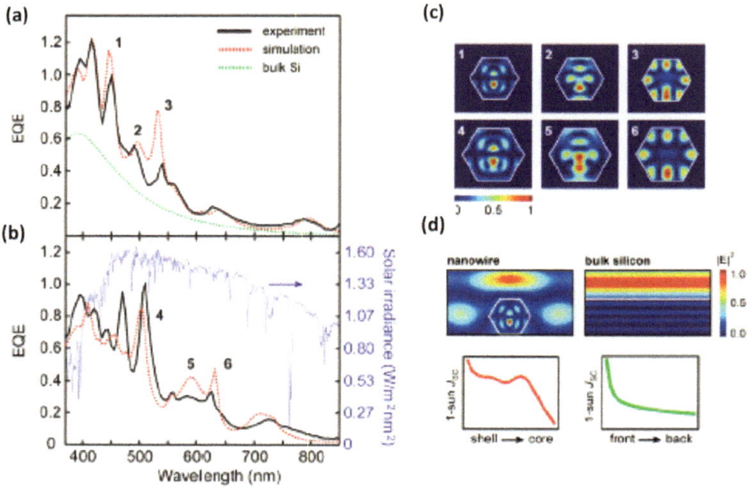

Figure 6.32 Experimental and simulated external quantum efficiency (EQE) and optical modes confined in NWs. (a) Experimental (black curve) and simulated (dashed red curve) EQE vs. Length of for a p/in NW solar cell with NW diameter (or height) as the only adjustable parameter used in simulation (the presented simulation uses a NW diameter of 240 nm). For comparison, a simulated EQE of a top 240 nm of planar bulk Si solar cell is presented (dashed green curve). (b) Experimental (black curve) and simulated (dashed red curve) EQE of a p/pin NW with a diameter of 305 nm, compared to the irradiance solar spectrum at the AM1.5G condition (dashed blue curve). (c) Spatial light absorption profiles of the resonant modes 1–3 and modes 4–6 obtained by FDTD simulation as labeled in (a) and (b), respectively. The spatial profiles correspond to TM polarizations. Modes labeled as 3 and 6 are whispering-gallery type modes and other modes Fabry–Perot resonances. (d) Profiles of electric field intensity for plane wave at λ = 445 nm for a NW (top left) and bulk Si (top right) systems. The white lines outline the NW and top surface of the bulk Si. JSC vs. distance from the shell into the core for the NW (bottom left) or from surface into bulk Si (bottom right) is also presented. Reproduced with permission from [82, 83].

Note that the linear color scale corresponds to light absorption rather than electric field intensity. To further shed light on the interaction, calculation is performed to show the distribution of electrical field intensity for an incident beam at a wavelength of 445 nm (Fig. 6.32d). Enhanced interaction between light and NW system is evidenced by the strong resonant modes excited within the NW and the featureless profile of electric field intensities in bulk Si with its thickness same as the value of the NW diameter.

Tandem Si single NW solar cells

The length scale of NWs enable to construct a tandem structure along the NWs. Kempa and co-workers reported the realization of axial p-i-n^+-p^+-i-n tandem photovoltaic device on a single Si NW, where n^+ and p^+ indicate heavily doped n- and p-type, respectively [47]. The schematic diagram and SEM (scanning electron microscopy) image of the Si single NWSC are shown in Fig. 6.33a,b, respectively. The tandem device yields an open-circuit voltage V_{oc} of 0.36 eV that is 57% larger than that of the single p-i-n junction device, as shown in Fig. 6.33c. The tandem cell displays a short-circuit current I_{sc} of 8 pA that is about 2.4 pA below that of the single junction device, indicating that the two subcells (in the tandem cell) are not current-matched. As shown by the blue curve in Fig. 6.33c, the overall conversion efficiency of the p-i-n+-p+-i-n tandem cell is still very low, due to many uncontrollable lost mechanisms.

The best single axial p-i-n junction nanopillar solar cell gave a conversion efficiency of 0.5% measured under one-sun conditions (AM 1.5G), corresponding to a V_{oc} of 0.29 V and a J_{sc} of 3.5 mA/cm^2. The 0.5% conversion efficiency was obtained for the insertion layer with an optimized thickness of 4 μm. The i-layer thickness has an evident impact on dark current, as shown in Fig. 6.34a where dark I–V characteristics of the p-i-n Si NW solar cell are presented with different i-layer thickness of 0, 2, and 4 μm (red, green, and black curves, respectively). Under illumination intensity of AM 1.5G, the devices with i-layer thickness = 0 (p–n junction), 2 and 4 μm give V_{oc} and J_{sc} (J_{sc}) values of 0.12 V and 3.5 pA (0.4 mA/cm^2), 0.24 V and 14.0 pA (1.5 mA/cm^2), and 0.29 V and 31.1 pA (3.5 mA/cm^2), respectively, as shown in Fig. 6.34b. Comparatively, the V_{oc}

and the J_{sc} values are 0.7 V and 50 mA/cm^2 for conventional Si-wafer solar cells. There are plenty of rooms for improvement in Si NW solar cells.

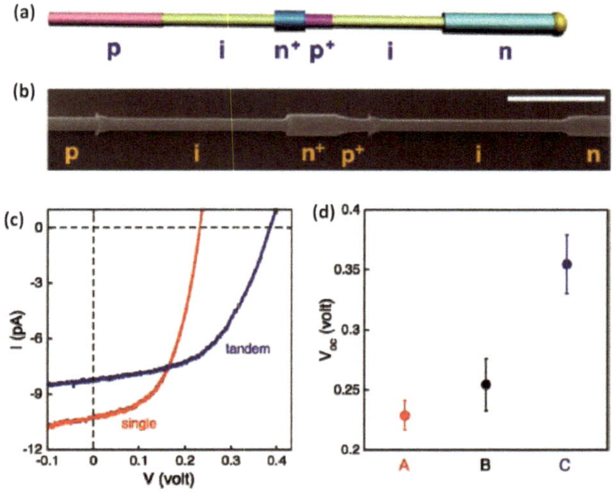

Figure 6.33 (a) Schematic of a tandem *p-i-n*$^+$*-p*$^+$*-i-n* Si NW solar cell; (b) SEM images of the NW solar cell prepared by selective etching (the scale bar is 1 μm); (c) *I–V* responses measured on *p-i-n* (red) and *p-i-n*$^+$*-p*$^+$*-i-n* (blue) tandem structures under AM 1.5G illumination; (d) V_{oc} for the axial Si NW devices with 2 μm *i*-layer *p-i-n* (red), 4 μm *i*-layer *p-i-n* (black), and 2 μm *i*-layer *p-i-n*$^+$*-p*$^+$*-i-n* (blue) structures. Reproduced with permission from [47].

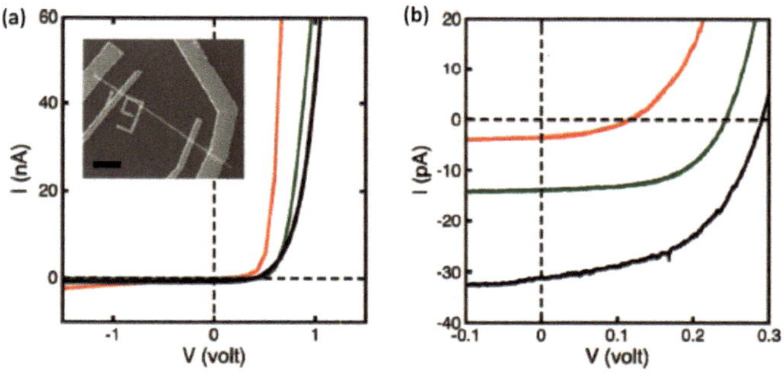

Figure 6.34 The dark (a) and light (b) *I–V* results for axial *p-i-n* junction of Si single nanowire solar cells with the *i*-layer thickness = 0 (red), 2 (green), and 4 (black) μm. The inset in (a) shows the SEM image of the Si NWSC with 2 μm *i*-layer length (the scale bar is 4 μm). The light *I–V* curves were measured at AM 1.5G. Reproduced with permission from [47].

6.5 Some Important Work Review

6.5.1 Solar Cells Based on Solution-Processed Core–Shell Nanowires

Nanowires for solar cells are in most cases fabricated by gas-phase epitaxial technologies including chemical vapor deposition (CVD) and molecular beam epitaxy (MBE) which are not cost-effective. Solution-processed fabrication is highly desired for lowering the cost of NWSCs and scaling up, in which a solution-based cation exchange reaction offers an interesting route to prepare various NWs [84]. The low-temperature fabrication also makes it easy to dope the NWs. II–VI CdS, CdSe, etc., are the kind of semiconductors that are fairly suitable for solution-based nanomaterial synthesis on which a great number of works have been reported.

Peidong Yang's group has prepared CdS/Cu_2S core–shell nanowires for solar cells and achieved 5.4% conversion efficiency [85]. Initially, CdS nanowires were fabricated by physical vapor transport route via a VLS procedure, with their lengths up to 50 μm and diameters in the range of 100 to 400 nm. The CdS nanowires are single crystalline with a hexagonal crystal phase. The CdS nanowires were then immersed in a 0.5 M CuCl solution and a Cu_2S shell formed on the surface of the CdS wires through cation exchange reaction in the solution. The solution temperature was kept at 50°C and the cation exchange reaction time was 5–10 s, which led to a Cu_2S shell layer with controlled thickness of 5–20 nm by tuning the reaction time. The nanowires were about 5 μm long and 260 nm in diameter. On the interface between CdS and Cu_2S, the lattice mismatch was less than 4%, small enough allowing high-quality material growth. The electronic contacts are made to the core and the shell. The preparation procedure is shown schematically in Fig. 6.35a. The CdS/Cu_2S junction prepared by cation exchange reaction in solution is atomically smooth and steep, which shows excellent performance in charge separation and low recombination rate for photogenerated carriers. Note that the world's first efficient thin-film solar cell was based on Cu_2S/CdS architecture, with Cu_2S being *p*-type and CdS intrinsic or weakly *n*-type [86].

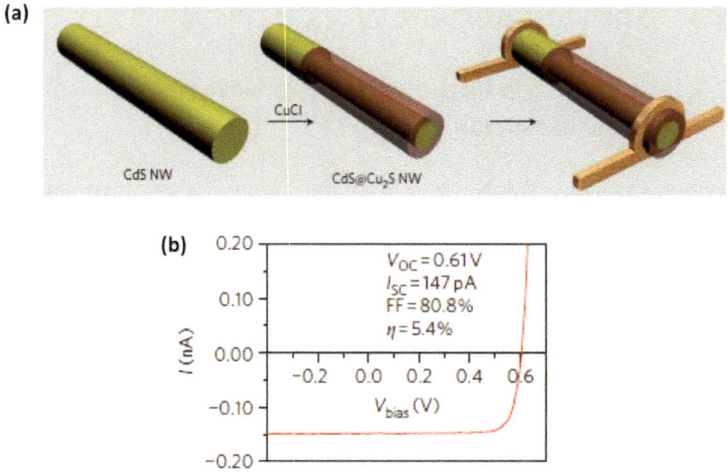

Figure 6.35 (a) Schematic preparation procedure of a CdS/Cu_2S core/shell nanowire, from initial CdS nanowire (yellow) to Cu_2S shell layer (brown) formation by selective ion exchange in CuCl solution, and finally the electric contacts are made to the Cd core and the Cu_2S shell. (b) Measured *I–V* curve of the nanowire solar cell under 1 AM 1.5G condition. Reproduced with permission from [85].

The high-quality material and excellent CdS/Cu_2S heterojunction prepared by the solution-based ion exchange reaction give relatively good solar cell parameters of V_{oc}, J_{sc}, and *FF*, as shown in Fig. 6.35b together with the measured *I–V* curve. The device has power conversion efficiency (PCE) of 5.4%. The efficiency of the NWSC is limited mainly by the thin Cu_2S shell layer that absorbs only 15–20% of the incident light. Thus, the use of a relatively thick Cu_2S shell could improve the conversion efficiency. In addition, the NWSCs show more efficient response to weak light levels than planar solar cells, suggesting that they could be used efficiently in cloudy days.

6.5.2 Si Nanowire Solar Cells with Axial and Radial *p–n* Junctions

Wafer-based planar silicon SCs present low light absorption efficiency in the vicinity of their *p–n* junctions and require highly purified Si material for long minority carrier lifetime and low carrier combination rate. Si NWs have the potential to overcome those limitations of crystalline Si planar solar cells by enhancing the light trapping and enabling the use of low purity Si material. Together with the large radial *p–n* junction area of the core–shell NW structure, the perpendicular light incident and carrier transport directions enable highly efficient charge separation and collection.

Si NWs have long been the focus of NSC research, but their PCEs were very low initially (below 1%). A breakthrough with 1.47% cell efficiency was achieved on devices in which nanowires were fabricated using an Au-film-assisted electrochemical etching process and the sizes of the nanowires were determined by that of the nanosphere diameters [80]. Later, device efficiency reached 2.95% with an improved J_{sc} of 14.2 mA/cm^2 [87], which was attributed to very efficient light harvesting and high junction performance. Conversion efficiency of more than 10%—even 15%—was achieved later [88, 89].

Degenerate doping is beneficial for charge collection where low resistive contacts can be made, while it is unfavorable for light absorption due to enhanced recombination introduced by high-density carrier recombination centers. The idea of selective-emitter technology has been used by wafer-based Si solar cells, where multiple doping steps are adopted with the use of photomasks. By decoupling heavily doped contact region (for charge collection) from lightly doped emitter region (for photon absorption), charge recombination is significantly suppressed, leading to great enhancement of J_{sc} and V_{oc},. For Si NWSCs, however, the geometry of the NWs makes the selective-emitter technology a big challenge. Um and co-workers developed a unique approach in which self-aligned selective emitters was integrated onto degenerately doped Si NWs. Good Ohmic contacts were achieved via one-step thermal diffusion on lightly doped Si NWs. Due to improved contacts and reduced carrier recombination rate, the device gives a J_{sc} as high as 33.65 mA cm^{-2}. A significant increase in internal quantum efficiency (IQE) in almost the whole solar spectrum is observed, due mainly to nearly perfect light trapping by Si NWs together with highly efficient carrier collection due to the use of selective-emitters. Remarkably, it shows 40% enhanced photoelectronic response in blue region. The NW solar cells with selective emitters give a PCE of 12.8% [90].

The nanowire array is prepared by one-step metal-assisted chemical etching on a thin-film planar Si with the total thickness of ~0.8 μm, as shown in Fig. 6.36a. The *p–n* junction is 0.6 μm below the top surface of the thin-film and the top *n*-type layer (with a thickness of 0.6 μm) has a *n*-type doping profile as shown in the figure, the top surface being degenerately doped to the level of 10^{21} cm^{-3} that continuously varies to 10^{16} cm^{-3} at the junction (see Fig. 6.36b). Such a *p–n* junction and doping profile were formed by phosphorus doping carried out using a spin on-doping method. Ag nanodots were prepared on the top surface of the planar thin-film and Si NWs were formed by one-step metal-assisted chemical etching,

$$Si + H_2O_2 + 6HF \rightarrow 2H_2O + H_2SiF_6 + H_2 \quad (6.3)$$

$$2Ag + H_2O_2 + 2H+ \rightarrow 2Ag+ + 2H_2O \quad (6.4)$$

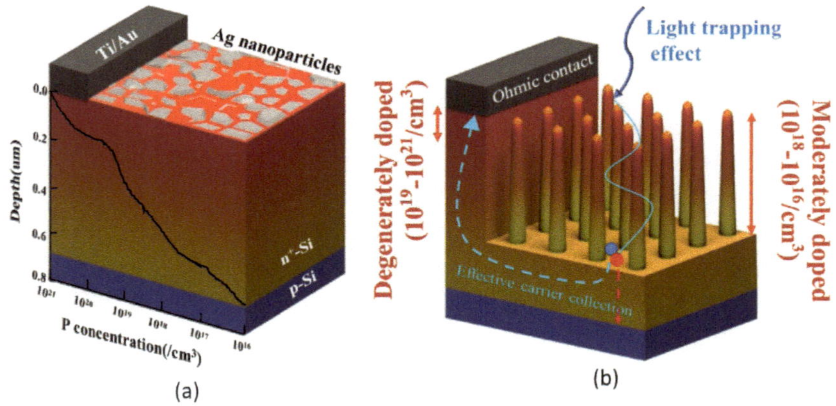

Figure 6.36 Schematic images of the preparation procedure a Si NW solar cell with selective emitter. (a) Device structure of a planar thin film solar cell with Ag nano particles on top and metal grids, prior to metal-assisted chemical etching. The doping profile is also shown schematically. (b) A Si NW solar cell with selective emitters: formation of Si NWs with moderate doping profile using one-step metal-assisted chemical etching, which is used as the light absorption area and the degenerately doped regions remain underneath the metal grids. Reproduced with permission from [90].

with a detailed preparation process as described previously. Note that degenerately doped top ends of the Si NWs are also etched away, left only with the moderately doped parts with a doping profile from 10^{18} cm^{-3} (at the tips of the NWs) to 10^{16} cm^{-3} (at the bottoms of the NWs). Si NWSCs with the selective emitter were thus constructed. Underneath the Ti/Au metal grids as shown in Fig. 6.36b, the heavily doped regions guarantee low resistance Ohmic contact and efficient carrier collection, while the moderately doped regions on NWs reduces the probability of carrier recombination.

6.5.3 GaAs Nanowire Array Solar Cells with Axial *p–i–n* Junctions

GaAs is direct-bandgap semiconductor that is more suitable for various optoelectronic devices than Si with indirect bandgap, due to its very large light absorption coefficient, superior carrier mobility, and well-established material preparation techniques. Furthermore, GaAs has the bandgap value of 1.41 eV, which matches better to the solar spectrum than Si with a bandgap of 1.12 eV. GaAs solar cells present a high open–circuit voltage V_{oc} close to 1 V. There are quite a few reports on GaAs nanowire solar cells with the nanowires prepared by either top-down lithographical patterning or various bottom-up methods, with either radial or axial junction geometries. In 2011, a PCE of 2.54% was achieved on SCs based on top-down nanorod arrays and radial junctions [91]. Afterwards, device PCEs have been raised to 5%, even to 10%. Lately, Gutsche and co-workers fabricated a single nanowire solar cell with the structure of GaAs/InGaP/GaAs and a radial *p-i-n* junction [92]. They achieved a PCE of 4.7%. An important part of PCE loss is due to the surface recombination of the carriers. Passivation of GaAs nanowires uses lattice-matched wide bandgap materials, which guarantees light penetration into nanowires. Reduction in surface recombination led to enhanced short-circuit current. In fact, InGaP passivation gives rise to a quantum efficiency of nearly 100% and this, in turn, results in enhanced device efficiency close to 7.6% [93].

Yao and co-workers fabricate solar cells configured by uniformly patterned GaAs nanowire arrays, using metal–organic chemical vapor deposition (MOCVD) [93]. The devices have axial *p-i-n* junctions that are prepared by selective area growth (SAG) and a short-circuit current density J_{sc} of up to 23.28 mA/cm^2 was achieved even without any surface passivation treatment on the wires. The device structure of their GaAs nanowire array solar cells is schematically depicted in Fig. 6.37, showing vertically aligned GaAs nanowire array on a GaAs substrate. The *p-i-n* junctions are grown along the nanowires, as indicated in the figure. The NWs were prepared by using MOCVD with selective area growth. All the NWs are embedded in BCB polymer that is transparent and insulating. ITO is used as front contact and the back contact is the n^+ GaAs substrate. Under sunlight, electrons and holes are excited by photons and swept by the built-in potential in the junction area toward the *n*- and *p*-regions, respectively. A photovoltaic potential is then established. The power conversion efficiency depends strongly on the light absorption ability of the device, which is one of the great advantages of the NWSC device where a great part of the incident light was absorbed. A numerical simulation was performed for the determination for optimal device structure.

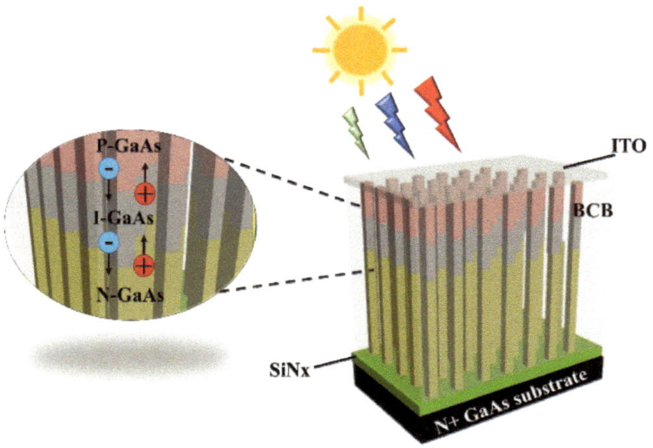

Figure 6.37 Schematic illustration of a GaAs NW solar cell with an axial junction and carrier transport when the device is illuminated (the zoomed-in graph). Reproduced with permission from [93].

Yao and co-workers fabricated GaAs NWSCs, achieved a PCE of 7.58%, and carried out a numerical simulation and compared the performances of both radial and axial junction NWSCs [93].

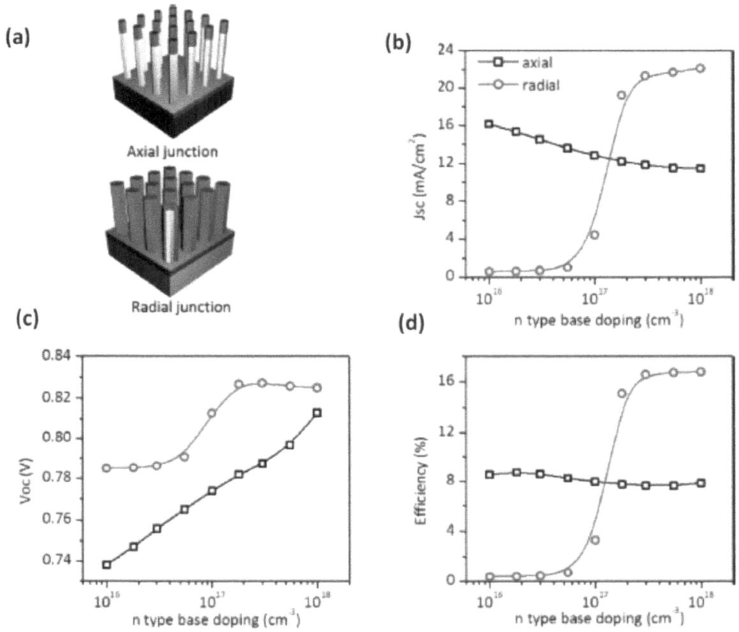

Figure 6.38 (a) Schematic of the vertically aligned GaAs NWs with axial (top) and radial (bottom) junctions (a), as well as J_{SC} (b), V_{OC} (c), and η (d) of the GaAs NWSCs with axial (black curves) and radial (red curves) junctions as a function of doping concentration in the *n*-type base. Reproduced with permission from [93].

Figure 6.38 presents the schematic illustration of GaAs NWs with axial and radial junctions (a), and measured device parameters of J_{SC} (b), V_{OC} (c), as well as PCE (d) as functions of *n*-type base doping concentration in the axial or radial junctions. The doping concentration of the *p*-type emitters is fixed at 10^{18} cm^{-3} to ensure Ohmic contact with a low resistance. As shown in the Fig. 6.38b, increase of the doping concentration leads to the turning-on of significant increase of J_{SC} in the case of the radial junction, while J_{SC} for the axial junction solar cells decreases gradually. According to semiconductor physics, increasing doping in the *n*-type base of axial junction gives rise to the reduction of width of the space charge region (junction depletion region). The built-in potential increases accordingly, which is beneficial for exciton splitting but; as a trade-off, the carrier drift length is also reduced, which is unfavorable for the extraction of electrons and holes. In addition, increasing doping concentration is likely to generate more carrier scattering centers and traps that act to kill the carrier mobility. The carrier mobility is essential, since it determines carrier diffusion. The enhanced built-in potential as a result of increasing doping concentration gives rise to large open-circuit voltage V_{oc}, as shown in Fig. 6.38c. However, short-circuit current J_{sc} drops with increasing doping concentration due to the trade-off factors such as reduced carrier mobility. As a result, PCE does not gain too much with increased doping (see Fig. 6.38d).

In the case of the radial junction, J_{sc} increases with doping concentration and it enhances quickly near the base doping concentration of 10^{17} cm^{-3}, as shown in Fig. 6.38b. The corresponding V_{oc} also increase with base doping, though the increase is not as significant as the increase of J_{sc} (see Fig. 6.38c), which gives rise to a PCE enhancement similar to J_{sc}. Apparently, the performance of radial junction NWSCs is more sensitively dependent on doping concentration than that of axial junction NWSCs. In other words, axial junction NWSCs are relatively inert to doping.

6.5.4 InP Nanowire Array Solar Cells with 13.8% Efficiency

Many NW materials and device architecture have been proposed. However, the NWSCs still suffer from low PCE due to high density of defects and surface states, as well as poor junction performance, in comparison to solar cells based on bulk materials. Wallentin and co-workers worked on the construction of NWSCs based on InP, a III–V semiconductor compound. Their device has the structure of axially defined *p-i-n* InP nanowires as building block for NW array SCs that were epitaxially prepared on *p*-type InP substrate.

InP is a semiconductor with a direct bandgap of 1.34 eV that is more suitable than Si for solar radiation capture. Gold particles used as seeds are prepared by nanoimprint route on InP substrates and the epitaxial growth occurs at the sites of Au particles. Figure 6.39a presents the SEM images at 0° and 30° (inset) tilted views [94]. The NW diameters of 130 to 190 nm and the pitch sizes of 470 or 500 nm were controlled by Au particle sizes and pattern distributions. The length of the NWs was determined by epitaxial growth time and in the work for solar cell devices the length of ~1.5 μm was selected.

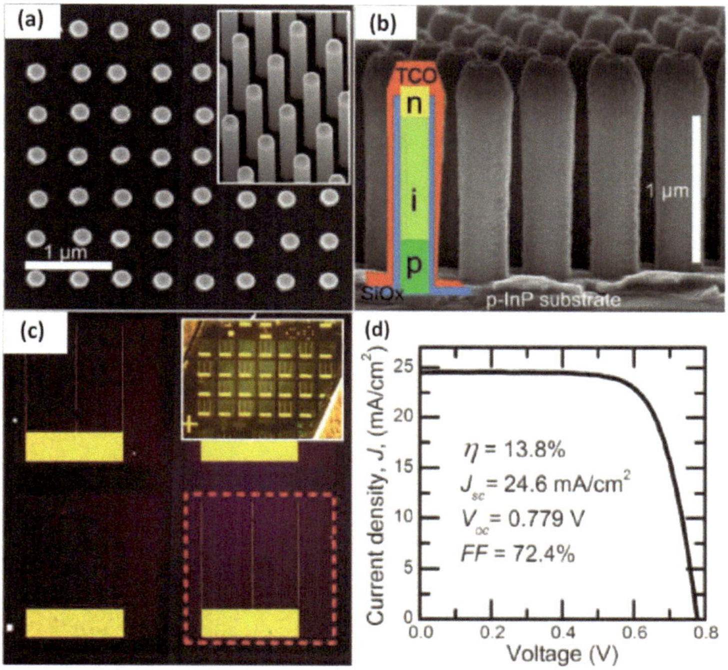

Figure 6.39 Scanning electron microscopy (SEM) and Optical microscope images, together with an *I–V* curve of InP NW solar cells: (a) SEM image of top surface of an as-grown NW array with 12% surface coverage and 30° tilt SEM image (inset); (b) SEM image of side view of the NW array. The color schematic shows the TCO (red), silicon oxide—SiOx (blue), as well as *p*-region (green), *i*-region (yellow-green), and *n*-region (yellow) in the NWs. (c) Optical microscope image of InP NW solar cells with metal contacts where the red dashed line highlights a cell with an area of 1 × 1 mm^2 and a sample with 4 × 7 cells are shown (top right); (d) *I–V* curve measured at AM 1.5G condition for InP NW cell with the highest efficiency. Reproduced with permission from [94].

The device performance is critically dependent on the geometrical parameters of the NW array. The Au seed particles are removed by chemical etching prior to putting the TCO contact layer. Figure 6.39b presents the side view SEM image of the NW array solar cell, where the colors are used to highlight the TCO (red), SiO_x (blue), as well as *p*-region (green), *i*-region (yellow-green), and *n*-region (yellow) in the NWs. Figure 6.39c shows the optical microscope image of the InP NW solar cells after putting metal contacts, where the red square (dashed line) defines a cell with an area of 1 × 1 mm^2. A sample with 4 × 7 cells on a substrate is shown on top right.

The highest PCE of 13.8% was achieved on the InP NW solar cell with axial *p-i-n* junctions using the NW diameter of 180 nm and the NW length of 1.5 μm and the NW array coverage of only 12% on the InP substrate. Their study shows that such a low NW coverage is able to convert 71% solar radiation into photocurrent, a sixfold increase in efficiency for light collection. The 13.8% PCE is largely attributed to the very high open-circuit voltage of 0.906 V, which exceeds that of wafer-based InP solar cells. The good passivation and isolation made by SiO_x layers are also expected, considering that the surface-to-volume ratio of the NW solar cell is about 30 times higher than its planar counterpart.

6.5.5 Three-Dimensional Nanopillar-Array Solar Cells Based on *n*-Type CdS Nanopillars Embedded in *p*-Type CdTe

II–VI semiconductors represent a kind of material with their bandgap covering a wide energy range from ultraviolet to infrared. The method of liquid precursors is also suitable for their preparation. Fan and co-workers developed a method to synthesize crystalline *n*-type CdS nanopillar (short nanowire) arrays on aluminum substrate and the arrays were embedded in *p*-type CdTe (see Fig. 6.28 in Section 6.4), forming junctions with radial geometry [78]. The bandgap of CdS is 2.42 eV and that of CdTe is 1.41 eV. Hence, most of the solar radiation is absorbed by CdTe. The CdS nanopillars form a highly ordered pattern for controlled optical properties, as depicted in Fig. 6.28. As shown in Fig. 6.28a, the heterojunction of CdTe/CdS act as a *p–n* junction to separate photogenerated electrons and holes. Such a coaxial vertical nanopillar arrays have the advantage that light absorption and carrier separation and collection are in orthogonal directions (see Fig. 6.28b). In this device configuration, carriers need to travel only a short distance before they are collected by contact electrodes. This is beneficial for SCs based on II–VI materials that usually have rather short carrier minority carrier lifetime and diffusion length. In addition, very thin devices can be designed since the thickness of the CdTe/CdS film only needs to be comparable to optical absorption depth which is very small for the II–VI materials with direct bandgaps.

CdS nanopillars were prepared by template-assisted VLS growth on aluminum substrates that were patterned with periodic alumina membranes as the template prepared by anodization. The template-assisted VLS is a versatile approach suitable for the growth of many semiconductors and device structures. The nanopillars are highly ordered and single-crystalline. The VLS preparation of CdS nanopillars was followed by deposition of a *p*-type CdTe thin film of ~1 μm in thickness using chemical vapor deposition. The CdTe layer is used to absorb solar radiation with its ideal bandgap (1.42 eV) for solar energy harvesting. The CdTe layer covers completely the *n*-type CdS nanopillar array and the top contact of Cu/Au bilayer (1 nm/13 nm) was made directly on the CdTe layer by thermal evaporation. The Au/*p*-CdTe interface exhibits good property of relative low resistance Ohmic contact due to the high work function of metal Au. Unfortunately, the good electric contact is achieved at the expense of optical transparency—only ~50% of overall solar energy goes through the Cu/Au bilayer. The aluminum substrate for the growth of the CdS nanopillars is used directly as the back electrode.

Figures 6.40a,b present the experimentally measured *J*–*V* curves, as well as the device PCE and V_{OC}, respectively [78]. The measurements are carried out at several solar radiation powers of 0 (dark), 17, 24, 35, 55, 78, and 100 mW; V_{OC} does not vary too much with the radiation power, J_{SC} decrease quickly with the reduction of radiation power, as shown in the figures. Device PCE remains essentially unchanged with radiation power. This suggests that the device might be used in low radiation circumstances such as indoor case. On the other hand, the device performance is sensitive to the geometric configuration of the nanopillars. Figure 6.41a presents the theoretical simulation of the solar cell efficiency versus H, the height of the CdS nanopillars imbedded in the CdTe thin film while the overall thickness of the CdTe layer is maintained at 1 μm. Apparently, the simulated result agrees with experimental data in general tendency.

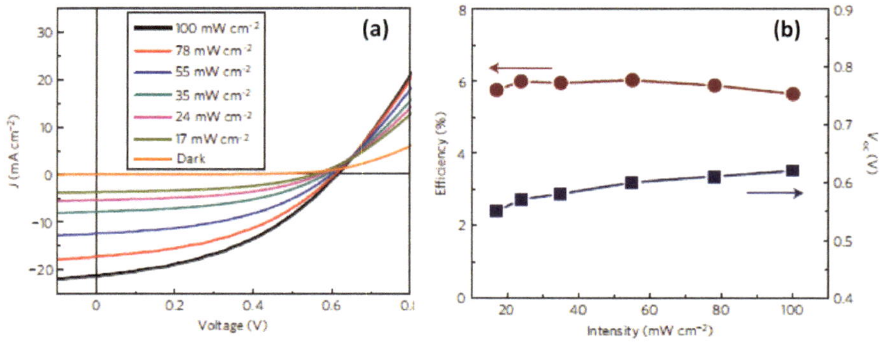

Figure 6.40 Measured J–V curves (a) and corresponding power conversion efficiencies and open-circuit voltages (b) at different radiation powers including dark. Reproduced with permission from [78].

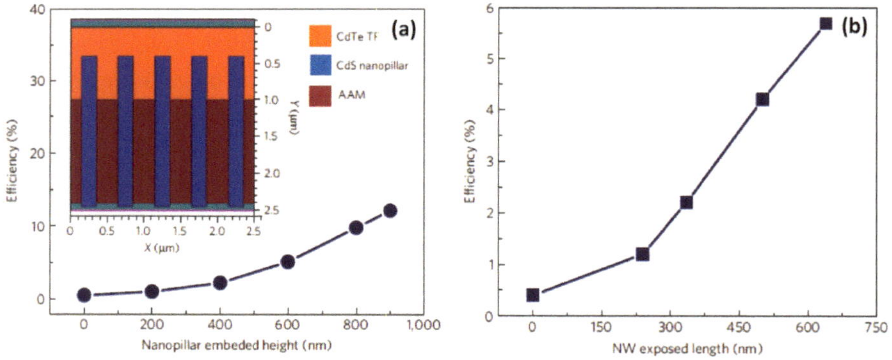

Figure 6.41 (a) Schematic side view of device cross section which shows the template (red, AAM—anodic alumina membrane), CdS nanopillars (blue), and CdTe thin film (orange). (b) PCE as a function of nanopillar length imbedded in CdTe thin film. Reproduced with permission from [78].

The best device gives a PCE of ~6% under AM1.5G illumination, as shown in Fig. 6.41b, with V_{oc} ~ 0.62 V, J_{sc} ~ 21 mA/cm^2, and FF ~ 0.43. It should be noted that the major PCE loss is due to the poor transparency of the front Cu/Au bilayer electrode. It is expected that an excellent front electrode might give rise to significant increase of short-circuit current and in turn the PCE to above 10%.

6.5.6 ZnO Core/Shell Nanowire Solar Cells

Nanostructured ZnO is highly attractive II–VI oxide semiconductor with extensive application in many fields. ZnO is a low-cost and well-studied material with a large direct bandgap of 3.37 eV. Nano-sized ZnO serves as building blocks in many applications, such as nanoelectronic and nano-optoelectronic devices, including solar cells. ZnO can also be grown with a large variety of geometries among which the preparation of ZnO nanowires is simple and mature. Similar to bulk ZnO, as-grown ZnO nanostructures are highly conductive and intrinsically *n*-type due to intrinsic defects. In the application for energy harvesting, ZnO is in most case used for photoanodes due to its highly conductive and transparent properties in large portion of the solar radiation spectrum. Light absorption of ZnO is only limited to UV photons with the energy of 3.37 eV and higher. Although ZnO itself cannot be used as absorber of sunlight, it can incorporate with other lower bandgap materials in the construction of solar cells. For instance, the radial or axial type-II heterojunction can be prepared with a ZnO NW core and another semiconductor shell. This kind of heterostructures are usually quite effective for charge separation where the shell layer serves as a photon absorber to generate electrons–hole pairs and the ZnO core as an electron transporter. In addition, ZnO has good anti-oxidant capacity, as a metal oxide.

Tak and co-workers investigated the ZnO/CdS core/shell NW solar cell with a PCE of 3.53% [54]. The ZnO/CdS core/shell NW heterostructure was fabricated by a solution-based approach on Si, TCO, and Ti substrates with ZnO buffer layers prepared by sputtering. The growth of the ZnO NW arrays was performed in zinc ammonia complex solution. The follow-up growth of the CdS shell layers was performed by successive ion layer adsorption and reaction during which the ZnO NW array were immersed in an aqueous solution containing Cd^{2+} cations (50 mM $Cd(NO_3)_2$) and another aqueous solution containing S^{2-} anions (50 mM

Na_2S) sequentially for 20 s, between which a thorough rinse by de-ionized water was carried out to rinse away left-over ions. The immersion-rinsing-immersion process was repeated 120 times to obtain the ZnO/CdS core/shell NWs.

One of the advantages of the growth route is the easiness in controlling the thickness of the CdS layer. Since the CdS shell acts as the light absorption layer for photocarrier generation, it plays an important role in determining device performance. For instance, J_{sc} of the NWSC enhances from 0.37 mA/cm^2 with bare ZnO NWs to 7.23 mA/cm^2 with CdS shell layers. On this device architecture, a PCE of 3.53% is obtained.

6.6 Existing Problems

Currently, the power conversion efficiency of NWSCs is still lower than that of the commercial wafer-based Si SCs. Although the nanowire geometry is intrinsically superior for excellent light trapping and could offers large room of designing various architectures of NWSCs, several issues limit the device performance of NWSCs. The major problem stems from the high density of surface states due to the large surface areas of the NWs. The defect states act as carrier trapping centers and/or carrier recombination centers, limiting the efficiency of carrier collection. Another issue is related to material quality. NWs for solar cells are generally grown by chemical methods in liquids at low temperature. This leads to high impurity concentration due to various contaminations and defect densities in NWs, which could significantly reduce the carrier mobility and lifetime of minority carriers. Other problems are mostly related to junction geometries. For NWSCs with *n*-type NWs grown on *p*-type substrate, for instance, the low efficiency is mainly caused by small junction area. This can be overcome theoretically by constructing junction with radial geometry. Despite of the currently existing issues in NWSCs, the advantages render them hopeful nanomaterial candidate for solar power conversion devices. One could see the way in which the conversion efficiency of NWSCs raises steadily.

References

1. L. Cao, P. Fan, A. P. Vasudev, J. S. White, Z. Yu, W. Cai, J. A. Schuller, S. Fan, and M. L. Brongersma, Semiconductor nanowire optical antenna solar absorbers, *Nano Lett.,* **10**, 439–445 (2010).
2. H. Sasaki, Scattering suppression and high-mobility effect of size-quantized electrons in ultrafine semiconductor wire structures, *Jpn. J. Appl. Phys.*, **19**, L735 (1980).
3. X. Zhao, C. M. Wei, L. Yang, and M. Y. Chou, Quantum confinement and electronic properties of silicon nanowires, *Phys. Rev. Lett.*, **92**, 236805 (2004).
4. J. Li, H. Y. Yu, and Y. Li, Solar energy harnessing in hexagonally arranged Si nanowire arrays and effects of array symmetry on optical characteristics, *Nanotechnology*, **23**, 194010 (2012).
5. E. Garnett and P. Yang, Light trapping in silicon nanowire solar cells, *Nano Lett.,* **10**, 1082–1087 (2010).
6. O. Ergen, D. J. Ruebusch, H. Fang, A. A. Rathore, R. Kapadia, Z. Fan, K. Takei, A. Jamshidi, M. Wu, and A. Javey, Shape-controlled synthesis of single-crystalline nanopillar arrays by template-assisted vapor–liquid–solid process, *J. Amer. Chem. Soc.,* **132**, 13972–13974 (2010).
7. S. A. Dayeh, C. Soci, X.-Y. Bao, and D. Wang, Advances in the synthesis of InAs and GaAs nanowires for electronic applications, *Nano Today,* **4**, 347–358 (2009).
8. B. Tian, X. Zheng, T. J. Kempa, Y. Fang, N. Yu, G. Yu, J. Huang, and C. M. Lieber, Coaxial silicon nanowires as solar cells and nanoelectronic power sources, *Nature,* **449**, 885–889 (2007).
9. K. Peng, Y. Xu, Y. Wu, Y. Yan, S.-T. Lee, and J. Zhu, Aligned single crystalline Si nanowire arrays for photovoltaic applications, *Small,* **1**, 1062–1067 (2005).
10. K. Ikejiri, J. Noborisaka, S. Hara, J.Motohisa, and T. Fukui, Mechanism of catalyst-free growth of GaAs nanowires by selective area MOVPE, *J. Cryst. Growth*, **298**, 616–619 (2007).
11. M. D. Kelzenberg, D. B. Turner-Evans, B. M. Kayes, M. A. Filler, M. C. Putnam, N. S. Lewis, and H. A. Atwater, Photovoltaic measurements in single-nanowire silicon solar cells, *Nano Lett.,* **8**, 710–714 (2008).
12. C. Colombo, D. Spirkoska, M. Frimmer, G. Abstreiter, and A. Fontcuberta i Morral, Ga-assisted catalyst-free growth mechanism

of GaAs nanowires by molecular beam epitaxy, *Phys. Rev. B: Condens. Matter,* **77**, 155326 (2008).

13. W. Wei, X.-Y. Bao, C. Soci, Y. Ding, Z.-L. Wang, and D. Wang, Direct heteroepitaxy of vertical InAs nanowires on Si substrates for broad band photovoltaics and photodetection, *Nano Lett.,* **9**, 2926–2934 (2009).
14. R.-Q. Zhang, Y. Lifshitz, and S.-T. Lee, Oxide-assisted growth of semiconducting nanowires, *Adv. Mater.,* **15**, 635–640 (2003).
15. P. Mohan, J. Motohisa, and T. Fukui, Fabrication of InP/InAs/InP core-multishell heterostructure nanowires by selective area metalorganic vapor phase epitaxy, *Appl. Phys. Lett.,* **88**, 133105-3 (2006).
16. J. A. Czaban, D. A. Thompson, and R. R. LaPierre, GaAs core–shell nanowires for photovoltaic applications, *Nano Lett.,* **9**, 148–154 (2008).
17. C. Colombo, M. Heib, M.Gratzel, and A. Fontcubertai Morral, Gallium arsenide *p-i-n* radial structures for photovoltaic applications, *Appl. Phys. Lett.,* **94**, 3125435 (2009).
18. M. T. Borgstrom, E. Norberg, P. Wickert, H. A. Nilsson, J. Tragardh, K. A. Dick, G. Statkute, P. Ramvall, K. Deppert, and L. Samuelson, Precursor evaluation for in situ InP nanowire doping, *Nanotechnology,* **19**, 445602-6 (2008).
19. D. Stichtenoth, K. Wegener, C. Gutsche, I. Regolin, F. J. Tegude, W. Prost, M. Seibt, and C. Ronning, P-type doping of GaAs nanowires, *Appl. Phys. Lett.,* **92**, 163107-3 (2008).
20. M. S. Gudiksen, L. J. Lauhon, J. Wang, D. C. Smith, and C. M. Lieber, Growth of nanowire superlattice structures for nanoscale photonics and electronics, *Nature,* **415**, 617–620 (2002).
21. R. S. Wagner and W. C. Ellis, Vapor-liquid-solid mechanism of single crystal growth, *Appl. Phys. Lett.,* **4**(5), 89–90 (1964).
22. R. Muhammad, Z. Othaman, S. Sakrani, and Y. Wahab, Vapor-liquid solid mechanism using gold colloids for the growth of GaAs nanowires, *J. Fundamental Sci.,* **4**, 363–367 (2008).
23. J. H. Lee and R. E. Geer, Templated Si-based Nanowires via Solid-Liquidsolid (SLS) and Vapor-Liquid-Solid (VLS) Growth: Novel Growth Mode, Synthesis, Morphology Control, Characteristics, and Electrical Transport, *Cutting Edge Nanotechnology*, Dragica Vasileska, IntechOpen, DOI: 10.5772/8844. Available from: https://www.intechopen.com/chapters/10097.

24. M. Paulose, O. K. Varghese, and C. A. Grimes, Synthesis of gold-silica composite nanowires through solid-liquid-solid phase growth, *J. Nanosci. Nanotechno.*, **3**(4), 341–346 (2003).

25. J. Zhu and Y. Cui, Photovoltaics: More solar cells for less, *Nat. Mater.*, **9**, 183–184 (2010).

26. X. Ji, X. Yang, and T. Yang, Self-catalyzed growth of vertical GaSb nanowires on InAs stems by metal-organic chemical vapor deposition, *Nanoscale Res. Lett.*, **12**, 428 (2017).

27. G. Koblmüller, S. Hertenberger, K. Vizbaras, M. Bichler, F. Bao, J-P. Zhang, and G. Abstreiter, Self-induced growth of vertical free-standing InAs nanowires on Si(111) by molecular beam epitaxy, *Nanotechnology*, **21**, 365602 (2010).

28. Y. Dong, B. Tian, T. J. Kempa, and C. M. Lieber, Coaxial group III nitride nanowire photovoltaics, *Nano Lett.*, **9**, 2183–2187 (2009).

29. L. J. Lauhon, M. S. Gudiksen, D. Wang, and C. M. Lieber, Epitaxial core–shell and core-multishell nanowire heterostructures, *Nature*, **420**, 57–61 (2002).

30. S. A. Dayeh, E. T. Yu, and D. Wang, Surface diffusion and substrate nanowire adatom exchange in InAs nanowire growth, *Nano Lett.*, **9**, 1967–1972 (2009).

31. S. Plissard, K. A. Dick, X. Wallart, and P. Caroff, Gold-free GaAs/GaAsSb heterostructure nanowires grown on silicon, *Appl. Phys. Lett.*, **96**, 121901-3 (2010).

32. M. Paladugu, J. Zou, Y.-N. Guo, G. J. Auchterlonie, H. J. Joyce, Q. Gao, H. H. Tan, C. Jagadish, and Y. Kim, Novel growth phenomena observed in axial InAs/GaAs nanowire heterostructures, *Small*, **3**, 1873–1877 (2007).

33. Y. N. Guo, J. Zou, M. Paladugu, H. Wang, Q. Gao, H. H. Tan, and C. Jagadish, Structural characteristics of GaSb/GaAs nanowire heterostructures grown by metal–organic chemical vapor deposition, *Appl. Phys. Lett.*, **89**, 231917-3 (2006).

34. T. Martensson, J. Wagner, E. Hilner, A. Mikkelsen, C. Thelander, J. Stangl, B. Ohlsson, A. Gustafsson, E. Lundgren, L. Samuelson, and W. Seifert, Epitaxial growth of indium arsenide nanowires on silicon using nucleation templates formed by self-assembled organic coatings, *Adv. Mater.*, **19**, 1801–1806 (2007).

35. H. D. Park, S. M. Prokes, M. E. Twigg, R. C. Cammarata, and A.-C. Gaillot, Si-assisted growth of InAs nanowires, *Appl. Phys. Lett.*, **89**, 223125 (2006).

36. W. M. Bullis, Properties of gold in silicon, *Solid State Electron.*, **9**, 143–168 (1966).

37. B. Mandl, J. Stangl, T. Mårtensson, A. Mikkelsen, J. Eriksson, L. S. Karlsson, G. Bauer, L. Samuelson, and W. Seifert, Au-free epitaxial growth of in as nanowires, *Nano Lett.*, **6**(8), 1817–1821 (2006).

38. M. Mattila, T. Hakkarainen, H. Lipsanen, H. Jiang, and E. I. Kauppinen, Catalyst-free growth of In(As)P nanowires on silicon, *Appl. Phys. Lett.*, **89**, 063119 (2006).

39. A. Wolfsteller, N. Geyer, T.-K. Nguyen-Duc, P. Das Kanungo, N.D. Zakharov, M. Reiche, W. Erfurth, H. Blumtritt, S. Kalem, P. Werner, U. Gösele, Comparison of the top-down and bottom-up approach to fabricate nanowire-based silicon/germanium heterostructures, *Thin Solid Films*, **518**, 2555–2561 (2010).

40. X. Li, Metal assisted chemical etching for high aspect ratio nanostructures: A review of characteristics and applications in photovoltaics, *Curr. Opin. Solid St. M.*, **16**(2), 71–81 (2012).

41. A. Pan, R. Liu, M. Sun, and C.-Z. Ning, Spatial composition grading of quaternary ZnCdSSe alloy nanowires with tunable light emission between 350 and 710 nm on a single substrate, *ACS Nano*, **4**, 671–680 (2010).

42. Z. Wu, J. B. Neaton, and J. C. Grossman, Charge separation via strain in silicon nanowires, *Nano Lett.*, **9**, 2418–2422 (2009).

43. A. Bouazzi, Y. Marfaing, and J. Mimila-Arroyo, Optimization of graded band gap CdHgTe solar cells, *Rev. Phys. Appl.* (Paris), **13**, 145 (1978).

44. G. F. Brown, J. W. Ager III, W. Walukiewicz, and J. Wu, Finite element simulations of compositionally graded InGaN solar cells, *Sol. Energy Mater. Sol. Cells*, **94**, 478–483 (2010).

45. L. Yang, T. Zhang, H. Zhou, S. C. Price, B. J. Wiley, and W. You, Solution-processed flexible polymer solar cells with silver nanowire electrodes, *ACS Appl. Mater. Interfaces*, **3**, 4075–4084 (2011).

46. J. Bae, H. Kim, X.-M. Zhang, C. H. Dang, Y. Zhang, Y. J. Choi, A. Nurmikko, and Z. L. Wang, Si nanowire metal–insulator–semiconductor photodetectors as efficient light harvesters, *Nanotechnology*, **21**, 095502 (2010).

47. T. J. Kempa, B. Tian, D. R. Kim, J. Hu, X. Zheng, and C. M. Lieber, Single and tandem axial *p-i-n* nanowire photovoltaic devices, *Nano Lett.*, **8**, 3456–3460 (2008).

48. S. Hoffmann, J. Bauer, C. Ronning, Th. Stelzner, J. Michler, C. Ballif, V. Sivakov, and S. H. Christiansen, Axial *p–n* junctions realized in silicon nanowires by ion implantation, *Nano Lett.,* **9**, 1341–1344 (2009).
49. N. Huang, C. Lin, and M. L. Povinelli, Limiting efficiencies of tandem solar cells consisting of III–V nanowire arrays on silicon, *J. Appl. Phys.,* **112**, 064321 (2012).
50. K. Wang, J. Chen, W. Zhou, Y. Zhang, Y. Yan, J. Pern, and A. Mascarenhas, Direct growth of highly mismatched type II ZnO/ZnSe core/shell nanowire arrays on transparent conducting oxide substrates for solar cell applications, *Adv. Mater.,* **20**, 3248–3253 (2008).
51. Y. Zhang, L.-W. Wang, and A. Mascarenhas, 'Quantum coaxial cables' for solar energy harvesting, *Nano Lett.,* **7**, 1264–1269 (2007).
52. K. A. Dick, S. Kodambaka, M. C. Reuter, K. Deppert, L. Samuelson, W. Seifert, L. R. Wallenberg, and F. M. Ross, The morphology of axial and branched nanowire heterostructures, *Nano Lett.,* **7**, 1817–1822 (2007).
53. C. Y. Wen, M. C. Reuter, J. Bruley, J. Tersoff, S. Kodambaka, E. A. Stach, and F. M. Ross, Formation of compositionally abrupt axial heterojunctions in silicon-germanium nanowires, *Science,* **326**, 1247–1250 (2009).
54. Y. Tak, S. J. Hong, J. S. Lee, and K. Yong, Fabrication of ZnO/CdS core/shell nanowire arrays for efficient solar energy conversion, *J. Mater. Chem.,* **19**, 5945–5951 (2009).
55. V. Sivakov, G. Andrä, A. Gawlik, A. Berger, J. Plentz, F. Falk, and S. H. Christiansen, Silicon nanowire-based solar cells on glass: Synthesis, optical properties, and cell parameters, *Nano Lett.,* **9**, 1549–1554 (2009).
56. N. Huang and M. L. Povinelli, Design of passivation layers on axial junction GaAs nanowire solar cells, *IEEE J. Photovol.*, **4**(6), 1511–1517 (2014).
57. L. Hu and G. Chen, Analysis of optical absorption in silicon nanowire arrays for photovoltaic applications, *Nano Lett.,* **7**, 3249–3252 (2007).
58. B. M. Kayes, H. A. Atwater, and N. S. Lewis, Comparison of the device physics principles of planar and radial *p–n* junction nanorod solar cells, *J. Appl. Phys.,* **97**, 114302-10 (2005).
59. K. Sun, A. Kargar, N. Park, K. N. Madsen, P. W. Naughton, T. Bright, Y. Jing, and D. Wang, Compound semiconductor nanowire solar cells, *IEEE J. Sel. Top. Quant.,* **17**, 1033–1049 (2011).

60. E. C. Garnett and P. Yang, Silicon nanowire radial *p*–*n* junction solar cells, *J. Amer. Chem. Soc.*, **130**, 9224–9225 (2008).

61. C.-Y. Huang, Y.-J. Yang, J.-Y. Chen, C.-H. Wang, Y.-F. Chen, L.-S. Hong, C.-S. Liu, and C.-Y. Wu, P-Si nanowires/SiO_2/*n*-ZnO heterojunction photodiodes, *Appl. Phys. Lett.*, **97**, 013503 (2010).

62. M. D. Kelzenberg, S. W. Boettcher, J. A. Petykiewicz, D. B. Turner-Evans, M. C. Putnam, E. L. Warren, J. M. Spurgeon, R. M. Briggs, N. S. Lewis, and H. A. Atwater, Enhanced absorption and carrier collection in Si wire arrays for photovoltaic applications, *Nat. Mater.*, **9**(4), 239–244 (2010).

63. K.-Q. Peng, X. Wang, X.-L. Wu, and S.-T. Lee, Platinum nanoparticle decorated silicon nanowires for efficient solar energy conversion, *Nano Lett.*, **9**, 3704–3709 (2009).

64. M.-S. Hu, H.-L. Chen, C.-H. Shen, L.-S. Hong, B.-R. Huang, K.-H. Chen, and L.-C. Chen, Photosensitive gold-nanoparticle-embedded dielectric nanowires, *Nat. Mater.*, **5**, 102–106 (2006).

65. Y. A. Akimov, W. S. Koh, S. Y. Sian, and S. Ren, Nanoparticle-enhanced thin film solar cells: metallic or dielectric nanoparticles? *Appl. Phys. Lett.*, **96**, 073111-3 (2010).

66. C. Pan, Z. Luo, C. Xu, J. Luo, R. Liang, G. Zhu, W. Wu, W. Guo, X. Yan, J. Xu, Z. L. Wang, and J. Zhu, Wafer-scale high-throughput ordered arrays of si and coaxial Si/$Si_{1-x}Ge_x$ wires: fabrication, characterization, and photovoltaic application, *ACS Nano*, **5**(8), 6629–6636 (2011).

67. J. F. Geisz, D. J. Friedman, J. S. Ward, A. Duda, W. J. Olavarria, T. E. Moriarty, J. T. Kiehl, M. J. Romero, A. G. Norman, and K. M. Jones, 40.8% efficient inverted triple-junction solar cell with two independently metamorphic junctions, *Appl. Phys. Lett.*, **93**, 123505 (2008).

68. W. Gutera, J. Schöne, S. P. Philipps, M. Steiner, G. Siefer, A. Wekkeli, E. Welser, E. Oliva, A. W. Bett, and F. Dimroth, Current-matched triple-junction solar cell reaching 41.1% conversion efficiency under concentrated sunlight, *Appl. Phys. Lett.*, **94**, 223504 (2009).

69. Y. B. Tang, Z. H. Chen, H. S. Song, C. S. Lee, H. T. Cong, H. M. Cheng, W. J. Zhang, I. Bello, and S. T. Lee, Vertically aligned *p*-type single-crystalline GaN nanorod arrays on *n*-type Si for heterojunction photovoltaic cells, *Nano Lett.*, **8**(12), 4191–4195 (2008).

70. K. Haraguchi, T. Katsuyama, K. Hiruma, and K. Ogawa, GaAs *p*–*n* junction formed in quantum wire crystals, *Appl. Phys. Lett.*, **60**(6), 745–747 (1992).

71. C. Colombo, M. Heiβ, M. Grätzel, and A. Fontcuberta, I. Morral, Gallium arsenide *p-i-n* radial structures for photovoltaic applications, *Appl. Phys. Lett.*, **94**, 173108–3 (2009).

72. M. Yamaguchi, A. Yamamoto, N. Uchida, and C. Uemura, A new approach for thin film InP solar cells, *Sol. Cells,* **19**(1), 85–96 (1986).

73. H. Goto, K. Nosaki, K. Tomioka, S. Hara, K. Hiruma, J. Motohisa, and T. Fukui, Growth of core–shell InP nanowires for photovoltaic application by selective-area metal organic vapor phase epitaxy, *App. Phys. Exp.*, **2**, 035004 (2009).

74. Z.-Q. Liu, R. Guo, G.-R. Li, Q. Bu W.-X. Zhao, and Y.-X. Tong, Facile electrodeposition of large-area CdO and $Cd_{1-x}Co_xO$ curved nanowires, *Electrochimica Acta,* **59**, 449–454 (2012).

75. K.-S. Cho, D. V. Talapin, W. Gaschler, and C. B. Murray, Designing PbSe nanowires and nanorings through oriented attachment of nanoparticles, *J. Am. Chem. Soc.*, **127**(19), 7143 (2005).

76. S. Y. Jang, Y. M. Song, H. S. Kim, Y. J. Cho, Y. S. Seo, G. B. Jung, C.-W. Lee, J. Park, M. Jung, J. Kim, B. Kim, J.-G. Kim, Y.-J. Kim, Three synthetic routes to single-crystalline PbS nanowires with controlled growth direction and their electrical transport properties, *ACS Nano,* **4**, 2391–2401 (2010).

77. S.-M. Yong, P. Muralidharan, D. S. Kim, and D. K. Kim, One-step hydrothermal synthesis of carbon-coated PbTe nanowires for thermoelectric applications, *Rev. Adv. Mater. Sci.*, **28**, 13–16 (2011).

78. Z. Fan, H. Razavi, J.-W. Do, A. Moriwaki, O. Ergen, Y.-L. Chueh, P. W. Leu, J. C. Ho, T. Takahashi, L. A. Reichertz, S. Neale, K. Yu, M. Wu, J. W. Ager, and A. Javey, Three-dimensional nanopillar-array photovoltaics on low-cost and flexible substrates, *Nat. Mater.*, **8**, 648 (2009).

79. R. Kapadia, Z. Fan, and A. Javey, Design constraints and guidelines for CdS/CdTe nanopillar based photovoltaics, *Appl. Phys. Lett.*, **96**, 103116-3 (2010).

80. X. Wang, K. L. Pey, C. H. Yip, E. A. Fitzgerald, and D. A. Antoniadis, Vertically arrayed Si nanowire/nanorod-based core–shell *p–n* junction solar cells, *J. Appl. Phys.*, **108**, 124303 (2010).

81. T. J. Kempa, J. F. Cahoon, S.-K. Kim, R. W. Day, D. C. Bell, H.-G. Park, and C. M. Lieber, Coaxial multishell nanowires with high-quality electronic interfaces and tunable optical cavities for ultrathin photovoltaics, *Proc. Natl. Acad. Sci. U.S.A.*, **109**, 1407–1412 (2012).

82. L. Novotny and N. van Hulst, Antennas for light, *Nat. Photon.*, **5**, 83–90 (2011).

83. L. Cao, J. S. White, J.-S. Park, J. A. Schuller, B. M. Clemens, and M. L. Brongersma, Engineering light absorption in semiconductor nanowire devices, *Nature Mater.*, **8**, 643–647 (2009).

84. D. H. Son, S. M. Hughes, Y. Yin, and A. Paul Alivisatos, Cation exchange reactions in ionic nanocrystals, *Science*, **306**, 1009–1012 (2004).

85. J. Tang, Z. Huo, S. Brittman, H. Gao, and P. Yang, Solution-processed core–shell nanowires for efficient photovoltaic cells *Nat. Nano.*, **6**, 568–572 (2011).

86. A. Rothwarf and A. M. Barnett, Design analysis of thin-film CdS–Cu_2S solar-cell, *IEEE Trans. Electron. Dev.*, **24**, 381–387 (1977).

87. Z. Li, J. Wang, N. Singh, and S. Lee, Optical and electrical study of core–shell silicon nanowires for solar applications, *Opt. Exp.*, **19**(S5), A1057–A1066 (2011).

88. J. Chen, T. Subramani, W. Jevasuwan, and N. Fukata, Improvement of silicon nanowire solar cells made by metal catalyzed electroless etching and nano imprint lithography, *Jap. J. Appl. Phys.*, **56**(4S), 04CP03 (2017).

89. X. Yan, C. Zhang, J. Wang, X. Zhang, and X. Ren, A high-efficiency Si nanowire array/perovskite hybrid solar cell, *Nanoscale Res. Lett.*, **12**, 14 (2017).

90. H.-D. Um, K.-T. Park, J.-Y. Jung, X. Li, K. Zhou, S.-W. Jee, and J.-H. Lee, Incorporation of a self-aligned selective emitter to realize highly efficient (12.8%) Si nanowire solar cells, *Nanoscale*, **6**, 5193–5199 (2014).

91. G. Mariani, P.-S. Wong, A. M. Katzenmeyer, F. Léonard, J. Shapiro, and D. L. Huffaker, Patterned radial GaAs nanopillar solar cells, *Nano Lett.*, **11**, 2490–2494 (2011).

92. C. Gutsche, A. Lysov, D. Braam, I. Regolin, G. Keller, G. Z.-A. Li, M. Geller, M. Spasova, W. Prost, and F.-J. Tegude, *n*-GaAs/InGaP/*p*-GaAs core-multishell nanowire diodes for efficient light-to-current conversion, *Adv. Funct. Mater.*, **22**, 929–936 (2012).

93. M. Yao, N. Huang, S. Cong, C.-Y. Chi, M. A. Seyedi, Y.-T. Lin, Y. Cao, M. L. Povinelli, P. D. Dapkus, and C. Zhou, GaAs nanowire array solar cells with axial *p*–*i*–*n* junctions, *Nano Lett.*, **14**, 3293–3303 (2014).

94. J. Wallentin, N. Anttu, D. Asoli, M. Huffman, I. Åberg, M. H. Magnusson, G. Siefer, P. Fuss-Kailuweit, F. Dimroth, B. Witzigmann, H. Q. Xu, L. Samuelson, K. Deppert, and M. T. Borgström, InP nanowire array solar cells achieving 13.8% efficiency by exceeding the ray optics limit, *Science,* **339**, 1057 (2013).

Chapter 7

Hybrid Nano Solar Cells

7.1 About Hybrid Solar Cells

Hybrid solar cells (HSCs) specifically refer to those using both inorganic and organic semiconductors, so that they combine the advantages of both organic and inorganic materials. In a hybrid nano solar cell (NSC), the photoactive layer is very often made by mixing semiconductor nanomaterials into a polymer matrix. Hybrid NSCs have demonstrated a great potential for flexible devices using cost-effective solution processing techniques. With a variety of architectures due to a great deal of organic materials one could select, a HSC features the characteristics that the conjugated polymer mostly acts as light absorber, donor material, as well as hole transporter, when inorganic material functions as both the acceptor and the electron transport layer. The selection of the materials for HSCs follows the process which requires a band alignment to guide the motion of photoexcited electrons and holes.

Figure 7.1 presents the band alignment of a typical HSC. The two materials form heterojunction where ZnO nanorod array is blended with PCBM polymer. The band profile of the device is shown in Fig. 7.1b. Similar to the conduction band (CB) and the valence band (VB) semiconductors, the "CB" and "VB" in organic semiconductors are termed as lowest unoccupied molecular orbital (LUMO) and highest occupied molecular orbital (HOMO), respectively. Together with the band alignment, the junction

Introduction to Nano Solar Cells
Ning Dai

ISBN 978-981-4877-49-7 (Hardcover), 978-1-003-13198-4 (eBook)
www.jennystanford.com

offers a driving force that drives the separation of photoexcited electrons from holes and the carrier transfer toward respective electrodes. Note that the two materials with different bandgaps are beneficial for the absorption of wide light spectrum. With a large variety of material choices, HSCs are potentially lower cost and easier to incorporate with a roll-to-roll processing for scaling up the production.

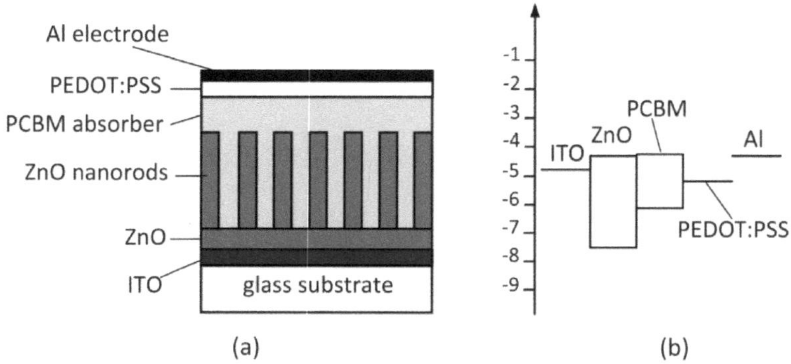

Figure 7.1 Schematics of a typical HSC (a) and its band alignment (b).

The price distribution in *c*-Si solar cells is roughly 1/3 Si material + 1/3 device fabrication + 1/3 others. The expense in Si comes from the requirement of high-purity (6N) crystalline material that is achieved through complicated purification process and high-temperature crystal growth. The device fabrication includes the formation of a *p–n* junction that usually uses expensive ion implantation followed by high-temperature annealing to reinstall the crystal quality. Studies on hybridized solar cells (organic-inorganic solar cells) aim to cut down the cost by using inexpensive materials (or at least less crystalline Si) and Schottky junction formed by conducting-polymer/Si, instead of conventional *p–n* junction, for carrier extraction.

Conventional solar cells built from polycrystal silicon wafers have high conversion efficiency around 20%, but they are expensive and not environmentally friendly since the material purification and crystal growth processes requires high-temperature treatment that is energy intensive. The motivation of using hybrid materials includes simplifying material processes and device preparation procedures and, more importantly, lowering down preparation temperature. HSCs such as photo-electrochemical cells (typically dye-sensitized SCs) are expected to be the cheap alternatives for conventional wafer-based Si SCs currently dominating the PV market. Hybrid could refer to devices or structures using very different properties of materials for active parts, including device concepts such as solid state dye-sensitized SCs and SCs based on the bulk heterojunctions formed by the combinations of semiconducting polymers and a variety of different nanoparticles such as Si and ZnO. In some literature, SCs consisting of carbon nanotubes and conventional semiconductors, for instance, are referred to as hybrid ones. In this book, the so-called hybrid solar cells refer to those formed by the combination of organic and inorganic materials for the active parts.

The hybrid concept for SCs has become more and more interesting in recent years. Both the nano-sized inorganic semiconductors and organic materials are cost effective and can be prepared and processed at low temperature. Being blended intimately together, the HSCs combine the unique photoelectric properties of inorganic nanosemiconductors with the ease of the thin-film technologies of conjugated polymers. In addition, physical parameters of the inorganic nanosemiconductors can be tuned by their sizes and functionalities of organic polymers can be tailored by molecular design. The major advantage for HSC devices lies in the versatility in the device manufacturing.

7.2 Fundamental Material Properties

There are limited choices for inorganic semiconductors for the wide solar spectrum with the major radiation energy covering the range from ultraviolet to mid infrared. Particularly, the nature lacks suitable narrow-bandgap semiconductors for energy conversion in the infrared solar radiation. The existing narrow-bandgap semiconductors are expensive and require high-cost equipment to fabricate. Thus, the organic narrow-bandgap materials are highly demanded for HSCs. It is also essential for the organic materials to form good contact with semiconductors and metals in the devices, since high contact resistance easily gives rise to low filling factor and conversion efficiency. In addition, relatively low photon absorption coefficient requires thick layer of organic semiconductors. The carrier mobility in the organic semiconductors needs to be high enough for reducing the carrier recombination. Other requirements for organic semiconductors include high conductivity, ease of fabrication, and enabling large area and quick production, as required for inorganic SC materials.

7.2.1 Optical and Electrical Properties of Polymers

Apparently, the suitable band alignment is the fundamental requirement in order to guide the transport of electrons and holes toward respective electrodes to complete the photovoltaic effect, as soon as excitons are generated upon photon absorption. In addition, the distinct difference between an HSC and a conventional one lies in the interface formed by organic and inorganic materials. An electron, when traveling across the interface, would experience sudden changes in energy, carrier lifetime, effective mass, conductivity, diffusion length, wave packet size, scattering intensity, etc. If the sudden changes are severe, the interface applies a strong reflection to the electron, which should be avoided. Therefore, it is essential to smoothen the changes as much as possible, which can be done through careful material design and device construction.

HSCs offer a broad selection for the combination of organic and inorganic materials, such as polymer–Si combination, polymer–nanorod composite, and fullerene–quantum dot composites. Interestingly, those nanomaterials have their size in the nanometer range at least in one dimension. Their optical properties, including the photoabsorption wavelength, can be varied by their sizes in nanometer range. The very large surface-area-to-volume ratio of the organic/nanosemiconductor composites is beneficial to both the photon absorption and carrier separation.

In most cases, the organics are used as hole transport materials and the semiconductors of various nano forms as electron transport ones. The 3D heterojunction layer in a HSC should provide efficient pathways for electrons to travel to the electrode. Thus, the nanomaterial is often in the form of nanorods to provide the pathways since, otherwise, the electrons have to transfer through hopping mechanism in polymers with low conductivity. The conversion efficiency of the HSCs relies on the geometry of the nanomaterials such as size, aspect ratio, volume fraction, etc. To a large extent, the geometry of the nanomaterials affects the nanoparticle dispersion and pathways for electron transport.

The development of HSCs is made possible due large to the successful preparation of various conducting polymers that have been studied intensively in the recent 20 to 30 years. The Nobel Prize in Chemistry was awarded to Heeger, MacDiarmid, and Shirakawa to honor their discovery and creative work on conducting polymers in 2000. Not only materials, but composites and structures add ways toward the realization of many optoelectronics devices, including SCs. The important conducting polymers include poly(3,4-ethylenedioxythiophene): sulfonated polystyrene, Phenyl-C_{61}-butyric acid methyl ester, polythiophenes, poly(*p*-phenylene vinylene), etc. The conductivity of some organic semiconductors can reach 1000 S/cm, about 2 to 3 orders of magnitudes lower than the most conductive metals such as silver and copper. Iodine-doped poly(3-dodecylthiophene) has an electric conductivity ~1000 S/cm.

7.2.2 Some Electron and Hole Transport Polymers

PEDOT:PSS

PEDOT:PSS with respective molecular formulas of $[C_6H_6O_2S]_m$ and $[C_8H_8SO_3]_n$ is a mixture of two ionomers, namely, poly(3,4-ethylenedioxythiophene) and sulfonated polystyrene, with their molecular structures and chemical formulas as shown in Fig. 7.2. PEDOT is a conjugated polymer carrying positive charges, while part of the sulfonyl groups are deprotonated so that they are negatively charged. Adding some organic materials, such as dimethyl sulfoxide and ionic liquids, into PEDOT:PSS might change its conductivity by orders of magnitude, making the doped PEDOT:PSS a transparent and conductive material suitable for the use as a transparent electrode. Some post-treatments with cosolvents and geminal diols, for instance, are also proven to be effective for the enhancement of material conductivity. Xia and co-workers have demonstrated that the conductivity of PEDOT:PSS, when treated with sulfuric acids, is comparable to those of indium tin oxide (ITO) and Al-doped zinc oxide (AZO), the popular transparent electrode materials [1].

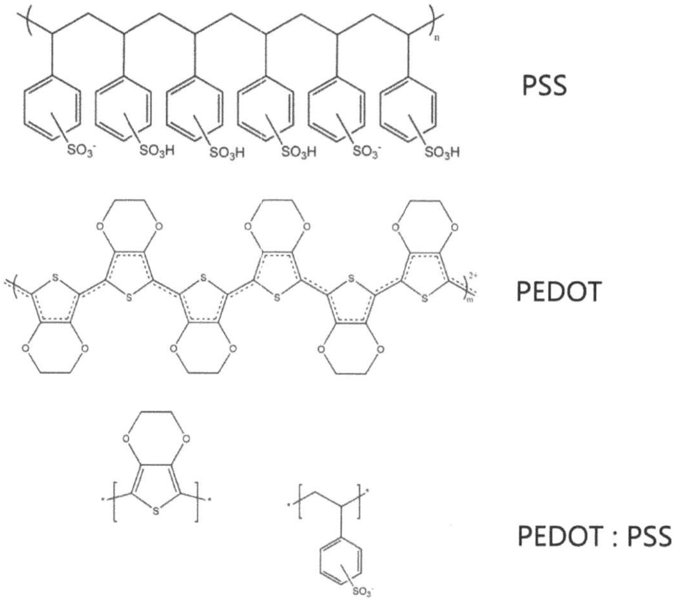

Figure 7.2 Structural formulas of PSS, PEDOT, and PEDOT:PSS. PEDOT: PSS is a mixed monomer of PEDOT and PSS.

Phenyl-C_{61}-butyric acid methyl ester

[6,6]-Phenyl-C_{61}-butyric acid methyl ester ($PC_{61}BM$) is a fullerene derivative of the C_{60} buckyball with the structural formula shown in Fig. 7.3, widely used in the fabrication of organic solar cells. $PC_{61}BM$ is a *p*-type organic semiconductor and soluble in chlorobenzene, allowing for solution process, useful for the fabrication of "printable" solar cells.

Polythiophenes

Polythiophenes (polymerized thiophenes, PTs) are a sulfur heterocycle, with an unsubstituted polythiophene monomer as shown in Fig. 7.4. One of the interesting phenomena of the material is that its optical properties are tunable by varying the temperature and applying electric field. The color of the material can even be changed by attachment of other molecules. PTs can be made conducting via doping by adding electrons (*n*-type doping) or removing electrons (*p*-type doping) from the conjugated π-orbitals, accompanied by forming a bipolaron structure. The enhanced electrical conductivity of the material is due to the delocalization of electrons along the polymer backbones where the overlapping of π-orbitals gives rise to extended energy band.

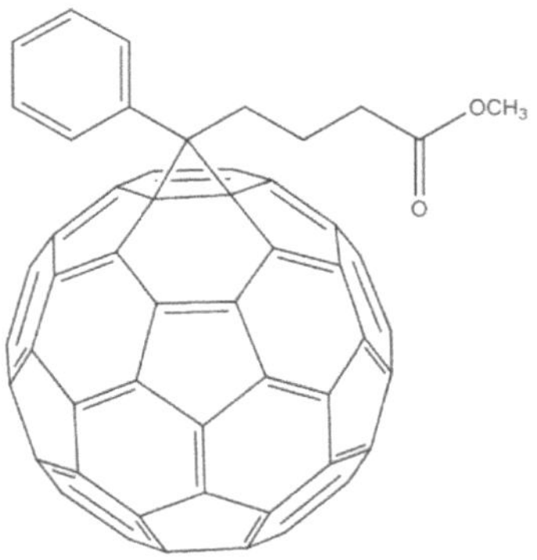

Figure 7.3 Structural formula of polymerized thiophenes.

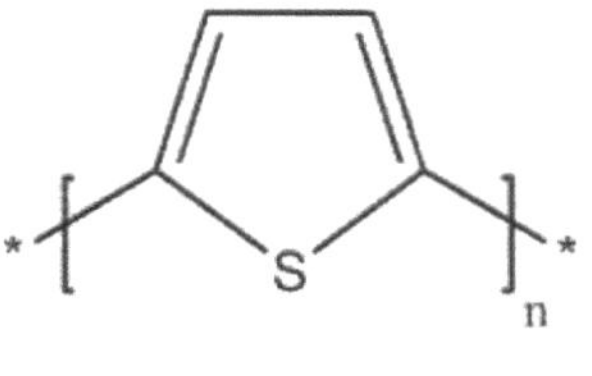

Figure 7.4 The monomer unsubstituted polythiophene (PT).

Usually, organic semiconductors can be doped with high concentration of dopants up to 30%, while it is usually less than 1% in inorganic semiconductors. Thus, doping in an organic semiconductor very often gives rise to another material. A variety of reagents have been used to dope PTs and the PTs of highest conductivity are doped by iodine and bromine. For PTs, *n*-type doping appears to be more difficult than *p*-type doping, manifested by the fact that both the element selectivity and the efficiency for *n*-type doping are less than *p*-type doping. The *p*-type doping can be realized by using counterions during material growth, though doping an as-grown PT film is possible. In electrochemical preparation, for instance, counterions in a solvent are doped directly in the polymer while it is growing on the electrode under oxidization condition. In this way, thick PT films can be growth with uniform doping.

The conductivity of PTs is high, but below 1000 S/cm. PTs have been synthesized either chemically or electrochemically. When chemically preparing PTs, oxidants or catalysts are usually used. Electrochemical polymerization uses a voltage applied across the solution of thiophenes in an electrolyte. During the electrochemical process, the PT film forms on the anode.

Poly(*p*-phenylene vinylene)

Poly(*p*-phenylene vinylene) (polyphenylene vinylene PPV), with its structural and chemical formulas shown in Fig. 7.5a, is also a conducting polymer with many derivatives. PPV can be synthesized into thin films with crystalline quality and high purity. The crystalline PPV films synthesized by the polymeric precursor route have a monoclinic unit cell with c (chain axis) = 0.658, a = 0.790, b = 0.605 nm, and the monoclinic angle = 123° (see Fig. 7.5b). There are quite a few methods that have been adopted for the preparation of the material. One of them are the so-called step growth coupling reactions to fabricate PPV with 5–10 repeat units. PPV is insoluble in water, but after it is incorporated with alkoxy and phenyl, for instance, its water-solubility can be greatly improved.

PPV and its derivatives have been used in photoelectronic applications as active materials. PPV became well-known after the fabrication of the first polymer-based light-emitting diode (LED) in 1989 in which PPV was used as the material for the emission layer [2]. PPV is a diamagnetic material that can be prepared with high purity. Pure PPV is, however, highly resistive.

Conductive PPV can be obtained by doping and the common dopants include iodine, alkali metals, and some acids. In fact, PPV is one of the very few polymers that can be synthesized into ordered crystalline structure.

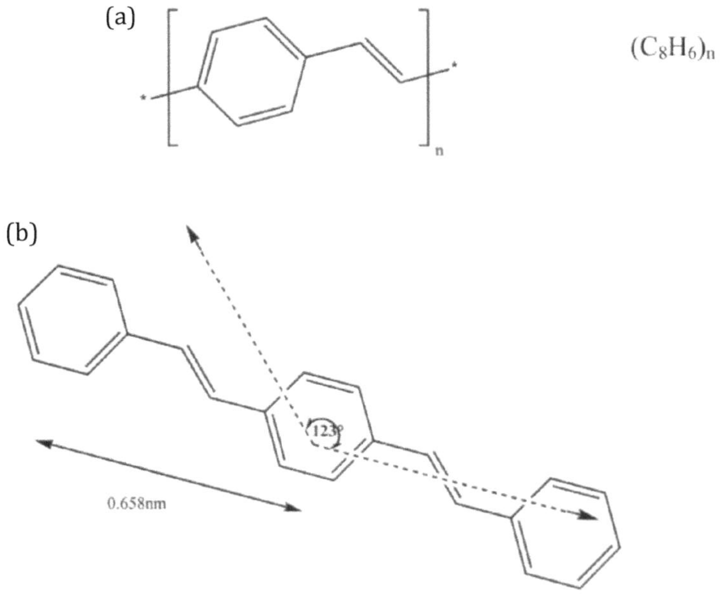

Figure 7.5 Structural and chemical formulas of PPV (a) and a short PPV chain (b).

In HSCs, the role of donor/acceptor interface is to split electrons and holes excited by photons, similar to the built-in electric field of *n*-type/*p*-type interface in conventional wafer-based Si SCs. High-quality donor/acceptor interface is required not only for low charge trapping defect density, but also for proper bandgap for efficient solar light absorption and energy level arrangement to direct transport of electrons and holes. For instance, the bandgap of CdSe is 2.10 eV, with an electron affinity ranges from 4.4 to 4.7 eV. When CdSe nanocrystal and MEH-PPV, with an electron affinity of 3.0 eV, are blended intimately together, the 1.5 eV difference in the electron affinities is large enough to drive electron transferring from CdSe into MEH-PPV polymer. Other material properties such as carrier mobilities and carrier lifetimes are also critical for the device performance. CdSe, for instance, has a high electron mobility of 600 $cm^2 \cdot V^{-1} \cdot s^{-1}$, while, unfortunately, polymers usually have very low carrier mobility. Structural and chemical formulas of MEH-PPV are presented in Fig. 7.6a.

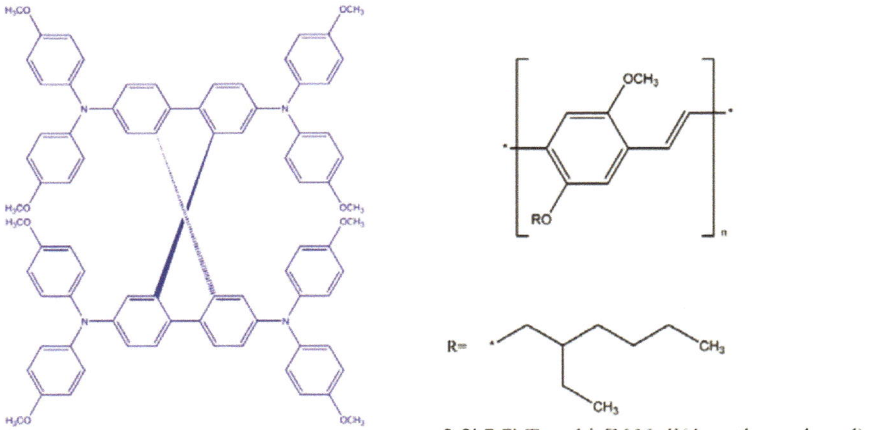

poly[2-methoxy-5-(2-ethylhexyloxy) phenylenevinylene-1,4-diyl]

2,2',7,7'-Tetrakis[N,N-di(4-methoxyphenyl) amino]-9,9'-spirobifluorene

(a) (b)

Figure 7.6 Structural and chemical formulas of MEH-PPV (a) and spiro-OMeTAD (b).

Spiro-OMeTAD, an organic non-crystalline, is a *p*-type conductive polymer that has been widely used as the hole transport material in solid state dye-sensitized solar cells. Figure 7.6b shows the structural and chemical formulas of spiro-OMeTAD. Spiro-OMeTAD has a high glass-transition temperature and good solubility in organic solvents since, as an organic material of small molecules, it infiltrates well into the nano-sized spaces and forms close contact with nanomaterials on substrates, which is the fundamental base for the nano solar cell architecture. There are many organic materials that have been used for SCs. Figure 7.7 presents structural and chemical formulas of some of them.

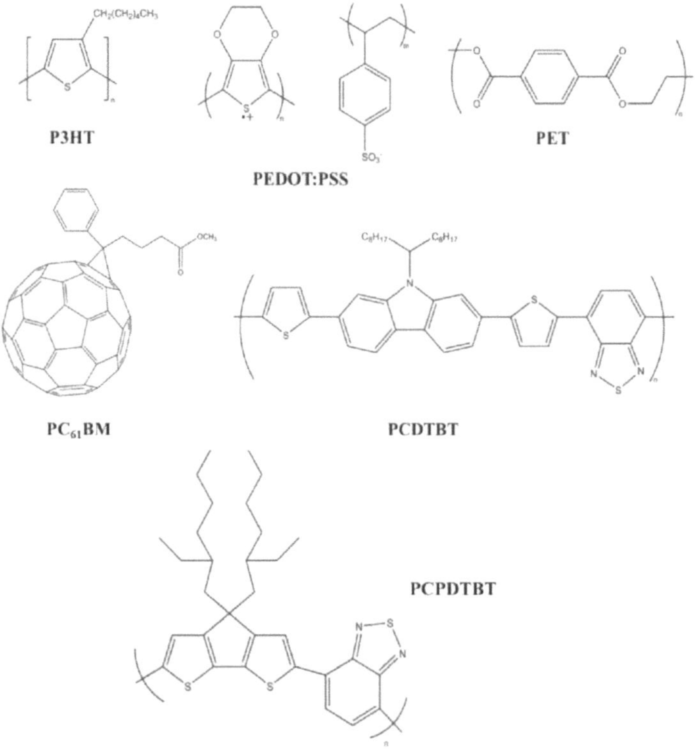

Figure 7.7 The structural and chemical formulas of poly-3-hexylthiophene (P3HT), poly(3,4-ethylenedioxythiophene) polystyrene sulfonate (PEDOT:PSS), polyethylene terephthalate (PET), [6,6]-phenyl-C_{61}-butyric acid methyl ester ($PC_{61}BM$), poly-[[9-(1-octylnonyl)-9H-carbazole-2,7-diyl]-2,5-thiophenediyl-2,1,3-benzothiadiazole-4,7-diyl-2,5-thiophenediyl] (PCDTBT), poly[2,6-(4,4-bis-(2-ethylhexyl)-4H-cyclopenta [2,1-b;3,4-b′]dithiophene)-alt-4,7(2,1,3-benzothiadiazole)] (PCPDTBT).

7.3 Device Architectures and Working Principles

7.3.1 Device Work Principle

The photovoltaic process relies on energy band profile in a SC that directs the motion of carriers. The key part of a conventional Si is the *p*–*n* junction with band profile schematically shown in Fig. 7.8a.

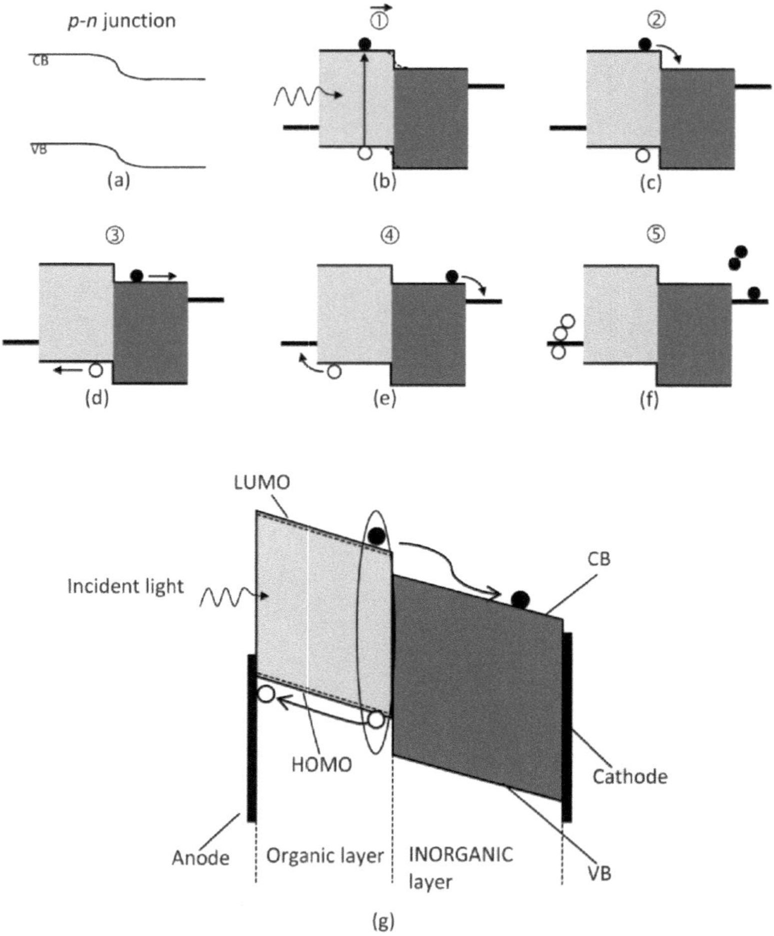

Figure 7.8 An organic/inorganic hybrid heterojunction can be constructed and used for solar cells. Schematic band profile in the conduction band (CB) and the valence band (VB) for a conventional *p-n* junction (a). A self-consistent process (charge redistribution) smears out the band discontinuities at the hetero-interface between organic and inorganic materials as shown by the dashed curves at the interface in (b), which leads to a band alignment similar to that in (a). A complete photovoltaic process for electrons in CB and holes in VB (b–f). Typical band alignment of HSCs with the organic material acts as donor and hole transport layers and the inorganic material as the acceptor and the electron transport layer (g). Note that the exciton binding energy in the inorganic layer are ignored.

The p–n junction functions to disassociate photoexcited electrons and holes. The corresponding key part in a HSC is the heterojunction consisting of a donor and acceptor materials. Note that here a donor material refers to the material that contributes electrons in the photovoltaic process. Thus, the donor material acts as a hole transport layer, similar to a p-type doped semiconductor layer of a p–n junction. An acceptor material contributes holes and acts as electron transport layer. As shown in Fig. 7.8b, the self-consistent charge redistribution upon the formation of the heterojunction smears out the band discontinuities at the hetero-interface, as shown by the dashed curve in the figure. Therefore, the band arrangement of heterojunction is very similar to that of a p–n junction.

Assuming that the acceptor material is an inorganic semiconductor and donor material is an organic semiconductor, Figs. 7.8b–f depict the working principle of a hybrid solar cell. An incident photon excites an exciton and the exciton moves toward the hetero-interface through diffusion (Figs. 7.8b, step ①). Near the junction region, the electron and the hole are separated by the potential due to the band discontinuity (Figs. 7.8c, step ②). In fact, the HOMO (LUMO) and the VB (CB) levels are tilted as shown in Fig. 7.8g, owing to the self-consistent process featuring the charge redistribution after formation of the heterostructure. In the organic layer, an exciton has a limited diffusion length—the electron and the hole will recombine after traveling a diffusion length. The diffusion length in the organic layer is on the order of 5–10 nm, much shorter than that in the inorganic semiconductor layer. The tilted band alignment is helpful for carrier diffusion but, still, excitons excited in the vicinity of the hetero-interface within the diffusion length make the major contribution to the photovoltaic effect. As shown in Fig. 7.8, the inorganic material needs to have a suitable amount of offset with respect to the exciton-binding energy in the organic material, i.e.,

$$E_{LUMO}^{D} - E_{CB}^{A} > U_b \tag{7.1}$$

for favorable change separation, assuming that the Coulombic energy corresponding to attraction force between separated electron and hole is negligible [3]. Here V_d is the exciton-binding energy in the organic layer, and E_A^D (organic layer in Fig. 7.8) and E_A^A (inorganic layer) are the electron affinities of the donor and the acceptor layers, respectively. Most polymers used for SCs have the exciton-binding energy ranging from 0.3 eV to 1.4 eV. V_d of MEH-PPV is 0.6 eV. The energy offset between the LUMO level in organic layer and the CB in the inorganic layer offers the energy to separate the electron and the hole in the exciton. In addition, the interface should be smooth and with low defect density to reduce charge trapping effects.

After disassociation of the exciton, the electron is transported to the cathode and the hole to the anode (see Fig. 7.8d step ③), driven by the potential in the device. The photoexcited carriers might be annihilated via recombination with defects or impurities. The diffusion lengths of electrons in the inorganic layer and holes in the organic layer are also critical parameters on which device performance relies. Apparently, using thin organic and inorganic layers with their thickness that does not exceed the diffusion lengths of holes and electrons, respectively, would be helpful for achieving high device performance. However, a thin active layer is unfavorable for sufficient light absorption. The effective way to overcome the problem of short exciton diffusion length is to use bulk heterostructures, instead of phase separation bilayers as described in Chapter 5. For bulk heterostructures, the interface between the organic (polymer) and the inorganic materials is modified to form a 3D interpenetrative form to increase the interface area by using appropriate geometric distribution of the interface profile. As shown in Fig. 5.3 of Chapter 5, the 3D bulk heterojunction facilitates the photogeneration of excitons in a distance to the interface within the diffusion length. It also shortens the distance for the electrons and holes to be transported throughout the device, which is helpful for reducing the probability of carrier recombination.

After migrating through the donor organic layer and the acceptor semiconductor, the electrons and the holes are extracted by their electrodes, as shown in Fig. 7.8e (step ④). Macroscopic photovoltaic voltage is established when a large number of electrons and holes are accumulated on the contact electrodes (see Fig. 7.8f for step ⑤).

7.3.2 Junction Potentials in Devices

There are a number of organic materials that have been used for HSCs, mostly being donors, including PPHT, P3HT, P3OT, P3BT, MEH-PPV, MDMO-PPV, and MOPPV, as well as some of their mixtures. The inorganic materials such as Si, CdSe, CdTe, CdS, PbSe, PbTe, PbS, ZnO, TiO_2, CuPc, and SnS, are usually in their nano forms of nanoparticle, nanowires, nanorods, nanotubes, etc. In HSCs, the inorganic nanomaterials act as acceptors, though some of them, like ZnO and TiO_2, are *n*-type semiconductors. The device performance relies strongly on the band alignment of materials in a HSC, which should have the junction energy profile as shown in Fig. 7.8, where the potential arrangement makes it possible for electrons and holes to be split and extracted by contact electrodes. Thus, the band alignment in a SC determines device performance to a large extent and the band alignment relies on the energy band positions of the organic materials also to a large extent. Figure 7.9 presents the ground state and first excited state level (HOMO and LUMO) positions of some of the commonly used materials prior to the formation of heterojunctions. A proper energy level alignment is essential for HSCs to work efficiently. The use of semiconductor nanomaterials allows one to customize the bandgaps that can be easily tuned to match to the solar spectrum. In addition, semiconductor nanomaterials offer possibility for multiple-exciton generation upon the absorption of a high-energy photon.

There are a large number of organic materials from which one can choose to construct hybrid SCs, based on their energy band structures with respect those of inorganic semiconductors. Energy level positions of some important organic materials used for hybrid solar cells are depicted in Fig. 7.9. The energy levels are scaled with respect to the vacuum level.

A built-in electric field is formed in the junction area of organic and inorganic materials. When a photon is absorbed, an exciton, a bound electron–hole pair, is created in the vicinity of the junction area (the donor/acceptor (D/A) interface in Fig. 7.10a) [4]. Two factors have to be taken into consideration. First, the exciton needs to be split to form a free electron and a free hole. Energy is thus required to split the electron–hole pair; therefore, a strong built-in electric field needs to be established to disassociate the electron and the hole. Second, the exciton diffusion length in a conjugated polymer material is around 10–20 nm. Therefore, an exciton should better be created within 10–20 nm in the vicinity of the donor/acceptor (D/A) interface for efficient separation of the electron and the hole. The free electron and the free hole can then be driven to their respective electrodes by the built-in electric field [5]. In order for efficient exciton disassociation and charge transport, the device structure of Fig. 7.10b was proposed, where an inter-digital donor/acceptor interface configuration dramatically increases the interface area that is very helpful for the exciton disassociation and the reduction of charge transfer distance. In addition, the inter-digital structure is beneficial for effective photon trapping. The fabrication of inter-digital structure requires photolithography, while a practical device structure can be formed by blending together the donor and the acceptor materials, as shown in Fig. 7.10c [6]. The so-called bulk-heterojunction configuration is a practical similarity to the inter-digital device, characterized by irregular donor/acceptor interface geometry and pathways of donor and acceptor phase that could direct transport of the photon-excited electrons and holes toward their respective electrodes.

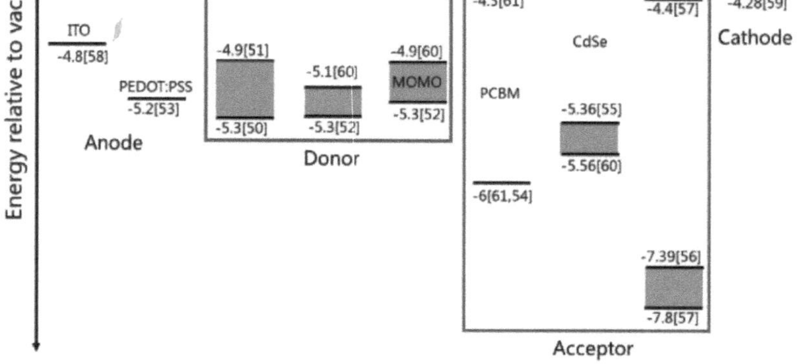

Figure 7.9 Energy level positions with respect to the vacuum level for some materials used for hybrid solar cells.

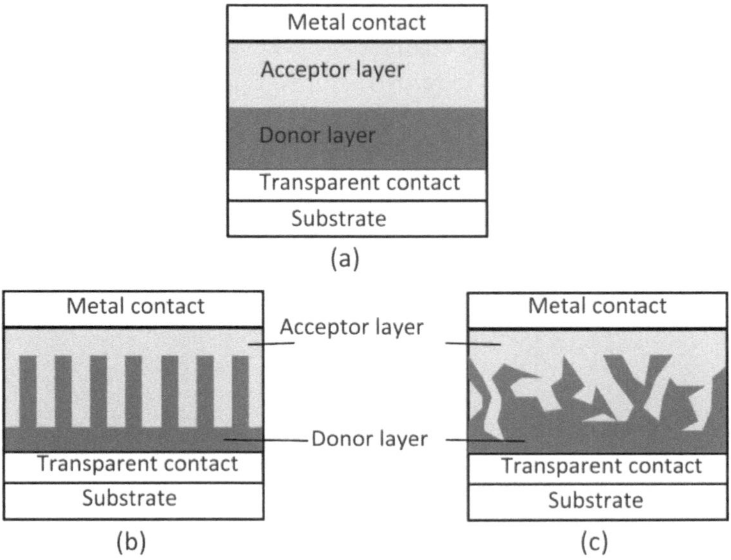

Figure 7.10 Schematic device architectures with planar (a), finger-crossing (b), and bulk heterojunction (c) interfaces between a donor and an acceptor layers.

Geometrically, HSC configuration can be arranged in planar heterojunction, a bulk heterojunction, or an ordered heterojunction. In the case of planar junction, both the inorganic and the organic materials are grown parallel on FTO substrate. For the bulk heterojunctions, the organic and inorganic semiconductors form an interpenetrating network that is either disordered or ordered. Usually, the bulk heterojunctions are formed by inorganic nanosemiconductors (nanowires, nanorods, etc.) embedded in the polymer matrix, or vice versa, by inorganic mesoporous semiconductor nanotube/nanorod array infiltrated with hole-transporting organic materials. Although both the planar and the bulk heterojunctions provide large potential discontinuity that allows high-efficiency exciton disassociation, the carrier collection process, in which electrons are transported to the cathode and holes to the anode, is rather inefficient. The inefficiency is caused by poor conductivity of both the organic and the nanostructured inorganic materials and by the random interpenetration and phase segregation of the organic and the inorganic materials. In the phase segregation case, the carriers have to migrate across the interfaces many times. Nano forms of materials offer wide choices of device configuration for 3D heterostructures. Some of them, with the nanomaterials being quantum dots (QDs), nanorods, and porous semiconductor, are schematically depicted in Fig. 7.11.

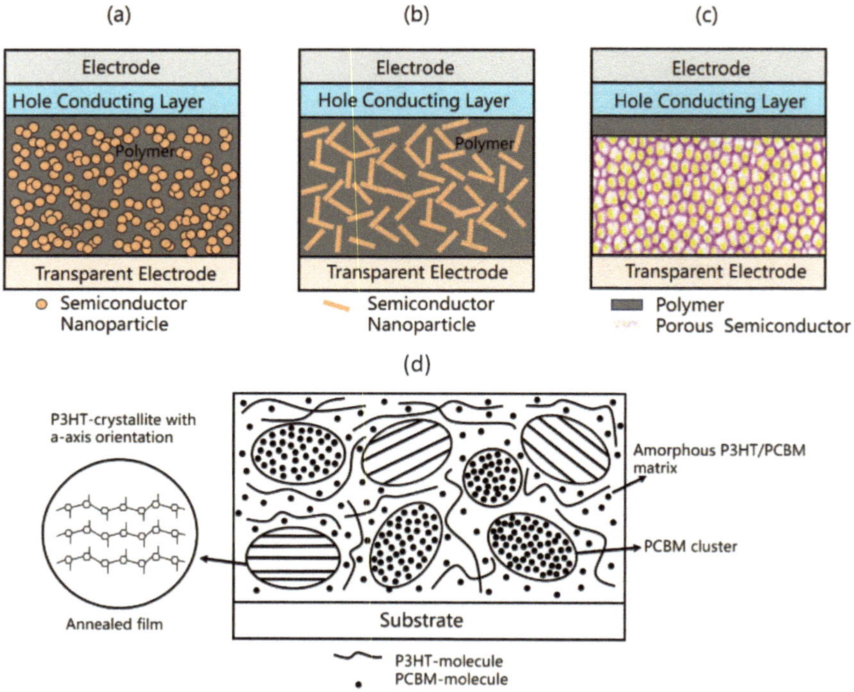

Figure 7.11 3D heterostructures formed with nano materials: (a) quantum dots (QDs), (b) nanorods, (c) porous semiconductor, and the schematic architecture of a 3D heterojunction solar cell based on P3HT/PCBM. Figures (a)–(c) were drawn following those in *Materials* **7**, 2747–2771 (2014).

The fundamental structure of organic-inorganic HSCs consists of a photoactive layer(s) formed by a donor and an acceptor layers, a metal contact and a transparent contact. The whole structure is prepared on a flexible (e.g., polyimide) or solid substrate (e.g., glass). The metal contact as cathode uses metals, e.g., Au, Ag, Al, Ca/Al, and LiF/Al, with high work function in order to extract holes and the transparent contact uses conductive indium tin oxide (ITO) as anode. The photoactive layer(s) is comprised of an organic part that is usually a conjugated polymer and an inorganic part that is usually semiconducting nanocrystals. Simplified configurations of hybrid solar cells are illustrated in Fig. 7.10a. The basic form of the active layer of a hybrid SC is a bilayer structure formed by a donor layer and an acceptor layer. The variations of the planar SC include the inter-digital form (Fig. 7.10b) and the bulk heterojunction form (Fig. 7.10c). Obviously, both the inter-digital form and the bulk-heterojunction form are microscopically planar SCs, as shown in the figure. The reported organic materials for hybrid SCs includes MEH-PPV (M3H-PPV) (poly(2-methoxy-5-(2-ethyl-hexyloxy)-phenylene vinylene)), P3HT (poly (3-hexylthiophene)), PCPDTBT (poly(cyclopentadithiophene-alt-benzothiadiazole)), as well as PCBM and the inorganic semiconductors are Si, GaAs, ZnO, Sb_2S_3, TiO_2, PbS, etc. The most common polymers used are P3HT, and M3H-PPV. P3HT has a bandgap of 2.1 eV and M3H-PPV has a bandgap of ~2.4 eV. The candidate materials for the *n*-type nanosemiconductors are usually Si, GaAs, ZnO, etc. The organic donor layer materials must be hole conductive, which are MEH-PPV, P3HT, PCPDTBT that have suitable HOMO and LUMO energy levels and conductivities for hole transport. The device configuration can also be arranged in a reverse way where inorganic semiconductors act as donor layers and the organic materials (e.g., PCBM) as acceptor layers.

7.4 Device Structure of Hybrid Solar Cells

7.4.1 3D Bulk Heterojunctions

A 3D bulk heterojunction has the advantages of large interface area, ease of carrier separation, and minimized distance that excitons travel to reach the interface for dissociation. A typical 3D heterojunction can be constructed by using polymer and fullerene. This 3D bulk heterojunction configuration for SCs was investigated dated to as early as 1995 and then remained as a hot system afterwards [7]. The general attracting and the expected room for performance improvement continue to stimulate the research effects that push the current conversion efficient from initially below 1% to currently beyond 10%.

The configuration is regarded as a model one since, based on it, other similar configuration can be constructed by replacing fullerenes by colloidal quantum dots, Si nanorods, etc. In this section, some architectures of hybrid SCs will be presented, emphasizing innovation concepts on new structures.

The device configuration of the 3D heterojunction solar cell is shown in Fig. 7.11d. The fullerenes and P3HT are mixed in a solution and the P3HT/PCBM precursor is spin-coated on a substrate, followed by annealing (solvent evaporation). Appropriate annealing temperature is required to choose and, in most cases, only a low annealing temperature is needed to increase the conductivity. Prolonged annealing and over high temperature could result in polymer domain sizes that are larger or comparable to carrier diffusion length, leading to low device efficiency.

Depending on the use of materials and the interface configurations, many architectures of HSCs based on the 3D bulk heterojunctions have been proposed and studied. In this section, three device architectures are presented and discussed.

7.4.2 Double Heterojunctions

Mor and co-workers proposed a HSC with the device configuration as shown in Fig. 7.12a and band structure profile as shown in Fig. 7.12b [8]. The device consists of a mesoporous TiO_2 nanotube array vertically grown on the FTO glass substrate (a structure widely used in DSSCs), a mixture of electron-donating P3HT [poly(3-hexylthiophene)] and an electron-accepting PCBM [[6,6]-phenyl-C_{71}-butyric acid methyl ester], a hole-injecting polymer of PEDOT [poly(3,4-ethylenedioxythiophene)]: PSS [poly(styrenesulfonate)] layer, and an Au top electrode. PCBM is a fullerene derivative with relatively high electron mobility (see the inset in Fig. 7.12b). The active part of the structures is the mesoporous TiO_2 nanotube array layer infused with the blend of the P3HT and the PCBM and the PEDOT: PSS layer.

The organic material absorbs light to generate excitons and the electrons and holes are separated at the interface between the mesoporous TiO_2 and the organic mixture of P3HT and PCBM.

Electrons then transfer to transparent conducting oxide layer electrode (the cathode) and holes to a metal back electrode (the anode). HSCs based on the mesoporous inorganic layer infused with organic donor materials have relatively high conversion efficiencies.

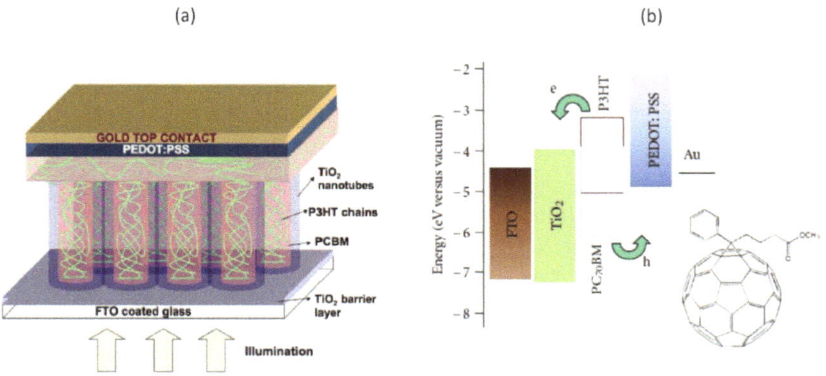

Figure 7.12 (a) Schematic device architecture (a) and the energy band alignment (b) of the 3D heterojunction solar cell with the structure of $FTO/TiO_2/PC_{70}BM$: P3HT/PEDOT:PSS/Au. The inset in (b) shows the molecular structure of $PC_{70}BM$. Reproduced with permission from [8].

7.4.3 Ordered Lamellar Architecture

An interesting format of hybrid photovoltaic materials is the ordered lamellar structure consisting of alternating nanoscale donor and acceptor layers, a periodical sequence of organic and inorganic semiconductors [9]. The lamellar structure can be fabricated using solution chemistry, where the inorganic semiconductor layer acts to accept electrons and organic surfactants sandwiching the inorganic layer absorb strongly the visible light. Such a structure has been found to show strong built-in potential for exciton splitting. Figure 7.13 presents the schematic diagram of the lamellar structures on substrates. Apparently, for maximum effect, the multi-layer planes of the lamellar structures should be oriented parallel to the substrate surface to allow for efficient charge transport to the respective anode and cathode electrodes. This, however, presents challenge for material growth.

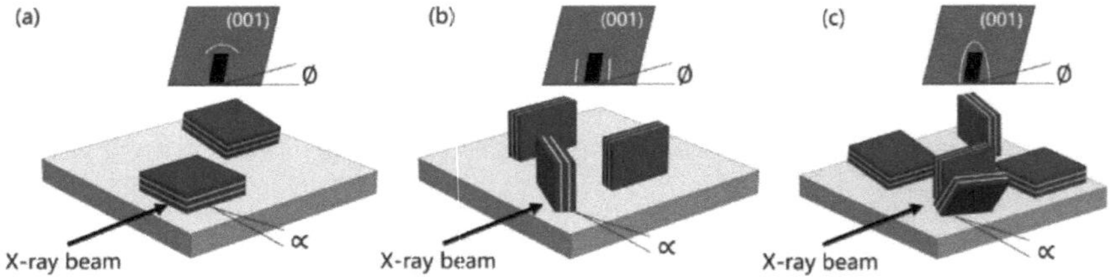

Figure 7.13 (a) Schematic geometric relationship among lamellar structures and the substrate: lamellar structures parallel to (a), perpendicular to (b), oriented randomly with respect to the substrate.

Herman and co-workers reported the preparation of hybrid periodic lamellar structures consisting of ZnO and 1-pyrenebutyric acid (PyBA) as the conjugated surfactant using electrodeposition. The orientation and density of the alternating inorganic and organic lamellar structures are controllable through solution and substrate chemistry [10]. The preparation used a solution of $Zn(NO_3)_2 \cdot 6H_2O$ and PyBA surfactant in a mixed solvent of water/dimethyl sulfoxide that is helpful for the solubilization of the surfactant. The reaction takes the following form:

$$NO_3^- + H_2O + 2e^- \rightarrow NO_2^- + 2OH^- \quad (7.2a)$$

$$Zn_2^+ + 2OH^- \rightarrow Zn(OH)_2 \quad (7.2b)$$

The ZnO layers are formed from $Zn(OH)_2$ after annealing, with the surfaces attached by the surfactants. The periodic ZnO/surfactant lamellar structure is then obtained.

7.4.4 Ordered Nanowire (Nanorod) Structures

Hybrid nanostructures, consisting of inorganic nanotubes and nanowires surrounded by organics, are another form of architectures for HSCs, where inorganics are used as electron acceptors and the organics as electron donors. The inorganics and the organics offer pathways for electron and hole transport, respectively. The *p–n* junctions form on the boundaries [11]. On the other hand, the hybrid nanostructures such as nanowires, for instance, form morphologies that reduce light reflection through multiple scattering. The major advantage of the nanostructures is that the small size of the hybrids tolerates relatively large lattice mismatch and makes the growth of high-quality materials easy.

The hybrid nanostructures have been prepared with controllable material qualities and geometric configurations based on a cost-effective self-organizing process on low-cost substrates. Recently, nanowire SCs have been emerging as one of the major nanostructures to construct SCs. A number of important works have been reported and the device efficiency increases steadily. One of the device configurations designed by Takanezawa and co-workers is presented in Fig. 7.14a for the field emission scanning electron microscopy (FESEM) cross-sectional image and Fig. 7.14b for the schematic device structure [12]. The energy band alignment of the device with the structure of ITO/ZnO/PCBM:P3HT/VO_x/Ag is depicted in Fig. 7.4c. The ZnO-organic hybrid solar cell with a VO_x buffer layer displays a short-circuit current density of 10.4 mA·cm^{-2}, an open-circuit voltage of 0.58 V, a fill factor of 65%, and a power conversion efficiency of 3.9%, under standard AM1.5 G condition. All parameters are relatively balanced.

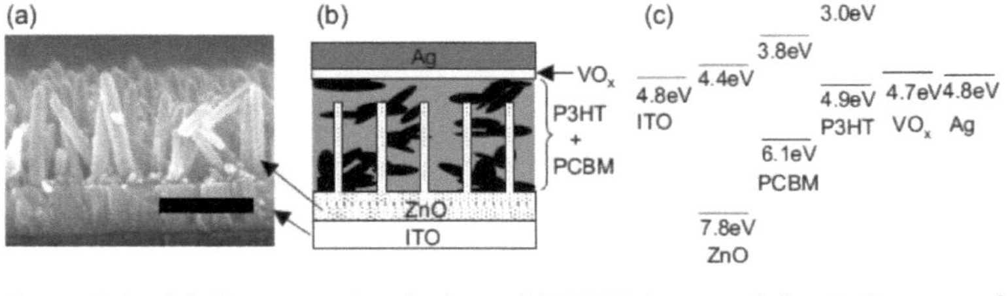

Figure 7.14 (a) Cross-sectional view of FESEM image of the ZnO nanorod arrays (scale bar: 300 nm) on ITO substrate; (b) schematic device structure of the ITO/ZnO/PCBM:P3HT/VO_x/Ag; and (c) energy band alignment of the device. Reproduced with permission from [12].

7.5 Some Hybrid Solar Cells

7.5.1 A Hybrid Cell with Si Nanowires on Pyramid-Textured Surface

An architecture of a completed Si/PEDOT hybrid solar cell was proposed with the cell surface of Si NWs on pyramid-texture [13].

More specifically, the HSC has a Si-NW/poly(3,4-ethylene-dioxythiophene):polystyrenesulfonate (PEDOT:PSS) configuration. The structure of the Si-NWs-on-pyramid-texture was fabricated by a two-step process. A pyramid textured surface was obtained by etching an *n*-type Si(100) wafer using a solution of potassium hydroxide and isopropanol solution. Randomly distributed Si-NWs were then formed by etching the textured surface using the mixture of hydrofluoric acid and silver nitrate. The length of the formed Si-NWs is 0.4–0.8 μm, which offers good light trapping. The HSC was constructed by spin-coating PEDOT:PSS onto the Si-NWs-covered pyramid-textured surface, followed by the preparation of the front and the back electrodes. The front electrode was a silver grid and the back electrode an Al layer prepared by thermal evaporation.

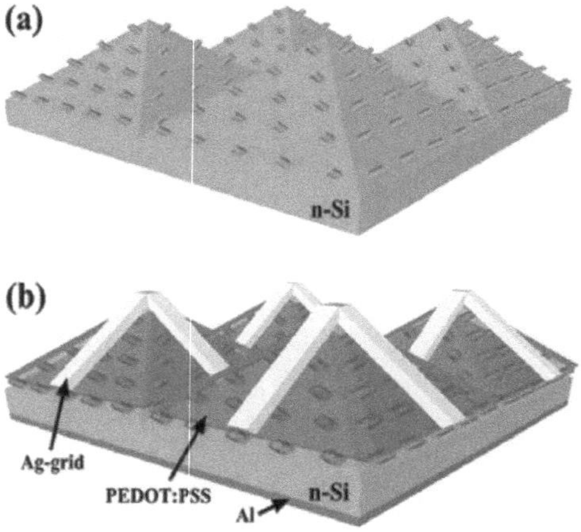

Figure 7.15 (a) The surface of the Si NWs on pyramid-texture and (b) the structure of Si-NWs-on-pyramid-texture/PEDOT:PSS hybrid solar cell.

Figure 7.15 presents the schematic surface and the device structure of the hybrid solar cell. Apparently, the Si-NWs on a pyramid texture device have a large Si/PEDOT:PSS heterojunction area and the binary structure of Si-NWs-on-pyramid-texture offers good light trapping than Si-NWs-on-planar-surface. With the use of the Si binary structure, short Si NWs with a small aspect ratio are able to achieve the similar light-trapping effect as long Si NWs on planar Si, which is beneficial for the reduction of carrier recombination. The overall advantages of such a hybrid solar cell lead to a maximum power conversion efficiency of 9.9%. The main improvement is the high short-circuit current density J_{sc} of 31.9 mA/cm^2.

Si-NW/polymer structures represent hybrid material systems that have been investigated intensively for hybrid solar cells. Table 7.1 lists performance (PCE and J_{sc}) of the devices based on several Si-NW/polymer structures [13, 14]. Clearly, more than 10% of conversion efficiency seems to be at hand with the structures.

Table 7.1 Summary of some reported HSCs based on Si NWs and organic materials

Si structure	Organics	Group/refs.	PCE [%]	J_{sc} (mA/cm^2)
SiNW	Poly(3-octylthiophene) (P3OT)	Kalita et al. [14a]	0.61	7.85
SiNW	PEDOT:PSS	Shiu et al. [14b]	5.08	19.28
SiNW	Poly(3-hexylthiophene) (P3HT)	Zhang et al. [14c]	5.9	26.2
SiNW	Spiro-OMeTAD/ PEDOT:PSS	He et al. [14d]	10.3	30.9
SiNW	PEDOT:PSS	He et al. [14e]	9.0	26.3
SiNW	PEDOT:PSS	Ozdemir et al. [14f]	5.3	30.2
SiNW/ Pyramid	PEDOT:PSS	He et al. [13]	9.9	31.9

7.5.2 Nano-CdSe/Polymer Hybrid Cells

CdSe and CdS colloidal nanoparticles were among the earliest that were used for nano solar cells and a great number of works have been reported. An early version of device design was illustrated in Fig. 7.16a [15]. The polymer used in the architecture is PCPDTBT that has the molecular structure as shown in Fig. 7.7. The CdSe tetrapods, with TEM image shown in Fig. 7.16b, were used as sensitizing material. The CdSe tetrapods present strong absorption at the wavelength range below 600 nm, while the PCPDTBT shows absorption band within 600–800 nm (see Fig. 7.16c). In the major solar spectrum range of 350–800 nm, the hybrid solar cell exhibits a high external quantum efficiency

(EQE) of >30%. The device shows an EQE of ~55% in 630–720 nm, mainly due to the strong absorption band of PCPDTBT. The device efficiency of 3.1% was certified by National Renewable Energy Laboratory (NREL) of USA, with V_{oc} = 0.6740 V, J_{sc} = 9.0151 Am/cm^2, and a fill factor = 51.47%.

(a) (b) (c)

Figure 7.16 Schematic device architecture of the CdSe-PCPDTBT hybrid solar cell (a), TEM image of the CdSe tetrapods (the Scale bar: 50 nm) (b), and light absorbance of films of CdSe tetrapods (dot-dashed), PCPDTBT (dotted, red), and CdSe-tetrapod/PCPDTBT mixture (solid, black) (c). Reproduced with permission from [15].

The device shows the open-circuit voltage V_{oc} of 0.674 V that is comparable to that of wafer-based Si SCs. However, the short-circuit current density J_{sc} (9.0151 Am/cm^2) and the fill factor are rather poor. The fact that light absorption of PCPDTBT makes the major contribution to high EQE indicates that the CdSe tetrapods are not very efficient. In addition, the electrical connection between quantum dots could be rather poor. Seo and co-workers demonstrated an improvement in device efficiency of polymer/nano-CdSe solar cells [16]. The device has the hybrid structure of CdSe nanocrystals functionalized by poly(3-hexylthiophene)(P3HT). In the work, the lengths of the ligands between the CdSe nanocrystals and between the polymer matrix and the CdSe nanocrystal were shortened, using thermal deprotection strategy. A significant increase in short current density was observed due to the improved electrical connection in the nano-CdSe/polymer hybrids.

Early in this century, Huynh and co-workers demonstrated that semiconductor quantum dots and polymers could be put together to fabricate solar cells [17]. Since the electron mobilities of polymers are typically below 10^{-4} cm^{2}V^{-1}s^{-1}, they used semiconductors to transfer electrons. In their work to fabricate the HSC with a CdSe-nanorods/P3HT hybrid structure as shown in Fig. 7.17, a precursor of mixture solution of 90 wt % CdSe nanorods in P3HT was spin-coated onto an PEDOT:PSS-coated ITO glass substrate. The front electrode is an aluminum thin film, as shown schematically in the figure. Although the conversion efficiency was only 1.7% measured under AM1.5 condition, this work pioneered the investigation on hybrid solar cells using inorganic semiconductors and organic materials.

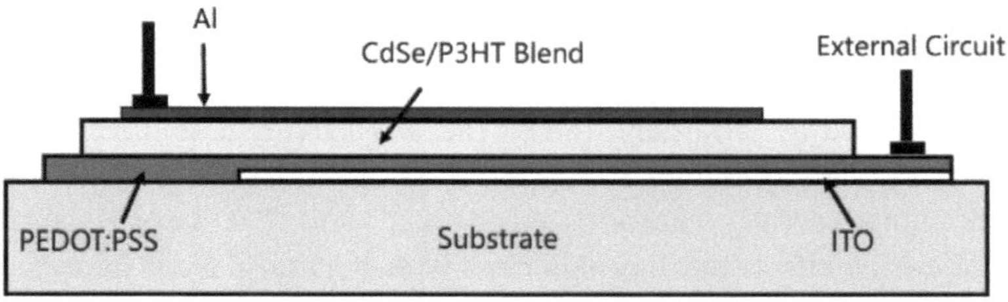

Figure 7.17 The schematic illustration of a CdSe-nanorods/P3HT/PEDOT: PSS hybrid cell.

The P3HT and CdSe form a band alignment such that both the conduction and the valence bands in CdSe are lower in energies than the LUMO and the HOMO levels of P3HT, respectively. Thus, CdSe nanorods are the electron-accepting material and P3HT the hole-accepting one. With light incident from the bottom of the device, photoexcited excitons are created due to photon absorption. The excitons in the vicinity of the CdSe/P3HT interface are split by the interface potential discontinuity and the electrons and the holes are transported to their electrodes. In this device configuration, electrons are transferred to the Al electrode (the top electrode) and the holes to the ITO electrode (the bottom electrode). The CdSe/P3HT blend has a broad absorption spectrum extending from 300 to 700 nm, since the absorption spectra of CdSe and P3HT are complementary in the visible spectrum. This leads to a broad photon-induced current spectrum. Furthermore, the use of CdSe nanorods has the advantage that the size of the nanorods can be varied to achieve wide visible spectrum absorption, which supplements the problem of insufficient light absorption capacity of P3HT.

The hybrid cells using the CdSe nanorods with 7 nm in diameter and 7, 30, and 60 nm in length were fabricated and the best result was obtained on the cell with nanorods of 7 nm diameter and 60 nm length. The best hybrid cell has an open-circuit voltage $V_{oc} \sim 0.7$ V determined by the difference between the work functions of ITO, PEDOT:PSS, Al, as well as the conduction band energy in CdSe nanorods and the HUMO level in P3HT. The open-circuit voltage of 0.7 V is comparable to that of wafer-based Si SCs. The short-circuit current density J_{sc} is, however, below 6 mA/cm^2, while it is typically ~30 mA/cm^2 for wafer-based Si SCs. Low J_{sc}, together with rather high internal resistance of the device, gives rise to poor filling factor FF = 0.4. The poor J_{sc} value is largely due to the low carrier mobilities in the organic materials of P3HT and PEDOT:PSS caused by inefficient hopping carrier transport mechanism and the strong carrier capturing effect (there is high density of structural traps in organic materials). The slow transportation of electrons and holes in the organic materials results in high chance for the carriers to recombine on their way to be transferred to the electrodes. The inorganic nanorods does offer easy paths for electron transport, as demonstrated in Fig. 7.18 where the cells with longer CdSe nanorods show relatively high external quantum efficiencies in visible [17]. The 1D feature of the longer nanorods is more desirable with respect to the short ones (like quantum dots) since the long rods themselves are the path for carrier transport. In the ideal case that the length of the nanorods equals to the thickness of the device, electrons might behave like free electrons with very high mobility.

A number of studies have suggested that combining organic polymers with inorganic nanomaterials be very effective to improve the charge transport efficiency and such a hybrid strategy has been widely adopted for various photoelectric devices. In fact, the combination of inorganic materials of high electron affinity and organic polymers of low ionization potential offers great help for the enhancement of carrier transport, if the chemical binding between organic and inorganic molecules gives good electronic contact [3]. Additionally, attention should also be paid to improving the quality of the organic materials themselves.

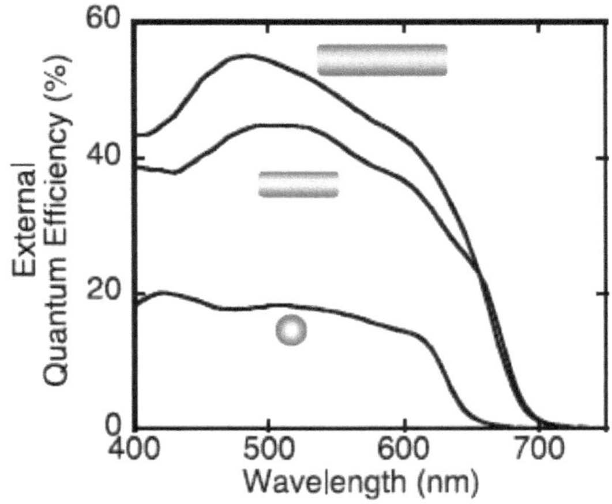

Figure 7.18 External quantum efficiencies vs. wavelength of hybrid SCs with different nanorod length of 7, 30, and 60 nm. Reproduced with permission from [17].

Bulk heterojunction (BHJ) structures have been proven effective in increasing nano solar cell conversion efficiency, due to the large area of heterojunction. Imran and co-workers designed a BHJ CdS-conjugated hybrid polymer solar cells fabricated by solution method [18]. In the fabrication of the devices, a 30 nm PEDOT:PSS thin layer was prepared on ITO-coated glass substrate by spin-coating. A photoactive layer was then spin-casted on the PEDOT:PSS thin layer. A 0.3 nm LiF buffer layer and 100 nm-Al contact were high-vacuum prepared. Four samples were fabricated, each having a photoactive layer with different CdS QD contents made from the solutions of P3HT:PCBM:CdS at several weight ratios of 1:0.8:0, 1:0.8:0.2, 1:0.8:0.4, and 1:0.8:0.6 dissolved in 1:1 chlorobenzene and dichlorobenzene. The hybrid solar cells, with a device structure of ITO/PEDOT:PSS/P3HT-CdS-PCBM/LiF/Al, have the band alignment as shown in Fig. 7.19a. Figure 7.19b presents the SEM image of the CdS nanostructure. The CdS nanoparticles form a percolation network, which composes a bulk heterostructure with the organic matrix.

The hybrid SC presents an efficiency of 4.41% due to the use of the CdS nanostructure, as compared to a reference solar cell that has the same device structure except for the absence of the quantum dots. The improved conversion efficiency is attributed to the formation of the bulk heterojunction structure and the enhancement in light absorption in CdS, which gives rise to the increase in short-circuit current density J_{sc} and the fill factor FF. Apparently, such a percolation network is helpful for the enhancement of J_{sc} by offering good electrical connections.

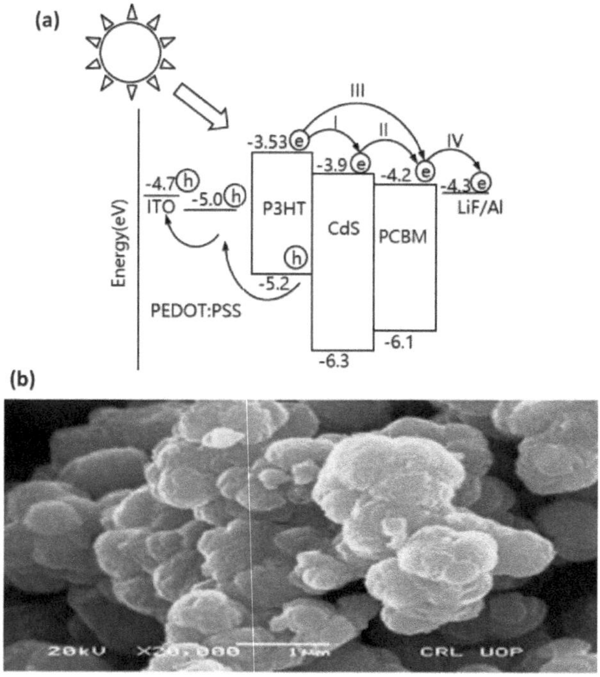

Figure 7.19 (a) Schematic energy band profile of the CdS/P3HT:PCBM hybrid SC (b) and SEM image of the CdS nanostructure. Reproduced with permission from [18].

7.5.3 Hybrid Solar Cells Using GaAs Nanopillars

GaAs has a bandgap of 1.41 eV, which is almost a perfect match to the solar spectrum for single-bandgap solar cells. It forms good electric contact with many organic polymers; therefore it is a good inorganic candidate to construct hybrid solar cells with organic materials. The key part of the hybrid solar cells is the GaAs-nanopillars/P3HT structure. The nanopillars were grown by metal organic chemical vapor deposition (MOCVD) and the ordered array was formed via selective growth on *n*-type GaAs (111)B substrate covered by patterned SiO_2 layer prepared by e-beam lithography. Prior to the deposition of P3HT layer, AuGe/Ni/Au bottom contact was prepared, followed by annealing the GaAs-nanopillars/substrate at 400°C for 45 s. The P3HT layer was deposited by spin coating the P3HT precursor on the ordered array of GaAs nanopillars. Immediately after spin-coating the P3HT layer, transparent ITO electrode was deposited using RF sputtering. Figure 7.20a presents the schematic diagram of the GaAs-nanopillars/P3HT hybrid solar cell, with the SEM image of the nanopillar array shown in Fig. 7.20b. The details of the device preparation are provided in Ref. [19].

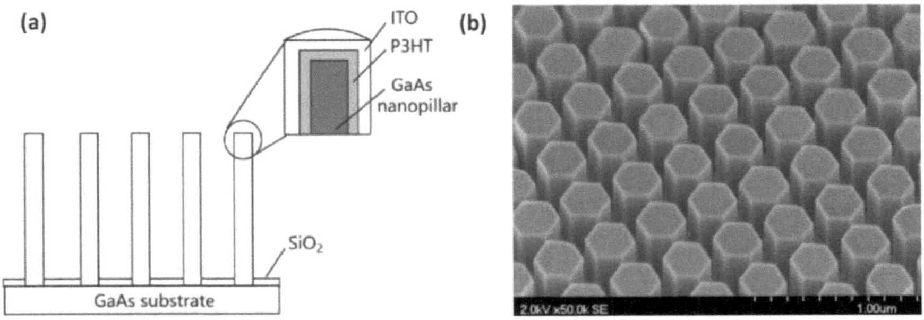

Figure 7.20 (a) Schematic illustration of the GaAs-nanopillars/P3HT hybrid solar and (b) SEM image of the ordered as-grown GaAs nanopillar array cell. Reproduced with permission from [19].

The GaAs nanopillars are *n*-type doped at the level of 5×10^{17} cm^{-3} for electron transport. When photons with energy greater than or equal to the bandgaps of GaAs and P3HT are incident on the device, excitons are generated and disassociated at the interface between P3HT and GaAs nanopillars, as evidenced by a photoluminescence quenching experiment [19]. Electrons and holes are transferred toward the AuGe/Ni/Au bottom and the ITO top electrodes, respectively, to complete the photovoltaic process. For comparison, two other cells with the similar structure but having either no GaAs nanopillars (only the planar GaAs substrate) or no P3HT layer were also fabricated. The GaAs-nanopillars/P3HT hybrid cell displays device parameters of V_{oc} = 0.2 V, J_{sc} = 8.7 mA/cm^2, FF = 32%, and PCE = 0.6% measured under the AM 1.5 condition.

Figure 7.21 presents the *I–V* curves for the three solar cells. The cell without a P3HT layer shows no photovoltaic effect, as understood by the fact that there is actually no *p–n* junction. Therefore, excitons cannot be disassociated, though they can be photogenerated. The cell without GaAs nanopillars displays very weak photovoltaic effect, as shown by the very small photocurrent response. This is expected since there exist some small P3HT/GaAs interfacial areas in patterned SiO_2 at which *p–n* junction is located. The cell with GaAs-nanopillars/P3HT junction displays the strongest photovoltaic effect among the three cells, benefited by both the GaAs nanopillars and P3HT layer. The advantages of the GaAs-nanopillars/P3HT structure over the other two are twofold at least. One is the very large area of *p–n* junction—the GaAs-nanopillars/P3HT interface. The other is the superior light trapping due to the nanopillar configuration. In addition, the P3HT in the GaAs-nanopillars/P3HT structure could be thinner than the P3HT layer in the planar GaAs/P3HT structure, allowing photogenerated excitons to diffuse to the heterojunction over a shorter distance.

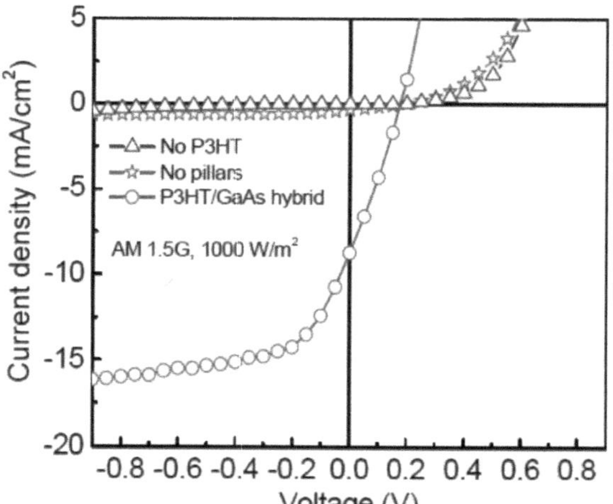

Figure 7.21 Measured *I–V* curves for a GaAs-nanopillars/P3HT hybrid SC and other two with the similar structure but either without the GaAs nanopillars or without the P3HT layer. Reproduced with permission from [19].

The GaAs-nanopillars/P3HT hybrid cell still shows low PCE less than 1%. Noticing the low open-circuit voltage of only 0.2 V, it is apparent that the energy band alignment, determined by

GaAs, P3HT, and the contacts materials, is far from ideal. One of the limitations is the high density of interface states in the P3HT/GaAs interface that acts as traps for carriers and intermediate states for electron hole recombination. It has been demonstrated that surface passivation using ammonium sulfide ($(NH_4)_2S$) reduces the density of the interface traps effectively. The passivation gave rise to a significant increase of PCE from 0.6 to 1.44% [20]. Another limitation is the unmatched mobilities of electrons in GaAs-nanopillars and holes in P3HT layer. The much lower hole mobility in P3HT than electron mobility in GaAs is responsible for the poor filling factor of 32%.

7.5.4 A Hybrid Tandem Solar Cell Consisting of a-Si:H Top Cell and a Dye-Sensitized Bottom Cell

One of the main reasons for PCE of the current SCs to be far from 100% lies in the solar radiation spectrum being broadband and the solar cell materials being single bandgaps. Tandem SCs are thus designed to harvest a broad spectrum of solar radiation by combining two or more solar cells with different absorption bands. A hybrid tandem solar cell is constructed using both inorganic and organic semiconductors with two or more heterojunctions. In constructing an inorganic tandem SC, the semiconductors need to be nearly lattice-matched for high-quality material preparation, in addition to the requirement that the semiconductors need to have suitable bandgaps matching to the solar spectrum. This is a severe challenge since really there are not many combinations of semiconductors to choose from the limited number of materials. For organic materials, however, the requirement for close lattice matching is relaxed since an organic/inorganic interface offers ease of strain relaxation. The large organic material family makes it easy for one to choose suitable materials for hybrid tandem SCs.

Hao et al. proposed an architecture of a hybrid tandem solar cell based on hydrogenated amorphous silicon (a-Si:H) top cell and a dye-sensitized solar cell (DSSC) as the bottom cell [21]. Figure 7.22 presents the schematic diagram of the device architecture. The a-Si:H top cell is comprised of *p*-type, *n*-type, and a nominally intrinsic a-Si:H three layers. The top cell was prepared in a multi-chamber plasma-enhanced chemical vapor deposition (PECVD) system. The DSSC bottom cell consists of an electrolyte layer and a dye-sensitized TiO_2 layer. The platinized zinc oxide (Pt/ZnO) is used both as the bottom electrode of the top cell and as the top electrode of the bottom cell. Here the sensitized dyes use N-719 [$RuL_2(NCS)_2$, L=4,4′-dicarboxylate−2,2′-bipyridine] and N-749 [$RuL'(NCS)_3$, L′=2,2′:6′,2″-terpyridine-4,4′,4″-tricarboxylate]. For details of the device fabrication process, see Ref. [21].

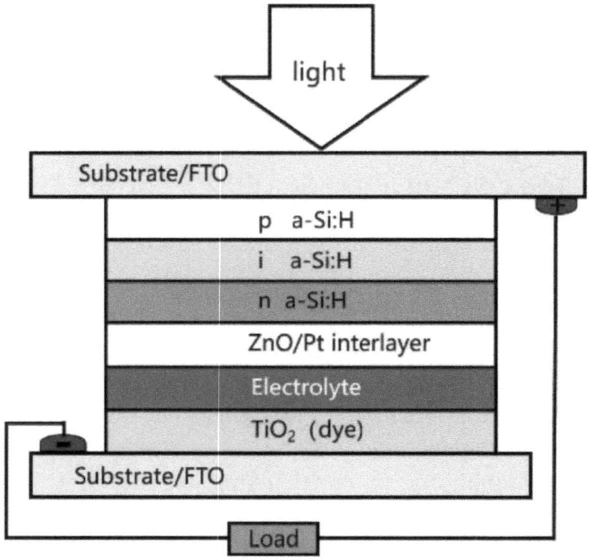

Figure 7.22 Architecture of a hybrid tandem solar cell based on an a-Si:H top cell and a dye-sensitized TiO_2 bottom cell.

The complete band alignment of the hybrid solar cell is shown in Fig. 7.23. The incident photons from top penetrate through the top glass substrate and FTO electrode into the a-Si:H top cell where the photons, with energies equal to and higher than the bandgap of a-Si:H, are absorbed and the excitons are excited. The excitons photon-generated in the intrinsic a-Si:H layer experience a strong built-in potential, where electrons and holes in the excitons are disassociated and swept toward the *n*-type a-Si:H layer and the *p*-type a-Si:H layer, respectively. Photons with lower energies penetrate through the top cell and are absorbed by the dye molecules where the electrons are driven into TiO_2 by potential discontinuity at the dye-molecule/TiO_2 interfaces and then transferred into the FTO counter electrode through load. The holes remain in the ground state of dye molecules and are reduced after the holes are transferred to the Pt layer through the I^-/I_3^- redox electrolyte where they combine with the electrons. The carrier migration gives rise to a photovoltaic voltage cross the top and the bottom electrodes. The photovoltaic voltage of the tandem solar cell relies on the interrelation of Fermi level potentials of the three a-Si:H layers, TiO_2, and the electrolyte.

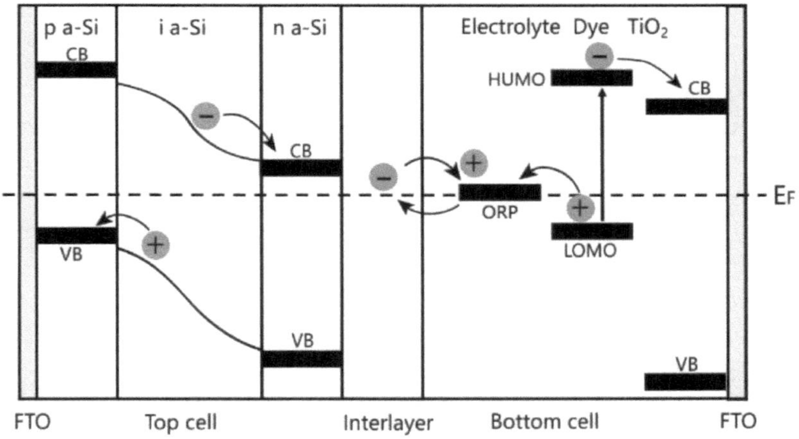

Figure 7.23 Energy band alignment of the hybrid tandem solar cell with an a-Si:H top cell and a dye-sensitized TiO_2 bottom cell.

The hybrid tandem SC should have higher PCE than the individual a-Si:H top cell and the individual dye-sensitized bottom cell. At their optimized device configuration, i.e., the device with the thicknesses of a 235 nm a-Si:H layer, 100 nm ZnO/Pt interlayer, and 8.5 μm dye-sensitized bottom cell, Hao et al. obtained photon conversion efficiency of 8.31%, with 1.45 V open-circuit voltage V_{oc} and 10.61 mA/cm^2 short-circuit current density, measured under AM1.5 condition. V_{oc} of 1.45 V is extremely good, while J_{sc} of 10.61 mA/cm^2 is much smaller than that of wafer-based Si SCs. The high V_{oc} is attributed to the fantastic band alignment of the materials making up the hybrid tandem SC, particularly the large band edge offset between the top and the bottom cells. V_{oc} of the hybrid tandem SC is essentially the sum of the open-circuit voltages of the individual subcells since they are connected in series, i.e., V_{oc} of the cell is approximately equal to the sum of the Fermi potential difference between the TiO_2 layer and the electrolyte and between the three a-Si:H layers. One of major technical challenges in fabricating organic-based tandem SCs is the preparation of a high-quality interlayer connecting the neighboring subcells but separating the subcells physically [22].

7.5.5 A Hybrid Tandem Cell Consisting of an a-Si:H Front Cell and a Polymer-Based Organic Back Cell

Seo et al. worked on another architecture of a hybrid tandem SC that consists of an a-Si:H front cell and a polymer-based organic back cell [23]. Figure 7.24 shows the schematic tandem structure of AZO/a-Si:H/ITO/PEDOT:PSS/PBDTTT-C:PCBM/Al, together with its cross-sectional image of high-resolution transmission electron microscopy (TEM) that shows well-defined layers and distinct interfaces without noticeable inter-diffusion between layers and physical damage. The front cell has the structure of *p*-type-a-Si:H/intrinsic-a-Si:H/*n*-type-a-Si:H configuration that was prepared on AZO-coated glass using PECVD. The active layer of the organic back cell consisted of a mixture of poly(4,8-bis-alkyloxybenzo(1,2-b:4,5-b′)dithiophene-2, 6-diyl-alt-(alkyl thieno(3,4-b) thiophene-2-carboxylate)-2,6-diyl) (PBDTTT-C) and [6,6]-phenyl-C_{61}-butyric acid methyl ester (PCBM). Prior to the fabrication of the back cell, an ITO layer was deposited on *n*-type a-Si:H layer using magnetron sputtering, followed by deposition of a conductive poly(3,4-ethylenedioxylenethiophene)- polystylene sulfonic acid (PEDOT:PSS) layer prepared by spin-coating. The highly conductive and transparent ITO/PEDOT:PSS double layers separate the organic and the inorganic subcells. The two cells are connected in series. ITO acts as the electron transport layer for the a-Si:H front cell and the PEDOT:PSS as the hole transport layer for the organic back cell. The highly transparent ITO/PEDOT:PSS double layers allow for the photons, especially those with energies below the bandgap of a-Si:H (1.7 eV), to reach the back cell.

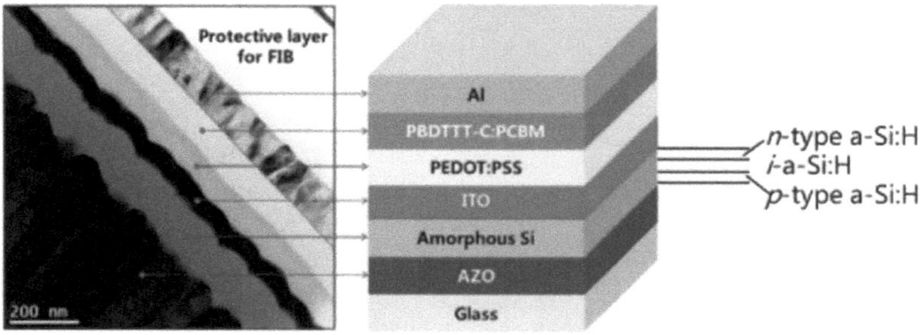

Figure 7.24 Cross-sectional SEM image and schematic structure of a hybrid tandem cell consisting of an a-Si:H front cell and a polymer-based organic back cell. Reproduced with permission from [23].

The photons incident into the front cell generate excitons in the intrinsic a-Si:H layer. The excitons disassociate into electrons and holes driven by the built-in potential in the intrinsic a-Si:H layer (see Fig. 7.24). The holes are transferred to the AZO electrode through the *p*-type a-Si:H layer in the front cell, while the electrons are swept into the ITO/PEDOT:PSS double layers, where they meet with holes from the back cell. The photons with energy below 1.7 eV (equal to the bandgap energy of the a-Si:H layer) penetrate through the front cell and the ITO/PEDOT:PSS double layers and are absorbed by PBDTTT-C:PCBM layer. The photon-generated excitons are disassociated by the discontinuity potential on PEDOT:PSS/PBDTTT-C:PCBM interface. The electrons and the holes are transferred to Al electrode and PEDOT:PSS, respectively.

Figure 7.25 presents the *J–V* curves of the individual a-Si:H front cell, the organic PEDOT:PSS/PBDTTT-C:PCBM back cell, and their combination—the hybrid tandem SC. The measurements were performed under AM1.5 simulated illumination. The measured device parameters, including power conversion efficiency PCE extracted from the measured *J–V* curves, open-circuit voltage V_{oc}, short-circuit current density J_{sc} and fill fact FF are listed in Table 7.2. One of the benefits for using the a-Si:H front cell and an organic PBDTTT-C:PCBM back cell is their compensative photocurrent response spectra of the two individual subcells. As shown in Fig. 7.25b, the response spectrum of a-Si:H cell is rather narrowly centered around 400 to 600 nm, while the organic subcell has a wide response range from 300 to 750 nm.

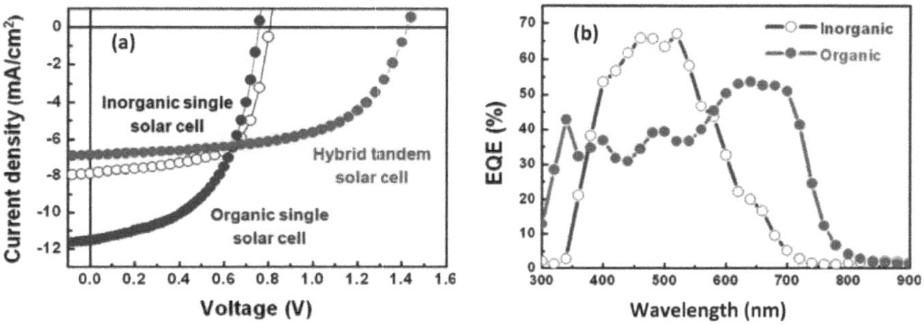

Figure 7.25 (a) the *J*–*V* curves of a a-Si:H front cell, organic PBDTTT-C: PCBM back cell, and the hybrid tandem solar cell measured under AM1.5 simulated illumination, and (b) external quantum efficiencies of the a-Si:H and the organic cells vs. wavelength from 300 to 900 nm. Reproduced with permission from [23].

Table 7.2 Measured device parameters of the a-Si:H front cell, organic PEDOT:PSS/PBDTTT-C: PCBM back cell, and the hybrid tandem solar cell

	PCE (%)	V_{oc}(V)	J_{sc} (mA/cm^2)	FF
a-Si:H front cell	4.11	0.80	7.86	0.64
Organic back cell	4.65	0.75	11.54	0.53
Hybrid tandem cell	5.72	1.42	6.84	0.58

The hybrid cell has a PCE value of 5.72%, higher than 4.11% of the a-Si:H subcell and 4.65% of the organic cell, demonstrating the effectiveness of the tandem combination. As expected, the V_{oc} value of the hybrid tandem solar cell reaches 1.42 V, which is almost the sum of the V_{oc} values of the two individual subcells, a result that the a-Si:H and the organic subcells are electrically connected in series. Another evidence for the "good" electrical connection of the two subcells is the J_{sc} value of the hybrid cell, which is the smallest among all three cells. Small J_{sc} is one of the major limitations for tandem SCs. J_{sc} value of a tandem cell is always smaller than (at most equal to) that of a subcell in the tandem cell, i.e., the short-circuit current value of a tandem solar cell is limited to that of a subcell with smallest J_{sc}. The best result will be given for a tandem solar cell in which all the subcells have the same short-circuit current values.

As mentioned above, the interlayer that electrically connects the two subcells plays an essential role in determining the performance of a tandem cell. Other than being highly transparent to the solar spectrum and highly conductive to carriers, the fundamental requirements for the interlayer involve in forming good interfaces with the neighboring layers in subcells. Seo et al. studied in detail how the interlayer affected device performance by preparing tandem solar cells without an interlayer, with an interlayer of either only ITO or only PEDOT:PSS, and with an interlayer consisting of ITO/PEDOT:PSS double layers. With no interlayer, a tandem cell shows very poor performance of a J_{sc} of 2.22 mA/cm^2, a V_{oc} of 0.87 V, a FF of 0.15, and a PCE of 0.30%. The poor results are mainly due to the bad electric connection of the two subcells. Adding an interlayer of either ITO or PEDOT:PSS gives rise to jumps in J_{sc}, V_{oc}, FF, and PCE to nearly 4.3 mA/cm^2, 1.4 V, 0.37, and 2.0%, respectively. Using an interlayer of ITO/PEDOT:PSS double layers, the device parameters of J_{sc}, FF, and PCE further increase to nearly 5.0 mA/cm^2, 0.5, and 3.5%, respectively, except V_{oc} that remains approximately the same. The improvement is mainly due to the decrease in series resistance from ~124 Ω·cm^2 with the use of either an ITO or a PEDOT:PSS interlayer to ~30 Ω·cm^2 with the use of an ITO/PEDOT:PSS interlayer. Better Ohmic contact leads to a remarkable increase in FF from 0.35 to 0.49 by offering efficient recombination for electrons from the a-Si:H subcell and holes from the organic subcells. Note again that the benefit of using an interlayer is due to the good electrical conductivity and contact, as well as excellent optical transparency.

An earlier but similar device configuration was proposed by Kim et al., where the hybrid tandem SC consists of a organic back cell, with the device structure of HTL/PCPDTBT:PC$_{70}$BM/TiO$_2$/Al, and a a-Si:H front cell, as shown in Fig. 7.26 [24]. While the performance of the a-Si:H subcell is nearly the same as the front cell in the hybrid cell fabricated by Seo et al., the organic back cell shows relatively poor J_{sc}, as shown in Fig. 7.27. This leads to overall low PCE of only 1.84%, though the device has a high V_{oc} of 1.50 V that is nearly the sum of the open-circuit voltages of the front and the back cells. In the work by Kim et al., the interlayer was a 30 nm-thick PEDOT:PSS or a 5 nm-thick MoO$_3$ films, prepared by thermal evaporation or spin-coating using a solution precursor, respectively. Apparently, the performance of the hybrid tandem cells relies on the interlayer, as shown by the better *J–V* response of the MoO_3 film than that of the PEDOT:PSS film. MoO_3 is an efficient hole transporting layer compared to conventional conducting polymers. As indicated in Fig. 7.27 and Fig. 7.25a, the electrical performance of the single MoO_3 and PEDOT:PSS interlayer is as good as the ITO/PEDOT:PSS interlayer.

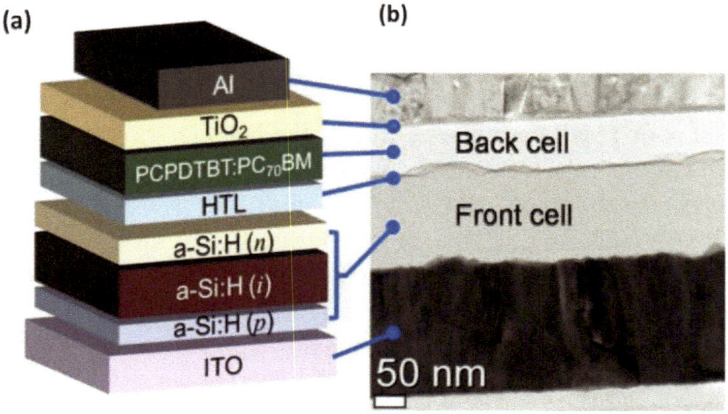

Figure 7.26 (a) Device architecture of the hybrid SC composed of a-Si:H and PCPDTBT:$PC_{70}BM$ organic subcells, (b) SEM image of the hybrid cell. Reproduced with permission from [24].

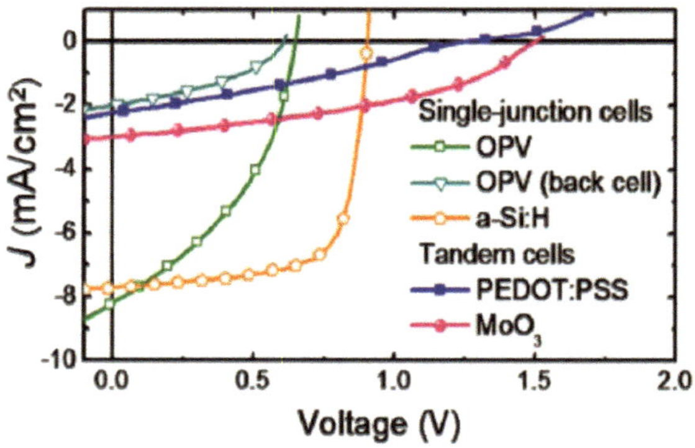

Figure 7.27 *J–V* curves of the front, back, and hybrid tandem solar cells measured under AM1.5 illumination. For comparison, a single junction organic SC and hybrid tandem SCs using PEDOT:PSS and MoO_3 interlayers are also shown. Reproduced with permission from [24].

The major advantage of the a-Si:H/organic hybrid SCs lies in their very high open-circuit voltage V_{oc} and the technical feasibility for large-scale device fabrication. The high V_{oc} is an indication that the material combination and the structural arrangement are well-designed. The existing low short-circuit current density J_{sc} suggests the details of the device need to be further optimized. The further improvement in the electrical connection between the front and the back cells, for instance, should give a positive impact on PCE. Other improvements include reducing carrier recombination during charge transport and increasing overall absorption of solar radiation.

7.6 Several Hybrid Nano Solar Cells with Efficiency ~10%

Limited by material properties and device structures, most HSCs do not have very high power conversion efficiencies, compared to conventional wafer-based Si SCs. Quite a few HSCs have been reported to achieve a PCE close to or even over 10%. In this section, several high-efficiency HSCs are presented. Some of them might have a future for productions.

7.6.1 Polymer/Nano-Si Hybrid Cells

Hybrid silicon nanocone–polymer solar cells

The device performance of an organic/Si hybrid solar cell relies on organic/Si Schottky junction, broadband light absorption, etc. Jeong and co-workers prepared a Si-nanocone/PEDOT:PSS hybrid solar cell and were able to achieve a PCE of 11.1% [25]. In their device, the Si nanocones were prepared by nanosphere lithography using SiO_2 nanoparticles. The SiO_2 nanoparticles were synthesized in a modified Stober process [26]. A uniform coverage of a SiO_2 nanoparticle monolayer was then formed on an *n*-type Si substrate via Langmuir–Blodgett assembly, where the size of the SiO_2 nanoparticles can be reduced by oxygen (O_2) and trifluoromethane (CHF_3) plasma etching. With the SiO_2 nanoparticle monolayer as template, a pattern is formed after etching the *n*-type Si substrate with chlorine (Cl_2) and hydrogen bromide (HBr) plasma. Figure 7.28 shows the schematic diagram and the measured TEM image of the Si nanocones. The

Si/polymer Schottky junction was prepared by spin-coating PEDOT:PSS on Si-nanocone-patterned substrate in air and at low temperature. It is easier to form conformal coating on nanocone-patterned substrate than on nanowire-patterned one. Note that prior to the PEDOT:PSS deposition, the patterned substrate needs to be cleaned by the hydrofluoric (HF) acid for removal of the native SiO_2 layer and possible remaining SiO_2 nanoparticles (see Ref. [25] for details), which is a critical step. In fact, the conformal PEDOT:PSS coating also act to passivate the Si surface [27]. Finally, a gold (Au) finger grid, with 80 μm width and the 450 μm inter-finger distance, is deposited on top of the PEDOT:PSS-covered Si-nanocone pattern as electrode. Although conductive and transparent, the PEDOT:PSS layer alone is not suitable for the use as electrode since it has much higher resistivity than Au.

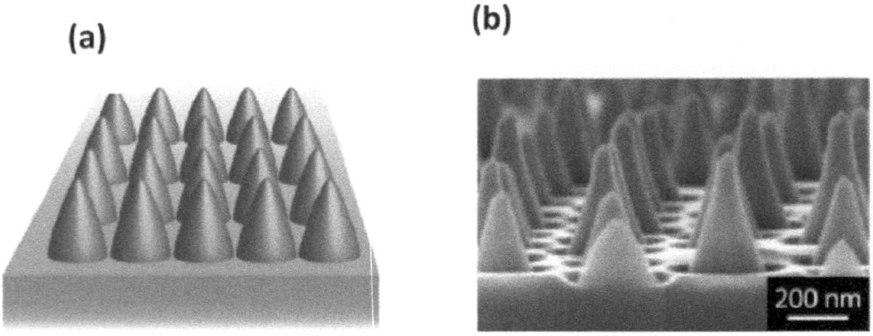

Figure 7.28 The schematic diagram (a) and TEM image of Si nanocone coverage on Si substrate (b). The Si nanocones in (b) is fabricated with SiO_2 nanoparticles 400 nm in diameter. Reproduced with permission from [25].

Figure 7.29 depicts the measured *J–V* characteristic responses of the Si-nanocones/PEDOT:PSS hybrid SCs with and without Au electrode, together with those of planar-Si/PEDOT:PSS with and without Au electrode for comparison. The extracted device parameters from Fig. 7.29 are listed in Table 7.3. As described in Chapter 3, the nanocone-patterned surface has nearly perfect light-trapping properties. With a change of device architecture from planar-Si/PEDOT:PSS to Si-nanocones/PEDOT:PSS, more than 50% increases in J_{sc} are obtained, as shown in Fig. 7.29a and Table 7.3. The significant increase of J_{sc} is due to superior anti-reflection property of the nanocone-patterned surface of the Si-nanocones/PEDOT:PSS cell, which leads to the enhancement in external quantum effect (EQE), as evidenced in Fig. 7.29b. Excellent anti-reflection and light-trapping surface was found for the nanocones having an aspect ratio (height:diameter) around 1. Apparently, the overall EQE of the Si-nanocones/PEDOT:PSS cell without Au grid is close to 90% in the spectrum range of 500 to 950 nm, not very sensitive to the wavelength. The EQE of the planar-Si/PEDOT:PSS cell (without Au grid) is, however, less than 70% and becomes low and low at the wavelength above 600 nm, reaching only 50% at 950 nm. The similar behavior was found for the hybrid cells with the Au electrode. Thus, the anti-reflection nanocones are especially beneficial for improving the device performance at longer wavelengths.

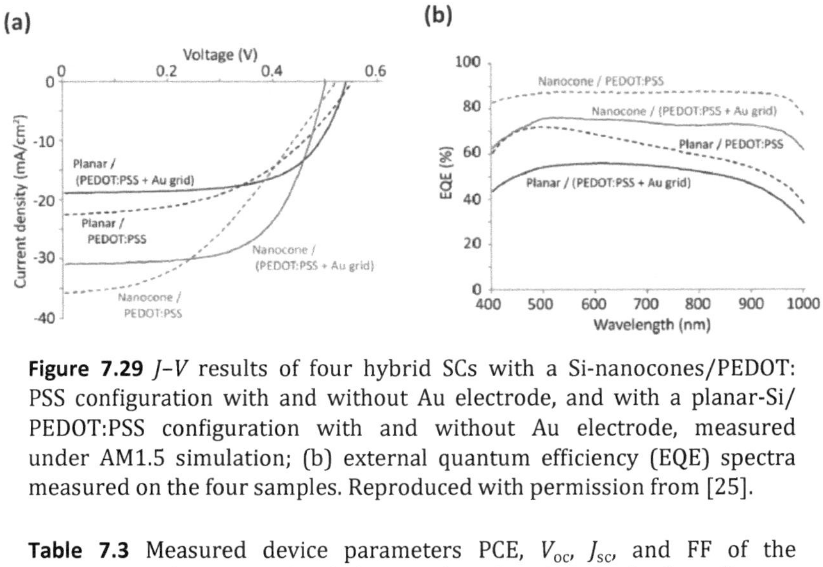

Figure 7.29 *J*–*V* results of four hybrid SCs with a Si-nanocones/PEDOT:PSS configuration with and without Au electrode, and with a planar-Si/PEDOT:PSS configuration with and without Au electrode, measured under AM1.5 simulation; (b) external quantum efficiency (EQE) spectra measured on the four samples. Reproduced with permission from [25].

Table 7.3 Measured device parameters PCE, V_{oc}, J_{sc}, and FF of the Si-nanocones/PEDOT:PSS and planar-Si/PEDOT:PSS hybrid solar cells

	PCE (%)	V_{oc} (V)	J_{sc} (mA/cm^2)	FF
Si-nanocones/PEDOT:PSS	7.54	0.51	35.6	0.413
Si-nanocones/PEDOT:PSS/Au	9.62	0.50	31.0	0.626
Si-nanocones/PEDOT:PSS/Au+doping	11.1	0.55	29.6	0.677
Planar-Si/PEDOT:PSS	5.92	0.55	22.5	0.480
Planar-Si/PEDOT:PSS/Au	6.47	0.54	18.8	0.641

The use of Au finger grid electrode results in a great enhancement in device FF. The sheet resistance of the Au finger grid is 0.83 $\Omega \cdot cm^2$, instead of 4.55 $\Omega \cdot cm^2$ for the PEDOT:PSS layer. The nearly fivefold reduction in resistance gives rise to more than 50% enhancement in FF, as shown in Fig. 7.29a for the improved rectangular-like *J–V* response of the device with Au electrodes. There is a trade-off—more than 10% decrease in J_{sc} attributed to reduced light absorption due to the block to light by the Au electrode. However, the overall improvement in device performance, with the use of Au electrode, is still significant. The only negative impact of using the nanocone surface is the 7.3% decrease of V_{oc}. The increase in area from the planar to nanocone surface could be significant. It has good side in terms of offering optoelectronic process for light from all directions. Yet the large area also results in an increasing chance for carrier recombination.

The Si-nanocones/PEDOT:PSS/Au exhibits a PCE of 9.62%. A slight improvement could kick the PCE into above 10%. It is noticed that V_{oc} of the device is only 0.5 V, even lower than that of planar-Si/PEDOT:PSS/Au cell. According to device physics of solar cells, a heavily doped back side layer (n^+ layer) gives rise to improved device performance through reducing contact resistance and enhancing V_{oc} by forming a built-in electric field. The additional built-in field on the back side of the device blocks the minority carriers, leading to reduction of the recombination rate at the back side of the device. In the work by Jeong and co-workers [25], the back side of the *n*-type Si substrate was further heavily *n*-type doped via phosphorus (P) diffusion in phosphoryl chloride ($POCl_3$) gas environment at the high temperature of 950°C. The n^+ doping at the level of 8×10^{20} cm^{-3} increases V_{oc} from 0.50 to 0.55 V and FF from 0.626 to 0.677, due to the formation of additional built-in electric field in the back side of the cell. The PCE of the cell increases from 9.62% to 11.1% accordingly.

Hybrid solar cell based on organic/inorganic Si nanowire arrays

A similar work was done by Shen and co-workers, who fabricated a hybrid solar cell using the Schottky junction formed by depositing spiro-OMeTAD onto Si nanowire arrays (Si NWs) standing on a Si substrate [28]. The device was slightly modified by adding other layers such as PEDOT:PSS, metal Cu and using core–shell or imbedded device configurations. The Si NWs on an *n*-type Si substrate was prepared by metal-assisted chemical wet etching [29]. To form Si-NWs/spiro-OMeTAD heterojunction, a precursor was prepared by mixing lithium bis(trifluoromethylsulfonyl)imide salt with spiro-OMeTAD dissolved in chlorobenzene. The junction was then constructed by depositing the precursor on the surface of the Si-NW-covered substrate followed by spin-coating. The fabrication was continued by depositing PEDOT:PSS and/or Cu layers for some of the devices. The back and the front electrodes were In:Ga alloy and Cu:Ag (50 nm:100 nm in thickness), respectively. For details of the device fabrication, see Ref. [28].

Compared to planar surfaces, the spiro-OMeTAD-covered Si-NWs on the planar Si substrate have enhanced light-trapping capability and the Si-NWs/spiro-OMeTAD core–shell structure is highly efficient for carrier collection. In Shen and co-workers' work, the Si NWs have diameters of ~20–100 nm and a density of $\sim 10^8$–10^9 cm^{-2}. Standing on the planar Si substrate, the *n*-type Si NWs form cores covered by spiro-OMeTAD shells. The Si NWs and the sheath-like spiro-OMeTAD act as the electron and the hole transport layers, respectively. The organic spiro-OMeTAD exhibits narrow absorption spectrum in visible light so that most light is absorbed by Si. The spiro-OMeTAD layer needs to be very thin in order for photogenerated holes to be collected efficiently by the front electrode, since the material has rather low carrier mobility. Additionally, the non-crystalline spiro-OMeTAD forms good contact with Si NWs.

Device parameters of three Si-NWs/spiro-OMeTAD hybrid SCs fabricated with different front electrodes with or without inserting layers are listed in Table 7.4 and the *J*–*V* responses and the EQE spectra of the devices are presented in Figs. 7.30a,b. The device with a configuration of In:Ga/SiNWs/spiro-OMeTAD/Cu:Ag shows V_{oc}, J_{sc}, FF, and PCE of 0.484 V, 25.5 mA/cm^2, 0.365, and 4.50%, respectively. Inserting a PEDOT:PSS layer between the spiro-OMeTAD layer and the Cu:Ag front grid electrode leads to significant increases of 12% for V_{oc}, 13% for J_{sc}, 48% for FF, and 88% for PCE. The increase in V_{oc} is apparently an indication of improvement of the organic/inorganic interface quality. The improvement may have the mechanism of the blocking of shunt paths formed during metal deposition that causes the possible electronic connection between the metal electrode and Si. A 13% increase in J_{sc} suggests the improved hole collection efficiency and better electrode contact after the insertion of the PEDOT:PSS layer. The enhancement both in V_{oc} and J_{sc} gives rise to dramatic increases of 48% in FF and 88% in PCE.

Table 7.4 The hybrid cell parameters of three Si-NWs/spiro-OMeTAD HSCs fabricated with different configurations of the hole transport layers

	PCE (%)	V_{oc} (V)	J_{sc} (mA/cm^2)	FF
Si-NWs/spiro-OMeTAD/Cu:Ag	4.50	0.484	25.5	0.365
Si-NWs/spiro-OMeTAD/PEDOT:PSS/Cu:Ag	8.47	0.542	28.9	0.541
Si-NWs/spiro-OMeTAD/Cu/PEDOT:PSS/Cu:Ag	9.70	0.527	31.3	0.588

Note: The PCE of the devices were calculated from EQE curves.

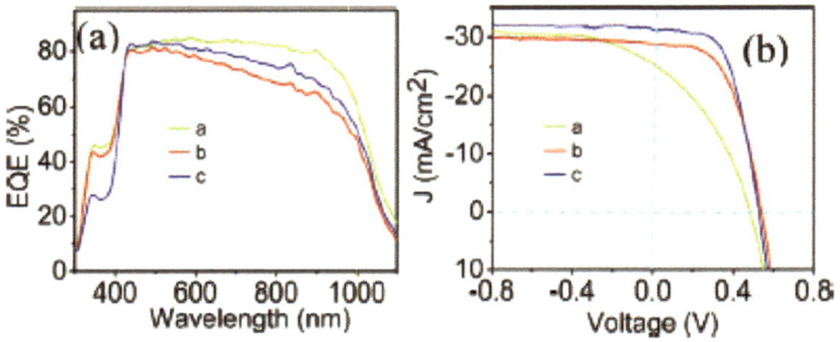

Figure 7.30 EQE (a) and *J–V* (b) responses of three SiNWs/spiro-OMeTAD hybrid SCs fabricated with different front electrodes with or without inserting layers, measured under AM1.5 simulation. The measurements were performed on the cell a, b, and c with the following device configurations: Cell a: In:Ga/SiNWs/spiro-OMeTAD/Cu:Ag; Cell b: In:Ga/SiNWs/spiro-OMeTAD/PEDOT:PSS/Cu:Ag; Cell c: In:Ga/SiNWs/spiro-OMeTAD/Cu/PEDOT:PSS/Cu:Ag. Reproduced with permission from [28].

The hole transport region of the device was further improved by inserting a Cu layer between the spiro-OMeTAD and the PEDOT:PSS layers, which resulted in the device architecture of In:Ga/SiNWs/spiro-OMeTAD/Cu/PEDOT:PSS/Cu:Ag. The hybrid cell then displayed the device parameters of 0.527 V, 31.3 mA/cm^2, and 0.588% for V_{oc}, J_{sc}, and FF, respectively, and a PCE of nearly 10% (9.70%).

He et al. also reported the fabrication of hybrid SCs based on the same Si-NWs/spiro-OMeTAD heterojunction configuration via wet chemical etching followed by a spin-coating process [30]. Prior to the deposition of Ag grid front electrode by electron-beam evaporation, a highly conductive PEDOT:PSS layer was spin-coated on spiro-OMeTAD that covered the Si-NWs. The PEDOT:PSS layer was prepared using precursor of PEDOT:PSS (Clevios PH500) mixed with 5 wt % dimethyl sulfoxide (DMSO) by spin-coating followed by 110°C annealing for 10 min. The back contact is metal Al. Si NWs of various lengths were prepared and tested in devices. A maximum PCE of 9.92% was obtained with the use of Si NWs with the length of 0.35 μm in cells. The details of the device fabrication process are described in Ref. [30].

Figure 7.31 shows the top-view SEM images of the as-prepared Si NWs coated by spiro-OMeTAD (a–d) and the corresponding cross-sectional views after the deposition of the PEDOT:PSS layer (e–h). The top-view images depict clearly uniform distribution of the Si-NWs standing on the Si substrate. The cross-sectional views of the SEM images present distinctly evidence that spiro-OMeTAD infiltrates deep into the pore regions among the Si NWs and fills the pores well so that it forms close contact with the sidewalls of the Si NWs, which is one of the advantages for spiro-OMeTAD being a small-molecule organic.

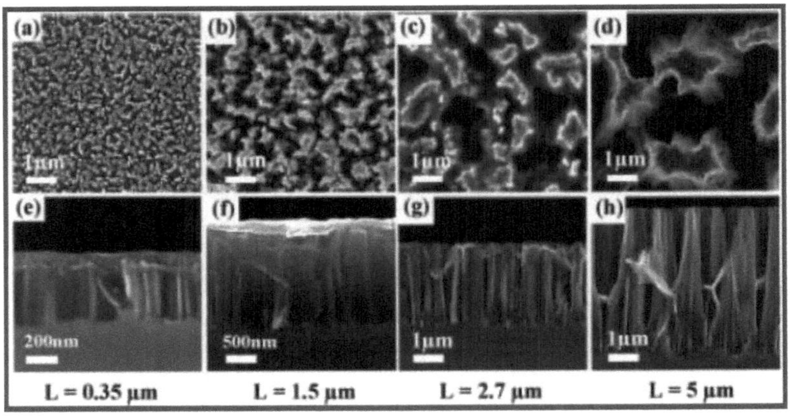

Figure 7.31 Top views of SEM images of the spiro-OMeTAD-covered SiNWs standing on the Si substrate with the NW length L = 0.35, 1.5, 2.7, and 5 μm (a–d) and corresponding cross-sectional views (e–h) after further coated with PEDOT:PSS layers. Reproduced with permission from [30].

The incident photon-to-current conversion efficiency (IPCE) measurement was performed on the Al/Si-NWs/spiro-OMeTAD/PEDOT:PSS/Ag hybrid SCs with the NW lengths of 0.15, 0.35, 0.9, 2.7, and 5.0 μm and the results are presented in Fig. 7.32. The cell with 5 μm Si NW length shows an IPCE curve with a response maximum of above 50% in 800–900 nm. The IPCE response decreases as the wavelength becomes shorter, even in the range of 500 to 600 nm where the sun exhibits the strongest radiation under AM1.5. This is attributed to the absorption of short-wavelength photons at top of the Si NWs where electrons have to travel long distances so that they have large probability to recombine with holes before reaching the electrode. Photons with long wavelength have better chance to reach the bottom of the NWs.

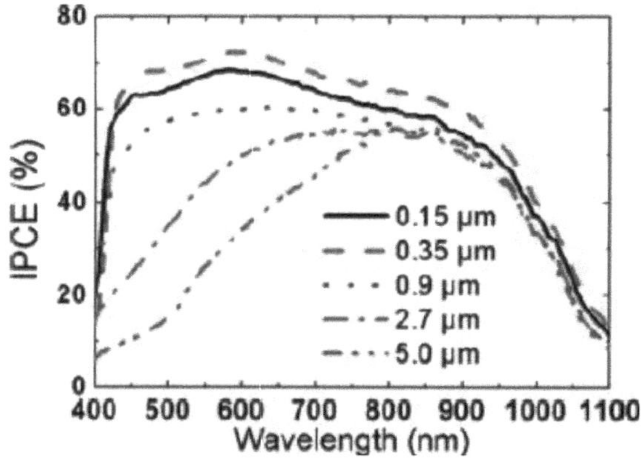

Figure 7.32 IPCE characteristics of Si-NWs/spiro-OMeTAD heterojunction hybrid cells with the Si NW lengths of 0.15, 0.35, 0.9, 2.7, and 5.0 μm. Reproduced with permission from [30].

Decreasing the Si NW length leads to overall increasing IPCE response. As depicted in Fig. 7.32, the increase is more significant in the long-wavelength region. The IPCE increases up to the maximum value of 72.4% at 600 nm for the Si NW length of 0.35 μm. This is attributed to the fact that the optoelectronic process occurs in the NWs more and more close to the Si substrate for short NW lengths. The HSC with Si NW of 0.35μm in length has the strongest overall IPCE response, which displays an open-circuit voltage V_{sc} of ~ 0.57 V, a short-circuit current density J_{sc} of ~30.7 mA.cm^{-2}, an FF of 0.57, and a PCE of 9.92%. The device performance with the 0.15 μm Si NWs degrades as shown in Fig. 7.32, which is attributed to the weakening light-trapping effect for short NWs. It should be noted that the geometry for the Ag grid electrode is not optimized. The 9.92% PCE was achieved on the grid electrode with a shading area fraction of 12%. Very likely, decreasing the shading area fraction could result in a PCE above 10%.

7.6.2 Efficient Hybrid Heterojunction Solar Cells Containing Perovskite Compound and Polymeric Hole Conductors

In recent years, perovskite compounds have been merging as hot solar cell materials due to their simple fabrication processes and quickly increasing power conversion efficiencies over 20%. A perovskite refers to any material with the same type of crystal structures as calcium titanium trioxide ($CaTiO_3$), which is known to have the perovskite structure shown in Fig. 7.33a. A perovskite material has a chemical formula of ABX_3, where "A" and "B" are two cations with very different sizes and "X" is an anion bonding to both cations. The A-atom is the group-II metal cation with larger size, such as Ca^{2+}, and the B-atom the smaller group-IV metal cations, such as Ti^{4+}. The X-atom is usually oxygen. The crystal unit cells of the perovskites can be cubic, orthorhombic, tetragonal, or trigonal, depending the way how the cubic structure is distorted. A metal halide perovskite compound has a general chemical formula of $(RNH_3)MX_3$ with R being C_nH_{2n+1}; X being halogen I, Br, Cl; M being Pb, Cd, Sn, etc. Perovskite compounds can be fabricated by many technical routes including self-assembly [31]. Perovskite-based hybrid materials have self-assembly characteristics and can be synthesized by simple and cheap methods.

Heo et al. prepared HSCs based on an inorganic-organic metal halide perovskite compound, $CH_3NH_3PbI_3$, with its molecule structure shown in Fig. 7.33b [32]. The $CH_3NH_3PbI_3$ was prepared by drying a mixture of 40 wt% $CH_3NH_3PbI_3$ in γ-butyrolactone solution at 100°C. An XRD measurement revealed characteristic lines at 14.08°, 14.19°, 28.24°, and 28.50°, corresponding to the diffraction of the (002), (110), (004), and (220) planes, respectively. The XRD measurement confirmed the tetragonal perovskite structure with the unit cell parameters of $a = b = 8.88$ and $c = 12.68$ Å.

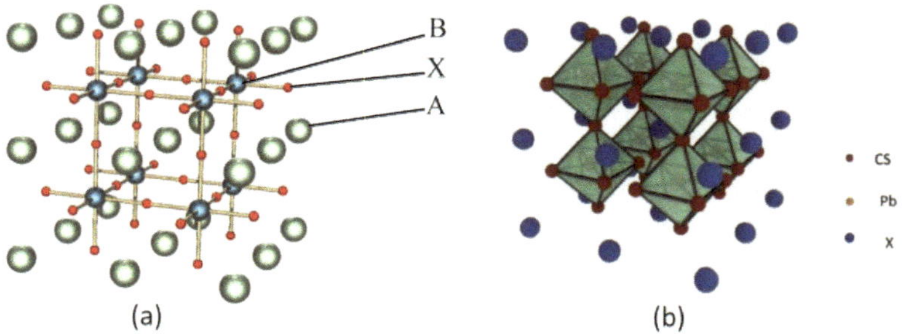

Figure 7.33 (a) The crystal structure of a perovskite with a chemical formula ABO_3 where the X atoms are mostly oxygens, the A-atoms the group-II metal cations such as Ca^{2+}, and the B-atoms smaller metal cations, such as Ti^{4+}. (b) Molecule structure of $CH_3NH_3PbI_3$.

The materials used in constructing the cells are $CH_3NH_3PbI_3$, mesoporous (mp)-TiO_2, hole-transport materials, including P3HT, PCPDTBT, PCDTBT, PTAA, spiro-OMeTAD, as well as fluorine-doped tin oxide (FTO) for transparent conductive front electrode and Au for back electrode. The device has a configuration of Au/HTM/$CH_3NH_3PbI_3$/mp-TiO_2/FTO. Figure 7.34 presents the schematic diagram of the perovskite HSC, together with the SEM image, where the hole-transport material is the PTAA layer. The best hybrid cell presents the device parameters of V_{oc} = 0.997 V, J_{sc} = 16.4 mA·cm^{-2}, and FF = 0.727, corresponding to a PCE value of 12.0%.

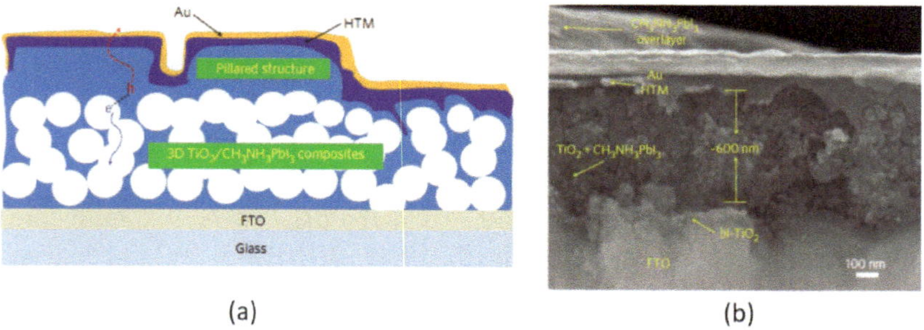

Figure 7.34 (a) The schematic picture of the hybrid SC with an architecture of Au/HTM/$CH_3NH_3PbI_3$/mp-TiO_2/FTO and (b) the cross-sectional SEM image of the device. Reproduced with permission from [32].

The architecture displays a band alignment, as shown in Fig. 7.35, suitable for photoelectronic conversion. Here the $CH_3NH_3PbI_3$ is used as the light harvester and the light is mainly absorbed at the $CH_3NH_3PbI_3$/mp-TiO_2 interface where the photon-generated excitons were disassociated and electrons and holes were swept towards FTO and Au electrodes, respectively, to complete the photovoltaic process. In addition to the suitable band alignment of the materials, the high device performance benefits from the facts that $CH_3NH_3PbI_3$ is good light harvester and P3HT, PCPDTBT, PCDTBT, PTAA, as well as spiro-OMeTAD are good hole-transport materials. The mp-TiO_2 is a good electron transport material. Particularly, the high-quality combination of the materials gives rise to high-quality heterojunctions and, in turn, a high PCE. Although $CH_3NH_3PbI_3$ does not form a continuous and conformal coverage on the mp-TiO_2 as shown in Fig. 7.34, this architecture does offer large $CH_3NH_3PbI_3$/mp-TiO_2 junction area for exciton disassociation. Figure 7.34 showed that the perovskite $CH_3NH_3PbI_3$ infiltrated deep into the pore space in the mp-TiO_2 layer, as confirmed by energy dispersive spectrometer (EDS) mapping and X-ray photoelectron spectroscopy (XPS) depth profile measurement of the elements (see the supplementary information of Ref. [32]. The three-dimensional composites of $CH_3NH_3PbI_3/TiO_2$ form pillared structure. Such a bicontinuous nanocomposite structure and the formation of densely packed small $CH_3NH_3PbI_3$ crystalline domains on surface and top of the mp-TiO_2 layer offer effective extraction of the charge carriers. Such an architecture also makes it easy for excitons, having short diffusion length, to diffuse to TiO_2/CH_3NH_3 PbI_3 interface where they were disassociated. Electrons can then be easily extracted by the extremely close mp-TiO_2 layer.

In another work, Noh and co-workers investigated a hybrid solar cell using perovskite $CH_3NH_3Pb(I_{1-x}Br_x)_3$ as the light-absorbing layer [33]. $CH_3NH_3Pb(I_{1-x}Br_x)_3$ has the property that its absorption covers almost the whole solar irradiation spectrum by varying the contents of I and Br. They demonstrated a conversion efficiency of 12.3% at around $x = 0.06$.

As a hole transport material, the high-molecular-weight PTAA layer is, in contrast to low-molecular-weight spiro-OMeTAD, only partially infiltrated into mp-TiO_2 and it is located near the surface of the $TiO_2/CH_3NH_3PbI_3$ composite layer. PTAA is an polymer with a very high hole mobility of 2×10^{-2} to 4×10^{-2} $cm^2 \cdot V^{-1} \cdot s^{-1}$, compared to the typical hole mobilities at around 1×10^{-4} $cm^2 \cdot V^{-1} \cdot s^{-1}$ for other hole transport polymers [34]. The potential discontinuity at the PATT/TiO_2-$CH_3NH_3PbI_3$-composite interface renders holes the easiness to be extracted by the Au electrode, which is essential for excellent device performance.

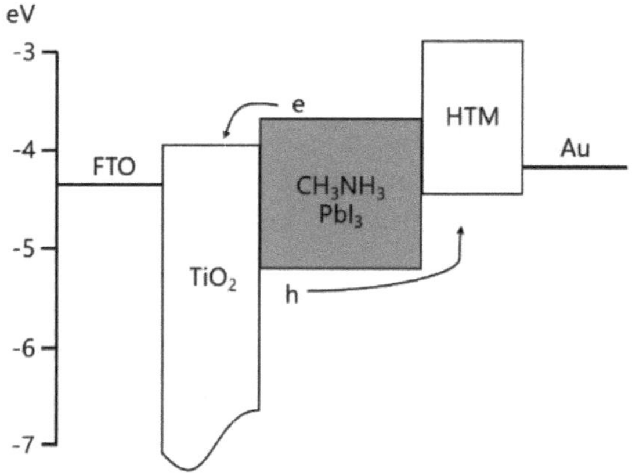

Figure 7.35 The band alignment of the Au/HTM/$CH_3NH_3PbI_3$/mp-TiO_2/FTO hybrid SC.

From the double junction configuration of the hybrid cell, photovoltaic process can also perform at the junction between the PTAA and the $CH_3NH_3PbI_3$ layers. The $CH_3NH_3PbI_3$ thus displays an ambipolar feature, since it acts as both *n*- and *p*-type semiconductors. Whether it is an *n*-type or a *p*-type depends on the band structure of the material with which they form the heterojunction. In other words, $CH_3NH_3PbI_3$, the inorganic-organic metal halide perovskite compound, is used as the electron transport material, if excitons are generated in the vicinity of the PTAA/$CH_3NH_3PbI_3$ interface. In the work by Heo et al., light is incident into the device from the FTO side and the highly absorptive $CH_3NH_3PbI_3$ absorbs most of the light at CH_3NH_3 PbI_3/TiO_2 interface, as evidenced by their studies which show that the $CH_3NH_3PbI_3$ behaves predominantly as a *p*-type material. In many cases, the ambipolar behavior of a light harvesting material is essential for high performance of a double junction solar cell.

For comparison, Heo and co-workers fabricated HSCs of the same device architecture with other HTMs such as P3HT, PCPDTBT, PCDTBT and without a HTM. Figure 7.36 shows that the performance of the device relies strongly on the use of the HTMs. V_{oc} of the SCs increases in the order of 0.73 V for P3HT, 0.77 V for PCPDTBT, and 0.92 V for PCDTBT, matching with the HOMO energy levels of these polymers: −25.2 eV for P3HT [35], −25.3 eV for PCPDTBT [36], and −25.45 eV for PCDTBT [37]. For the device without a HTM layer, V_{oc} is only 0.68 V. Although the cell with a PTAA layer as HTP shows a V_{oc} value of 0.90 V, which is not the highest among those in Fig. 7.36, it does display a J_{sc} value of 16.4 mA·cm^{-2} far superior to those of the other devices. This is owing to the fact that PTAA is well-matched to the $TiO_2/CH_3NH_3PbI_3$ composite as a hole-transport material.

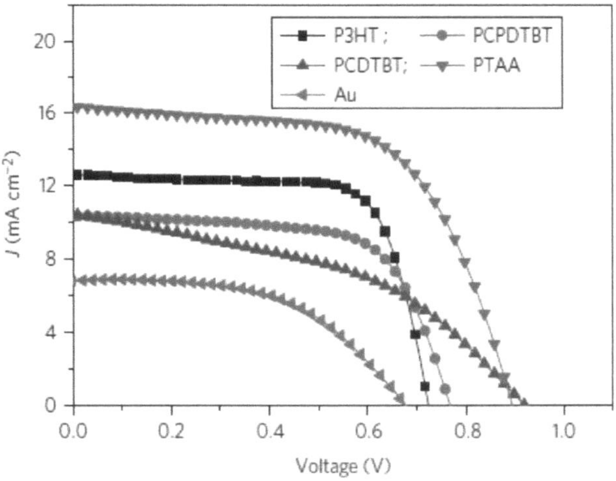

Figure 7.36 The measured *J–V* characteristics of the Au/HTM/$CH_3NH_3PbI_3$/mp-TiO_2/FTO hybrid SCs with the HTM layer being P3HT, PCPDTBT, PCDTBT, and PTAA, or without a HTM layer. Reproduced with permission from [35–37].

7.6.3 Highly Efficient Si-Nanorods/Organic Hybrid Core–Sheath Heterojunction Solar Cells

He and co-workers fabricated HSCs using *n*-type Si nanorods (NRs) and *p*-type spiro-OMeTAD and the devices have the configuration of Al/Si-NRs/spiro-OMeTAD/PH500/Ag, with PH500/Ag-grid being the front electrode and Al-film the back electrode [38]. The spiro-OMeTAD was prepared by a simple low-cost solution route. The Si-NRs/spiro-OMeTAD forms a core–sheath hybrid hetero-junction and the spiro-OMeTAD coverage on Si NRs is complete due to the small molecule size of spiro-OMeTAD. The Si-NRs/spiro-OMeTAD hybrid cell yields a PCE of 10.3%. The Si NRs arrays were prepared by etching *n*-type (100) Si wafers using a solution of 0.46 M HF (hydrofluoric acid) and 0.02 M $AgNO_3$ (silver nitride). Vertically aligned, single-crystal Si NRs have the diameters ranging 30–150 nm and the spacing between the neighboring NRs in the range 20–80 nm. The hybrid junction was formed by spin-coating a solution of 50 mg spiro-OMeTAD dissolved in chlorobenzene containing 4.17 mM Li on the Si NRs array. PH500 was deposited by spin-coating an aqueous solution of PEDOT:PSS with a solid content of 1% mixed with 5 wt.% dimethyl sulfoxide (DMSO) on the Spiro-OMeTAD layer, followed by annealing at 110°C for 10 min. Both the silver grid front electrode and the 250 nm-thick Al back electrode were deposited by e-beam evaporation. Note that Ag-grid shadowing gives rise to 12% power loss due to blocking the incident light. The device has the configuration and energy band alignment shown in Fig. 7.37. For comparison, planar Si hybrid cells were also prepared using the same fabricating procedure with a 25 nm spiro-OMeTAD layer in thickness.

The schematic energy band profile of the device shows that the HOMO level of spiro-OMeTAD is at 4.9 eV, about 0.27 eV above the valence band edge of Si, while the LUMO level of spiro-OMeTAD is at 1.9, 2.15 eV above the conduction band edge of Si. The potential discontinuities at the Si-NRs/spiro-OMeTAD interface disassociate the photon-generated excitons efficiently and drive the electrons and the holes towards the back and the front electrodes, respectively. Hybrid cells with different Si NR lengths have been fabricated and tested for performance. It was found that a cell with Si NRs of 0.35 μm in length gave the best device property.

J–*V* characteristics of the hybrid cells with 0.35 μm Si NRs and planar Si were measured at AM 1.5 condition and the results are presented in Fig. 7.38a. From the *J*–*V* data, V_{oc}, J_{sc}, FF, and PCE were extracted and presented in the figure.

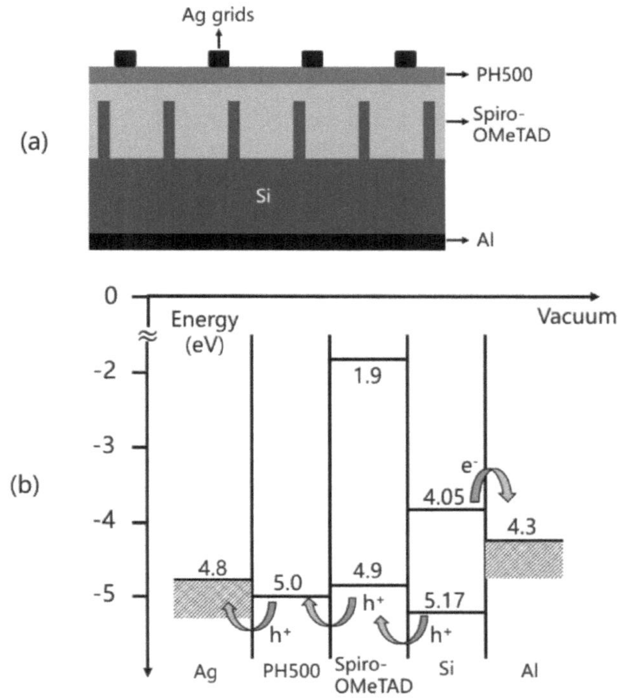

Figure 7.37 Schematic diagram of (a) the Si-NRs/spiro-OMeTAD hybrid solar cell and (b) the corresponding energy band profile of the device.

Replacing planar Si by Si NRs leads to a substantial increase of PCE from 5.6% to 10.3%, due to the increases in V_{oc} (from 0.54 to 0.57 V), J_{sc} (24.8 to 30.9 mA/cm^2), as well as FF (42.2% to 58.8%). The optical reflectance measurement indicates that the surface of the Si NRs coated with spiro-OMeTAD/PH500 exhibits much better photon trapping property than that of the spiro-OMeTAD/PH500-coated planar Si. Additionally, the increased junction area in the Si NR cell enhances the carrier separation and the reduced travel distance of minority carriers make it easy for the charged carriers to be transported to the electrodes. Thus, the increases in J_{sc} and FF are beneficial due to the use of Si NRs.

It appears that the use of Si NRs leads to improvement in device performance in almost the whole solar spectrum range except at very short wavelength (shorter than 450 nm), which can be seen clearly in Fig. 7.38b that presents the IPCE results of both the 0.35 μm Si NRs cell and the Si planar cell. The Si NRs cell has a maximum IPCE value above 70% at 600 nm and the planar cell shows a maximum IPCE of ~55% at 650 nm.

Among various HSCs, it could be expected that the HSC based on organic materials and nano silicon will still play a leading role [39].

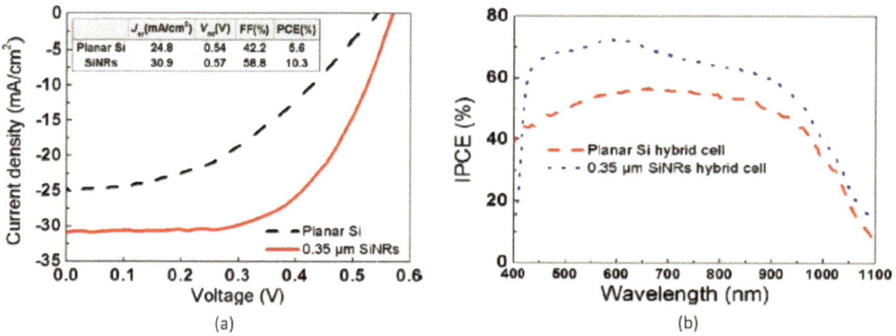

Figure 7.38 *J–V* curves (a) and IPCE spectra (b) of the hybrid cell with 0.35 μm Si-NRs and the hybrid cell with planar Si. The measurements were performed at AM1.5 condition. Reproduced with permission from [38].

7.7 The Existing Problems

Most reported hybrid SCs display PCE below 10%, so that the major existing problem for hybrid SCs is their low efficiencies. On the other hand, unlike inorganic semiconductors, organic materials are often unstable over time, aging being a serious problem. Another critical issue is the mismatching of some fundamental properties, which could negatively affect device performance very seriously. The carrier mobilities, for instance, are the typical material parameter regarding which one often feels helpless in the construction of hybrid SCs, considering the fact that the inorganic materials are often highly superior to organic materials. The electron mobility in silicon, for instance, is typically the order of 500 $cm^2 \cdot V^{-1} \cdot s^{-1}$. The hole mobility in MEH-PPV is typically below 1×10^{-4} $cm^2 \cdot V^{-1} \cdot s^{-1}$. Carriers in organic materials undergo various scatterings so that the speed of carrier transport is significantly slowed down. Those scatterings originate from not only various defects and impurities but also atomic arrangement without long-range order. In fact, the absence of lattice translation invariance is a key reason for the low mobility in organic materials. The low carrier mobility and, in turn, short diffusion length (around 10 nm) leads to low device PCE. To overcome these obstacles, a number of straightforward ways can be adopted and some of them have been proven to be effective, such as choosing suitable annealing temperature and annealing time, controlling the film thickness, and using right donor/acceptor materials and work function electrodes.

For the realization of high conversion efficiency, it is essential to control the quality of the interface where the processes of exciton disassociation and charge transfer occur. The interface is a place where there is high density of defects, dislocation, etc., that act as charge trapping centers, or as non-radiative recombination centers. In addition to constructing the heterojunction for potential alignment to drive the electrons and holes, organic materials are often used as coating material for the passivation of inorganic semiconductors. The smaller the mobility of the carriers, the slower they migrate. In addition, the poor conductivity of nano and organic materials is one of the main factors for the low efficiency of HSCs.

In HSCs, one of the typical material forms is inorganic nanomaterial imbedded in organic matrix. Thus, controlling the nanomaterial aggregation is a critical issue. It requires that the nanomaterials be dispersed in a suitable form to achieve maximize interface area. The complex material structure also form networks offering for effective electron transport. A number of methods were proposed to fabricate this kind of materials. A typical way is integrating semiconductor nanowires in organic polymers so that the nanowires offer channels for electron conduction. One of the major advantages of using nanosemiconductors is that the bandgap nanosemiconductors can be tuned to match to the solar spectrum. However, another important thing that needs to be taken serious consideration is the band edge alignment with those of polymers for optimized exciton disassociation and carrier transport.

For cost effectiveness, nanosemiconductors used for SCs are usually fabricated in liquid phases. The colloidal nanomaterials are thus stabilized by various ligands and the carrier transport is via the ligands. The fundamental requirements of the ligands include good current carrying property (good conductivity), suitable energy level position, and good passivation for the nanomaterials. Obviously, if the ligands are electrically insulators, they are harmful for device performance since they block the way of carrier transport. Using suitable pyridine or short-chain ligand has proven to be effective for enhancing the device performance [40]. More recently, a HSC with a polyvinylpyrrolidone (PVP)/nano-SnO_2 structure has gained a PCE near 19% (18.98%) [41].

References

1. Y. Xia, K. Sun, and J. Ouyang, Solution-processed metallic conducting polymer films as transparent electrode of optoelectronic devices, *Adv. Mater.*, **24**, 2436–2440 (2012).
2. J. H. Burroughes, D. D. C. Bradley, A. R. N. Brown, R. N. Marks, K. D. Mackay, R. H. Friend, P. L. Burns, and A. B. Holmes, Light-emitting diodes based on conjugated polymers, *Nature*, **347**(6293), 539–541 (1990).
3. D. S. Ginger and N. C. Greenham, Photoinduced electron transfer from conjugated polymers to CdSe nanocrystals, *Phys. Rev. B*, **59**(16), 10622–10629 (1999).
4. S. E. Gledhill, B. Scott, and B. A. Gregg, Organic and nano-structured composite photovoltaics: An overview, *J. Mater. Res.*, **20**(12), 3167–3179 (2005).
5. J. J. M. Halls, K. Pichler, R. H. Friend, S. C. Moratti, and A. B. Holmes, Exciton diffusion and dissociation in a poly(*p*-phenylenevinylene)/C-60 heterojunction photovoltaic cell, *Appl. Phys. Lett.*, **68**(22), 3120–3122 (1996).
6. J. J. M. Halls, C. A. Walsh, N. C. Greenham, E. A. Marseglia, R. H. Friend, S. C. Moratti, and A. B. Holmes, Efficient photodiodes from interpenetrating polymer networks, *Nature*, **376**(6540), 498–500 (1995).
7. G. Yu, J. Gao, J. C. Hummelen, F. Wudl, and A. J. Heeger, Polymer photovoltaic cells: Enhanced efficiencies via a network of internal donor-acceptor heterojunctions, *Science*, **270**(5243), 1789–1791 (1995).
8. G. K. Mor, K. Shankar, M. Paulose, O. K. Varghese, and C. A. Grimes, High efficiency double heterojunction polymer photovoltaic cells using highly ordered TiO_2 nanotube arrays, *Appl. Phys. Lett.*, **91**(15), 152111 (2007).

9. K. M. Coakley, Y. Liu, C. Goh, and M. D. McGehee, Ordered organic-inorganic bulk heterojunction photovoltaic cells, *MRS Bull.*, **30**, 37–40 (2005).

10. D. J. Herman, J. E. Goldberger, S. Chao, D. T. Martin, and S. I. Stupp, Orienting periodic organic-inorganic nanoscale domains through one-step electrodeposition, *ACS Nano*, **5**(1), 565–573 (2011).

11. J. Weickert, R. B. Dunbar, W. Wiedemann, H. C. Hesse, and L. Schmidt-Mende, Nanostructured organic and hybrid solar cells, *Adv. Mater.*, **23**, 1810 (2011).

12. K. Takanezawa, K. Tajima, and K. Hashimoto, Efficiency enhancement of polymer photovoltaic devices hybridized with ZnO nanorod arrays by the introduction of a vanadium oxide buffer layer, *Appl. Phys. Lett.*, **93**, 063308 (2008).

13. L. He, D. Lai, H. Wang, C. Jiang, and Rusli, High-efficiency Si/polymer hybrid solar cells based on synergistic surface texturing of Si nanowires on pyramids, *Small*, **8**(11), 1664–1668 (2012).

14. (a) G. Kalita, S. Adhikari, H. R. Aryal, R. Afre, T. Soga, M. Sharon, W. Koichi, and M. Umeno, Silicon nanowire array/polymer hybrid solar cell incorporating carbon nanotubes, *J. Phys. D: Appl. Phys.*, **42**, 115104 (2009); (b) S.-C. Shiu, J.-J. Chao, S.-C. Hung, C.-L. Yeh, and C.-F. Lin, Morphology dependence of silicon nanowire/poly (3,4- ethylenedioxythiophene): Poly(styrenesulfonate) heterojunction solar cells, *Chem. Mater.*, **22**, 3108 (2010); (c) F. Zhang, B. Sun, T. Song, X. Zhu, and S. Lee, Air stable, efficient hybrid photovoltaic devices based on poly(3-hexylthiophene) and silicon nanostructures, *Chem. Mater.*, **23**, 2084 (2011); (d) L. He, C. Jiang, Rusli, D. Lai, and H. Wang, Highly efficient Si-nanorods/organic hybrid core–sheath heterojunction solar cells, *Appl. Phys. Lett.*, **99**, 021104 (2011); (e) L. He, Rusli, C. Jiang, H. Wang, and D. Lai, Simple approach of fabricating high efficiency Si nanowire/conductive polymer hybrid solar cells, *IEEE Electron Device Lett.*, **32**, 1406 (2011); (f) B. Ozdemir, M. Kulakci, R. Turan, and H. E. Unalan, Silicon nanowire-poly(3,4-ethylenedioxythiophene)-poly(styrenesulfonate) heterojunction solar cells, *Appl. Phys. Lett.*, **99**, 113510 (2011).

15. S. Dayal, N. Kopidakis, D. C. Olson, D. S. Ginley, and G. Rumbles, Photovoltaic devices with a low band gap polymer and CdSe nanostructures exceeding 3% efficiency, *Nano Lett.*, **10**, 239–242 (2010).

16. J. Seo, W. J. Kim, S. J. Kim, K.-S. Lee, A. N. Cartwright, and P. N. Prasad, Polymer nanocomposite photovoltaics utilizing CdSe nanocrystals capped with a thermally cleavable solubilizing ligand, *Appl. Phys. Lett.*, **94**, 133302 (2009).

17. W. U. Huynh, J. J. Dittmer, and A. P. Alivisatos, Hybrid nanorod-polymer solar cells, *Science*, **295**, 2425 (2002).

18. M. Imran, M. Ikram, A. Shahzadi, S. Dilpazir, H. Khan, I. Shahzadi, S. Amber Yousaf, S. Ali, J. Geng, and Y. Huang, High-performance solution-based CdS-conjugated hybrid polymer solar cells, *RSC Adv.*, **8**, 18051 (2018).

19. R. B. Laghumavarapu, G. Mariani, B. Tremolet de Villers, J. Shapiro, P. Senanayake, A. Lin, B. J. Schwartz, and D. L. Huffaker, Hybrid solar cells using GaAs nanopillars, *IEEE*, 000943 (2010).

20. G. Mariani, R. B. Laghumavarapu, B. T. de Villers, J. Shapiro, P. Senanayake, A. Lin, B. J. Schwartz, and D. L. Huffaker, Hybrid conjugated polymer solar cells using patterned GaAs nanopillars, *Appl. Phys. Lett.*, **97**, 013107 (2010).

21. S. Hao, J. Wu, and Z. Sun, A hybrid tandem solar cell based on hydrogenated amorphous silicon and dye-sensitized TiO_2 film, *Thin Solid Films*, **520**, 2102–2105 (2012).

22. J. Yang, R. Zhu, Z. Hong, Y. He, A. Kumar, Y. Li, and Y. Yang, A robust inter-connecting layer for achieving high performance tandem polymer solar cells, *Adv. Mater.*, **23**, 3465 (2011).

23. J. H. Seo, D.-H. Kim, S.-H. Kwon, M. Song, M.-S. Choi, S. Y. Ryu, H. W. Lee, Y. C. Park, J.-D. Kwon, K.-S. Nam, Y. Jeong, J.-W. Kang, and C. S. Kim, High efficiency inorganic/organic hybrid tandem solar cells, *Adv. Mater.*, **24**(33), 4523–4527 (2012).

24. T. Kim, J. H. Jeon, S. Han, D.-K. Lee, H. Kim, W. Lee, and K. Kim, Organic-inorganic hybrid tandem multijunction photovoltaics with extended spectral response, *Appl. Phys. Lett.*, **98**, 183503 (2011).

25. S. Jeong, E. C. Garnett, S. Wang, Z. Yu, S. Fan, M. L. Brongersma, M. D. McGehee, and Y. Cui, Hybrid silicon nanocone–polymer solar cells, *Nano Lett.*, **12**, 2971–2976 (2012).

26. G. H. Bogush, M. A. Tracy, and C. F. Zukoski IV, Preparation of monodisperse silica particles: Control of size and mass fraction, *J. Non-Cryst. Solids*, **104**, 95–106 (1988).

27. M. J. Sailor, E. J. Ginsburg, C. B. Gorman, A. Kumar, R. H. Grubbs, and N. S. Lewis, Thin films of *n*-Si/poly-$(CH_3)_3$Si-cyclooctatetraene:

Conducting-polymer solar cells and layered structures, *Science,* **249**, 1146–1149 (1990).

28. X. Shen, B. Sun, D. Liu, and S.-T. Lee, Hybrid heterojunction solar cell based on organic-inorganic silicon nanowire array architecture, *J. Am. Chem. Soc.,* **133**, 19408–19415 (2011).
29. K. Peng, Y. Xu, Y. Wu, Y. Yan, S. T. Lee, and J. Zhu, Aligned single-crystalline Si nanowire arrays for photovoltaic applications, *Small,* **1**, 1062 (2005).
30. L. He, C. Jiang, H. Wang, D. Lai, and Rusli, Si nanowires organic semiconductor hybrid heterojunction solar cells toward 10% efficiency, *ACS Appl. Mater. Interfaces,* **4**(3), 1704–1708 (2012).
31. Z. Cheng and J. Lin, Layered organic–inorganic hybrid perovskites: Structure, optical properties, film preparation, patterning and templating engineering. *Cryst. Eng. Comm.,* **12**(10), 2646–2662 (2010).
32. J. H. Heo, S. H. Im, J. H. Noh, T. N. Mandal, C.-S. Lim, J. A. Chang, Y. H. Lee, H.-J. Kim, A. Sarkar, M. K. Nazeeruddin, M. Grätzel, and S. I. Seok, Efficient inorganic–organic hybrid heterojunction solar cells containing perovskite compound and polymeric hole conductors, *Nat. Photon.,* **7**(6), 486–491 (2013).
33. J. H. Noh, S. H. Im, J. H. Heo, T. N. Mandal, and S. I. Seok, Chemical management for colorful, efficient, and stable inorganic–organic hybrid nanostructured solar cells, *Nano Lett.,* **13**(4), 1764–1769 (2013).
34. W. Zhang, J. Smith, R. Hamilton, M. Heeney, and I. McCulloch, Systematic improvement in charge carrier mobility of air stable triarylamine copolymers, *J. Am. Chem. Soc.,* **131**(31), 10814–10815 (2009).
35. L. Shen, G. Zhu, W. Guo, C. Tao, X. Zhang, C. Liu, W. Chen, S. Ruan, and Z. Zhong, Performance improvement of TiO_2/P3HT solar cells using CuPc as a sensitizer, *Appl. Phys. Lett.,* **92**, 073307 (2008).
36. M. Morana, M. Wegscheider, A. Bonanni, N. Kopidakis, S. Shaheen, M. Scharber, Z. Zhu, D. Waller, R. Gaudiana, and C. Brabec, Bipolar charge transport in PCPDTBT-PCBM bulk heterojunctions for photovoltaic applications, *Adv. Funct. Mater.,* **18**, 1757–1766 (2008).
37. B. Kim, H. R. Yeom, M. H. Yun, J. Y. Kim, and C. Yang, A selenophene analogue of PCDTBT: Selective fine-tuning of LUMO to lower the bandgap for efficient polymer solar cells, *Macromolecules,* **45**, 8658–8664 (2012).

38. L. He, C. Jiang, R., D. Lai, and H. Wang, Highly efficient si-nanorods/organic hybrid core–sheath heterojunction solar cells, *Appl. Phys. Lett.*, **99**(2), 021104 (2011).

39. G. Lin, Z. Gao, T. Gao, Y. Chen, Q. Geng, Y. Li, L. Chen, M. Li, Research progress in improving the performance of PEDOT:PSS/Micro- and Nano-textured Si heterojunction for hybrid solar cells, *J. Materiomics* **7**(5), 1161–1179 (2021).

40. B. R. Saunders, Hybrid polymer/nanoparticle solar cells: Preparation, principles and challenges, *J.* Colloid. Interface *Sci.*, **369** (1), 1–15 (2012).

41. L. Qiu, D. Mei, W.-H. Chen, Y. Yuan, L. Song, L. Chen, B. Bai, P. Du, and J. Xiong, Organic-inorganic hybrid electron transport layer of PVP-doped SnO_2 for high-efficiency stable perovskite solar cells, *Sol. Energy Mater. Sol. Cells*, **248**, 112032 (2022).

Chapter 8

Some Advanced Ideas for Enhancing the Conversion Efficiency

8.1 Overview of Strategies for Improving Photovoltaic Cell Efficiency

A single-junction bulk solar cell has a maximum theoretical efficiency that is limited to 33%, the Shockley–Queisser (SQ) limit. The SQ limit is mainly limited by two factors: (1) below-bandgap photons that cannot be absorbed, so that the corresponding photon energies cannot be transformed into electric power; (2) above-bandgap photons that excite hot electron/hole pairs that decay to give away excess energies to lattice via emitting phonons and the excess energy is wasted. Some advanced ideas, including upconversion and downconversion, have been proposed to achieve high solar cells efficiency beyond the SQ limit.

As addressed in Chapter 2, the major lost mechanisms lead to, for instance, typical 19–24% power conversion efficiency for wafer-based Si solar cells (SCs). The loss mechanisms are associated with optical, electrical, material, and device features. There are many proposed approaches aiming at overcoming the limitations to achieve high conversion efficiency. This chapter discusses some major themes that have been demonstrated to be effective for the construction of solar cells with high efficiency. An important advantage for nanostructured solar cells is that

Introduction to Nano Solar Cells
Ning Dai

ISBN 978-981-4877-49-7 (Hardcover), 978-1-003-13198-4 (eBook)
www.jennystanford.com

they can be used to incorporate new physical mechanisms which allow solar power conversion efficiency (PCE) greater than that of a one-junction solar cell.

Nanomaterials and nanostructures have a large number of varieties. Nanotechnologies provide great possibilities to modify materials and nanostructures, offering various routes towards applications by constructing great number of devices with their properties and parameters tunable simply through modifying material and structure sizes. The most expected application of nanomaterials and nanostructures lies in the utilization of their practical advantages over conventional devices, together with their applications using new physical mechanisms for the enhancement of device performance. Through many avenues, nanomaterials and nanostructures can be used to construct nano solar cells with efficiencies exceeding those of conventional solar cells such as Si-wafer-based ones. For increasing conversion efficiency, a number of approaches have been suggested to overcome the SQ limit, which can be clarified into several groups and some of them are effective for future applications. In terms of their different physical mechanisms, the reported effective approaches include enhanced light trapping via surface nano textures and surface plasmon oscillation using nano metal particles, bandgap modification of existing materials, introduction of intermediate bands, hot carrier utilization, full solar spectrum absorption, multi-exciton generation, utilization of low-energy photons, etc. These are the approaches to exceed the Shockley–Queisser limit. One expects that great benefits could be gain from solar cells with the configuration of nanomaterials and nanostructures to implement the approaches, the benefits depending on the specific device configuration of choices of nanomaterials and nanostructures.

In multi-junction solar cells, the devices are configured to match the solar radiation spectrum. The matching can also be done in an alternative way: instead of configuring the devices, the solar spectrum is converted to match the solar cells whose conversion efficiencies are wavelength-dependent. The spectrum conversion has two major advantages: (1) It is applicable to existing solar cells, and (2) SCs and the converter can be optimized separately due to their mechanical independence. Spectral splitting techniques can be divided into three categories.

1. Spectral upconversion

   Spectral upconversion aims at the utilization of below-bandgap photons that cannot be absorbed by solar cells with a bandgap larger than the energies of the photons. In an upconversion process, two or more below-bandgap photons are combined to generate an above-bandgap photon.

2. Spectral downconversion (or quantum cutting)

   The downconversion is featured to make the best use of a photon with energy at least two times the bandgap energy of the solar cell. In the downconversion process, a high-energy photon is split into at least two above-bandgap photons, i.e., multiple excitons are produced by the absorption of one photon.

3. Spectral downshifting (or luminescence)

   It transforms a photon into another with lower energy with which the solar cell has a higher quantum efficiency.

According to theoretical modeling of those approaches in which realistic physical limits have been taken into consideration, a three-junction tandem cell could have efficiency reaching 66%, a thermophotonic one to 50%, a hot-carrier one to 67%, an intermediate energy band one up to 62.3%, and a multiple exciton cell to 75%. Of these approaches, the one receiving the most attention and focus for experimental implementation is the multi-junction tandem solar cell.

8.2 Multi-Junction Tandem SCs

A tandem solar cell has multiple junctions that provide an effective way to harvest the broad solar radiation spectrum using two or more subcells with different absorption bands. Thus, in the multi-junction tandem cell, multiple materials are needed to match the broad solar radiation spectrum from infrared to ultraviolet.

8.2.1 Principle of Multi-Junction Solar Cells

For a given semiconductor used as the active layer of a solar cell, a photon with energy larger than its bandgap ($h\nu > E_g$) will be absorbed and an electron–hole pair is excited in the conduction and the valence bands that might contribute to electrical energy. However, the excess part of energy, $h\nu - E_g$, is wasted as heat through the electron–lattice interaction; the larger $h\nu - E_g$ is, the more the photon energy is wasted. A photon with its energy $h\nu < E_g$, on the other hand, cannot be absorbed and cannot make contribution to electrical energy at all. Henry calculated the maximum work per absorbed photon, which is less than 100% [1]. A Si wafer solar cell is unable to use the photon with energy below 1.12 eV (the bandgap of crystalline Si) and the fraction of the energy above 1.12 eV.

A single-junction solar cell would have been highly efficient for power conversion had the sun radiated monochromatic light. Unfortunately, the solar spectrum covers a broad photon energy range of about 0 to 4 eV. Intuitively, one might use semiconductors of different bandgaps to match the broad solar spectrum. In Fig. 8.1, the AM1.5 solar spectrum of 0.25 to 2.75 μm is divided into three spectral regions of 0.25 ~ 0.75, 0.75 ~ 1.10, and 1.10 ~ 1.85 μm and, for each region, the photon energy is converted by solar cells whose active semiconductor layers have the bandgaps E_g = 0.67 eV (Ge), 1.18 eV (InGaAs), and 1.70 eV (GaInP), respectively. Thus, a tandem solar cell in which different subcells are stacked on top of each other has different energy thresholds for the photon absorption in a wide spectral range. The achievement of high efficiencies of tandem solar cells relies on the utilization of subcells of different bandgap energies for relevantly different parts of the solar spectrum. In other words, the lower subcell use the photons penetrating through the upper subcell. By splitting the solar spectrum into narrow spectral range, each *p–n* junction in subcells converts a relevant narrow spectral region efficiently.

Concerning the way of connection between different subcells, the tandem devices could be assembled in two-terminal or four-terminal configurations. In the two-terminal case, the subcells in a tandem SC are connected in the back-to-back series way.

In practice, the neighboring subcells are connected with tunnel diodes to allow efficient tunneling of photon-excited current. Due to the series configuration of the subcells, the current of the tandem solar cell is limited by the subcell with the lowest current, which is one of the disadvantages of a tandem SC. Current matching is rather difficult in a practical situation. In the case of four-terminal (two-subcell tandem SC) or six-terminal (three-subcell tandem SC) configurations requires at least three or five transparent electrodes, respectively. Those electrodes are likely to introduce unwanted light absorption and scattering.

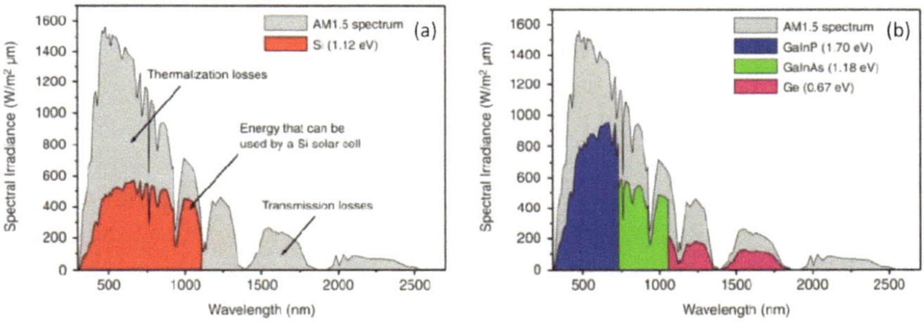

Figure 8.1 (a) The solar irradiation spectrum at AM1.5 and the part of the spectrum usable by Si solar cells; (b) the part of the spectrum usable by $Ga_{0.35}In_{0.65}P/Ga_{0.83}In_{0.17}As/Ge$ tandem solar cells.

Studies show that graded-bandgap materials present a broadened spectral absorption and, in turn, a high conversion efficiency. A graded-bandgap solar cell has a wide bandgap at the top surface used as the window for light incidence.

8.2.2 Thin-Film Multi-Junction Tandem Solar Cells

Most multi-junction SCs are based on III–V compound semiconductors. They are excellent materials with excellent photoelectric performance for the fabrication of multi-junction solar cells. III–V compounds are a large family which offers large room of material choices. Multi-junction SCs have the highest conversion efficiency among SCs. Spectrolab in the United States demonstrated an efficiency of 40.8% based on a triple-junction architecture with a concentrated sunlight of 326 suns in 2008 [2].

Later, the Institute of Solar Energy System in Germany set a new record of conversion efficiency at 41.1% measured at 454 suns and, in 2014, Dimroth and co-workers raised the conversion efficiency to 44.7% using a four-junction GaInP/GaAs/GaInAsP/GaInAs concentrator solar cell [3]. Later at the end of 2014, NREL (National Renewable Energy Laboratory, US) announced the new conversion efficiency result of 45.7% based on a four-junction device at 234 sunlight concentration [4]. With the advantages of high light absorption ability due to direct bandgaps, tunable bandgaps via ternary composition, relatively large bandgaps beneficial for low excess reversed saturation current, as well as robust materials suitable for space uses, most record-breaking tandem SCs are based on III–V compounds.

The tandem SCs based on both III–V and II–VI semiconductors have been reported. In a tandem configuration, each subcell is grown on top of the other directly by MBE or MOVCD. MBE or MOVCD are very expensive thin-film growth technologies. During the growth, in situ monitoring technologies are used to monitor the material and structural parameters of the subcells, including crystal qualities, interface smoothness, layer thickness, etc. Solar energy harvesting requires solar cells of larger area. A tandem solar cell is therefore very complicated in its structure and very expensive. For the reduction of cost, concentrators are usually used to concentrate the light, so as to reduce the required size of the solar cells for light collection. Hence, light-concentrating systems with sunlight-tracking facility are usually used to cut down cost and raise efficiency. Figure 8.2 presents a typical optical and mechanical system for the tracking and concentrating sunlight.

The lens-like concentrators are Fresnel lenses that focus the light onto the solar cells as shown in Fig. 8.3a. The use of Fresnel lens is for the reduction of lens thickness since the concentrators are large in sizes. There are several different forms of Fresnel lenses that focus sunlight into point, line, and a small area. In the case of point focus, the facets of the Fresnel lens are circularly symmetric with respect to its axis, while in linear focusing case it has a constant cross section along a given axis. In the point and the linear focusing cases, one small cell and a linear array of cells are set behind corresponding lens at their focal point and focal line, respectively. The mirror-like concentrators are concave mirrors that make dense of light on solar cells (see Fig. 8.3b). Using a lens or mirror, the intensity concentration can be up to 500 suns. On the other hand, the sun rises in the east and sets in the west, the sun radiation changing the direction continuously. The concentrators have to adjust the position in terms of the sun's movement so that the sunlight is kept to focus on solar cell surfaces. Sensors and mechanical system are thus needed to track the sun position.

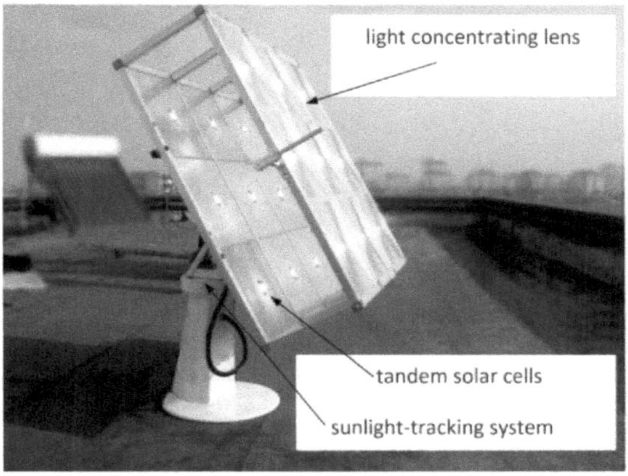

Figure 8.2 An optical and mechanical sunlight tracking system for multi-juction solar cells.

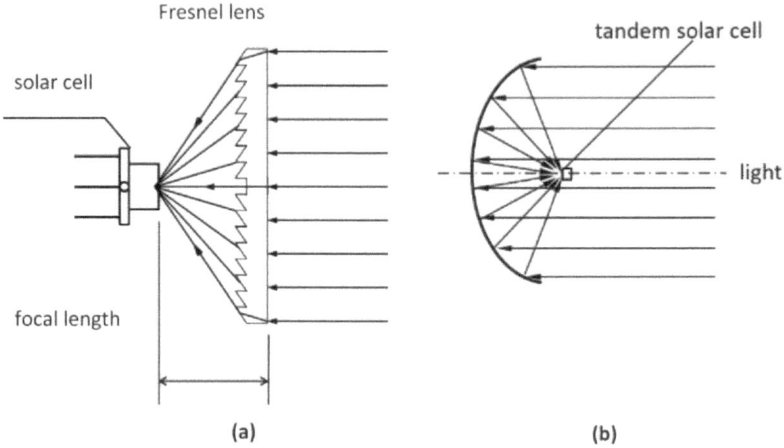

Figure 8.3 (a) A Fresnel lens (left) and a concave mirror for focusing the light onto solar cells; (b) a concave mirror as a concentrator to make dense of light.

For low cost and easy maintenance, the Fresnel lenses or concave mirrors, as well as the mechanical sunlight-tracking system need to have excellent weather stability and long-term durability. Acrylic plastic (polymethyl methacrylate, or PMM) is good candidate for Fresnel lenses due to its low cost, although the lifetime of the material is currently still a concern.

Multi-junction solar cells are composed of several single-junction solar cells. Theoretically, one might prepare single-junction solar cells separately and then mechanically stack them together. However, the electric quality of the multi-junction cells is rather poor, since it is extremely difficult to obtain atomic smooth interfaces between the single-junction cells. The widely adopted technical scheme is to grow all the single-junction cells on a substrate epitaxially and consecutively. One major challenge for the approach is the requirement that the semiconductors with suitable bandgaps need to be nearly lattice-matched for high-quality material growth.

Modern epitaxial crystal growth techniques, such as MBE and MOCVD, make it possible to grown complicated multi-layer material structures with high crystal quality. Semiconductor layers of different bandgaps can be grown in a stacked configuration, as shown schematically in Fig. 8.4. The tandem solar cell combines subcells made with semiconductors of different bandgaps in one stack. The bandgap of each subcell matches the relevant part of the solar spectrum, leading to very high conversion efficiency. In the device configuration, the subcell with highest bandgap is on top of the tandem solar cell so that the high-energy photons are absorbed and the photons with energy below its bandgap penetrate through the top subcell. The subcell with the second highest bandgap is underneath the top subcell, which absorbs the photons of relevant energies. In practice, a tandem solar cell have three or four subcells due to requirements on close matching mainly associated with bandgaps and lattice constants of the subcell materials. In a three-subcell tandem SC made by semiconductors with the bandgaps of E_g^1, E_g^2, and E_g^3 (see Fig. 8.4), for instance, the device behaves optically like low-pass photon energy filters. Assuming $E_g^1 > E_g^2 > E_g^3$, high-energy photons with energy $h\nu \geq E_g^1$ will be absorbed by the top subcell with the bandgap

E_g^1, when sunlight is incident from the top. The rest photons with energies below the top bandgap penetrate through the top subcell and consecutively absorbed by lower subcells. Photons with energies $E_g^2 \le hv < E_g^1$ are absorbed by the subcell with bandgap E_g^2, and photons with $E_g^3 \le hv < E_g^2$ are absorbed by the subcell with bandgap E_g^3.

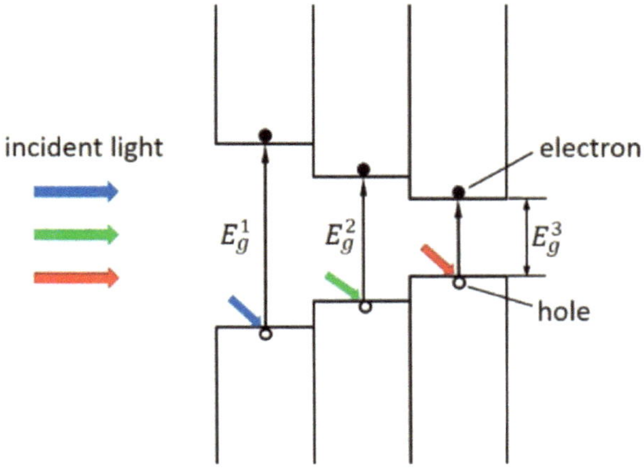

Figure 8.4 The ideas for the conversion of full solar spectrum energies: semiconductors with different energy gap are sequentially stacked together so that different energy ranges of photons are nearly resonantly absorbed by corresponding semiconductors. The blue, green, and red arrows represent the colors of light, showing that the blue photons are absorbed by the wide bandgap E_g^1 semiconductor, the green photons by the medium bandgap E_g^2 semiconductor, and the red photons by the narrow bandgap E_g^3 one, sequentially.

Figure 8.5 presents device configurations of two 3-junction tandem solar cells based on III–V semiconductor compounds, the right-side being a metamorphic $Ga_{0.44}In_{0.56}P/Ga_{0.92}In_{0.08}As/Ge$ three-junction solar cell and the left-side a lattice-matched three-junction solar cell of the same material system. For the metamorphic cell, a record of 40.7% conversion efficiency has been realized at concentration of 240 suns under the concentrator terrestrial spectrum of the standard AM1.5 condition. For the lattice-matched cell, a 40.1% conversion efficiency has been achieved at 135 suns. This metamorphic three-junction tandem cell, the first solar cell reaching over 40% conversion efficiency, was a milestone for solar cell research.

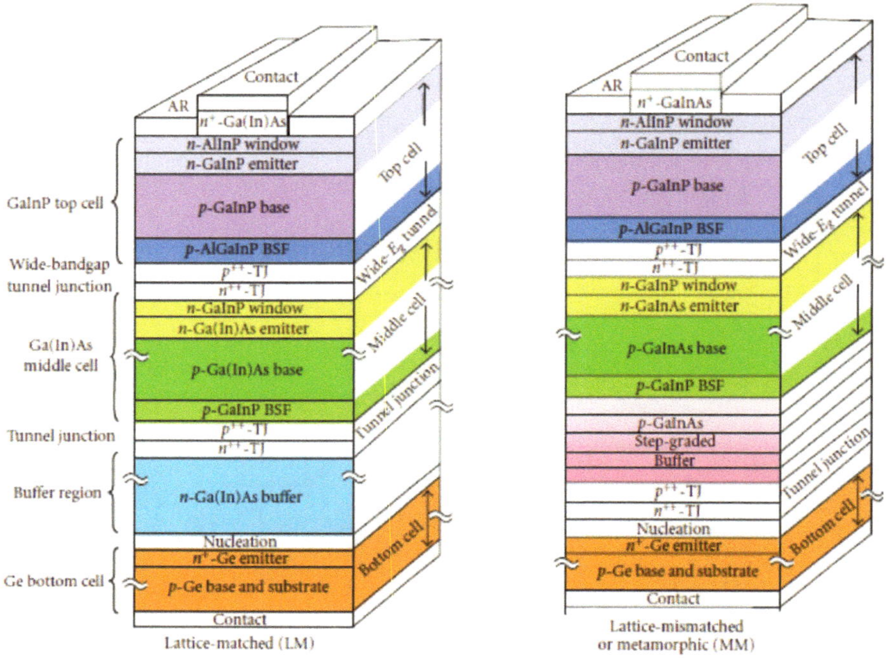

Figure 8.5 Schematic view of the GaInP/GaInAs/Ge triple junction tandem solar cells with lattice-matched (left) and metamorphic (right) configurations.

Effectiveness of the stacked configuration for tandem solar cells has been demonstrated by their very high conversion efficiencies higher than 40%. Recently, conversion efficiency of 47.6%, more than double of that of the crystalline Si solar cell, has been achieved on a tandem solar cell consisting of four subcells under about 665 suns concentration [5]. There is still a big room for improving the efficiency considering that a tandem cell with a infinite number of subcells has a theoretical maximum efficiency of ~87% [6]. Another configuration of a lattice-matched GaNPAs/Si tandem solar cell is shown in Fig. 8.6 [7].

So far, tandem SCs have achieved ~48% and PCE can exceed 80% theoretically; the large number of junctions is troublesome due to the increasing complexity associated with structure and materials. Other challenges include controlling interface quality and interlaminar strain accumulation.

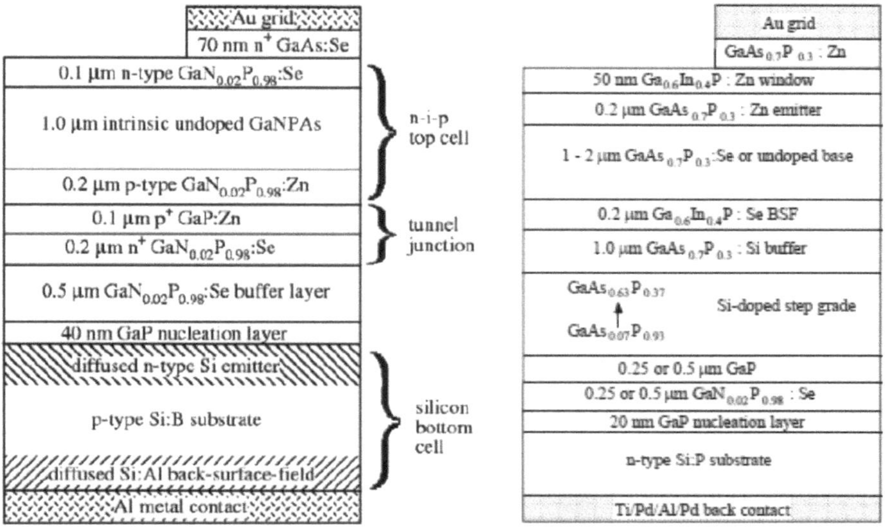

Figure 8.6 Schematic illustration of the GaNPAs/Si tandem solar cell. Reproduced with permission from [7].

8.2.3 Nano Multi-Junction Tandem Solar Cells

The fabrication of the monolithic multi-junction solar cells is a big challenge, due to the requirement for close lattice matching imposed by the growth methods such as MBE and MOCVD, while the semiconductor materials used in the devices are usually not lattice-matched. Additional constraint comes from the limited choices of substrates that are compatible in their lattice constants with those of the material system composed of the SCs. On the other hand, the growth of a buffer layer or wafer peeling-off is not always viable techniques for practical fabrication.

Nanomaterials offer alternative approaches for the fabrication of tandem solar cells by largely relaxing those constraints. For axial junction nanowires, for instance, the junction interface is a few hundred square nanometers or smaller. It then allows for large lattice mismatch and mismatch strain is more or less relaxed via nanowire sidewalls. Another benefit is that the growth of high-quality nanowires does not really need high-quality substrates. Yao and co-workers fabricated nano tandem SCs with a device structure as shown schematically in Fig. 8.7a [8].

The nano SC consists of a top cell and a bottom cell connected in series. The top cell is a GaAs nano subcell consisting of nanowires (NWs) with the structure of p^+-GaAs-emitter/i-GaAs-layer/n-GaAs-base/n^+-GaAs-back. The bottom cell is a wafer-based Si subcell with the structure of p^+-Si-emitter/n-Si-base/n^+-Si-BSF (BSF: back surface field). In fabricating the top cell, the wafer-based Si solar cell was used as a substrate and the growth of the NW solar cell was initiated from the n^+-GaAs-back.

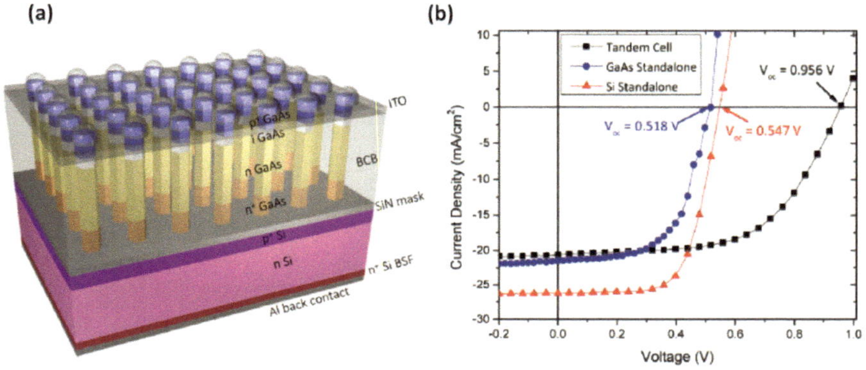

Figure 8.7 (a) Schematic structure of a nano tandem SC consisting of GaAs NWs on Si. The GaAs nanowire has a structure of p^+-GaAs-emitter/i-GaAs-layer/n-GaAs-base/n^+-GaAs- bottom on a wafer Si subcell with the three-layer structure of p^+-Si-emitter/n-Si-base/n^+-Si-BSF (back surface field). The top contact is made by ITO and the bottom contact by Al. (b) Measured J–V curves of the GaAs-NW/Si-wafer tandem cell (black), the standalone GaAs-NW subcell (blue), and the standalone Si-wafer subcell (red). Reproduced with permission from [8].

It is of utmost importance for the top and the bottom subcells to have low junction resistance, otherwise photocurrent is largely stopped by the high-resistance junction. In addition, the current of the two subcells should be in a close matching. Figure 8.7b presents the J–V curves measured on the GaAs-NW/Si-wafer solar cell, the standalone GaAs-NW subcell, and the standalone Si-wafer subcell. The open-circuit voltage V_{oc} of the tandem cell is almost equal to the addition of those of the two subcells (see the arrows in the figure). The short-circuit current I_{sc} of the tandem cell, is, however, the smallest among the three cells, close to that of the GaAs standalone cell. This is a clear indication that it is critical to match the current of the subcells. In Yao and co-workers' work, the most efficient tandem cell has the device performance of V_{oc} = 0.956 V, J_{sc} = 20.64 mA/cm^2, FF = 0.578, and PCE = 11.4%. The cell is not great in terms of its performance. However, the idea of constructing tandem solar cell using nanomaterials opens a way toward highly efficient solar cells based on low-cost and low quantity of materials. A theoretical simulation showed that an optimal tandem cell based on a 1.1/1.7 eV double junction could have cell efficiency above 40% [9].

Both nanomaterials and organic materials have great flexibility for the construction of tandem solar cells. Kim and co-workers studied a hybrid device architecture with two subcells based on quantum dots and polymers [10]. In fabricating two separate cells, the PbS colloidal quantum dot (CQD) cell is deposited on the fluorine-tin-oxide (FTO) coated glass substrate. The TiO_2 nanoparticle layer is formed by spin-coating the solution of diluted TiO_2 in ethanol. The PbS QD layer was deposited from the solution of PbS CQDs in octane (50 mg per ml) and the whole QD layer is grown four times by the layer-by-layer process in order to achieve an targeted layer thickness. The QDs form a depleted heterojunction with the polymers. The fullerene cell was fabricated on indium-tin-oxide (ITO)-coated glass substrate in a vacuum chamber. The ZnO layer, for electron transport, was evaporated on the ITO-coated substrate using ZnO powder as the source material. A conjugated polyelectrolyte, poly[(9,9-bis(3′-(*N*,*N*-dimethylamino)propyl)-2,7-fluorene)-alt-2,7-(9,9-dioctylfluorene)] (PFN solution), was then spin-coated on the ZnO layer. Next, the $PC_{71}BM$, PBDTTT-C-T, and PTB7 layers are all deposited by spin-coating, followed by the deposition of MoO_x/Ag electrode. Two kinds of fullerene cells are prepared for the study: one using PC71BM/PBDTTT-C-T and the other adopting a PC71BM/PTB7 bilayer.

The studied hybrid tandem solar cell consists of above two single-junction cells with the band alignment shown in Fig. 8.8 where the two subcells, the PbS colloidal QD and the fullerene ones, are in series. In the tandem device architecture of FTO-glass/TiO_2/PbS-QD/MoO_x/ZnO/PFN/polymer:fullerene/MoO_x/Ag, the PbS-QD and polymer:fullerene active layers were fabricated in the same ways as the respective layers in above-mentioned single-junction cells. The MoO_x/ZnO bilayer in the middle of the device is the recombination layer that must offer good electrical connection for the two subcells. Good connection for the subcells in tandem SCs is a big challenge for achieving high conversion efficiency. While the PbS QD, PC71BM/PBDTTT-C-T, and PC71BM/PTB7 single-junction cells present maximum conversion efficiency η (open-circuit voltages V_{oc}) of 3.87% (0.573 V), 7.43% (0.758 V), and 7.87% (0.745 V), respectively, the hybrid tandem cells only achieve η (V_{oc}) of 5.33% (1.252 V), i.e., V_{oc} of the hybrid tandem cell is close to the summation of those of the two subcells, but conversion efficiency η is lower. Some mechanisms are responsible for the low efficiency. One is the fact that the spectral absorption of the PbS QDs and the polymer:fullerene system is not complementary, leading to low short-circuit current of the hybrid tandem cell. A significant part of visible light is absorbed by the PbS QD layer, since light first incidents on the QD cell. Another is the unmatched current of the two subcells.

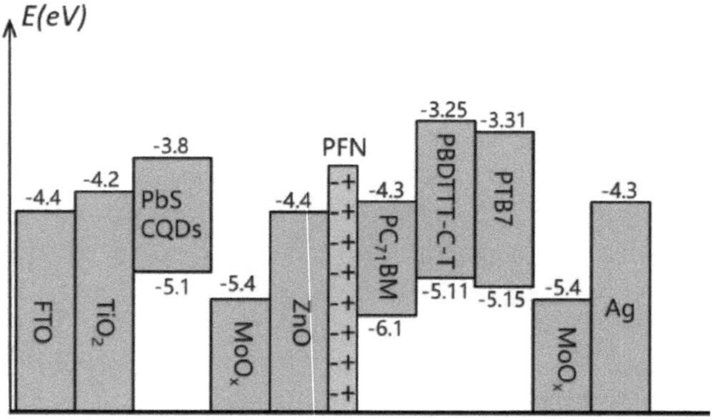

Figure 8.8 Schematic band alignment of the QD/polymer tandem solar cell [10].

Alternatively, a tandem SCs composed of semiconductor QDs was proposed and studied by Crisp and co-workers [11]. The monolithic tandem solar cell uses CdTe and PbS QD thin films as the active layers, with both the CdTe and the PbS layers being cost-effectively solution-prepared, as shown schematically in Fig. 8.9a. By tuning the sizes of CdTe and PbS QDs, spectral absorption of the two active layers can be made complementary to absorb a wide range of solar spectrum. A theoretical simulation gives a conversion efficiency of 45% for a bandgap of 1.5 eV for CdTe and that of 0.75 for PbS, assuming ideal structural design and device technology. The device has the structure of FTO-glass-cathode/CdS-QDs/CdTe-QDs/ZnTe-ZnO/PbS-QDs/MoO_x/Al-anode (see Fig. 8.9a). The thickness of the CdTe-QD and the PbS-QD layers is 250 and 500 nm, respectively. The ZnTe-ZnO is the recombination bilayer, serving as a tunnel junction. For comparison, two separate CdTe-QD and PbS-QD standalone cells were fabricated following the identical parameters and processes as their respective parts in the tandem cell.

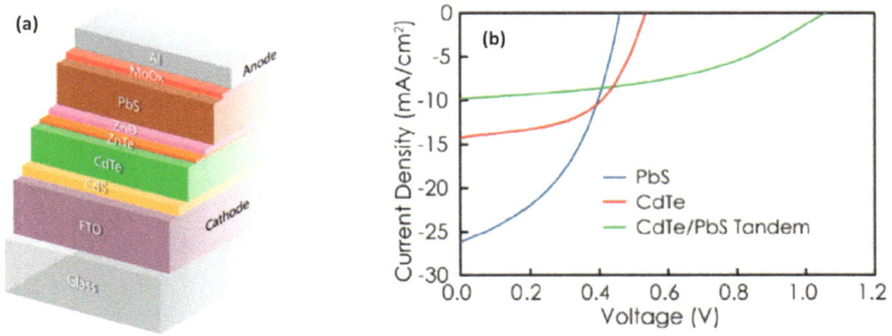

Figure 8.9 Schematics of two QD tandem solar cells. (a) an architecture of and (b) *J–V* curves of the CdTe-QD tandem cell (green) and the standalone PbS-QD (blue) and CdTe (red) cells. Reproduced with permission from [11].

It was found that the ZnTe–ZnO bilayer plays a critical role for effective series connection of the two subcells in the tandem cell. The open-circuit voltage of the tandem cell is even slightly higher than the sum of the open-circuit voltages of the two CdTe-QD and PbS-QD standalone cells due to the use of the ZnTe-ZnO tunnel junction, as shown by the measured *J–V* curves of the CdTe-QD(red), CdTe-QD (blue), and the tandem cells presented in Fig. 8.9b. Contrary to the obvious benefit of open-circuit voltage V_{oc} obtained for the tandem cell, the large loss in short-circuit current density J_{sc} for the tandem is definitely a problem that needs to be solved. While the complementarity of the spectral absorption of the two subcells is satisfactory, current mismatch in the two subcells connected in series and a large number of carrier-trapping interface and defects are the major reasons for the low J_{SC}.

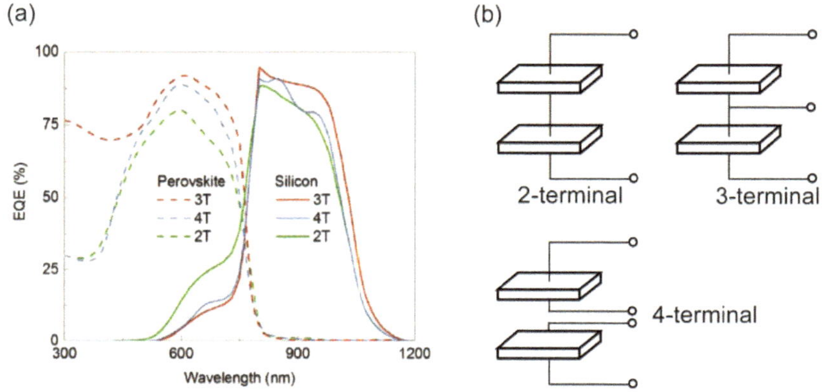

Figure 8.10 (a) Calculated EQE of perovskite (dashed lines) and Si (solid lines) solar cells for two-terminal, four-terminal, and three-terminal (IBC) configurations. The perovskite layer thickness in the cases of two-terminal, four-terminal, and three-terminal (IBC) cells is 300, 950, and 550 nm, respectively. (b) Methods of two-terminal, three-terminal, and four-terminal connections for two-junction solar cells. Reproduced with permission from [12].

Perovskites have been the star photoelectric conversion materials in the past decade and perovskite solar cells have achieved conversion efficiencies over 20%. Figure 10a presents the external quantum efficiency (EQE) of Si and perovskite solar cells, showing that their spectral absorption curves are almost perfectly complementary over the whole solar irradiative spectrum [12]. One could thus expect to achieve high PCE on the perovskite–silicon tandem solar cells. There are three ways of power output connections for the two-junction solar cells, which have two, three, or four terminals as shown in Fig. 8.10b. Based on Si and perovskite as two absorber layers, Garnett's group designed a novel tandem SC in which nanoscale contacts were sandwiched between planar perovskite and Si layers, as shown in Fig. 8.11. The Si subcell has an inter-digitated back contact (IBC) configuration and the tandem SC has three terminals. The novelty of this back-to-back connected tandem SC lies in the fact that its efficiency does not require current matching, so that the requirement on electronic quality of the whole tandem cell is reduced.

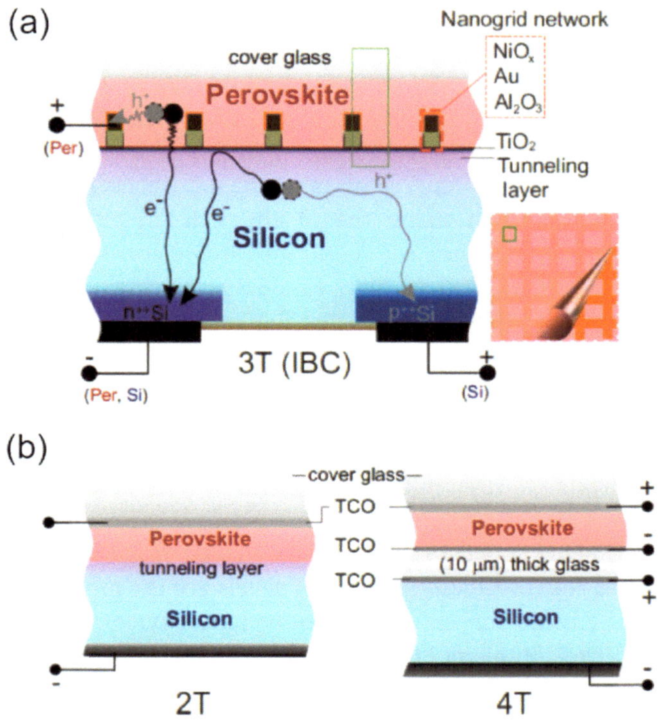

Figure 8.11 Schematic architectures of three two-subcell perovskite/silicon tandem SCs. (a) Back-to-back connected perovskite/Si-IBC tandem SC with 3 terminals (3T). The nanoscale contacts consist of Au metal contact (dark brown) covered by Al_2O_3 (light brown) and NiO_x (orange) insulating layers. The inset is the top view of the device with the unit cell for 3D simulation highlighted by the light-green box (half-pitch). (b) Architectures of perovskite tandem cells with two terminals (2T) and four terminals (4T). In the simulation, thickness of the perovskite layer varies from 0.05 to 1 μm and that of the silicon is fixed at 180 μm. Reproduced with permission from [12].

Light incidence is from the top of the device. High-energy photons are absorbed by the perovskite active layer and holes are collected by the positive electrode on the right side while electrons are picked up by the n^{++} Si back electrode on the bottom left side of the Si IBC subcell. Photons with energy below the bandgap of the perovskite material penetrate into the Si subcell where excitons are optically excited. Holes are collected by the p^{++} positive electrode of the Si IBC subcell and the electrons by the same negative electrode shared by perovskite subcell.

Device simulation indicates that the device in Fig. 8.11a has an efficiency reaching 32.9%, assuming an 18% PCE of the Si IBC subcell and a 10 μm thickness of the perovskite layer. Comparatively, the two- and four-terminal tandem cells, with the structure shown in Fig. 8.11b, display PCEs of 24.8% and 30.2%, respectively, using the same amount of perovskite material. Furthermore, when using the Si IBC cell of 25% conversion efficiency, the tandem cell with the architecture of Fig. 8.11a could achieve a PCE as high as 35.2%. Further improvement of the device performance is possible with better light trapping and device passivation.

An alternative device architecture of perovskite/Si tandem cells was designed and studied by Hossain and co-workers [13]. Based on nanophotonic design, the perovskite/silicon tandem cell shows a short-circuit current density J_{sc} of and a conversion efficiency of 30%. The high J_{sc} of 20 mA·cm^{-2} is achieved due to the minimized optical losses using the nanophotonic structure.

Solar cells with multiple energy levels are the most intensively studied scheme for high-PCE photovoltaic devices. The multiple energy levels implement multiple absorption paths, which allows wide solar light spectrum absorption. A triple-junction solar cell, for instance, uses three semiconductors, each with its bandgap matching to corresponding one third of the whole solar spectrum. If the part is implemented so that it is an optimized device for its given part of the solar spectrum, the three-junction solar cell will give the best performance if they are in series with perfect connection. In the case when multiple energy levels are constructed in each semiconductor (intermediate energy band, for instance), the single semiconductor then allows multiple absorptions. Multiple path absorption can be readily configured using doping, quantum dots and quantum wells.

8.2.4 Spectrum Splitting

The solar spectrum that covers the photon energy range of 0–4 eV can be split by a number of ways. The devices that realize the splitting include filters, prism, mirrors, diffraction gratings, etc. Each split part can be introduced into a solar cell of relevant bandgap. Since different split parts of the solar spectrum do not have to follow the same path, the solar cells for different split parts do not require a stack configuration. The main advantage of spectrum splitting is to avoid the stack configuration which often suffers from extremely strict technical requirements for the sub-SC interconnection and the very narrow allowed ranges of material parameters for the sub SCs, such as the lattice constants and bandgaps. The spectrum splitting offers great freedom in choosing solar cell materials.

Spectrum splitter splits the broad solar spectrum into portions and separates them spatially. Theoretically, one could divide the solar spectrum into more fractions and use more solar cells to achieve high power conversion efficiency. Henry calculated the limiting terrestrial PCEs with 1, 2, 3, and 36 single solar cells, which gave the conversion efficiencies of 37, 50, 56, and 72%, respectively [1]. Apparently, the improvement on conversion efficiency is significant by going from single bandgap to two and three bandgaps. It becomes less and less significant when more and more semiconductors are added even though right bandgaps are chosen for them. For the construction of a practical solar cell for photoelectric conversion in broad solar spectrum, incident photons of different energies need to be directed onto junctions of corresponding semiconductor layers, which requires a proper device architecture. As shown in Figs. 8.12a,b, the simplest way is the grating (prism) approach in which a grating (prism) spatially disperses the incident light so that photons of different energies are collected by solar cells of proper energy gaps. However, the architectures suffer from the complexities both mechanically and optically, which makes it impractical. In addition, the complex design has the problems of stability and cost. A useful tool is the device that has the functions of both a spectrum splitter and an optical concentrator. Goetzberger et al. proposed a practical system of a spectrum splitter and a fluorescent concentrator, enabling the use of diffuse light [14].

Splitting the wide solar irradiation spectrum into parts and using individual solar cells of different absorption bandgaps to perform photovoltaic conversion offers an effective way to bypass the obstacle that electrical connection between subcells in a tandem solar cell is usually hard to achieve. In this case, each solar cell (subcell) is designed to execute photoelectric conversion for solar spectrum in a certain wavelength range that is much narrower than the whole solar spectrum. A spectrum splitter splits the incident spectrum into two beams, reflecting lower-energy photons and allowing higher-energy photons to pass through, or vice versa. A three-splitter configuration is presented in Fig. 8.13a [15]. Assuming the broad spectrum is divided into the low-, medium-, and high-energy ranges, the incident photons in the low-energy range are totally reflected by the first splitter at 45° with respect to the surface normal of the splitter and collected by the low-bandgap solar cell. The photons with energies in the high- and medium-energy ranges penetrate through the first splitter and the medium-energy photons is reflected by the second splitter and taken by the underneath solar cell with medium bandgap. Photons in the high-energy range go through the second splitter and directed by the third splitter into the solar cell with high bandgap. Instead of optimizing the solar cell parameters for the whole solar spectrum, the three solar cells in the figure just have to be designed to match their relevant parts of the solar spectrum.

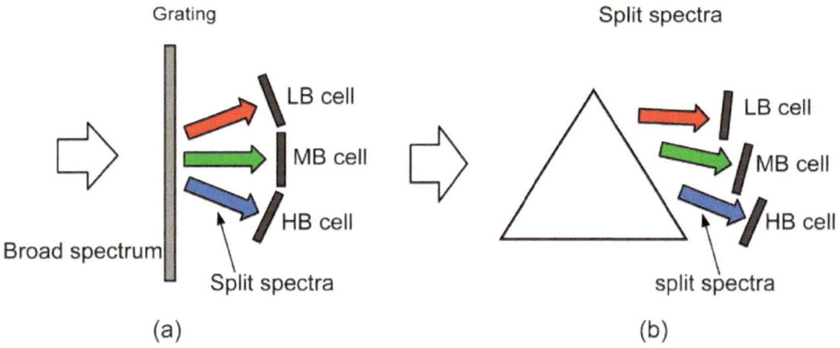

Figure 8.12 Spectrum splitting using grating (a) or a prism (b). The three split spectra are taken by the low-bandgap (LB), medium-bandgap (MB) and high-bandgap (HB) solar cells accordingly.

Instead of using spectral splitter, a nanophotonic splitter can also be designed to concentrating light through nanoscale resonance. Figure 8.13b presents schematic architecture of the nanophotonic splitter [15].

Using the straightforward spectrum splitting scheme, a-Si:H and CIGS solar cells are tested at five different wavelength splitters, where the long wavelength portion of the solar spectrum is directed to the CIGS cell and the short wavelength portion to the a-Si:H [16](a). Figure 8.14 presents the measured device efficiency of the combined a-Si:H/CIGS configuration, together with those of the individual a-Si:H and CIGS solar cells. Obviously, the efficiency of the a-Si:H cell is quite low at short splitting wavelength, since a large portion of solar irradiation is directed into the CIGS cell. On the other hand, the efficiency of the a-Si:H cell can then be raised by increasing its bandgap. In Fig. 8.14, the best PEC of 18.6% corresponds to the splitter at the splitting wavelength of 614 nm.

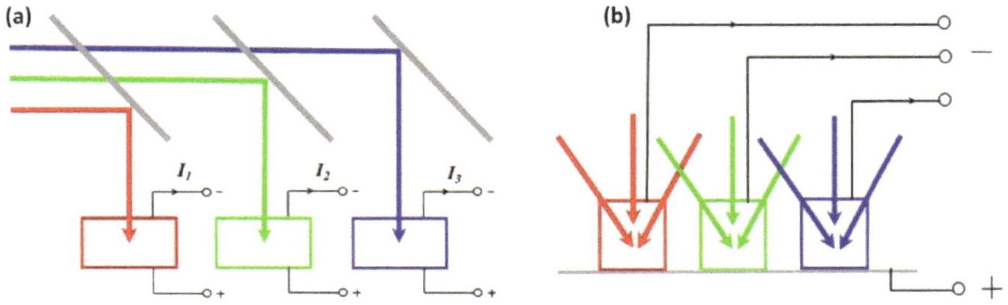

Figure 8.13 (a) A beam splitter approach in which the broad solar spectrum is split into three ranges that are directed into their respective solar cells and (b) a nanophotonic splitter approach where optical concentration is offered by nanoscale resonance.

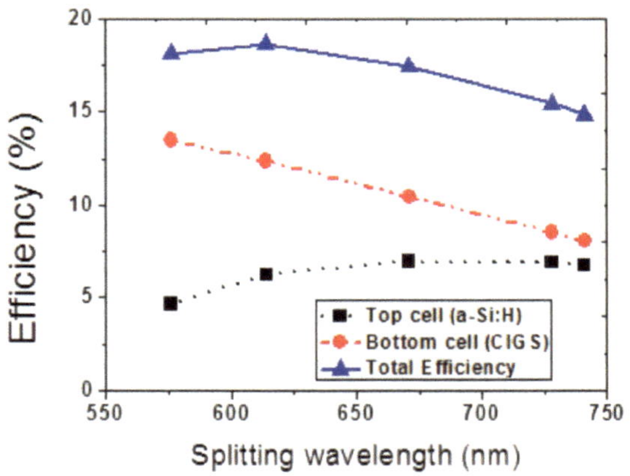

Figure 8.14 Efficiencies of the a-Si:H (black) and CIGS (red) cells, as well as the combined a-Si:H/CIGS configuration at five different splitting wavelengths. Reproduced with permission from [16].

A more dedicated spectrum splitting technique uses off-axis holographic lens, as shown in Fig. 8.15 [16] (b). Light is normally incident on the holographic lens from top. After passing the lens, solar irradiation is split into two portions: shorter-wavelength and longer-wavelength ones separated by a transition wavelength (green rays in the figure). The longer-wavelength ray is absorbed by the low-bandgap cell and the shorter-wavelength light by the high-bandgap cell.

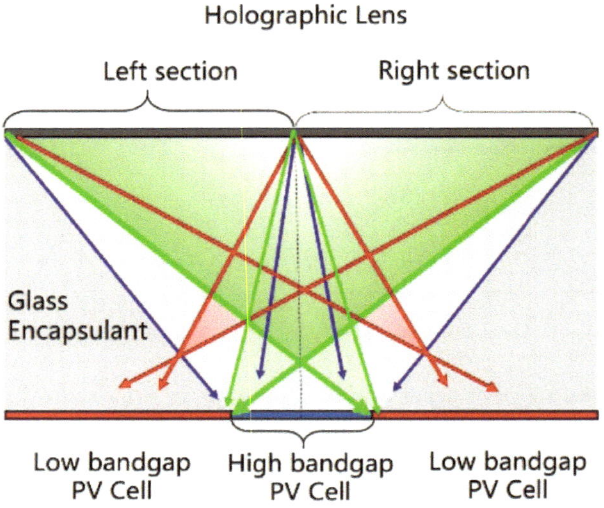

Figure 8.15 Schematic of a spectrum-splitting solar cell system, showing the ray trace of the transition wavelength (green), shorter wavelength (blue), and longer wavelength (red). Assuming normal incident light, the holographic lens directs the longer wavelength ray to low bandgap solar cell and the shorter wavelength ray to the high bandgap solar cell.

A NW solar cell architecture was proposed by Dorodnyy and co-workers, based on three NW array subcells working at different spectral ranges [17]. A theoretical simulation showed that the solar cell system could reach the conversion efficiency of 48.3%. The key part of the solar cell system are the three NW subarrays on a Si substrate, as shown in Fig. 8.16a with one unit cell highlighted where the tallest and the shortest NWs stand on one side and the medium-length NWs on the other side. Among three sets of NWs, the one with large bandgap is the tallest and that with low bandgap is the shortest, while the one with medium bandgap is in the middle. Each of the three NW subarrays, with a given NW material, performs the photoelectric conversion with respect to solar irradiation from one of the three spectral ranges of the whole solar spectrum, as depicted in Fig. 8.16b.

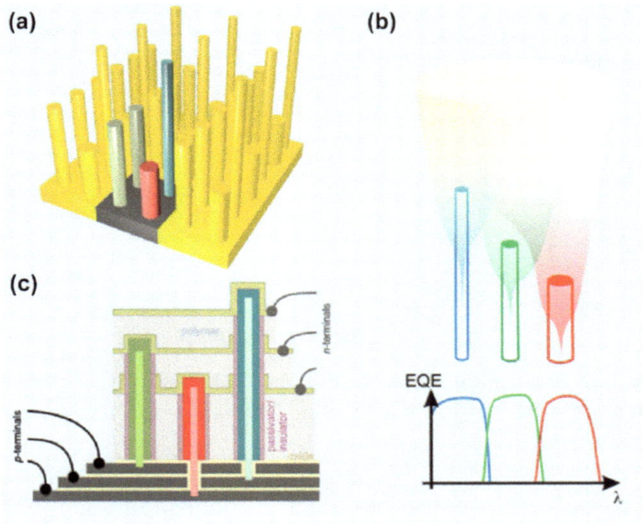

Figure 8.16 A conceptual NW solar cell designed with three NW subarrays: (a) schematic of the three NW subarrays on Si substrate, with each unit cell having NWs of three different bandgaps; (b) operating principle of spectrum splitting where each of the three different NWs absorbing photons in one of three spectral ranges of the whole solar spectrum correspondingly; (c) schematic of the six-terminal connection. Reproduced with permission from [17].

The materials for the three NW subarrays are $In_{0.37}Ga_{0.63}As$, GaAs, and $Al_{0.54}Ga_{0.46}As$, with the bandgaps of 0.93, 1.42, and 2.01 eV, respectively. Those material bandgaps are ideal combination to form the best match with the solar spectrum. The materials are all GaAs-based, so they can be easily grown on a single substrate. The spectrum splitting is based on the principle that when incident light reaches the height of the tallest NWs, high-energy photons will be absorbed by these NWs with large bandgap while the medium- and low-energy photons will not be absorbed. The medium-energy photons are absorbed by the medium-bandgap NWs and only the low-energy photons can reach the shortest NWs being absorbed there. Thus, the whole solar spectrum is actually divided into three spectral ranges in which photons are respectively absorbed by three NW subarrays in terms of photon energy.

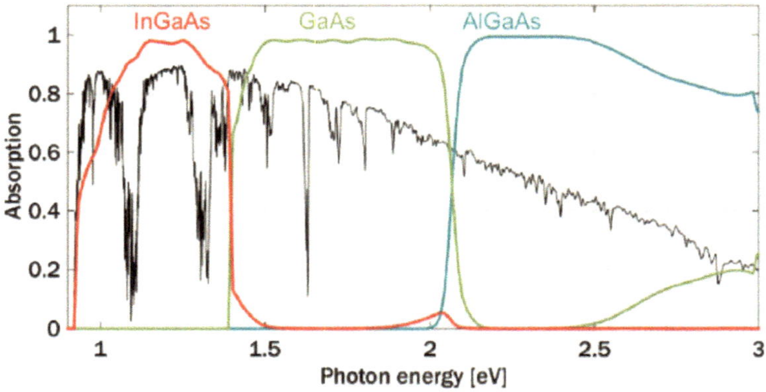

Figure 8.17 Absorption bands of InGaAs, GaAs, and AlGaAs which cover the dominant part of the whole solar irradiation spectrum. Reproduced with permission from [17].

Figure 8.16c presents the electric connection detail of the six-terminal device. It is clearly shown that the three NW subarrays are independently connected, which gives large device design room, such as material combinations and spectrum splitting. Theoretically, high solar energy conversion efficiency of 48.3% is achievable with such device architecture in optimized device parameters, mainly due to the utilization of almost whole solar spectrum, as shown by Fig. 8.17. In addition, the NW array enables excellent photon-trapping.

8.3 Hot-Carrier Capture

High-energy photons (with their energies larger than the bandgap) excites electron–hole pairs, i.e., electrons are excited to eigen states above the bottom of the conduction band and holes to states below the top of the valence band. In conventional solar cells, those hot electrons and hot holes relax to the band edge through phonon emission. Utilizing the energies of the hot carrier prior to emitting phonons could significantly improve power conversion efficiency of solar cells. The idea of utilizing extra energy of hot carriers is to collect electron–hole pairs before they get cooled to their respective band edges. There are essentially two processes to utilizing the hot carrier energies: The hot carriers remain "hot" before collected by contact electrodes and they give away their energies through impact ionization that excites second electron–hole pairs. Here collision ionization is the inverse of an Auger process, during which two electron–hole pairs annihilate to generate one electron–hole pair with higher energy. Thus, the rates of completing the process of photoexcited carrier separation, transport, and collection by the contacts must be faster than the cooling rate of the carriers. Simultaneously, the rate of impact ionization needs to be faster than the cooling rate of the hot carrier relaxation processes [18].

8.3.1 Hot-Carrier Generation in Semiconductors

One of the losses in solar power conversion efficiency is the hot electron generated in the conduction band by high-energy photons with energy larger than the junction bandgap, as shown in Fig. 8.18. The photons with below bandgap energies ($hc/\lambda < E_g$) cannot excite electrons and the photons with energy equal to the bandgap ($hc/\lambda = E_g$) are absorbed by the material and electrons are excited into the bottom of the conduction band. When photon energies are larger than the bandgap ($hc/\lambda > E_g$), electrons are excited to energy levels above the bottom of the conduction band. The excess energy, $hc/\lambda - E_g$, is usually lost through interaction with lattice, generating heat in cells. $hc/\lambda - E_g$ takes important amount of energy due to the single-bandgap characteristic in a solar cell. The type of optoelectronic device called hot carrier solar cell is designed to harvest the excess energy, $hc/\lambda - E_g$. The hot-electron relaxation is a fast process. Thus, the hot carrier solar cell should be designed to slow down the relaxation process, or extract charge carriers before relaxation.

It is, however, rather difficult for traditional wafer Si solar cells to harvest the excess energy, due to the long distance between the site of the electron being excited and the electrode. On the way being extracted by electrodes, the photogenerated hot electrons lose the hot energy to the lattice after going through many scattering interactions with the crystal material. In thin-film and nano solar cells, the distance between the sites of electron being excited and being collected is much shorter, making it possible to capture the extra energy $h\nu - E_g$. However, the effect in thin-film and nano solar cells is often overwhelmed by defects and interfaces in the devices.

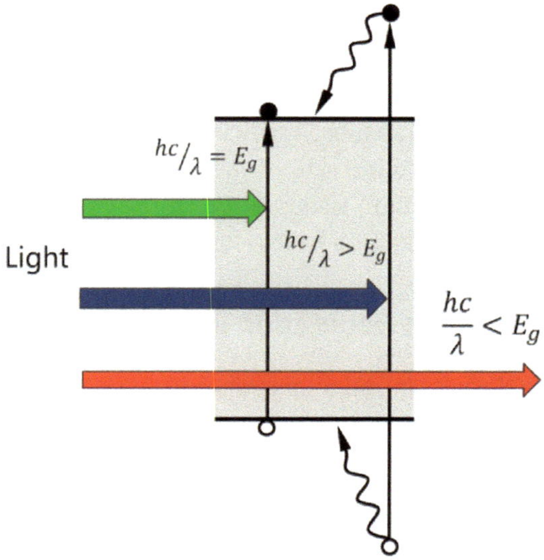

Figure 8.18 What happens when sunlight enters the surface of a solar cell: Photons penetrate through the cell at photon energy $hc/\lambda < E_g$ (red arrow), photons are absorbed and electrons are excited to the bottom of the conduction band at $hc/\lambda = E_g$ (green arrow), and photons are absorbed and the eletrons are excited to eigen energy levels above the bottom of the conduction band at $hc/\lambda > E_g$ (blue arrow).

8.3.2 Device Structures of Hot Carrier Solar Cells

A hot carrier solar cell consists of carrier-extracting layers, contact electrodes, and a key part—photon absorber in which hot electrons and hot holes are photon-excited with slow cooling-down rate, as shown for the fundamental device structure and band alignment in Fig. 8.19. The electron- and hole-extracting layers (EEL and HEL) present narrowly distributed energy bands that are electrically connected to the photon absorber on both sides. At the excitation of hot carriers by high-energy photons, hot electrons (hot holes) are excited into the conduction (valence) band. The hot electrons and holes are extracted by the EEL and

HEL layer, respectively, where the carriers are then cooled down to the EEL and HEL lattice temperature. Finally, the carriers are collected by the cathode and the anode, the contact electrodes of the SC. Apparently, the performance of the hot carrier SC relies largely on the efficiency of electron (hole) being extracted by the EEL (HEL) layer. Generally, the photon absorber is fundamentally required to possess the following requisite properties:

(1) A low bandgap to allow for photon absorption of a broad spectral range and appropriate energy positions of the top of the conduction band and the bottom of the valence band matching to the corresponding energy levels of the ΔE_e and ΔE_h bands of the carrier-extracting layers. Assuming 100% of photon absorption with $hc/\lambda \geq E_g$, the maximum PCE could reach 66% for the bandgap of active layer around 0.8 eV [19].

(2) Good material quality for high carrier mobilty for fast carrier transport to the electrodes.

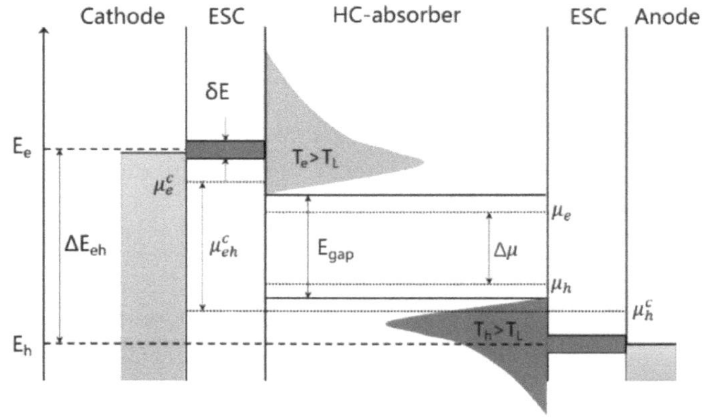

Figure 8.19 The potential profile of a designed hot carrier solar cell. The absorber layer absorbs incident photons and generates hot electrons and hot hole with their respective temperature T_e and T_h above the lattice temperature T_l. The carrier-extracting layers for both electrons and holes (EEL and HEL, respectively) have electric contact band with narrow energy distributions (ΔE_e and ΔE_h) to extract hot electrons and hot holes before the they are cooled down to the lattice temperature. The electrons and holes are then collected by cathode and anode, respectively.

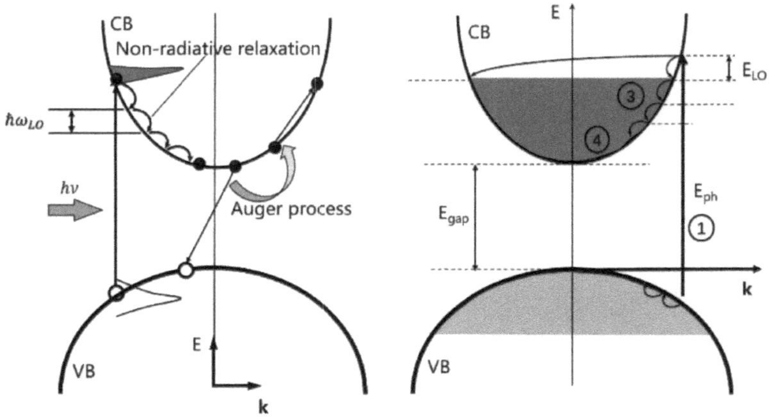

Figure 8.20 (a) Schematic of the non-radiative recombination of an electron/hole pair governed by an Auger process where the energy of the excited electron is given to another electron at low energy, accompanied by temperature increase in the conduction band. Also shown is the excitation of hot electron and hot hole that are relaxed to the bottom of the conduction band and the top of the valence band by phonon emission. (b) Schematic illustration of the cooling processes for electrons and holes in parabolic bands of unequal curvature.

(3) Wide photonic bandgap in order for the effective suppression of Klemens' decay to block the mechanism of an optical phonon splitting into two acoustic phonons and narrow energy dispersion of optical phonons to reduce Ridley process.

(4) Small energies for optical and acoustic phonons, helpful since more steps are needed for the electrons (holes) to relax to the bottom (top) of the conduction (valence) band, as shown in Fig. 8.20.

Unlike the quasi-continuous energy bands in bulk semiconductors, the eigen energy levels in nanocrystals are discrete, due to quantum confinement effect. When the energy spacing between the neighboring energy levels is much larger than the energy of LO phonons, the hot carrier cooling has to rely the much slower effect of multiple phonon emission. As a result, the carrier cooling rate is reduced by the "phonon bottleneck" effect due to the use of nanomaterials. In the case of polar materials, the initial energy loss is dominated by the emission of LO phonons.

8.3.3 Hot Carrier Solar Cells Based on Metal–Insulator–Metal and Metal–Semiconductor Structures

Simple designs are based on a metal–insulator–metal (MIM) structure, or a simpler metal–semiconductor (MS) Schottky barrier type, as shown in Fig. 8.21 for the band profiles. In both structures, photoexcited holes are collected by the cathode and electrodes by the anode on right side of the devices. Electrons have to overcome the barrier, either an insulating layer as in the MIM structure or a Schottky barrier as in the MS structure. Thus, it requires that the hot carriers are energetic enough, as shown in the figure. So far, those concepts have not been well-demonstrated experimentally for solar cells, though they have been theoretically proposed and considered promising device structures.

Cooling rates of hot carriers depend on many processes in a material, such as the sizes of nanomaterials, surface conditions, density of the photogenerated hot carriers, etc. However, it has been understood that hot electrons and hot holes cool at different rates, electrons cooling more slowly than holes due to their much smaller effective masses. Thus, hot electrons dominate the dynamical process of the photogenerated hot carriers.

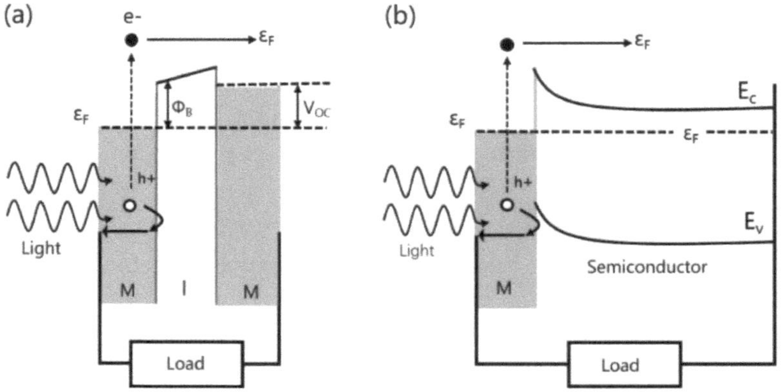

Figure 8.21 Schematic of a metal–insulator–metal (MIM) (a) and a metal–semiconductor (MS) (b) hot carrier solar cells.

In a hot carrier SC, the excess energy is extracted through impact ionization [20], for instance, before it is lost to lattice.

The time scale of thermalization is typically about 10 ps. However, it is rather difficult to find or design a structure in which the excess energy is extracted and used for photovoltaic effect within the time scale. An alternative way is to slow down the electron–phonon interaction. As pointed out by Conibeer and Green, a phonon bottleneck can be designed to increase the time scale, with the help of nanotechnology [21].

Even if the hot electrons are not relaxed in the semiconductor, they would lose energy as soon as they travel into metal contacts since there are a large number of electrons in metals. The hot carriers are then cooled down to the lattice temperature. Therefore, the charge carriers in the metal must be prevented from interacting with the hot carriers in the solar cell. A way to potentially solve this problem is to construct energy selective metal contact such as a tunneling junction, through which the hot carriers are extracted.

8.3.4 Hot Carrier Solar Cells Based on QDs

One of the practical difficulties in constructing a hot carrier SC is that hot electrons thermalize (loses excess energy to lattice) too fast. However, the time scale of thermalization can be prolonged by using materials or structures such as perovskite or quantum dots [21, 22]. Effective nanostructures can be designed to collect electron–hole pair before they get relaxed, both for slowing cooling down rate and, at the same time, enhancing the impact ionization. It has been shown that the cold electrons and holes at contact electrodes are the major mechanism for cooling down the hot carriers. Jiang and co-workers found that resonant tunneling contact electrodes based on quantum dots showed good anti-cooling property [23]. Hot electrons, due to their much smaller effective masses than holes, leading to more closely spaced eigen states, cooled down much slowly than holes. Typically, electrons cool to the band edge in the order of 10 ps while holes in the order of 0.2 ps. Besides, cooling rates depend upon hot carrier densities. There are a few hot carrier relaxation channels through which hot carriers give away their extra energies. Via an Auger process, for instance, a hot electron could give its excess kinetic energy to a hot hole and the hot hole, with much shorter relaxation time, then cools down quickly to the top of the valence band. For semiconductor quantum dots, hole could be trapped by the surface states. This blocks the Auger process due to the absence of holes in the cores of quantum dots, which in turn slows down electron cooling. With the use of the hot-carrier approaches, a solar cell efficiency could surpass 50% theoretically, under solar light concentration.

A number of theoretical works have shown the potential usefulness of the strategy, although it has been proven rather difficult to extract the excess energy of the hot electrons before they are cooled down to the lattice temperature. Extraction of the excess energy was experimentally demonstrated by Tisdale and co-workers [24], who observed the transfer of hot electrons from PbSe colloidal QDs to TiO_2 through time-resolved second harmonic response measured on TiO_2 surface coated with 1.5 monolayers of 3.3 nm PbSe QDs. The work suggests a useful route to construct a hot carrier solar cell by slowing down hot carrier cooling by means of nano-sized semiconductors.

Although high PCEs are predicted by theoretical analysis and simulation, the hot carrier solar cells have not been widely demonstrated by experiments. The major challenges stem from two key factors. The first is the fact that one could hardly find a semiconductor having both properties of a good absorber (with very low hot carrier cooling rate to allow for the extraction of excess energy) and a good SC active material (with a high quantum efficiency for photon absorption over a wide solar spectral range and an excellent carrier transport ability). The second is the carrier-extracting layer that is required to have good electric connection with the absorber and, at the same time, has a well-engineered narrow energy band at an appropriate energy position for high-efficiency carrier extraction. A quite few works have been devoted to designing the hot carrier SCs and preparing high-performance absorber and carrier-extracting layer [24–26].

8.4 Multiple Exciton Generation or Downconversion

8.4.1 Physical Background of Multiple Exciton Generation

In the hot carrier case, the extra part of the photon energy is used for photon-electric conversion, instead of being transformed into heat. In the case that photon–generated electrons are very hot, i.e., the photon energy is larger than two times of the bandgap, it is possible for the high-energy photon to excite two or more excitons. In a process called multiple exciton generation (MEG), the excessive part of photon energy that exceeds the bandgap is used to generate another or even more electron–hole pairs, if the photon is energetic enough. Utilizing MEG, the loss of a photovoltaic device could be reduced significantly. MEG, also called downconversion sometimes, is based on the idea of quantum cutting in which the energy of a high-energy photon is cut into two or more parts. In other words, a high-energy photon is transformed into two or more low-energy photons and the low-energy photons that still have sufficient energies larger than the bandgap, are absorbed by two or more electrons in the valence band, corresponding to 200% or higher quantum efficiencies.

Quantum cutting can be realized by several routes including intermediate band (levels). It is predicted theoretically that a single-bandgap solar cell facilitated with one intermediate level has a highest efficiency of 39.6% [27]. The most widely adopted scheme for quantum cutting uses luminescent materials. Figure 8.22 presents a conceptual solar cell using luminescent material to perform the quantum cutting. As shown in the figure, adding the luminescent material on the solar cell should not block low-energy photons and the luminescent low-energy photon should be collected by the underneath solar cell efficiently.

In a MEG process, the energy of an incident photon is two or more times of the semiconductor bandgap energy. In 2004, MEG was first observed on colloidal PbSe quantum dots [28]. The MEG effect was soon detected in other lead salt quantum dots (PbTe and PbS), as well as quantum dots of other material systems, including InAs, CdSe, CdS, Si [29], InP [30], as well as in semiconducting single-walled carbon nanotubes (SWNTs) [31]. Wang et al. demonstrated that absorbing a single photon with the energy of three times the energy gap of the SWNT leads to 130% quantum efficiency for exciton generation. MEG is considered due to several mechanisms. One is, as referred in bulk case, the impact ionization during which a high-energy photon excites an electron (exciton) into an energy level much higher than the bottom of the conduction band and the electron excites another electron (exciton), or other electrons (excitons), into the conduction band through impact interaction. This process requires high density of states (DOS) of the multi-excitons. The impact ionization is extremely inefficient in bulk semiconductors so that it makes ignorable contribution to the solar cell efficiency. It could, however, be greatly enhanced in nanosemiconductors since, unlike in bulk materials, the requirement for both energy and momentum conservation is relaxed in QDs.

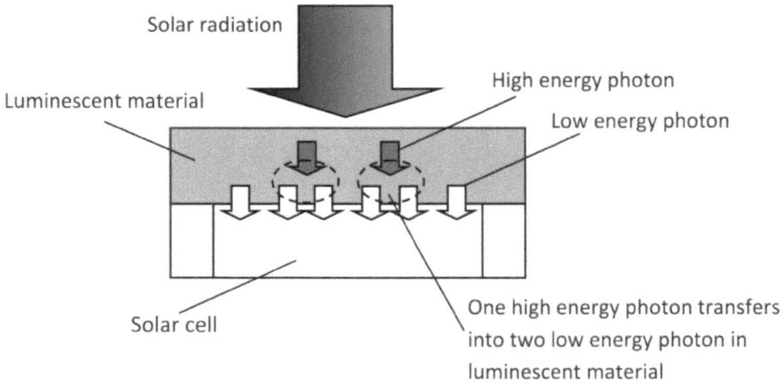

Figure 8.22 Schematic diagram of a solar cell using luminescent material to convert a high-energy photon into two or more low-energy photons.

For Si with a bandgap of 1.12 eV, photon energies above 1.12 eV would excite hot carrier with energy above the bottom of the conduction band. Theoretically, it is possible for a high-energy photon to excite multiple excitons and a photovoltaic process could has a quantum efficiency of $n \times 100\%$, where n is the number of excitons excited by the single photons. Correspondingly, the solar cells could have an open-circuit voltage ~ $n \times E_g/e$ evidenced by the experimental work on solar cells based on colloidal PbS and PbSe QDs [32]. Although a great deal of efforts have been devoted to develop techniques of utilizing MEG effect to raise the conversion efficiency, real evidence that demonstrates the enhancement of device efficiency due to MEG effect is insufficient. This is attributed to currently unclear MEG mechanism.

8.4.2 Multiple Exciton Generation in QDs

Similar to an atomic system, an Auger process can be configured through the construction of a quantum dot imbedded in a matrix semiconductor having an energy band profile as shown in Fig. 8.23. In the system, the ground state is within the bandgap and the first excited state above the conduction band of the matrix semiconductor. When an incident photon excites an electron to the energy level well above the bottom of the conduction band, the electron could give away its extra energy to the second electron and excite it to the ground confinement state of the QD, while the first electron relaxes to the same level. In the Auger process, one photon creates two electron–hole pairs. Materials with suitable bandgaps should be chosen together with an optimized device configuration designed to obtain the highest Auger efficiency.

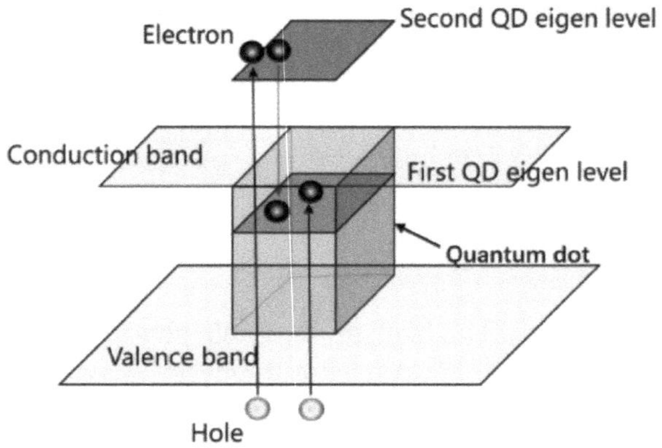

Figure 8.23 A schematic shows that a quantum dot is imbedded in a semiconductor matrix to form a structure that configures a multiple exciton solar cell.

As what can be seen obviously, this device architecture bears the challenge of carrier transport. The photon-excited carriers are confined in quantum dots in some degree and they need to be collected by electrodes of solar cells. Two ways can be adopted to overcome the problem. One is to excite the electrons to the continuum above the bottom of the conduction band via a certain kind of excitations, so that the carriers could travel through the device and collected by electrodes. One of the strategies for the excitation is to utilize mini-bands that locate within the thermal energy below the conduction band continuum. Another is to embed multiple quantum dots so that, instead forming energy levels, mini-bands must form in the material complex of semiconductor matrix and quantum dots to allow for easy carrier transport.

Apparently, MEG solar cells are only suitable for high-energy photons in solar spectrum. For photon energy equal to three to five bandgap energy, each photon could excite three to five excitons.

Kumar et al. investigated MEG in PbSe QDs of three different sizes with ground optical absorption peaks at 1100, 1200, and 1300 nm, by transient absorption spectroscopy based on a pump-probe route [33]. In the measurements, MEG threshold was determined by transitions associated with the electronic states at critical points along the Σ direction in the Brillouin zone. They observed reduction of the exciton cooling rate, which is beneficial for intensifying MEG process.

8.4.3 Solar Cell Architectures with Multiple Exciton Process

David and co-workers observed MEG in PbSe nano solar cells with over 120% external quantum efficiencies [34]. The reported solar cells have a device architecture as shown in Fig. 8.24a (see ref. [34] for specific technology of device fabrication) and the band profile as shown in Fig. 8.24b. Figure 8.25 presents the EQE curves measured on the SCs, showing that the EQEs' exceed 100% in the spectral range roughly 2.8–3.5 eV. The EQEs are 109 $\pm$ 3%, 113 $\pm$ 4% and 122 $\pm$ 3% for nanorods with E_g = 1.05, 0.95 and 0.80 eV, respectively. Clearly, a cell with a smaller bandgap is more likely to achieve a higher over-100%

external quantum efficiency. It was estimated that exceeding 150% EQE is achievable using narrow-bandgap nanomaterials with the similar-device architecture. The measurements for EQEs have been performed under different white light biases and essentially the same spectra were observed. The narrow EQE peaks around 1.0 eV correspond to the ground state excitons peak in all three tested devices and the fast decrease in EQE above 3.5 eV photon energy is due to the onset of ZnO layer absorption.

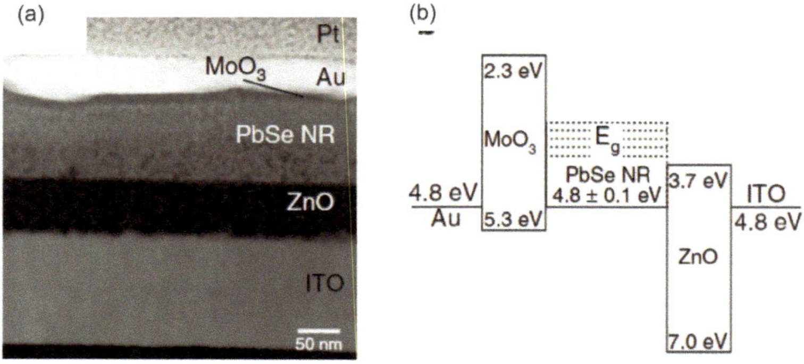

Figure 8.24 (a) TEM image of cross-sectional structure of the PbSe nanorod SC and (b) corresponding energy band profile. Reproduced with permission from [34].

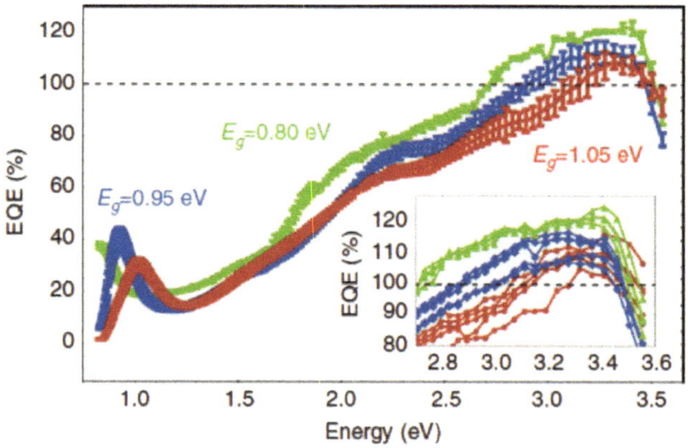

Figure 8.25 Measured EQE of the PbSe nanorod solar cells for E_g = 1.05, 0.95, and 0.80 eV, with the portions of the EQE curves exceeding 100% being depicted in the inset. The error bars are obtained from the standard deviation of those multiple solar cells. There are six cells for E_g = 1.05 eV, five cells for E_g = 0.95 eV, and five cells for E_g = 0.80 eV. Reproduced with permission from [34].

It is important to know the quantitative contribution to the photocurrent by the MEG effect. David and co-workers made estimation by weighting the experimental EQE by the fraction in excess of 100% in the EQE curves. This actually assumes that only the fraction in excess of 100% is contributed by MEG and the fraction is used as the internal quantum efficiency (IQE) in excess of 100%. The real IQE should be larger due to the surface light reflection (IQE = EQE/(1 – *R*), *R* is the surface light reflection). Results show that at AM1.5G illumination, the short-circuit current increases by 1.7%, 4.5%, and 5.8% for the solar cells with the bandgaps of 1.05, 0.95, and 0.8 eV, respectively. This demonstrates the potential usefulness of the MEG effect that could make fair amount of contributes to the solar cell efficiency. The MEG contribution to device efficiency could be further improved by optimizing the device architecture. It is worthy to note that the MEG effect is much stronger in nanostructures than in bulk materials.

Although different physical mechanisms are incorporated in the device architectures, both MEG and hot carrier SCs utilize the excess energies of the above-bandgap photons to achieve high device efficiency. A number of theoretical simulations have predicted high achievable device efficiency for the advanced device architectures. However, very few experimental works are devoted to constructed device that shows distinct efficiency enhancement contributed by those advanced ideas.

8.5 Upconversion

8.5.1 Principle of Spectral Upconversion

Due to the transmission of photons with energies below the bandgap, the photons make no contribution to photoelectric conversion. An upconverter produces one high-energy photon out of two or more below-bandgap photons, so that the energies of the low-energy photons can be used in the conversion.

A typical mechanism for the realization of spectral upconversion is through non-linear processes accompanied by the second- or higher order harmonic generations. However, this non-linear effect requires a very high light intensity (typically above 1000 suns). A high-efficiency upconversion is based on the energy transfer mechanism, referred to energy transfer from a sensitizer (an excited ion) to an activator (a neighboring ion) [35]. According to Trupke et al., the maximum PCE could potentially reach 47.6% under non-concentrated solar illumination if the upconversion effect is fully incorporated in a solar cell [36]. An energy upconversion process in the (Yb^{3+}, Er^{3+}) couple is illustrated in Fig. 8.26, where a 980 nm excitation is upconverted into light with shorter wavelengths as shown by the green and red arrows [37]. If this upconversion converter is attached to a solar cell, the solar cell is able to absorb low-energy photons for photovoltaic conversion. The upconverter, Gd_2O_2S doped with Yb^{3+} and Er^{3+}, for instance, is commercially available.

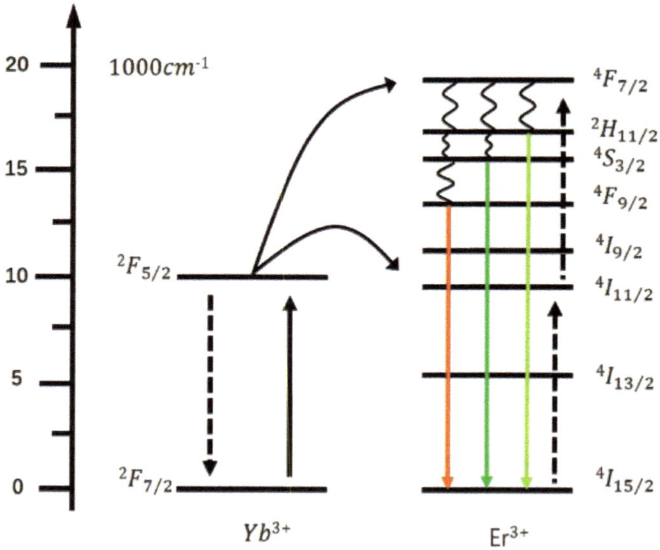

Figure 8.26 An energy upconversion process in the (Yb^{3+}, Er^{3+}) couple, the Yb^{3+} ion is excited at 980 nm from its $^2F_{7/2}$ level to the $^2F_{5/2}$ state and the energy is then transferred to the Er^{3+} ion, leading to the excitation of the Er^{3+} ion, as shown by the dashed arrows in the energy scheme of the Er^{3+} ion. The relaxation of the Er^{3+} ion gives rise to the emissions associated with the $^2H_{11/2}$ (light green arrow), the $^4S_{3/2}$ level (green arrow), and the $^4F_{9/2}$ level (red arrow) [37].

Technically, the upconversion can be realized by introducing an intermediate energy band (or level) so that the low-energy photons could use the energy spacings between the valence band and the intermediate band and between the conduction band and the intermediate band.

Thus, the fundamental idea for upconversion is to generate an above-bandgap photon through absorbing two or more sub-bandgap photons. The above-bandgap photon can then be absorbed by a solar cell for photoelectric conversion. The process uses an intermediate energy state into which a valence electron is excited by absorbing a sub-bandgap photon. The electron that is excited into the intermediate band and the excited electron is further excited into the conduction band by absorbing a second low-energy photon. The two (or more)–photon excited conduction band electron then recombines with a left-over hole in the valence band radiatively by emitting a high-energy photon. The emitted high-energy photon can then excite an electron–hole pair in a solar cell, as shown in Fig. 8.27. Thus, an upconverter is a device that outputs one above-bandgap photon with inputing of at least two low-energy photons.

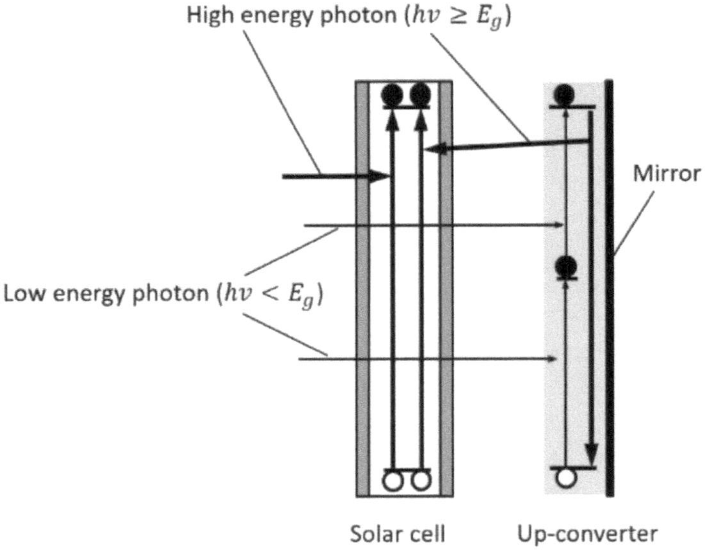

Figure 8.27 Schematic illustration of a bifacial solar cell with an attached upconverter on the back. Two below-bandgap photons penetrate through the cell and absorbed by the upconverter (thin arrows) where they are converted to an above-bandgap photon (thick arrows) that can be absorbed by the solar cell.

The upconversion process is likely to have high efficiency in nanomaterials and nanostructures, similar to the multiple exciton absorption due to the relaxing of momentum conservation. In the whole solar spectrum, about 20% of the solar energy is owned by photons with energies below the Si bandgap. The low-energy photons transmit directly through the device, making no contribution to the photovoltaic process. Thus, upconverting the low-energy photons in solar spectrum is an effective approach for raising the solar cell efficiency particularly for large-bandgap materials used for solar cells. Upconversion approach for photons with energies below the bandgap is technically feasible when nanotechnology is included.

8.5.2 Solar Cells with a Spectral Upconverter

Adding such an upconverter to a solar cell raises the power conversion efficiency of the solar cell. Theoretically, Trupke and co-workers showed that the original 30% conversion efficiency of wafer-based Si SCs can be enhanced to more than 40% at the illumination by non-concentrated light [36]. A solar cell and an upconverter can be designed separately so that there is essentially no modification needed to be made on the solar cell. As shown in Fig. 8.27, the upconverter is usually attached at the back of the solar cell so that low-energy photons can penetrate through the SC and absorbed by the upconverter. Only the back part of the solar cell need to be modified a bit and the design should let the output high-energy photons impinge on the solar cell. Thus, the upconversion device can incorporate with the commercial wafer-based SCs. As shown in Fig. 8.27, a solar cell with an upconversion part attached needs to deal with a solar spectrum that can be divided into three parts. The first part is the high frequency photons with the photon energy $h\nu_h \geq 1.12$ eV and the photons can be converted by the Si SC. In the second part, photons with their energy ~ 0.5 eV $\leq h\nu_m \leqq 1.12$ eV cannot be absorbed by the Si SC and the upconversion efficiency of the upconverter is low. The third part is the low-energy photons of $h\nu_l < \sim 0.5$ eV and the upconverter works efficiently.

Following the conceptual upconversion, Goldschmidt et al. designed a practical upconverter with spectral and geometric concentration [38]. As shown in Fig. 8.28, the upconversion

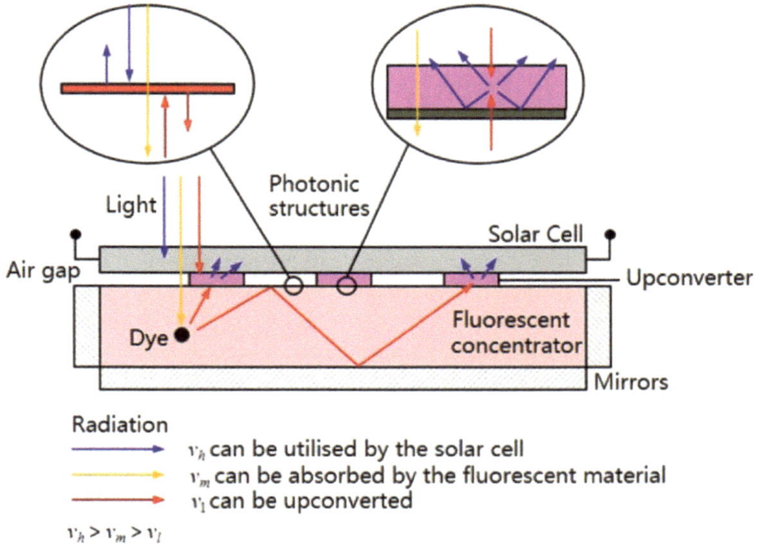

Figure 8.28 Potential setup of an advanced upconverter system. Assuming that the incident spectrum are divided into three frequency ranges: ν_h with photon energies above the bandgap of the solar cell, ν_m with photon energies below the bandgap and beyond the absorption range of the converter but in the absorption range of the dye molecules in the fluorescent waveguide concentrator, and ν_l having energies absorbable by the upconverter. The solar cell absorbs above-bandgap ν_h photons (blue arrows) and convert the photons into electric power. The below-bandgap ν_m and ν_l photons (yellow arrows and red arrows, respectively) penetrate through the cell. The upconverter converts the ν_l photons into ν_h photons that can be absorbed by the solar cell. The ν_m photons are absorbed by the dye molecules in fluorescent material where the dye molecule emits fluorescent ν_l photons absorbed by the upconverter. The waveguide is used to guide the dye-emitted to the upconverter.

elements are incorporated with a wafer-based Si SC and a fluorescent concentrator for high-efficiency photoelectric conversion, so that photons with energies in the range of 0.5 to 1.1 eV can be utilized efficiently. The whole system includes the conventional SC, the fluorescent concentrator consisting of luminescent nanocrystalline QDs imbedded in a transparent matrix, as well as the upconverter. The NQDs are PbSe and PbS QDs with strong absorption above their threshold photon energies and very efficient photoluminescence emission. The threshold and the photoluminescence (PL) emission energies depend on the sizes of the QDs through quantum confinement effect. For the PbSe- and PbS-QD concentrator in Fig. 8.28, the emission energies are tunable from 0.5 eV to 1.1 eV [39, 40]. For the use as a concentrator, it is required that QDs have high PL efficiency. In fact, the QDs of $PbSe/PbSe_xS_{1-x}$ core–shell structure (with a core of PbSe and a graded shell of $PbSe_xS_{1-x}$ alloy for lattice and dielectric matches) exhibit excellent photoelectric properties characterized by large cross section for optical absorption and high internal quantum efficiency of 50–80% [41]. In addition, the PbSe/PbSeS core–shell structure are of photochemical robustness, which is essential for the use in SCs.

When solar radiation is shining on the system, the high-energy photons ($h\nu_h$ > 1.12 eV) are absorbed and converted into electric power by the Si SC. The less energetic photons penetrate through the Si SC. The converter can convert the photons with low energies of around and less than 0.5 eV ($h\nu_l$), as shown in Fig. 8.28. There are photons with the energies between 0.5 and 1.12 eV ($h\nu_m$) and they are down–converted into low-energy photons by the concentrator so that they can be converted into high-energy photons by the converter. The photons with mediate energy are absorbed by the QDs in concentrator, PL emission of the QDs absorbed by the converter, and the high-energy photons converted by the converter absorbed by the Si SC. Especially, the concentrator does absorb the upconverted photons that should be taken by the Si SC. The upconverting device is designed to locate between the concentrator and the back of the solar cell. As shown in Fig. 8.28, the back of the Si SC is not completely covered by the upconverting device optically, in order for the photons with intermediate and low energies to incident into the concentrator, since the upconverter might reflect the incoming light. Thus, the system is designed in such a way that the QD concentrator absorbs the photons in the spectrum range of 0.5 eV $\leq h\nu_m \leq$ 1.12 eV and generates PL emission of photon energies $h\nu_l$ < 0.5 eV. The upconverter then converts the low-energy photons into high-energy photons of $h\nu_h \geq$ 1.12 eV. In order for the PL emission to be coupled efficiently into the upconverter, the QD spectral concentrator is designed so that it is also a geometric concentrator for light intensity. Figure 8.28 shows that the concentrator has also a waveguide configuration where PL emission is confined in the waveguide. The waveguide is designed in such a way that there is a dielectric matching between the waveguide-like concentrator and the upconverter so that the PL emission can be coupled into the upconverter efficiently, in addition to the low-energy photons that penetrate through the Si SC and incident on the upconverter. This leads to the increase of the upconversion efficiency. The dielectric discontinuity on the waveguide/air boundary (see Fig. 8.28), where upconverter is not attached, guarantees the total reflection for most photons in the PL emission.

As a relatively independent element, upconverters can be easily attached onto wafer-based silicon solar cells. Theoretically, adding an upconverter on a silicon solar cell increases the efficiency limit from 33% up to 40.2% at non-concentrated light condiction [36].

One of the important things in designing the system is to prevent the absorption of the high-energy photons, emitted by upconverter, by the QD concentrator. This is the reason why the upconverter and the QD concentrator are designed to be completely separated (a patent has been granted for this design) [42]. In fact, the use of photonic crystal structure between the QD concentrator and the upconverter is often an effective way to prevent the upconverted photons from being absorbed by the concentrator, as shown in top-right of Fig. 8.28. Another advantage of the design is the integration that combines spectral concentration and waveguide function, which makes it applicable to incorporate with lenses and mirrors as the external concentration. Thus, very high light intensity with relevant photon energies is achievable at the location of the upconverter, which is effective for increasing the conversion efficiency. In Fig. 8.28, selective mirrors are attached to the waveguide to increase the efficiency of fluorescent collection, where photons with the energies of $h\nu_m$ penetrate into the concentrator and the photons with the energies of $h\nu_l$ emitted by the QDs are reflected (see the left inset in Fig. 8.28).

An upconverter should be attached to a solar cell as close as possible. In conventional Si solar cells that use QD fluorescent concentrator, for instance, solar cells are usually attached at the side edges of concentrator plates. The PL photons emitted by the QDs may need to travel several centimeters before reaching the Si solar cells attached on the sides, and therefore, most of the light is expected to be lost. The upconverter needs to be attached on the front or back surface of the QD concentrator so that the PL photons only need to travel very short distance before hitting the upconverter, beneficial for the collection of PL photons by the upconverter.

8.5.3 An a-Si:H Thin-Film Solar Cell Attached with an Upconverter

Wild and co-workers reported an upconversion solar cell consisting of a a-Si:H thin-film solar cell attached with an upconverter [37, 43]. The upconverter uses Gd_2O_2S co-doped with Er^{3+} and Yb^{3+}, which has an energy scheme as shown in Fig. 8.26. When the upconverter is excited with 981 nm line of a Xenon lamp, strong emissions were observed within 520–570 nm and 650–680 nm. Figure 8.29 illustrates the measured *J–V* curves on the a a-Si:H SC with and without the upconverter, illuminated by 980 nm laser. The cell without the upconverter was used as a reference.

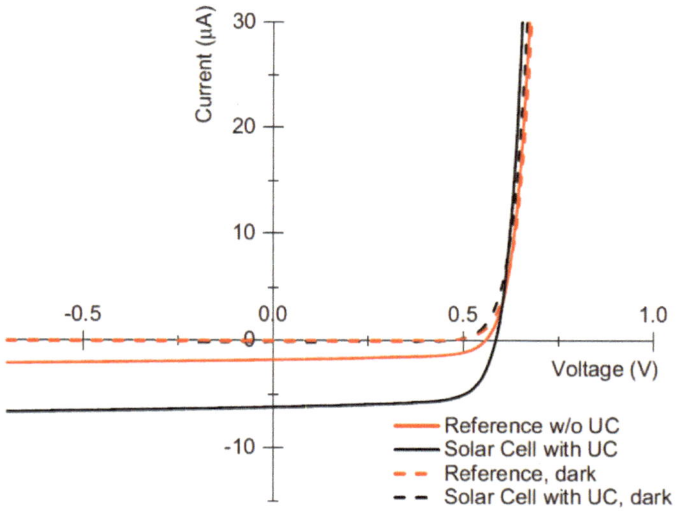

Figure 8.29 Measured *J–V* curves of a-Si:H solar cells with and without attached upconverter, together with those of the reference cell which has only the back reflector. Reproduced with permission from [37].

Short-circuit current J_{sc} extracted from the cell with the upconverter is three times of that of the cell without the upconverter. a-Si:H has a tail extending into the bandgap. But the band tail absorption and the upconversion response can be separated, since the former is linearly dependent on the excitation intensity while the latter increases quadratically. Experimentally, this was done by extracting slope from the data of EQE versus excitation power presented with a double logarithmic plot. The slope obtained on the cell with the upconverter is $\sim$ 0.72 and that on the cell without the cupconverter is $\sim$ 0.094, which is a clear indication that the upconverter enhance the power conversion efficiency through upconversion.

8.6 Plasmonic Solar Cells

8.6.1 The Plasmon Effect

Plasmons are quantized collective oscillations of the free electron gas density with respect to the fixed positive ion background. The plasmon effect is usually observed in materials with high free electron densities, such as metals. Plasmonic effect also shows up in heavily doped semiconductors that become degenerate. Metals show bright visible colors that are due to plasmon oscillation. Metal particles in nanometer sizes exhibit very interesting optical properties in addition to their unique bright colors, one of the optical properties being the surface plasmon oscillation. The electron mean free path in gold and silver is around 50 nm. Thus, for gold or silver nanoparticles with the particle diameter of a few nanometer, electrons receive essentially no scattering when travelling inside the particles, the only scattering being from the surface.

Light and electrons in material interact strongly through electric vector of light that drives electrons. Visible light has wavelength much larger than the mean free path of electrons in nano metal particles. Therefore, light drives electrons to more uniformly inside the metal particles, as schematically shown in Fig. 8.30 [44]. The plasmon oscillation exhibits a characteristic frequency or a frequency range. When the frequency of light matches to the characteristic frequency, a plasmon oscillation is in resonance with the incident light at the frequency and light energy is then coupled into the plasmon oscillation. The plasmon oscillation is characterized by the standing oscillation of electrons toward up and down surfaces inside the particles (see Fig. 8.30). Light interacts with plasmon system in different cross sections for absorption and scattering.

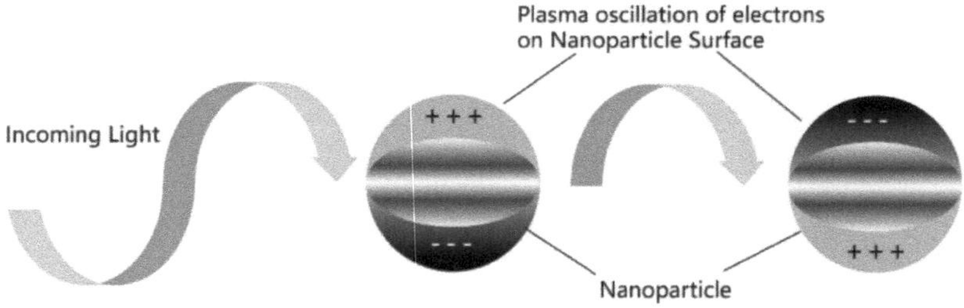

Figure 8.30 A schematic diagram showing the incident light drives the surface plasmon resonance due to resonant interaction of the electron density and light.

As shown in Fig. 8.30, the characteristic frequency of the plasmon oscillation is directional. Thus, oscillation frequency depends on not only the size but also the shape of the metal nanoparticles for a given light-incident geometry. This is due to the fact that change of surface geometry gives rise to variation in the surface electric field density, which in turn results in change of plasmon oscillation frequency of the charges. In addition, the dielectric constants of both the metal nanoparticles and the surrounding medium play important roles. All the factors can be used for comprehensive design of nanostructure systems for effective light trapping. Due to the fact that the charge oscillation occurs at the surface of the nanoparticles, it is named surface plasmon resonance (SPR), so that the oscillation condition is sensitive to surface conditions. As a result, in addition to frequency, the electric field distribution of the SPR also depends on light incidence geometry. A number of works have been reported on the use of metal nanoparticles for light trapping on solar cells. For a comprehensive review for plasmonic solar cells, see Ref. [45].

8.6.2 Working Principle of Plasmonic Solar Cells

Metallic nanostructures support surface plasmons. By proper design, light can be directed to propagate at the metal/semiconductor interface or be trapped in desired regions with nanoscale sizes. In other words, nanoplasmonics bears the characteristic of trapping, concentrating and localizing light. Solar cells incorporated with metallic nanostructures display surface plasmon effect to enhance the light absorption and, in turn, solar cell efficiency. Figure 8.31a shows the schematic device structure of a plasmonic organic solar cell, and Fig. 8.31b presents calculated electric field intensity distribution on x-y cross section, showing that the light is concentrated nearby the Ag nanoparticles due to localized surface plasmon resonance [46].

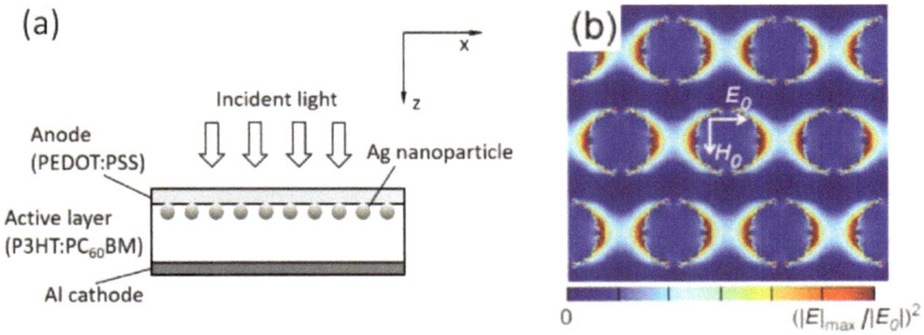

Figure 8.31 A schematic plasmonic organic solar cell with the structure of structure of PEDOT:PSS/Ag-particle/P3HT:$PC_{60}BM$/Al (a) and calculated electric field intensity distributed on x-y cross section cutting through Ag nanoparticles for *p*-polarized incidence at the wavelength of 604 nm (b). E_0 and H_0 are the polarization directions of electric field and magnetic field of the incident light, respectively. Reproduced with permission from [46].

Some advanced ideas require high light concentration to achieve high device PCE. The upconversion, for instance, is potentially useful to compensate for sub-bandgap photon transparency of solar cells. The effectiveness of upconversion has been demonstrated in a number of proof-of-principle experiments. However, in practical application on photovoltaic devices, many technological problems need to be solved. The effective way to enhance upconversion efficiency relies largely on the light concentration and conventional light concentration technologies are often not very practical for real application, since they need complicated optical system with lens to track solar irradiation. Aiming at the problems, some technologies were proposed to enhance low-energy photon absorption, including plasmonics and external sensitization routes [47, 48].

8.6.3 A Plasmonic Solar Cell with the $NaYF_4$:(Er^{3+}, Yb^{3+}) Upconverter

Chen and co-workers used the plasmonic effect to enhance the light concentration for upconverting sub-bandgap photons to super-bandgap ones [49]. They designed a composite module in which GaAs solar cells are incorporated with plasmonic effect. Improved 1-sun photovoltaic performance is due to the capture of sub-bandgap photons through spectral upconversion by the plasmonics. Figure 8.32 presents the device structure. The upconverter was composed of a hole-post hybrid silver nanostructure plasmonic reflector coated with $NaYF_4$:(Er^{3+}, Yb^{3+}) upconversion nanocrystals. A polymer layer (PMMA) was then coated on the upconverter as the waveguide, followed by attaching an ultrathin GaAs solar cell on the top. The conversion efficiency of the solar cell incorporated

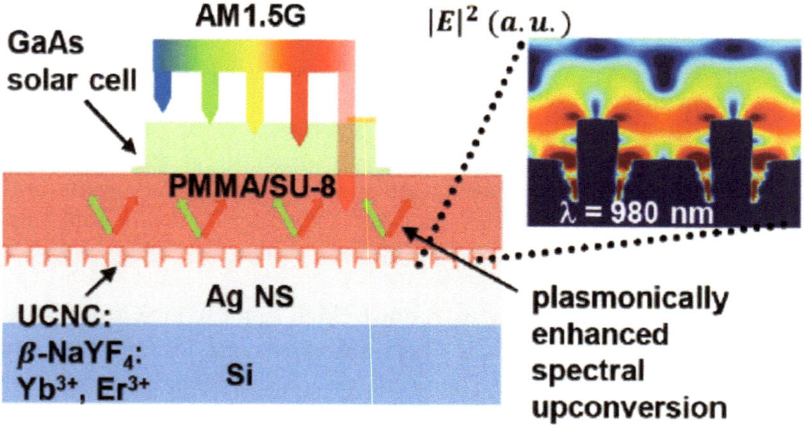

Figure 8.32 Schematic of a ultrathin GaAs solar cell with a $NaYF_4$: (Er^{3+}, Yb^{3+}) upconversion nanocrystal (UPNC) modified by plasmonic Ag nanostructure on a Si substrate. The calculated distribution of the square of electric field intensity ($|E|^2$) is plotted on the top right. Reproduced with permission from [49].

with the plasma-modified upconversion nanocrystal (UCNC in the figure), increases relatively by 6.4% and 11.7% compared to the cells on a nanostructured (NS) silver reflector and a plain silver reflector without the upconverter. As shown in Fig. 8.33, the *J–V* curves measured on the GaAs cells on a plain silver reflector are essentially the same with and without the upconverter. The GaAs cells on nanostructured (NS) Ag reflector with and without the upconverter show quite different *J–V* behaviors, indicating the significant effect of the plasmonic nanostructure in concentrating light. Note that there are three mechanisms that are contributed to the ECE enhancement: the spectral upconversion, light concentration, as well as the PMMA waveguide.

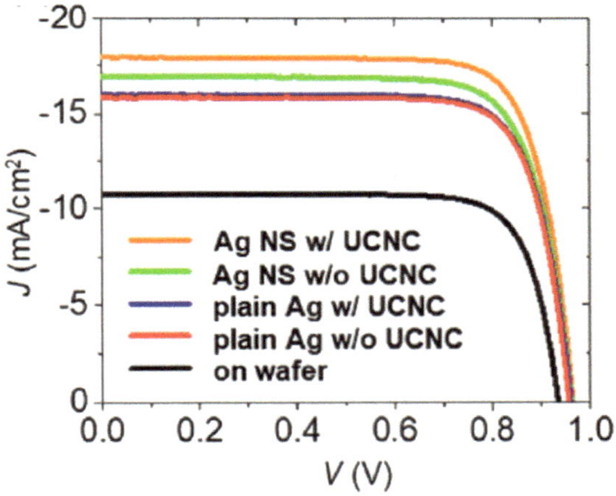

Figure 8.33 *J–V* curves of ultrathin GaAs solar cells at configurations: on nanostructured (NS) Ag reflector with (orange) and without (green) the upconverter, on a plain silver reflector with (red) and without (red) the upconverter, and on the GaAs solar cell only (black). Reproduced with permission from [49].

8.7 Intermediate Band Solar Cells

8.7.1 Working Principle of Intermediate Band Solar Cells

Intermediate band (IB) is a way to go beyond the efficiency limit of a single-junction SC by introducing an energy band in forbidden band, an alternative approach within one SC to achieve different energy thresholds through introducing an IB [50, 51]. In this device architecture, more electronic transition energies can be used for the absorption of photons of different energies, as shown in Fig. 8.34. This idea is more effective in infrared region of the solar spectrum since the infrared spectrum cover a large frequency range that cannot be used by Si SCs. On the other hand, conversion of infrared photons requires narrow-bandgap semiconductors that are expensive and technically amateur. Furthermore, photoelectric devices based on narrow-bandgap semiconductor often operate at low temperature. One of the advantages for the virtual bandgap is that it does not require narrow-physical-bandgap materials to achieve high conversion efficiency.

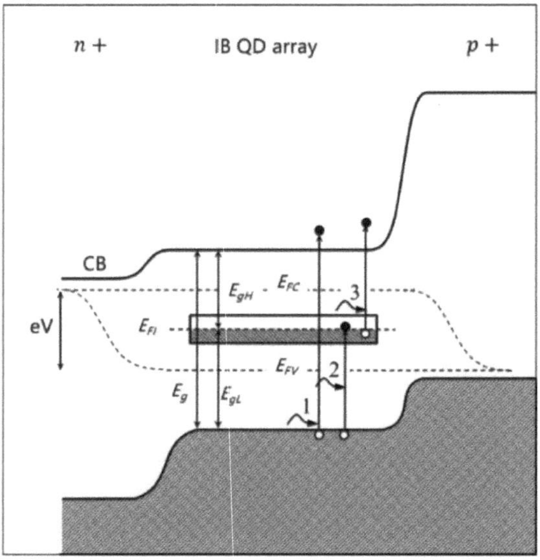

Figure 8.34 Schematic of energy band illustration for the IB solar cell.

In implementing intermediate band SCs, a half-filled band (or more half-filled bands) located within the bandgap is created. Such a structure offers more opportunities for lower energy photon absorptions. As shown in Fig. 8.34, an electron may jump from the valence band to the conduction band, as in conventional SCs and, in addition, an electron may reach the IB by absorbing a lower energy photon and then reach the conduction band by absorbing another low-energy photon. Of course, while creating more opportunities for photon absorption, the IBs also offer more possibilities for carrier recombination as loss of photovoltaic conversion. One of the challenges for IB solar cells is to provide efficient multi energy levels for photon absorptions and low probability for carrier recombination.

8.7.2 Methods of Introducing Intermediate Bands

There are several ways to obtain IBs by means of energy band engineering, such as doping, mini-band approach, and localized approach for virtual-bandgap SCs, etc. In material preparation, impurities and other nanomaterials can be incorporated into semiconductors, leading to the formation of multiple energy bands (MEBs) or multiple energy levels (MELs) in the bandgap of the matrix material. To exceeding the Shockley–Queisser limit, a system with MEBs or MELs provides with multiple channels for optoelectronic conversion. Efficient devices are then implemented in the way that photoelectric generation process is enhanced and the carrier recombination one is largely suppressed.

Two possible schemes for virtual-bandgap solar cells are shown in Fig. 8.35. At high density of the quantum dots, the eigen levels of quantum dots mix to form mini-bands, as shown in Fig. 8.35a. Similarly, nanomaterials can form localized energy levels if they are well separated and, very often, localized energy levels are implemented by quantum wells (see Fig. 8.35b). Both mini-bands and localized levels approaches require certain band offsets in the conduction (or valence), which calls for proper selection of materials.

Physically, efficient photoelectric conversion using the IB requires the simultaneous optimum radiative coupling between all the bands in Fig. 8.35 [52]. The effective radiative coupling is believed to be realizable by means of band engineering, a fast-developing area in both structures and materials, though it has not been demonstrated so far. Thus, referring to the working process of efficient photoelectronic conversion, the design of an ideal IB solar cell requires optimum material systems (mother semiconductor and nanomaterials) and related device configuration in which the photovoltaic process, particularly optical absorption and carrier transport, is completed in an optimized way. In well-designed conditions, the MEB or MEL is strongly coupled with solar radiation field simultaneously, so that optical absorption and recombination occur between those bands (levels).

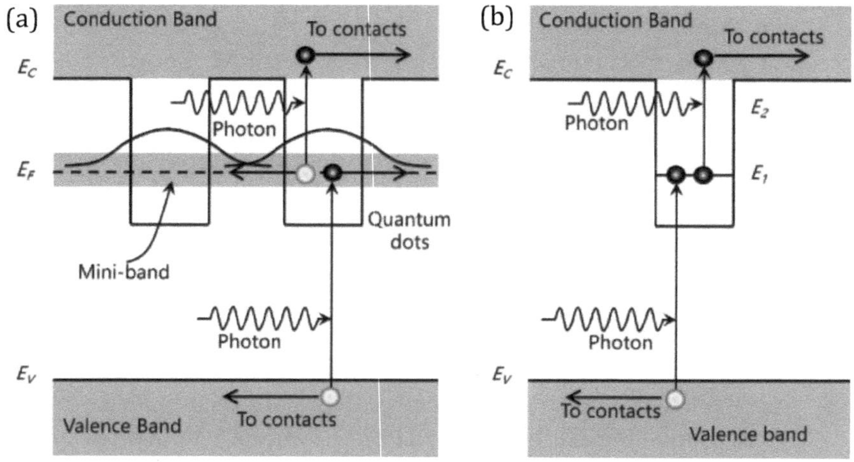

Figure 8.35 Two possible schemes for virtual-bandgap solar cells: (a) two-photon absorption via a mini-band (b) and two-photon absorption via a confined and localized state.

Intermediate bands formed by doping

Theoretically, the impurity-band photovoltaic devices could lead to the enhancement of the photoelectric conversion by providing additional energy bands allowing absorptions of photons with different energies, if the defect bands have optimum optoelectronic properties like the conduction and the valence bands in a semiconductor. The concept of MEBs (MELs) can be implemented in devices through introducing a defect or impurity band in the bandgap of a semiconductor. The essential requirement for the defect or impurity band is rigorous control of the defect (or impurity) level position and concentration. The requirement is often difficult to meet in current technology. Another challenge is that in this kind of devices, some important material parameters critical for device performance, such as carrier mobility and lifetime, are to low due to the crystal structure imperfection caused by doping and defects. The device is easy to design but finding the suitable defects and impurity is a big challenge.

The conceptually simple IB approach, however, encounters various problems. For the doping approach, the intermediate band (or energy levels) in the bandgap originates from deep-level-impurity doping that tend to degrade the quality of the semiconductor. The impurity bands have low carrier mobility. High light absorption coefficient associated with impurities requires very heavy doping, which is very often a disaster for semiconductor quality. Achieving IB by doping is thus unrealistic.

Intermediate bands formed by nanomaterials

The concept of MEBs (MELs) can also be implemented in devices through imbedding quantum dots (QDs), quantum wires (QWires), or quantum wells (QWells) in a semiconductor matrix. In the mini-band approach, QDs (or QWires, QWells) are introduced to form mini-bands within the semiconductor bandgap when the quantum dots are close enough to cause the wavefunction overlapping between neighboring quantum dots.

The mini-band could have high density of states that leads to efficient light absorption. The major challenge is the selection of quantum dot material having an appropriate bandgap and band alignment with the mother semiconductor. Quality requirements pose a big challenge to the preparation of quantum dots. For instance, the surface passivation is often a challenge for colloid quantum dots due to the requirement for suppression of high density of surface states, while cost is the biggest issue for the embedded quantum dots grown by self-assembly. Similarly, a local band can also be formed by other nanomaterials such as quantum wires and quantum wells.

A number of approaches have been demonstrated which used QDs to form energy levels or mini-bands in the bandgap. In reported works, the SCs with high efficiency based on the virtual-bandgap concept are those using existing solar materials as barrier layers that dominate the total photoelectronic conversion. The sizes, shapes, and distribution of nanomaterials offer great room for the implementation of optimized devices.

Still, challenges remain in proper selection of material system, device configurations and, more importantly, optimum radiative coupling between those bands that are very different in their physical properties. In addition, the issue associated with carrier mobility and interface trapping, for instance, is still problematic. The IBs implemented by mini-band and localized energy level approaches have demonstrated the high device efficiencies in principle [50].

Despite the technical challenges remaining in achieving high conversion efficiency, the virtual-bandgap concept has some practical advantages. The most important one is, with a clear physical picture of the idea aiming at wide spectrum absorption for both low and high-energy photons, that a large-bandgap semiconductor can be used and the interband spacing is tunable with nanotechnology. The use of large-bandgap materials keeps away the requirement of low-energy photon conversion that is always problematic for devices with a *p–n* junction due to thermaldynamic challenge. In addition, the virtual-bandgap concept can be incorporated into the current SC technologies.

Using doping, SCs with intermediate energy band functions like a three-junction tandem device, though only one material is used. For the quantum dot IB solar cells, two materials could result in efficiency similar to that of a three-junction tandem cell. Thermodynamical calculation indicates that the power conversion efficiency of an ideal IB solar cell could transcend the Shockley–Queisser limiting to ~47% at 1 sun and ~63% at full concentration [50].

The IB allows sub-bandgap photons be absorbed to generate additional photocurrent. In practice, the IB solar cells suffer from low carrier lifetime and absorptivity for transitions associated with the IB, due to low density of states in the IB band. The low carrier lifetime is mainly caused by fast non-radiative recombination. The concept of the "photon ratchet" (analogous to a mechanical ratchet) was proposed by Yoshida et al. [53]. As shown in Fig. 8.36, the ratchet band (RB) is a bit below the IB states in energy. It gains electrons irreversibly from the IB states through thermal transitions. If the RB energy states are non-emissive, i.e., the transition between RB and VB is forbidden due to the symmetries of their wavefunction, the carriers on RB will have very long lifetime, leading to the enhancement of absorptivity and, in turn, the conversion efficiency. Pusch and co-workers designed an implementation-independent thermodynamic model to evaluate the efficiency of IB solar cells due to absorptivity [54]. They found that the conventional IB solar cells cannot transcend the single-bandgap Shockley–Queisser limit due to the recombination through the IB. However, if the transitions from the VB to the IB and from the IB to the CB correspond to light absorptivity exceeding 36%, the conversion efficiency will break the Shockley–Queisser limit. This can be done by introducing a quantum ratchet into the IB solar cell.

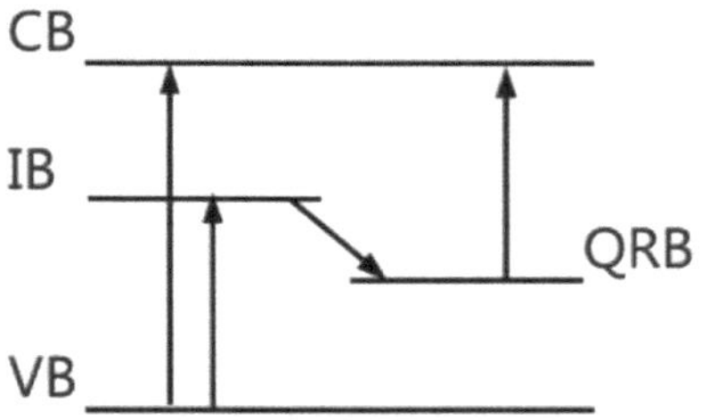

Figure 8.36 A quantum ratchet band (QRB) state close to the intermediate band (IB). The conduction band (CB) and the valence band (VB) are also shown.

8.7.3 Some Intermediate Band Solar Cells

A number of IB solar cells have been investigated, including 3D QDs and impurity bands. Yang and co-workers synthesized nano $CuIn_{1-x}Sn_xS_2$ and $CuGa_{1-x}Sn_xS_2$ thin films with $x = 0.02$ [55]. The Sn doping introduces a narrow IB 1.7 and 0.8 eV above the top of the valance band in $CuGaS_2$ and $CuInS_2$, respectively. Measured diffuse reflection and photoluminescence spectra show extra absorption and emission due to the IBs. The enhanced spectral response would intensify photovoltaics effect.

Vörös and co-workers investigated IB formation in colloidal nanoparticles and proposed the photovoltaic device architectures using the nanoparticles, based on the results from ab initio calculations [56]. There exists high density of surface states that are controllable to a large extent by modifying the nanoparticle surfaces using ligand exchange. The surface condition together with the quantum confinement could give rise to deep (intragap) energy states in the bandgap. Thus, nanoparticles could be promising materials for the construction of the IB solar cells. Figure 8.37a presents the calculated positions of the intragap states, lowest unoccupied molecular orbit (LUMO) and highest occupied molecular orbit (HOMO) energy levels. The intragap level is well-separated from the LUMO and HOMO levels, so that it can be used as IB energy level to construct IB solar cells. The absorption spectra of the three-level system is displayed in Fig. 8.37b, showing that the absorption spectra are well complementary.

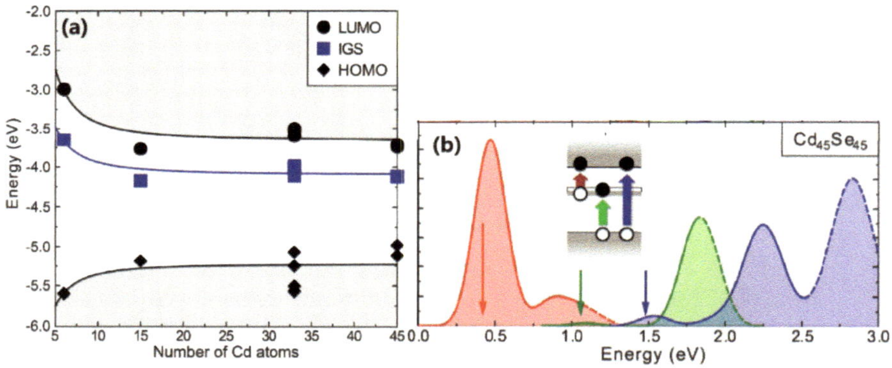

Figure 8.37 (a) The intragap, LUMO and HOMO energy levels (relative to the vacuum level) of a single QD as a function of the Cd-atom numbers (or QD size) in the QD and (b) calculated absorption spectra of the three-level system, with the related transitions shown in the inset. Reproduced with permission from [55].

Theoretically, the IB architecture can achieve solar cell efficiency far above the Shockley–Queisser limit and the IB approaches have been experimentally demonstrated, in the sense of physical principle and the operating principle of the IB solar cells. In practice, however, the Shockley–Queisser limit is not yet being transcended.

The research on solar cells with nanostructures is still on the way and it seems to have a long way to go. However, in closely viewing the whole picture of nano solar cell research, one could see that there are in fact a great number of ways to reach to the destination—more favorable than conventional solar cells in the combination of power conversion efficiency and cost.

References

1. C. H. Henry, Limiting efficiencies of ideal single and multiple energy gap terrestrial solar cells, *J. Appl. Phys.,* **51(8)**, 4494–4500 (1980).
2. J. F. Geisz, D. J. Friedman, J. S. Ward, A. Duda, W. J. Olavarria, T. E. Moriarty, J. T. Kiehl, M. J. Romero, A. G. Norman, and K. M. Jones, 40.8% efficient inverted triple-junction solar cell with two independently metamorphic junctions, *Appl. Phys. Lett.,* **93**, 123505-3 (2008).
3. W. Guter, J. Schone, S. P. Philipps, M. Steiner, G. Siefer, A. Wekkeli, E. Welser, E. Oliva, A. W. Bett, and F. Dimroth, Current-matched triple-junction solar cell reaching 41.1% conversion efficiency under concentrated sunlight, *Appl. Phys. Lett.,* **94**, 223504-3 (2009); F. Dimroth, M. Grave, P. Beutel, U. Fiedeler, C. Karcher, T. N. D. Tibbits, E. Oliva, G. Siefer, M. Schachtner, A. Wekkeli, A. W. Bett, R. Krause, M. Piccin, N. Blanc, C. Drazek, E. Guiot, B. Ghyselen, T. Salvetat, A. Tauzin, T. Signamarcheix, A. Dobrich, T. Hannappel, and K. Schwarzburg, Wafer bonded four-junction GaInP/GaAs//GaInAsP/GaInAs concentrator solar cells with 44.7% efficiency, *Prog. Photovolt: Res. Appl.* **22**, 277–282 (2014).
4. NREL demonstrates 45.7% efficiency for concentrator solar cell (2014, December 17) https://phys.org/news/2014-12-nrel-efficiency-solar-cell.html.
5. https://www.ise.fraunhofer.de/en/press-media/press-releases/2022/fraunhofer-ise-develops-the-worlds-most-efficient-solar-cell-with-47-comma-6-percent-efficiency.html.
6. M. A. Green, *Third Generation Photovoltaics—Advanced Solar Energy Conversion*, Springer Series in Photonics, ed., T. Kamiya, et al., Berlin, Heidelberg, New York, Springer, p. 160, vol. 12 (2003).
7. J. F. Geisz, J. M. Olson, M. J. Romero, C. S. Jiang, and A. G. Norman, Lattice-mismatched GaAsP solar cells grown on silicon by OMVPE, *Conference Record of the 2006 IEEE 4th World Conference on Photovoltaic Energy Conversion.* J. F. Geisz, J. M. Olson, D. Friedman, S. Kurtz, W. McMahon, M. Romero, R. Reedy, K. Jones, A. Norman, A. Duda, A. Kibbler, C. Kramer, and M. Young, III–V/Silicon Lattice-Matched Tandem Solar Cells, *the 2004 DOE Solar Energy Technologies Program Review Meeting*, October 25–28.

8. M. Yao, S. Cong, S. Arab, N. Huang, M. L. Povinelli, S. B. Cronin, P. D. Dapkus, and C. Zhou, Tandem solar cells using GaAs nanowires on Si: Design, fabrication, and observation of voltage addition, *Nano Lett.,* **15**, 7217–7224 (2015).
9. I. Tobías and A. Luque, Ideal efficiency of monolithic, series-connected multijunction solar cells, *Prog. Photovolt: Res. Appl.,* **10**, 323–329 (2002).
10. T. Kim, Y. Gao, H. Hu, B. Yan, Z. Ning, L. K. Jagadamma, K. Zhao, A. R. Kirmani, J. Eid, M. M. Adachi, E. H. Sargent, P. M. Beaujuge, and A. Amassian, Hybrid tandem solar cells with depleted-heterojunction quantum dot and polymer bulk heterojunction subcells, *Nano Energy,* **17**, 196–205 (2015).
11. R. W. Crisp, G. F. Pach, J. M. Kurley, R. M. France, M. O. Reese, S. U. Nanayakkara, B. A. MacLeod, D. V. Talapin, M. C. Beard, and J. M. Luther, Tandem solar cells from solution-processed CdTe and PbS quantum dots using a ZnTe–ZnO tunnel junction, *Nano Lett.,* **17**(2), 1020–1027 (2017).
12. G. W. P. Adhyaksa, E. Johlin, and E. C. Garnett, Nanoscale back contact perovskite solar cell design for improved tandem efficiency, *Nano Lett.,* **17**(9), 5206–5212 (2017).
13. M. I. Hossain, W. Qarony, V. Jovanov, Y. H. Tsang, and D. Knipp, Nanophotonic design of perovskite/silicon tandem solar cells, *J. Mater. Chem. A,* **6**, 3625–3633 (2018).
14. A. Goetzberger and W. Greubel, Solar energy conversion with fluorescent collectors, *Appl. Phys.,* **14**, 123–139 (1977).
15. S. A. Mann and E. C. Garnett, Resonant nanophotonic spectrum splitting for ultrathin multijunction solar cells, *ACS Photonics,* **2**, 816–821 (2015).
16. (a) M. Konagai, S. Kim, Y. Takiguchi, P. Sichanugrist, T. Kobayashi, and T. Nakada, Development of High-performance Solar Cells Using Solar Spectrum Splitting Technique, *2013 IEEE 39th Photovoltaic Specialists Conference* (PVSC) PART 2, p. 094-098; (b) S. D. Vorndran, B. Chrysler, B. Wheelwright, R. Angel, Z. Holman, and R. Kostuk, Off-axis holographic lens spectrum-splitting photovoltaic system for direct and diffuse solar energy conversion, *Appl. Opt.,* **55**(27), 7522–7529 (2016).
17. A. Dorodnyy, E. Alarcon-Lladó, V. Shklover, C. Hafner, A. F. I. Morral, and J. Leuthold, Efficient multiterminal spectrum splitting via a nanowire array solar cell, *ACS Photonics,* **2**, 1284–1288 (2015).

18. G. Conibeer, M. Green, R. Corkish, Y. Cho, E.-C. Cho, C.-W. Jiang, T. Fangsuwannarak, E. Pink, Y. Huang, T. Puzzer, T. Trupke, B. Richards, A. Shalav, and K.-L. Lin, Silicon nanostructures for third generation photovoltaic cells, *Thin Solid Films,* **654**, 511–512 (2006).

19. R. T. Ross, A. J. Nozik, Efficiency of hot-carrier solar energy converters, *J. Appl. Phys.,* **53**, 3813 (1982).

20. P. Würfel, Solar energy conversion with hot electrons from impact ionization, *Sol. Energy Mater. Sol. Cells,* **46**, 43–52 (1997).

21. G. J. Conibeer and M. A. Green, Phononic band gap engineering for hot carrier solar cell absorbers, in *Proceedings of the 19th European Photovoltaic Solar Energy Conference*, 2004, Paris, France, p. 270–273.

22. S. Kahmann and M. A. Loi, Hot carrier solar cells and the potential of perovskites for breaking the Shockley–Queisser limit, *J. Mater. Chem. C,* 2471–2486 (2019).

23. C. W. Jiang, E. C. Cho, G. Conibeer, and M. A. Green, Silicon quantum dots: Application for energy selective contacts to hot carrier solar cells, *Proc. 19th European Photovoltaic Solar Energy Conference*, WIP-Munich & ETAFlorence, Paris, p80 (June 2004).

24. J. Vura-Weis, S. H. Abdelwahed, R. Shukla, R. Rathore, M. A. Ratner, and M. R. Wasielewski, Crossover from single-step tunneling to multistep hopping for molecular triplet energy transfer, *Science,* **328**, 1543–1547 (2010).

25. G. Conibeer, S. Shrestha, S. Huang, R. Patterson, H. Xia, Y. Feng, P. Zhang, N. Gupta, M. Tayebjee, S. Smyth, Y. Liao, S. Lin, P. Wang, X. Dai, and S. Chung, Hot carrier solar cell absorber prerequisites and candidate material systems, *Sol. Energy Mater. Sol. Cells,* **135**, 124–129 (2015).

26. V. Svrcek, M. Kolenda, A. Kadys, I. Reklaitis, D. Dobrovolskas, T. Malinauskas, M. Lozach, D. Mariotti, M. Strassburg, and R. Tomašiūnas, Significant carrier extraction enhancement at the interface of an InN/p-GaN heterojunction under reverse bias voltage, *Nanomaterials,* **8**, 1039 (2018).

27. T. Trupke, M. A. Green and P. Würfel, Improving solar cell efficiencies by down-conversion of high-energy photons, *J. Appl. Phys.,* **92**(3), 1668–1674 (2002).

28. R. Schaller and V. Klimov, High efficiency carrier multiplication in PbSe nanocrystals: Implications for solar energy conversion, *Phys. Rev. Lett.,* **92**(18), 186601 (2004).

29. P. Yu, J. M. Luther, Q. Song, W. K. Metzger, R. J. Ellingson, A. J. Nozik, M. C. Beard, and K. P. Knutsen, Multiple exciton generation in colloidal silicon nanocrystals, *Nano Lett.*, **7**(8), 2506–2512 (2007).
30. S. K. Stubbs, S. J. O. Hardman, D. M. Graham, B. F. Spencer, W. R. Flavell, P. Glarvey, O. Masala, N. L. Pickett, and D. J. Binks, Efficient carrier multiplication in InP nanoparticles, *Phys. Rev. B*, **81**(8), 081303 (2010).
31. S. Wang, M. Khafizov, X. Tu, M. Zheng, and T. D. Krauss, Multiple exciton generation in single-walled carbon nanotubes, *Nano Lett.*, **10**(7), 2381–2386 (2010).
32. A. J. Nozik, Multiple exciton generation in semiconductor quantum dots, *Chem. Phys. Lett.*, **457**, 3–11 (2008).
33. M. Kumar, S. Vezzoli, Z. Wang, V. Chaudhary, R. V. Ramanujan, G. G. Gurzadyan, A. Bruno, and C. Soci, Hot exciton cooling and multiple exciton generation in PbSe quantum dots, *Phys. Chem. Chem. Phys.*, **18**, 31107–31114 (2016).
34. N. J. L. K. Davis, M. L. Böhm, M. Tabachnyk, F. Wisnivesky-Rocca-Rivarola, T. C. Jellicoe, C. Ducati, B. Ehrler, and N. C. Greenham, Multiple-exciton generation in lead selenide nanorod solar cells with external quantum efficiencies exceeding 120%, *Nat. Commun.*, **6**, 8259 (2015).
35. F. Auzel, Upconversion and anti-stokes processes with f and d ions in solids, *Chem. Rev.*, **104**, 139–173 (2004).
36. T. Trupke, M. A. Green, and P. Würfel, Improving solar cell efficiencies by up-conversion of sub-band-gap light, *J. Appl. Phys.*, **92** (7), 4117–4122 (2002).
37. J. de Wild, J. K. Rath, A. Meijerink, W. G. J. H. M. van Sark, and R. E. I. Schropp, Enhanced near-infrared response of a-Si:H solar cells with β-$NaYF_4$:Yb^{3+} (18%), Er^{3+} (2%) upconversion phosphors, *Sol. Energy Mater. Sol. Cells*, **94**, 2395–2398 (2010).
38. J. C. Goldschmidt, P. Löper, S. Fischer, S. Janz, M. Peters, S. W. Glunz, G. Willeke, E. Lifshitz, K. Krämer, and D. Biner, Advanced upconverter systems with spectral and geometric concentration for high upconversion efficiencies, in *Proceedings IUMRS International Conference on Electronic Materials*, Sydney, Australia, pp. 307–311 (2008).
39. I. Voitenko, J. F. Muth, M. Gerhold, D. Cui, and J. Xu, Tunable photoluminescence of polymer doped with PbSe quantum dots, *Mater. Sci. Eng. C*, **27**, 1078–1081 (2007).

40. M. Brumer, A. Kigel, L. Amirav, A. Sashchiuk, O. Solomesch, N. Tessler, and E. Lifshitz, PbSe/PbS and PbSe/PbSe$_x$S$_{1-x}$ core/shell nanocrystals, *Adv. Funct. Mater.*, **15**, 1111–1116 (2005).

41. A. Kigel, M. Brumer, G. Maikov, A. Sashchiuk, and E. Lifshitz, The ground-state exciton lifetime of PbSe nanocrystal quantum dots, *Superlattices Microstruct.*, **46**(1), 272–276 (2009).

42. J. C. Goldschmidt, P. Löper, and M. Peters, *Solar Element Mitgesteigerter Effizienz und Verfahrenzur Effizienzsteigerung*, in *Deutsches Patent*, Fraunhofer-Gesellschaftzur Förderung der angewandten Forschunge. V., Bundes republik Deutschland (2007).

43. J. de Wild, T. F. Duindam, J. K. Rath, A. Meijerink, W. G. J. H. M. van Sark, and R. E. I. Schropp Increased upconversion response in a-Si:H solar cells with broad band light, *IEEE J. Photovoltaics*, **3**, 17–21 (2013).

44. S. Eustis and M. A. El-Sayed, Why gold nanoparticles are more precious than pretty gold: Noble metal surface plasmon resonance and its enhancement of the radiative and nonradiative properties of nanocrystals of different shapes, *Chem. Soc. Rev.* **35**, 209–217(2006).

45. Y. H. Jang, Y. J. Jang, S. Kim, L. N. Quan, K. Chung, and D. H. Kim, Plasmonic solar cells: from rational design to mechanism overview, *Chem. Rev.*, **116**, 14982–15034 (2016).

46. J. Zhu, M. Xue, H. Shen, Z. Wu, S. Kim, J.-J. Ho, A. Hassani-Afshar, B. Zeng, and K. L. Wang, Plasmonic effects for light concentration in organic photovoltaic thin films induced by hexagonal periodic metallic nanospheres, *Appl. Phys. Lett.* **98**, 151110 (2011).

47. L. Feng, M. Niu, Z. Wen and X. Hao, Recent Advances of Plasmonic Organic solar cells: photophysical investigations, *Polymers* **10**, 123 (2018).

48. F. Farhadnia, A. Rostami, and S. Matloub, Plasmonic Solar Cells, a New Way to Enhance Energy Conversion Efficiency: Analysis and Modeling of Effect of Metal Geometry, Inter. *J. Opt. Photon.* **13**(1), 61–70 (2019).

49. H. Chen, S.-M. Lee, A. Montenegro, D. Kang, B. Gai, H. Lim, C. Dutta, W. He, M. L. Lee, A. Benderskii, and J. Yoon, Plasmonically enhanced spectral upconversion for improved performance of GaAs solar cells under nonconcentrated solar illumination, *ACS Photonics*, **5**(11), 4289–4295 (2018).

50. A. Luque and A. Martí, Increasing the efficiency of ideal solar cells by photon-induced transitions at intermediate levels, *Phys. Rev. Lett.*, **78**(26), 5014–5018 (1997).

51. D. Macdonald, K. McLean, J. Mitchell, P. N. K. Deenapanray, and S. De Wolf, An alternative scheme for implementing the impurity-photovoltaic effect incrystalline silicon, in *Proceedings of the 19th European Photovoltaic Solar Energy Conference*, Paris, France, pp. 88–91 (2004).

52. N. J. Ekins-Daukes, C. B. Honsberg, and M. Yamaguchi, Signature of intermediate band materials from luminescence measurements, *Proceedings of the 31st Photovoltaic Specialists Conference, Orlando*, 49–54 (2005).

53. M. Yoshida, N. J. Ekins-Daukes, D. J. Farrell, and C. C. Phillips, Photon ratchet intermediate band solar cells, *Appl. Phys. Lett.*, **100**, 263902 (2012).

54. A. Pusch, M. Yoshida, N. P. Hylton, A. Mellor, C. C. Phillips, O. Hess, and N. J. E. Daukes, Limiting efficiencies for intermediate band solar cells with partial absorptivity: the case for a quantum ratchet, *Prog. Photovolt: Res. Appl.*, **24**(5):n/a-n/a DOI: 10.1002/pip.2751 (2016).

55. M. Vörös, G. Galli, and G. T. Zimanyi, Colloidal nanoparticles for intermediate band solar cells, *ACS Nano*, **9**(7), 6882–6890 (2015).

56. C. Yang, M. Qin, Y. Wang, D. Wan, F. Huang, and J. Lin, Observation of an intermediate band in Sn-doped chalcopyrites with wide-spectrum solar response, *Sci. Rep.*, **3**, 1286 (2013).

Index

1,2-dichlorobenzene 241–242

α-phase 110
a-Si, *see* amorphous silicon
AAM, *see* anodic alumina membrane
absorption 63, 67, 134, 137, 142–144, 146, 148–150, 152, 158, 174, 179, 191, 208, 266, 279, 292, 326, 337, 377, 406, 419, 445, 452, 455, 469, 509, 512
  above-bandgap sunlight 190
  atmospheric 63–64
  balance light 122
  broadband light 445
  defect 94, 98
  dominant light 354
  efficient solar light 413
  two-photon 518
absorption coefficient 106, 119, 127, 328, 370, 385, 408, 519
absorption efficiency 105, 327, 382
absorption spectra 150, 191, 241–242, 273, 280, 293, 431, 522
absorptivity 520–521
Al-doped ZnO (AZO) 211–212, 215–217, 227, 410
ALD, *see* atomic layer deposition
amorphous silicon (a-Si) 56, 79, 105–107, 157, 188–189, 193–194, 437–444, 486–487, 510–511
  bandgap of 438, 440
anodic alumina membrane (AAM) 191–192, 392
atomic interdiffusion 23
atomic layer deposition (ALD) 164, 211, 215
atomic precipitation 334–335
Auger process 491, 494, 497, 500
AZO, *see* Al-doped ZnO
AZO backbone 215–217
AZO electrode 441
AZO H-P-G photoanode 216–217
AZO host 217

band alignment 28, 33, 37, 55, 70, 117, 240–241, 272, 300, 307, 346, 405–406, 408, 416, 419, 431, 433, 455–456, 479, 492, 519
  staged type-II 305, 346
  type-II 117–118, 305, 347, 350, 352, 372–373
band discontinuities 27, 119, 123, 127, 269, 304, 349, 354–355, 416–417
band dispersion 13
bandgap 1, 6, 9–11, 15, 69, 76–78, 96, 99–100, 102,

104–105, 116–117, 119, 174, 177, 185–186, 190, 210, 213, 271–272, 299, 302, 305–306, 309–310, 312, 316–318, 336, 339, 347–348, 359, 364–365, 368–370, 372–374, 390–391, 413, 419, 423, 434–435, 437, 469–470, 474–475, 481, 483, 485, 487–491, 493, 498–500, 503, 506–507, 516–519, 522
bulk 212–213, 309
electronic 177
indirect 140–141, 148, 385
tunable 116, 208, 310, 472
wide 55, 116–117, 151, 366–367, 372, 471, 475
bandgap loss 96
Basella alba 245
Basella alba dyes 245
Bloch electron 19
Blocking electron injection 102
Bohr radius 68, 270–271
boron atoms 18
boron diffusion 345, 361–362
Bragg scattering 176–177
Brillouin zone 9, 13, 178, 225, 501
brookite 214
bulk heterojunction colloidal QDSCs 299
bulk heterojunction configurations 289, 423
bulk materials 2, 53, 56, 331–332, 388, 499, 503
bulk semiconductor material 331
bulk semiconductors 2, 38, 326, 494, 499
carbon 54, 224, 247–248, 251
carbon atoms 224, 226–227, 251
carbon-based thin-film electrodes 222
carbon-based transparent conducting electrodes 222–223, 225
carbon nanomaterials 247–248
carbon nanotubes (CNTs) 37, 206, 210, 215, 218, 222–224, 247–248, 407
carbon-passivated porous silicon counter electrodes 252
carbonaceous materials 247
carrier collection 77, 91, 123, 246, 308, 316, 331, 339, 353, 355, 358–360, 394, 449
efficient 300, 383, 385
photon-generated 326
carrier concentrations 16, 24
carrier densities 24, 26, 79, 331, 356
carrier mobilities 90, 329, 331, 388, 394, 408, 460, 518, 520
carrier recombination 89, 151, 217–218, 289, 318, 352–353, 385, 408, 418, 429, 448, 517
carrier separation 354, 391, 408, 423, 459
catalysis 234, 311
catalyst 190, 208–209, 226, 233, 237, 247, 249, 311, 336, 340, 376–377, 412
CB, *see* conduction band
CBE, *see* chemical beam epitaxy
Cd 110, 209, 253, 292–293, 393, 453
Cd monolayers 293
Cd-terminated surfaces 291–292

CdS 3, 108–110, 113, 208, 221–222, 305, 370–372, 381, 419, 433–434, 499
  bandgap of 109, 390
CdS/CIGS junction 112
CdS/Cu_2S junction 381
CdS layers 108–109, 394
CdS nanopillars 371, 391–392
CdS nanowires 381
CdS QDs 253, 292–293
  as-prepared 292–293
CdS shell layers 373, 393–394
CdSe 55, 208, 221–222, 269, 289, 305, 370–371, 381, 413–414, 419, 431, 499
CdSe nanocrystals 8, 430
CdSe nanorods 431–432
CdSe QDs 267, 273–274, 289, 305, 307, 311
CdSe tetrapods 429–430
$CdSe/TiO_2$ junction 305
CdTe 3, 106, 108–109, 114, 208, 269, 370–371, 390, 392, 419, 480–481
CdTe/CdS junction 371
CdTe/CdS solar cells 109–110
CdTe layer 108–109, 391–392
CdTe QDs 481
CdTe thin films 371
cell efficiency 88–89, 115, 184, 186, 383, 479
cells
  crystalline CdTe 109
  high-bandgap 488
  low-bandgap 488
  micro-crystalline CdTe 109
  multi-junction 474
  multiple exciton 469
  organic 442
  photoelectrochemical 92, 105, 351, 407
  planar 459–460
  plasmonic 194
  prototype photovoltaic 206
  ribbon 362–363
  Schottky 295, 297
  single-junction 474, 479–480
cellulose sodium salt 248–249
charge carriers, photogenerated 101
charges, photon-generated 348
chemical beam epitaxy (CBE) 334–335
chemical etching, metal-assisted 384
chemical vapor deposition (CVD) 56, 223, 226, 334–335, 364, 366, 381, 391
chiralities 223
CIGS, *see* $CuIn_xGa_{1-x}Se_2$
CNT electrode 224
CNTs, *see* carbon nanotubes
colloidal quantum dots (CQDs) 54, 56, 125, 128, 266, 274, 279–280, 287, 290, 293–294, 300, 302, 315–316, 318, 424, 479
compound semiconductors 3, 100, 110, 370, 471
conduction band (CB) 5–6, 9, 11, 14–16, 18, 22–23, 27, 34, 65, 67, 69–70, 76, 78, 91, 96, 99, 102, 113, 115–117, 133, 206, 270, 272, 306, 346–347, 354, 373, 405, 416–418, 490–494, 499–501, 505, 516–517, 521
conductivity 5, 7, 16, 18, 20, 34, 175, 205–206, 213, 215, 227, 408–410, 412, 423–424

conversion efficiency 2, 77, 99–100, 107, 112, 115, 125, 141, 167, 185, 187, 194–195, 197, 218–220, 239–240, 246, 249, 289–291, 295, 297, 300, 310–313, 334, 351–352, 356, 358, 360, 363–364, 366, 368, 372, 379, 381–383, 394, 408–409, 429, 431, 455, 467–522
core/shell nanowires 339, 349–350, 357, 360
counter electrode materials 306, 312
  catalytic 247
counter electrodes 206–208, 217, 221, 245–255, 304, 311–312
  sensitized composite Pt-coated 219
  traditional Pt 249, 255
CQDs, *see* colloidal quantum dots
crystalline materials 98, 406
Cu_2S shell layer 381–382
Cu_2ZnSn 114, 253–255
$Cu_2ZnSn(S,Se)_4$ (CZTSSe) 114, 253–254
$CuIn_xGa_{1-x}Se_2$ (CIGS) 106, 110–114, 487
  cells 113, 486–487
  high-quality 111
  layer 113
  solar cells 112–115, 486–487
  solar cells on flexible stainless steel substrates 113
CVD, *see* chemical vapor deposition
CZTSSe, *see* $Cu_2ZnSn(S,Se)_4$

dangling bonds 53, 56, 105, 107
devices
  electrochromic 213
  four-junction 472
  nano-optoelectronic 393
  optoelectronic 3, 5, 25, 32, 37, 53, 62, 106, 145, 160, 185, 270, 330, 337, 374, 385, 409, 491
  photoelectric 432, 516
  photoelectric conversion 100
  photoelectronic 1, 8, 75
  photonic crystal 182
  radial junction nanowire 362
  upconverting 508
dielectric constant 94, 153, 156–157, 159–161, 174–175, 177, 180–181, 186, 188, 197, 331, 512
doping nanomaterials 19
Drude model for damped free electrons in metal 160
DSSCs, *see* dye-sensitized solar cell
Dye-sensitized solar cell (DSSCs) 93, 125, 128, 205–254, 414, 424, 437
dye-sensitized solar cells, photoanodes of 207

EEL, *see* electron-extracting layer
electrodeposition 108, 426
electron Bohr radius 7, 271
electron-extracting layer (EEL) 492–493
electron/hole pair 76, 78, 96, 270, 357, 372, 494
  photo-excited 348

photon-excited 115
electron mobility 7, 34, 111, 215, 217, 231, 304, 414, 430, 437, 460
electron transport layer 405, 416–417, 440
electron transport material 455–456
electrons
photo-excited 97, 221, 238
photo-generated 272, 295, 299, 308, 311
photon-excited 78, 115–117, 122, 218, 239, 268, 420
photon-generated 348–349, 367, 372
electrons and holes 15, 18–19, 24, 27–28, 30, 33–34, 46, 53, 55, 68–71, 76, 79, 91–93, 96, 100–104, 117, 119, 123, 127, 268, 288, 295–296, 345, 348–349, 352, 354, 358, 367–368, 372, 386, 408, 413, 418–419, 424, 432, 435, 438, 441, 455, 461, 493–494
photo-generated 272, 299, 305, 308, 311
photoexcited 93, 98, 101–102, 117, 349, 405, 417
photogenerated 29, 113, 346–347, 352, 390
photon-excited 76, 78, 420
photon-generated 79, 268, 347–349, 351, 369, 371
in semiconductors 8–9, 11–21
separation of 77, 79, 123, 345
electrostatic force 92, 348, 353
EQE, *see* external quantum efficiencies
exciton Bohr radius 37, 68, 270–272, 309
excitons 25–26, 29, 37, 68–71, 76–79, 92, 231, 266, 270–271, 293, 295, 304, 310, 316, 336, 354, 408, 417–420, 423–424, 431, 435–436, 438, 441, 455–456, 483, 498–499, 501
photon-generated 441, 455, 458
external quantum efficiencies (EQE) 112, 166–167, 297, 377–378, 430, 433, 442, 447, 450, 482, 501–503, 511

FDTD, *see* finite difference time domain
Fermi ball 20, 48–49
Fermi energies 13, 27
Fermi levels 11, 14–15, 28–31, 93, 103, 294, 296
FESEM, *see* field emission scanning electron microscopy
field emission scanning electron microscopy (FESEM) 427
filling factor 83–84, 104, 166, 194, 311–314
finite difference time domain (FDTD) 124, 137, 187, 377–378
fluorine-tin-oxide (FTO) 211–212, 215, 219, 221, 245, 298, 454–456, 479
Fresnel lenses 195, 472–474

FTO, *see* fluorine-tin-oxide
fullerene 224, 247, 408, 410, 423–424, 479
fullerene cells 479

GaAs 3, 7, 13, 27, 34, 269, 337, 339–340, 359, 364, 366, 369–370, 385, 423, 434–435, 437, 478, 489–490
GaAs core/shell nanowires 336
GaAs nanopillars 434–436
GaAs nanowire array solar cells 385–386
GaAs solar cells 167, 195–196, 385, 514–515
GaAs substrate 336, 386
GaN nanorods 366–368
graphene 59, 206, 210, 224–226, 247
  single layer 226
graphene oxide, reduced 225–227, 255
graphite 224, 226, 234, 245, 247–248

HB, *see* high-bandgap
HEL, *see* hole-extracting layer
hetero-interface 30, 116–118, 349, 354, 373, 416–417
heterojunction, bulk 269, 299–301, 407, 418, 421, 423–424, 433
heterojunction solar cells 109, 422, 424–425
heterojunctions 23, 27–28, 30, 34, 37–38, 76, 93, 103, 116, 119, 127, 137, 268–269, 294, 298–299, 301, 329–330, 340, 346–349, 354–357, 359–360, 374, 390, 405, 417, 419, 423, 433, 436–437, 456, 461
  depleted 297, 300–301, 479
  hybrid 268, 416
  nanowire 354
  phase separated bi-layer 268–269
  type II 93, 117, 374
  type II core/shell 117
high-bandgap (HB) 486, 488
high resolution transmission electron microscope (HRTEM) 366
highest occupied molecular orbital (HOMO) 105, 208, 405, 417, 419, 522
hole-extracting layer (HEL) 492–493
holes
  charged 16, 68, 272, 347
  hot 492–495, 497
  photoexcited 288, 495
  photon-excited 116, 118
  photon-generated 78
HOMO, *see* highest occupied molecular orbital
hot-carrier capture 490–491, 493, 495, 497
hot carrier solar cells 491–493, 495–497, 503
  device structures of 492
hot carriers 490–492, 494–497, 499
  photogenerated 495
hot electrons 65, 78, 96, 307, 490–497
  photogenerated 306, 491

HRTEM, *see* high resolution transmission electron microscope
HSCs, *see* hybrid solar cells
hybrid nano solar cells 405–462
hybrid solar cells (HSCs) 405–409, 413, 416–417, 419–420, 423–452, 454–457, 459–462
hybrid solar cells, device structure of 423, 425
hydrogen atoms 56, 68–69, 105, 107, 226, 271

impact ionization 307, 491, 495–496, 499
indium 5, 211–212
indium tin oxide (ITO) 188–189, 211–212, 222, 289, 294, 297–298, 303, 352, 364, 386, 410, 422, 432, 443, 478–479
inorganic materials 56, 405, 407–408, 416–419, 421, 432, 460
inorganic nanomaterials 419
inorganic semiconductors 407, 411, 419, 423, 425, 431, 460–461
intermediate band solar cells 515, 517, 519, 521
internal quantum efficiency (IQE) 295, 377, 383, 503, 508
ion etching
  deep reactive 333, 344, 361–362
  reactive 188–189, 345
IQE, *see* internal quantum efficiency
ITO, *see* indium tin oxide
ITO electrode 303–304, 431
  transparent 293, 435

Lambertian light trapping limit 141–142
leakage 81, 89, 375–377
ligand exchange 288, 297, 315–316, 522
light reflection 95, 101, 123, 131, 133, 135, 137, 139, 144, 153, 160, 174, 179, 186–187, 217, 426
  suppression of 143, 145, 154
light reflection reduction 101, 146
light scattering 145, 158, 332, 364
light scattering by metal nanoparticles 162
light trapping 95, 105, 107, 133, 138–140, 143, 146–147, 151, 153, 156–171, 173–174, 179, 185–186, 188, 190–191, 193, 195, 197, 308, 328, 331–332, 357, 359, 383, 512
  efficiency 151, 169
  enhanced 120, 124, 326, 355, 468
light trapping layer 171–173
light trapping methods 95, 138–139, 141–155
light trapping nanomaterials 190
liquid-phase processes 57–58
lithography
  electron beam 190, 237, 342
  substrate-conformal imprint 168

lowest unoccupied molecular orbital (LUMO) 208, 239, 405, 417, 419, 431, 522
LUMO, *see* lowest unoccupied molecular orbital

magnetron sputtering 210, 212, 440
materials
  acceptor 417, 420
  catalytic 248, 250, 253, 255
  composite 123–124
  compositional semiconductor 350
  counter-electrode 312
  electron-accepting 431
  graded-bandgap 471
  graphene-based 225–226
  hole transport 408, 414, 455
  iodide-treated 290–291
  large-bandgap 302, 506, 520
  lattice-matched wide bandgap 385
  linear isotropic 175
  luminescent 498–499
  narrow-physical-bandgap 516
  organic narrow-bandgap 407
  perovskite 453, 484
  photo-anode 213, 305
  photonic crystal 197
  popular transparent electrode 410
MBE, *see* molecular beam epitaxy
MEBs, *see* multiple energy bands
MEG, *see* multiple exciton generation
MELs, *see* multiple energy levels
metal organic chemical vapor deposition (MOCVD) 27, 103, 108, 196, 210, 274, 334–335, 338, 340–341, 368, 385–386, 434, 474, 477
metal organic vapor phase epitaxy (MOVPE) 334–335, 341, 369
metal oxide semiconductor field effect transistor (MOSFET) 32
metal oxides 205–206, 209, 213, 299, 305, 370, 393
metals, catalytic 236–237, 339
Mie scattering 135–137
MOCVD, *see* metal organic chemical vapor deposition
molecular beam epitaxy (MBE) 27, 43, 103, 196, 210, 274, 334–335, 338, 340, 369, 381, 472, 474, 477
MOSFET, *see* metal oxide semiconductor field effect transistor
MOVPE, *see* metal organic vapor phase epitaxy
multi-junction solar cells 471
multi-junction tandem solar cells 469–489
multi-wall carbon nanotubes (MWCNTs) 219
multiple energy bands (MEBs) 517–519
multiple energy levels (MELs) 484, 517–519
multiple exciton generation (MEG) 265, 317, 498–499, 501, 503
MWCNTs, *see* multi-wall carbon nanotubes

nano photovoltaic devices 268
nano photovoltaic materials 119–120

nano semiconductor materials 2
nano solar cells (NSCs) 30, 76, 83, 86, 88, 92–93, 115–119, 121–124, 127, 137, 142, 146, 163, 193, 205, 210, 218, 246, 265, 268, 284, 310–311, 313–314, 341, 356, 360, 363, 365, 369–370, 405, 429, 468, 491–492
nano TiO_2 209, 213, 227–229, 251, 304–305
nanobelts 232–233, 237
nanocone arrays 153–156, 188–189
nanocones 95, 153–155, 187–188, 325, 445–447
nanodevices 325, 330
nanoheterojunction colloidal QD solar cells 301
nanohole arrays 147, 152–153
nanoholes 147, 152–153
nanomaterial aggregation 461
nanomaterial arrays 165, 342–343
nanomaterials
  colloidal 461
  inorganic 432, 461
  sol-gel-grown 61
  solar cells based on 117, 121
nanoparticles, silica 188–189
nanophotonic splitter 486
nanopillar array
  CdS 371, 391
  multi-diameter 154–155
nanopillars 95, 156, 187, 299, 325, 351, 364, 391–392, 434
nanorods (NRs) 37, 45, 147, 152, 190, 228–229, 231–236, 266, 299, 305, 366–368, 372, 408, 419, 421–422, 424, 426, 431–432, 457–459, 501
nanosemiconductors 461, 499
  inorganic 407, 421
nanowire arrays 147–149, 151–152, 188–189, 191, 331–332, 342, 344, 365, 369, 384, 448
  optical absorption of 147
nanowire axes 341
nanowire solar cells (NWSCs) 92, 147, 326, 333–334, 345–385, 388, 394
  axial junction 352, 386, 388
  radial junction 358, 388
nanowire solar cells, device architectures of 345–359
nanowires (NWs) 4, 45, 51–52, 124, 143–145, 147–154, 170–173, 188, 190–192, 228, 232–235, 239, 305, 325–341, 343–364, 368–369, 371–372, 374–379, 381–386, 389–390, 394, 419, 421, 426–428, 448–449, 451–453, 461, 477–478, 488–489
NHE, *see* normalized hydrogen electrodes
non-electrostatic force 76, 93, 354
normalized hydrogen electrodes (NHE) 269–270
NRs, *see* nanorods
NSCs, *see* nano solar cells
NWs, *see* nanowires
NWSCs, *see* nanowire solar cells

PbS CQD solar cells 290–291
PbS CQDs 290–291, 479
PbS-QD standalone cells 481

PbS QDs 123, 290, 303, 480, 507
PCE, *see* photoelectric conversion efficiency
PCs, *see* photonic crystals
PECVD, *see* plasma enhanced chemical vapor deposition
perovskites 453–454, 482–483, 496
phonons 6–7, 15, 64–65, 78, 494
photo-generated carriers 210, 244, 300
photoanodes 101, 122, 206–208, 213–217, 219, 222, 228, 245, 251, 253–254, 304, 311, 393
photoelectric conversion efficiency (PCE) 54, 75, 77, 95, 195, 240, 242–246, 295, 299, 314, 326, 331, 346, 361, 363, 382–383, 385–388, 390, 392–394, 429, 435, 437, 442–443, 445, 447–450, 452–453, 458–459, 462, 468, 476, 479, 484
photoelectronic conversion 141, 300, 373, 455, 519
  efficient 517
photoelectronic conversion efficiency 205
photoelectrons 206–207, 219
photoexcited carriers 96, 109, 125, 418
photoexcited electrons 25, 68, 76–77, 92–93, 97, 99–100, 103, 109, 116–117, 123, 406
photogenerated carriers 122, 352–353, 356, 381
photogenerated holes 113, 357–358, 449
photoholes 207
photolithography 56, 190, 246, 334, 342, 345, 376, 420
photoluminescence 6–8, 62, 507
photoluminescence spectra 33, 292, 521
photon absorber 115–116, 122, 205, 294, 393, 492–493
photon absorption 2, 67–68, 75, 102, 107, 139, 300, 309, 325, 333, 345, 368, 383, 408, 431, 470, 493, 497, 516–518
photon absorption by semiconductors 76
photon conversion efficiency 439
photon emission 66–67, 78
photon energies 6, 15, 67, 76, 78, 96, 116, 125, 140, 148–150, 177, 272, 318, 373, 470, 490–492, 498–499, 501–502, 506–508
photon-excited carriers 350, 501
photon flux 64
photon management 138, 142–143
photon management design, efficient 142, 188
photon-to-electron conversion efficiency 311
photon trapping 131–196, 326, 353
photonic bandgaps 173–174, 177–178, 182, 185–186
photonic crystal structures 174, 509
photonic crystals (PCs) 142, 173–191, 195, 197, 213, 284
photonic grating 182–183
photons 1, 6, 64, 66–67, 69, 71, 75–79, 91, 96, 100, 115–119, 123, 127, 131,

133–134, 137–140, 142, 148–149, 177–178, 186, 190, 195, 206–208, 210, 239, 268, 272, 293, 295, 305, 309–310, 316, 318, 347, 353, 355, 435, 438, 440–441, 469–470, 474–475, 483, 485–486, 490–492, 498, 500–501, 503, 505–509, 516, 518
above-bandgap 467, 469, 503, 505
below-bandgap 467, 469, 503
low energy 25, 78, 309
single 499
upconverted 508–509
photoresist 342–343
photovoltaic conversion 310, 485, 504, 517
photovoltaic conversion efficiency 328
photovoltaic devices 1–2, 25, 64, 71, 75, 79, 83, 92, 99–117, 119, 121, 127, 139, 151, 222, 227, 293, 312, 369, 498, 513
high-PCE 484
illuminated 83
impurity-band 518
single-junction 314
tandem 379
thin-film 112
photovoltaic effect 1, 26, 29, 67, 75–79, 81–91, 93, 97, 102, 121, 124, 239, 288, 347, 349, 351, 408, 417, 435–436, 496, 521
photovoltaic materials 121, 127
hybrid 425
planar PEDOT thin film 153–154
plasma enhanced chemical vapor deposition (PECVD) 107, 334, 438, 440
plasmas 107, 342–343, 445
plasmonic light trapping 161, 192–193
plasmonic solar cells 168, 511–514
plasmonics 142, 168, 513–514
plasmons 157–158, 160, 162, 511
polarizability 136, 159–160
polarized lights 150, 331
polymers 227, 408–409, 412–414, 418, 423, 429–430, 456–457, 461, 479–480
electron-conducting 307–308
polythiophenes 227, 409–410
porous semiconductor 422
porphyrins 242–244
potential barrier 30, 32, 38–40, 102–103, 109, 270, 288, 347
power conversion efficiency 1–2, 83, 95, 105, 111, 119, 121–122, 125, 127–128, 143, 166, 194, 219, 239, 248, 295, 297, 301–302, 304, 310–311, 315, 317, 326, 346, 359, 382, 386, 394, 427, 445, 467–468, 485, 490–491, 506, 511, 520, 523
Pt counter electrode 208, 252–255

QD arrays 306–308
QD concentrator 508–510
QD film 294–296, 298–302, 314–317
colloidal 287, 298–299, 302
QD layers, PbS 304

QD-sensitized solar cells 304–306
QD solar cells 288–289, 293–311, 315–316
  device architectures of 293–309
QDs, *see* quantum dots
QDSCs, *see* quantum dot solar cells
QDSSCs, *see* quantum-dot-sensitized solar cells
quantum confinement 4, 7, 46–47, 68, 70–71, 121, 265, 273, 280, 347, 522
quantum-dot-sensitized solar cells (QDSSCs) 207–209, 239, 253–255
quantum dot solar cells (QDSCs) 30, 265–318
  colloidal 295, 298, 302
quantum dots (QDs) 4, 18, 30, 33, 37, 45–46, 53–54, 59, 71, 115–119, 121–124, 127, 173, 205, 208, 213, 239, 253, 265–286, 288–295, 300, 302, 304–309, 311, 316–318, 330, 363–364, 372, 422, 430, 432–433, 479, 484, 496–497, 499–501, 508–510, 517, 519–520, 522
  growth of 190, 276–277
  type-II 272
quantum efficiency 385, 499
quantum wells 4, 37–38, 43, 176–177, 484, 517, 519
quantum wires 37, 45, 115, 119, 127, 519
quasielectric fields 349–350
Queisser limit 468, 517, 521–522

Rayleigh scattering 136–137
reduced graphene oxide (RGO) 225–227, 255
reflectance 143, 148–149, 156, 165
resonance light trapping mechanisms 171–173
RGO, *see* reduced graphene oxide
ruthenium 239–240, 251
Rydberg energy 68–69, 271

SAED, *see* selected area electron diffraction
SAG, *see* selective area growth
scanning electron microscopy 379, 389
Schottky barrier 30–32, 293, 495
schottky junction 32, 268, 293, 301, 406, 448
Schottky junction solar cells 294, 296
Schrödinger equation 10, 12, 39, 44–45
SCs, *see* solar cells
selected area electron diffraction (SAED) 361, 366
selective area growth (SAG) 386
selective emitters 383–385
semiconductor bandgap energy 498
semiconductor bandgaps 10, 177, 519
semiconductor heterojunctions 27, 269
semiconductor junction 293
semiconductor materials 3, 7–8, 96, 99, 102, 333, 366, 477
  elementary 100
  light-sensitive 100

semiconductor nanomaterials 419
semiconductor nanowires 325–326, 329, 331, 335–336, 341, 461
semiconductor oxides 239, 266
semiconductor quantum dots 34, 45, 53, 115, 207–208, 265–274, 309–310, 317, 430, 480, 497
semiconductors
  crystalline 9, 119, 127
  direct-bandgap 119, 127, 141, 385
  doped 5, 27, 511
  electronic states in 5, 36
  electrons in 2–70, 76
  indirect bandgap 100, 105
  large-bandgap 520
  nano 2, 18, 34, 60
  nano-sized 2, 497
  narrow-bandgap 10, 12, 15, 39, 407, 516
  photovoltaic 119, 127
  wide-bandgap 10, 12, 15, 38, 151, 214, 228, 231, 299, 304
sensitizers 206, 208–209, 213–214, 216, 221, 238–246, 504
Shockley–Queisser limit 467–468, 517, 520–522
silicon 5, 56, 76, 96, 100, 107, 118, 139–140, 177, 251, 345, 460, 483
single nanowire solar cells 370, 374, 377, 379
SiNW PEDOT 429
SiO_2 nanoparticle monolayer 445
SiO_2 nanoparticles 445–446
SnO_2 206, 209, 211–212, 215, 217–218, 231
sol-gel growth 61
solar cells (SCs) 1–2, 32, 75–77, 79–80, 82–86, 88–97, 99–106, 113–117, 120–121, 123–125, 138–143, 145, 149–151, 162, 169–170, 179–181, 185–186, 188, 191–194, 206, 210, 213–214, 265–268, 287–291, 298–299, 301–302, 307–310, 313–316, 318, 325–326, 359–368, 371–372, 377–379, 381, 385, 393–394, 406–407, 413–414, 430, 432, 467–473, 476–478, 484–486, 490–492, 495–496, 498–501, 503–510, 512–514, 516–517, 519–522
  axial junction 387
  bionic 245–246
  nanowire array 348
  thin-film 105, 107, 113, 138–139, 142, 157, 180
solar cells
  conventional 116, 119, 122, 127, 347–348, 406, 468, 490, 523
  flexible 222, 227
  heterojunction colloidal QD 301
  hybrid tandem 437–439, 442, 444
  light trapping in nano 142, 193
  multi-junction 107, 366, 468, 470–471, 474
  nanoheterojunction CQD 302
  nanomaterial 30, 114–115, 118, 120, 122–123
  organic 114, 410, 513

perovskite 482
planar 365, 368, 382–383
ribbon 361–362
single-bandgap 434, 498
single-junction 470, 474
single nanowire 368, 374, 380, 385
two-junction 482
ultrathin GaAs 514–515
virtual-bandgap 517–518
solar irradiation 316, 487–489
solar radiation 1, 62–64, 77–78, 82, 86, 91, 100, 180, 191, 205, 318, 328, 364, 367, 369, 373, 388, 390–391, 437, 445, 508
solar radiation energy 1–2, 75, 143
solar radiation photons 97, 113, 116
solar radiation spectrum 62–63, 71, 111, 125, 166, 239, 309, 393, 437, 468
broad 469
solar spectrum 8, 102, 108–109, 115–116, 119, 127, 139, 150, 175, 210, 240, 302, 309–310, 368–369, 383, 385, 419, 434, 437, 443, 461, 468, 470, 474, 481, 484–486, 489–490, 501, 506, 516
space charge region 24, 33, 79, 90, 92–93, 96, 101, 326, 346, 387
spectral absorption 480–481
spectrum splitting 484–486, 489–490
spin-coating 246, 284, 433, 435, 440, 443, 449, 451, 458, 479
SPR, *see* surface plasmon resonance
superlattices 37, 43, 176–177, 229
supersaturation 57, 236, 276, 334, 337
surface ligand exchanges 53, 265, 287–289, 291
surface plasmon resonance (SPR) 157, 159–160, 512

tandem cell 105, 310, 379, 442–443, 476, 478, 481–482, 484
hybrid 441–442, 444, 480
multi-junction 114, 469
three-junction 469, 520
tandem devices 379, 470
three-junction 520
tandem solar cells 125, 196, 309–310, 439, 442, 469–472, 474, 476–477, 479, 485
hybrid 437, 439–441, 443–444
TCE, *see* transparent conducting electrode
TCFs, *see* transparent conductive films
TCO, *see* transparent conductive oxide
thin films 34–35, 50, 59, 105, 107, 115, 121, 124, 146, 148–149, 153–154, 157, 171–173, 179, 184, 188, 192, 206, 212, 219, 223, 228–229, 246–248, 266–267, 274, 283–289, 293–295, 297, 303–304, 309, 311, 313, 332, 335, 352, 358, 371, 391–392, 412, 431, 480, 521

TiO_2 140, 206, 208–209, 213–215, 217, 219–220, 228–229, 231, 240–241, 245, 249, 253, 269, 300, 304–305, 311, 419, 423, 439, 454, 479, 497
TiO_2 electrodes 213
TiO_2 nanomaterials 207, 227–228, 231, 253
TiO_2 nanoparticles 215–217, 228–229
transparent conducting electrode (TCE) 205–254
transparent conductive films (TCFs) 206, 209, 227–237
transparent conductive oxide (TCO) 107, 206, 209–214, 222–223, 227, 239, 285, 294–295, 298–299, 304–305, 351–352, 389–390, 393
transparent electrodes 223, 231, 352, 409

upconversion 467, 503–507, 509, 511, 513
upconverter 503–511, 514–515

vapor-liquid-solid (VLS) 190, 334, 338, 371, 507
vapor-phase fabrication 57–59
VLS, *see* vapor-liquid-solid

wet etching 342–344

X-ray photoelectron spectroscopy 303, 455

zinc blende 3–4, 108
ZnO 151, 206, 209, 211–213, 215, 221, 228, 231–232, 234, 236, 269, 294, 300, 304–305, 370–372, 393, 407, 419, 423
ZnO layer 113, 298, 426, 479
ZnO nanobelts 233–234, 237
ZnO nanomaterials 232, 234–238, 372
  as-grown 234
  patterned 237
ZnO nanostructures 220, 232–233
ZnO nanowire arrays 222, 393
ZnO nanowires 215, 220–221, 236–238, 373, 393